# *DESIGN AND PERFORMANCE OF GAS TURBINE POWER PLANTS*

VOLUME XI

HIGH SPEED AERODYNAMICS

AND JET PROPULSION

# *DESIGN AND PERFORMANCE OF GAS TURBINE POWER PLANTS*

EDITORS

W. R. HAWTHORNE

W. T. OLSON

PRINCETON, NEW JERSEY

PRINCETON UNIVERSITY PRESS

1960

*London:* Oxford University Press

L. C. Card 58-5027



Printed in the United States of America by
The Maple Press Company, Inc., York, Penna.

# *FOREWORD*

On behalf of the Editorial Board, I would like to make an acknowledgement to those branches of our military establishment whose interest and whose financial support were instrumental in the initiation of this publication program. It is noteworthy that this assistance has included all three branches of our Services. The Department of the Air Force through the Air Research and Development Command, the Department of the Army through the Office of the Chief of Ordnance, and the Department of the Navy through the Bureau of Aeronautics, Bureau of Ships, Bureau of Ordnance, and the Office of Naval Research made significant contributions. In particular, the Power Branch of the Office of Naval Research has carried the burden of responsibilities of the contractual administration and processing of all manuscripts from a security standpoint. The administration, operation, and editorial functions of the program have been centered at Princeton University. In addition, the University has contributed financially to the support of the undertaking. It is appropriate that special appreciation be expressed to Princeton University for its important over-all role in this effort.

The Editorial Board is confident that the present series which this support has made possible will have far-reaching beneficial effects on the further development of the aeronautical sciences.

Theodore von Kármán

# *PREFACE*

Rapid advances made during the past decade on problems associated with high speed flight have brought into ever sharper focus the need for a comprehensive and competent treatment of the fundamental aspects of the aerodynamic and propulsion problems of high speed flight, together with a survey of those aspects of the underlying basic sciences cognate to such problems. The need for a treatment of this type has been long felt in research institutions, universities, and private industry and its potential reflected importance in the advanced training of nascent aeronatucial scientists has also been an important motivation in this undertaking.

The entire program is the cumulative work of over one hundred scientists and engineers, representing many different branches of engineering and fields of science both in this country and abroad.

The work consists of twelve volumes treating in sequence elements of the properties of gases, liquids, and solids; combustion processes and chemical kinetics; fundamentals of gas dynamics; viscous phenomena; turbulence; heat transfer; theoretical methods in high speed aerodynamics; applications to wings, bodies and complete aircraft; nonsteady aerodynamics; principles of physical measurements; experimental methods in high speed aerodynamics and combustion; aerodynamic problems of turbo machines; the combination of aerodynamic and combustion principles in combustor design; and finally, problems of complete power plants. The intent has been to emphasize the fundamental aspects of jet propulsion and high speed aerodynamics, to develop the theoretical tools for attack on these problems, and to seek to highlight the directions in which research may be potentially most fruitful.

Preliminary discussions, which ultimately led to the foundation of the present program, were held in 1947 and 1948 and, in large measure, by virtue of the enthusiasm, inspiration, and encouragement of Dr. Theodore von Kármán and later the invaluable assistance of Dr. Hugh L. Dryden and Dean Hugh Taylor as members of the Editorial Board, these discussions ultimately saw their fruition in the formal establishment of the Aeronautics Publication Program at Princeton University in the fall of 1949.

The contributing authors and, in particular, the volume editors, have sacrificed generously of their spare time under present-day emergency conditions where continuing demands on their energies have been great. The program is also indebted to the work of Dr. Martin Summerfield who guided the planning work as General Editor from 1949–1952 and Dr. Joseph V. Charyk who served as General Editor from 1952–1956. The cooperation and assistance of the personnel of Princeton University Press and of the staff of this office has been noteworthy. In particular, Mr. H. S. Bailey, Jr., the Director of the Press, and Mr. R. S. Snedeker, who has supervised the project at the Press, have been of great help. The figures were prepared by Mr. Zane Anderson. Special mention is also due Mrs. E. W. Wetterau of this office who has handled the bulk of the detailed editorial work for the program.

Coleman duP. Donaldson
General Editor

# *PREFACE TO VOLUME XI*

This volume, XI, and its companion piece, Volume X, comprise a treatment of the aircraft gas turbine power plant. In particular, the two volumes discuss the major engine components—compressor, combustor, turbine, and controls—with a view toward describing, explaining, and interpreting the basic phenomena influencing their design and operation. The main subjects in Volume XI are combustion, mechanical and metallurgical aspects, and performance and control.

Preparation of drafts for this volume proceeded from 1952 to 1959, a period in which research and development activity on aircraft gas turbines was at maximum levels. This activity both kept the contributing authors very busy in the field and provided them a steady stream of information pertinent to the volume. The resulting problem, familiar to all editors of collected works in an active field, was to complete the volume with nearly uniform timeliness throughout. Revisions, re-writes, and re-assignments were required. Some dating remains inevitable. Nevertheless, the volume editors believe that the text now contains most of the major findings and ideas germane to the subjects covered.

Our sincere thanks go to all authors, both for their fine contributions and for their patience with one another and with us. We are also most grateful to Dr. Coleman duP. Donaldson and his staff for urging us on and for coordinating and completing our work.

W. R. Hawthorne
Walter T. Olson
Volume Editors

# CONTENTS

*PART 1. INTRODUCTION*

# PART ONE

## *INTRODUCTION*

# SECTION A

## *GENERAL CONSIDERATIONS*

W. R. HAWTHORNE

This volume and Vol. X of this series deal with some of the fundamental aspects of gas turbines for aircraft propulsion. The text follows the natural division of the gas turbine components. Vol. X is devoted to the aero- and thermodynamic aspects of compressors and turbines. The first portion of this volume contains sections on various aspects of the combustion chamber problem. The remainder of the volume comprises sections on materials, the aeroelastic aspects of blade vibration, and performance. The general aim of the authors has been to distil from the body of research results and development experience the more lasting contributions to our systematic understanding of gas turbine phenomena.

**A,1. Combustion Chamber Design.** In the development of the early turbojet engines the design of the combustion chamber offered considerable difficulty. The need for heat release rates ten or more times as great as those used in industrial practice, and the requirement for efficient and stable combustion with low pressure losses over a wide range of mixture strengths, followed by the rapid mixing of considerable quantities of diluent air, involved such extensions of, or departures from, existing practice that much time was consumed in combustion chamber development. In Germany, von Ohain foresaw this difficulty, and hydrogen was used in the early proof-running of the Heinkel turbojet. Whittle and others in Britain and America found that extensive development on separate combustion chamber test rigs was necessary. For a long time the duration of the test runs of the first Whittle engine was limited by the combustion chamber. It is not surprising, therefore, that a large research effort has been directed toward gas turbine combustion problems and that much development experience has been necessarily acquired.

There was some scientific support for the early development. At the end of the 1930's knowledge of the kinetics of flame reactions was growing, although theory scarcely covered adequately even the simplest hydrocarbon flame. Results of research on flames in homogeneous mixtures were available and included measurements of flame speeds in laminar flow and in detonations, as well as the effects of additives on detonations in piston engine cylinders. A little work on the combustion of nonhomogeneous streams of fuel and air had been done. This included some studies of diffusion and turbulent flames of gases and the burning of particles of solid and liquid fuels. The latter work proved the more stimulating since the physical rather than the chemical processes were found to control the rates of burning and flame stability. Stability, for instance, has not been achieved by chemical means as much as by aerodynamic and heat transfer processes.

The importance of the physical processes has been reflected in the emphasis put on research into atomization, evaporation, mixing, and the heat and mass transfer in wakes and stabilizing flow systems. The same emphasis will be apparent in the part of this volume devoted to combustion chamber design. Vol. II of this series on high speed aerodynamics and jet propulsion is devoted to the fundamentals of combustion processes and surveys the work on reaction kinetics and the basic physical processes in flames. Vol. XII reviews work on rockets, including the performance of liquid and solid rockets and heat transfer to the nozzle. It also contains a survey of ramjets and pulsating combustion chambers.

The part of this volume devoted to combustion deals only with combustion in the gas turbine engine. Some historical material will be found in Sec. H. Sec. B reviews the requirements for a gas turbine combustor and points from them to the essential techniques and to the processes into which the whole problem can be separated for detailed research.

In the research on, and in the development of, combustion systems time and effort have been devoted to experimental techniques. The measurement of temperature, pressure, and velocity in high speed streams, unmixed as regards composition and temperature, have presented certain difficulties. The measurement of gas composition in very lean as well as rich mixtures has received special attention. These problems and that presented by the measurement of time-dependent properties are discussed in Sec. C.

At one stage in the development of the early Whittle turbojet, the fuel was vaporized before injection. This system was soon replaced by pressure injection of the liquid fuel. Another system developed later has been based on the injection of mixtures of fuel vapor and air. Air blast injection of liquid fuel is also a possibility. The results of research on these various methods of injection are presented in Sec. D. This section also contains a discussion of some work on the use of solid fuels.

Sec. E deals with flame stabilization and includes material on the stabilization of flames of gaseous fuel-air mixtures and air-fuel spray mixtures as used in afterburners. In the majority of gas turbine combustors, flames are stabilized in perforated liners or flame tubes. This more complex phenomenon is discussed in this section, as is the relation between flame stability and combustion efficiency. Sec. E also surveys work on the mechanism of flame stabilization in boundary layers.

The mixing of heterogeneous streams of fuel and air has been found to be a controlling process both before, during, and after combustion. Sec. F deals with this phenomenon in its application to gas turbine combustors.

Sec. G deals with fuels for aircraft gas turbine engines from the point of view of their performance in the engine, their behavior in the fuel system, and their availability. The discussion is confined to hydrocarbon fuels for gas turbines and is not intended to cover slurries, fuels based on boron compounds, or liquefied gases.

In Sec. H the material presented in earlier sections is synthesized in a survey of the problems of combustion chamber design and development. The section emphasizes the factors that must be considered by the engine designer and surveys possible trends in design. This section completes the parts in Vol. X and XI which deal with the basic aerodynamic and thermodynamic processes in the gas turbine engine.

**A,2. Problems of Materials and Mechanical Strength.** The design of a successful turbine engine is by no means finished when the aero- and thermodynamic design of the components is completed. The major problems of mechanical design, selection of materials, matching of components, control, installation, and manufacture have still to be solved. Some of these problems are in the realm of conventional mechanical engineering. Perhaps *conventional* is not the correct word, for aeronautical engineering has been defined as mechanical engineering done better. It is, however, possible in the selection of material for a volume such as this that many design problems such as the stressing of rotating parts, the determination of blade and disk frequencies, the calculation of gyroscopic stresses and thermal expansions, and the peculiar features of aero-gas-turbine bearings and seals may be omitted not because they are unimportant, but because they can be considered to be suitably modified or extended developments of the normal design procedures used by the mechanical engineer. It seems appropriate therefore that the remainder of the volume should be devoted to those problems which experience has shown to be especially acute in aircraft gas turbines, and on which enough work has been done to make a reasonably lasting contribution possible.

A simple analysis of its thermodynamic cycle shows that the achievement of a useful gas turbine depended as much on the development of a high temperature material for the turbine blades as on obtaining good efficiencies in the components. The effort to develop the turbojet stimulated much of the development of high temperature materials. Not only have satisfactory materials been developed, but recent research has much improved the understanding of their behavior and properties. In Sec. I of Part 3, this research has been summarized and the insight it offers into the nature of the behavior of materials at high temperature has been explained.

The fluctuating aerodynamic loads acting on the blading of compressors and turbines have frequently caused fatigue failures. In the aircraft gas turbine this problem has been particularly acute because of the need to use light, highly stressed blading. Failures were experienced at first in the turbine. Then as the axial compressor was developed, failures of its blades in fatigue posed a serious problem. As might be expected, a considerable amount of work has been done on the vibration of blades and the factors which contribute to it. Some of it, such as that related to flutter, has been an extension of work done on aircraft wings. The vibration of a stalled blade is of particular importance in the compressor because several stages of blading are often stalled in the normal running range of an axial compressor. A remarkable feature of a stalled compressor is the phenomenon of propagating or rotating stall. The blades in a row do not all stall simultaneously, but the stall of one or more blades propagates along the row at a speed which is frequently found to be about half the rotor speed.

The aerodynamic features of the only recently revealed propagating stall (1953) are described in Vol. X. In this volume it is regarded as a source of excitation which, together with stall flutter phenomena, are of concern in the mechanical design of slender compressor blades.

Research on blade vibration is by no means finished but the subject warrants a summary of its fundamental aeroelastic aspects. It is hoped that Sec. J covers this and will serve as an introduction to further work.

**A,3. Performance.** Given a knowledge of the performance of its components, the calculation of the performance of a gas turbine is a straightforward, if laborious, operation. Interest is generally centered on the effects of the various factors, such as the efficiency of components, temperatures, and flight conditions, that affect the performance of the engine and on the part-load characteristics of the engine as influenced by the off-design performance of the components. The many theoretical calculations of the cycle and part-load performances which have been made in the development of the gas turbine have been strikingly confirmed by practical developments of engines such as the turboprop, the two-spool, and the bypass engines, which have behaved as predicted. In the history of the development of the engine, performance analysis has proved to be a satisfactory guide for future development.

In the development of controls for gas turbine engines an understanding of the part-load characteristics is fundamental. This aspect of the control problem has been covered in this volume; the more practical aspects are dealt with in Vol. XII. The effect of the performance of components on the performance of the engine has been emphasized in Part 4, Sec. K of this volume, which summarizes the extensive study of cycle and part-load performance which has presaged the successful development of the gas turbine.

# PART TWO

# *COMBUSTION CHAMBER DESIGN*

# SECTION B

## *REQUIREMENTS AND PROCESSES*[1]

P. LLOYD

**B,1. Basic Requirements.** The combustion system of a turbojet or turboprop engine must satisfy a wide range of requirements, not necessarily more arduous but certainly much more complex than those demanded of the other basic components, the compressor and the turbine. The primary purpose is simple enough, to raise the temperature of the compressed air by reasonably efficient burning of the supplied fuel. If this were the only thing that mattered, the task would be easy enough. However, in addition to the thermal efficiency there are also two kinds of aerodynamic efficiency to consider: (1) the loss of total pressure which obviously reduces the effective pressure ratio of the engine must not be excessive, and (2) while the air delivered from the compressor may be very far from uniformly distributed (a further complication is the fact that the inlet velocity distribution is not constant over the range of flight conditions), the hot gas supplied to the turbine must be at a substantially uniform temperature, preferably with a slight gradient from a lower temperature at the highly stressed roots of the rotor blades to a temperature rather above the mean at the blade tips.

Apart from efficiency there are various other requirements. Combustion must be stable over the whole range of operating conditions, including the transient states of acceleration and deceleration, i.e. the fire must not go out nor must there be any excessive pressure pulsations excited or amplified by the combustion system. Stability must also be maintained in various disturbed or abnormal conditions such as those created by the ingestion of ice particles or water resulting from the sudden shedding of built-up ice, or of water-methanol mixture introduced for thrust boosting. Ignition must be reliable and rapid both on the ground and in flight; the equipment must stand exposure to the flame temperature, to a pulsating air stream and to oxidizing gases; and there must be no substantial residue of carbon either in the form of smoke or as an internal deposit. All of these functions must be performed within a restricted space and without excessive weight, and the system must be mechanically reliable. Lastly the equipment should be capable of consuming a reasonably wide range of fuels in the distillate class.

[1] Manuscript received Sept. 1956.

The relative importance of these various factors differs somewhat between one class of engine and another, but they are always present in any application; the merit of a combustion system cannot be exactly assessed without taking all of them into account. The question of parameters to define aerodynamic and thermodynamic performance is examined in Art. 4, but it can be stated on quite general grounds that no single performance criterion can be expected to provide an effective basis for the comparison of alternative systems.

The difficulty of satisfying these various requirements is very largely associated with the wide range of conditions over which the system has to operate, and it may be well to consider this aspect of the matter first. Any engine must operate from starting to full rotational speed, from static conditions on the ground to maximum flight speed, and from sea level to a ceiling which will generally be in the stratosphere. Fig. B,1a, for instance, shows the range of operating conditions (pressure and temperature) that would be encountered by a hypothetical engine of 8:1 compression ratio at sea level and with representative intake and compressor efficiencies, over a range of flight speeds up to $M = 2.5$ and up to an altitude of 80,000 ft. At each altitude three flight regimes are considered: full engine speed over an appropriate range of flight speeds, engine idling at Mach 1 and 0.5 (these are, of course, purely notional speeds to indicate the approximate range of conditions), and relighting with a windmilling engine again at Mach numbers of 0.5 and 1. In this last condition it is assumed that the temperature and pressure rise contributed by the compressor are both zero. This diagram illustrates fairly clearly the versatility that is required, for the operating conditions range from the high pressures corresponding to high speed flight at sea level to the subatmospheric values corresponding to low engine and flight speeds at high altitude. The figures plotted in Fig. B,1a refer, of course, to a particular engine and intake, but they show a pattern that is quite general. It will be noticed, for instance, that at a given altitude the operating conditions could have been shown without much error by a single temperature-pressure line joining the extreme values.

The temperature rise required of the combustion system at equilibrium conditions does not, of course, vary over any comparable range, and even if the transient conditions of acceleration and deceleration are taken into account the picture is not changed very drastically. Exact values depend a good deal on engine characteristics, but Fig. B,1b gives a first impression of what happens. This graph shows the changes in measured gas temperature at turbine inlet with engine speed at several conditions including start to idling, idling to full speed on a fast acceleration, equilibrium running between idling and full speed, and full speed to idling. Excluding the starting case the temperature range covered is only 0.4:1, and, allowing for temperature rise in the compressor, the

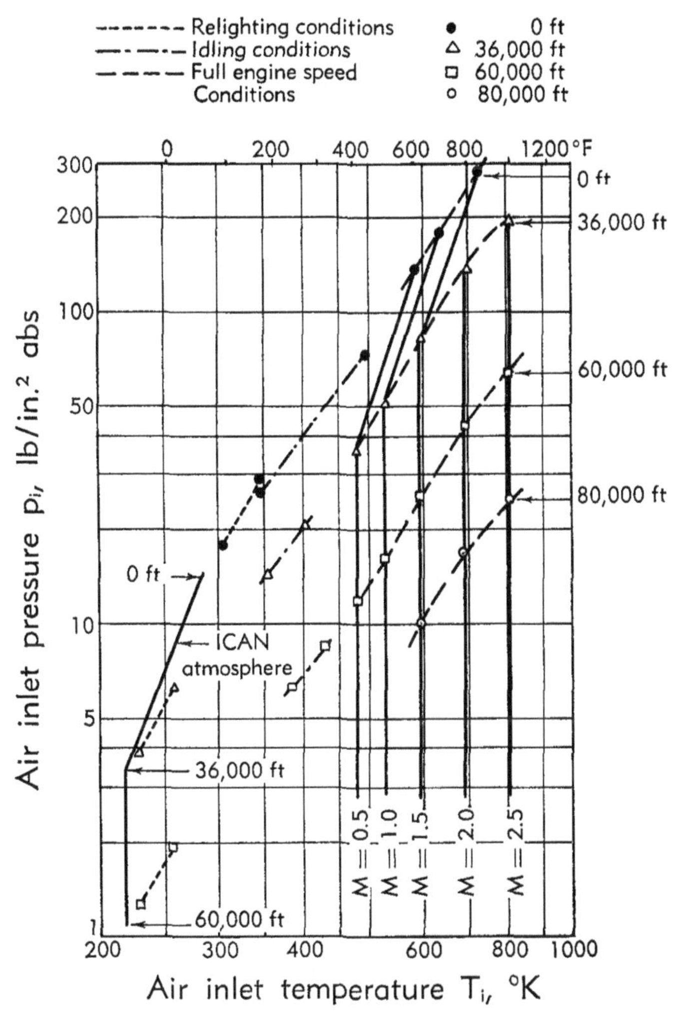


Fig. B,1a. Air temperature and pressure at the inlet to the combustion chamber of a turbojet engine under various conditions of flight. Assumptions: An 8:1 compression ratio engine having peak compressor efficiency 88.5 per cent polytropic; intake efficiency 100 per cent up to Mach 1, then as for a 2 shock center-body system. Relighting conditions are for a wind-milling engine giving no temperature or pressure rise above the ram value; flight speeds $M = 0.5$ and $M = 1.0$.

air-fuel ratio range will be rather less. There is, however, some scope for time lag between fuel flow changes and temperature response, and the true range of fuel-air ratios may be expected to exceed that deduced from these temperature values. The transient fuel flows and temperatures are, in fact, determined by fuel system settings; in acceleration, for instance, the control unit has to limit the combustion temperature rise to avoid the danger of engine surge. The data of Fig. B,1a imply that at the equilibrium conditions the air mass flow range from Mach number

1.5 at sea level to idling at 60,000 ft would be 33:1. Combining this with a fuel-air ratio range of 3:1 we deduce a fuel flow range of 100:1. In spite of the somewhat arbitrary form of calculation this is representative.

After the wide range of operating conditions the next most important feature is the spatial limitation imposed on the combustion system, which arises in these obvious ways: (1) in the typical straight-through engine layout, the length between the compressor outlet and the turbine annulus implies the length of shaft and of backbone, and must therefore be minimized to save weight and to avoid mechanical complications, (2) in all engines there is some restriction on the over-all diameter and hence

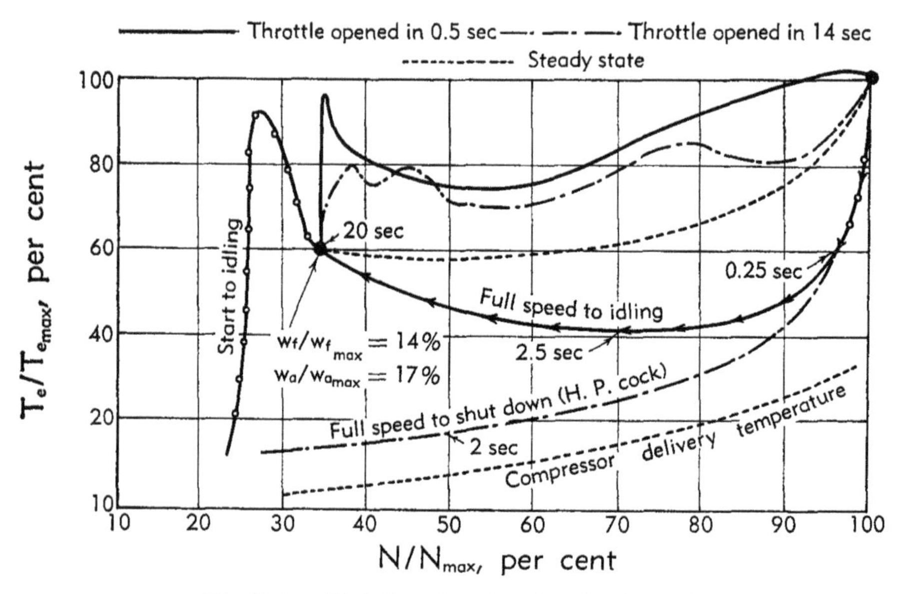


Fig. B,1b. Variation of combustion chamber exit temperature during transient operation.

on the cross-sectional area available for combustion space, and (3) when the thrust per unit frontal area is an important criterion of performance then the cross-sectional area of the combustion system may be restricted severely. In the original Whittle engine the limitation on the combustion chamber length was largely avoided by the use of the contra-flow layout, and while this arrangement implied some increase of pressure loss it was certainly a wise and useful provision at a time when combustion chamber performance was so uncertain. In later engines the straight-through arrangement has become universal and the effective length available for combustion equipment has generally fallen between the limits of 12 and 30 inches. This dimension, combined with the limit to the combustion chamber diameter derived from considerations of over-all engine diameter, thus defines the space available for the system.

These limitations bear on the functions of the combustion chamber in a number of ways. The limitation on cross-sectional area implies a minimum (idealized) air velocity based on the engine mass flow, the density at the combustion chamber inlet, and the aggregate cross-sectional area available. This velocity is generally in the range of 50 to 150 feet per second and is one characteristic of the combustion chamber loading. (This velocity may also be expressed as a Mach number, but the limitation on pressure loss keeps compressibility effects small, so that this is in no way essential.) It is to be noted that this velocity is greatly in excess of the normal flame speed of hydrocarbon-air mixtures. Next, a combination of length and cross-sectional area begins to characterize the shape of the combustion chamber and in this respect the length-to-diameter ratio (or in the annular case, the length-to-annulus width ratio) is a parameter of some importance in regard to the mixing and aerodynamic processes. Lastly, the volume available limits the time in which the various processes of mixing and burning must be completed. Since this time interval cannot be exactly deduced from the geometry of the system and from the initial and final states of the working fluid, and since there is often some doubt as to what is the rate-determining process, this characteristic has often been defined in terms of the "combustion intensity" as the rate of heat release per unit of volume and per unit of pressure

$$\frac{\text{CHU}}{\text{ft}^3\text{-hr-atm}} \text{ or } \frac{\text{BTU}}{\text{ft}^3\text{-hr-atm}}$$

This parameter is not uniquely related to residence time, but is a rough inverse measure of it, and for comparison with boiler and furnace practice it is worth making the point that turbojet intensities are in the region 2–10 million BTU/ft$^3$ hr atm (where the volume is that of the flame tube up to the plane of the nozzle guide vanes), whereas oil-fired boiler practice is in the region of 0.036–0.18 million.

However, combustion intensity is only a rough-and-ready index bearing no exact relationship to any of the component processes. To understand the significance of the volumetric limitation one must consider its effect on the separate processes of combustion and especially on that which ultimately sets the limit, namely the chemical reaction rate. This question is dealt with later.

Lastly, there is a special feature of gas turbine combustion which distinguishes it from all other heat engines, the high dilution of the working fluid which results from the relatively low cycle temperature. This dilution with excess air implies that the over-all air-fuel ratio is far beyond the inflammability limits of any liquid hydrocarbon fuels. Similarly the cycle temperature is below that at which the ignition of hydrocarbon fuels or of combustible gases can be considered instantaneous. These circumstances are responsible for various problems of stability and combustion efficiency.

To sum up: the gas turbine operates at air-fuel ratios outside the normal limits of inflammability and over a very wide range of air temperatures, pressures, and fuel flows. Combustion must take place in an air stream moving much faster than the normal flame speed of hydrocarbon/air mixtures and in a volume which implies the need for a high rate of heat release and allows but a short time for completion of the processes of burning and mixing. In spite of these limitations and handicaps, performance must be satisfactory in relation to stability, combustion efficiency, pressure loss, mixing, ignition, mechanical reliability, and other functions. These factors will be considered in greater detail later.

The requirements for afterburners are in some way similar to those for main combustion systems in that good aerodynamic and combustion efficiency, ignition on the ground and in flight, stability, and mechanical reliability are still necessary, but there are significant differences. Afterburning is essentially a high temperature process working not far from stoichiometric proportions. Although the fuel-air ratio may be varied to provide thrust modulation, mixtures are usually well within the inflammability limits. The high inlet temperature obviously favors carburation. Also, the range of operation is substantially less than for main combustion systems, the reheat being required only as a boosting function at full engine speed, so that the operating pressure level is determined by flight speed and altitude. Subsonic flight at high altitude implies pressures below atmospheric, although above the corresponding values for a main combustion system at idling condition. Spatial limitations are less rigid than in main combustion systems but are no less real. To avoid diffusion to a jet-pipe diameter greater than the rest of the engine, the gas velocity must be kept up (according to the engine pressure ratio) to the region Mach 0.2–0.3, and the jet-pipe length cannot be unrestricted either. In practice it turns out that afterburning requires combustion loadings which, in relation to maximum reaction rates, can be just about as high as those in main combustion systems.

There is also the related case of the ramjet engine which again raises characteristic combustion problems very similar to those of the turbojet, and especially afterburning. From the combustion aspect there is, in principle, no real difference between the two types of engine. The fact that in a ramjet the whole of the compression, and not merely a part of it, is accomplished by the intake means only that there is a slight difference in compression efficiency; for the same altitude and over-all compression ratio the operating conditions (of pressure and temperature) will inevitably be similar. Fig. B,1c gives a general impression of the pressure and temperature values by displaying the combustion chamber entry conditions for the case of a two-shock, two-dimensional intake at flight

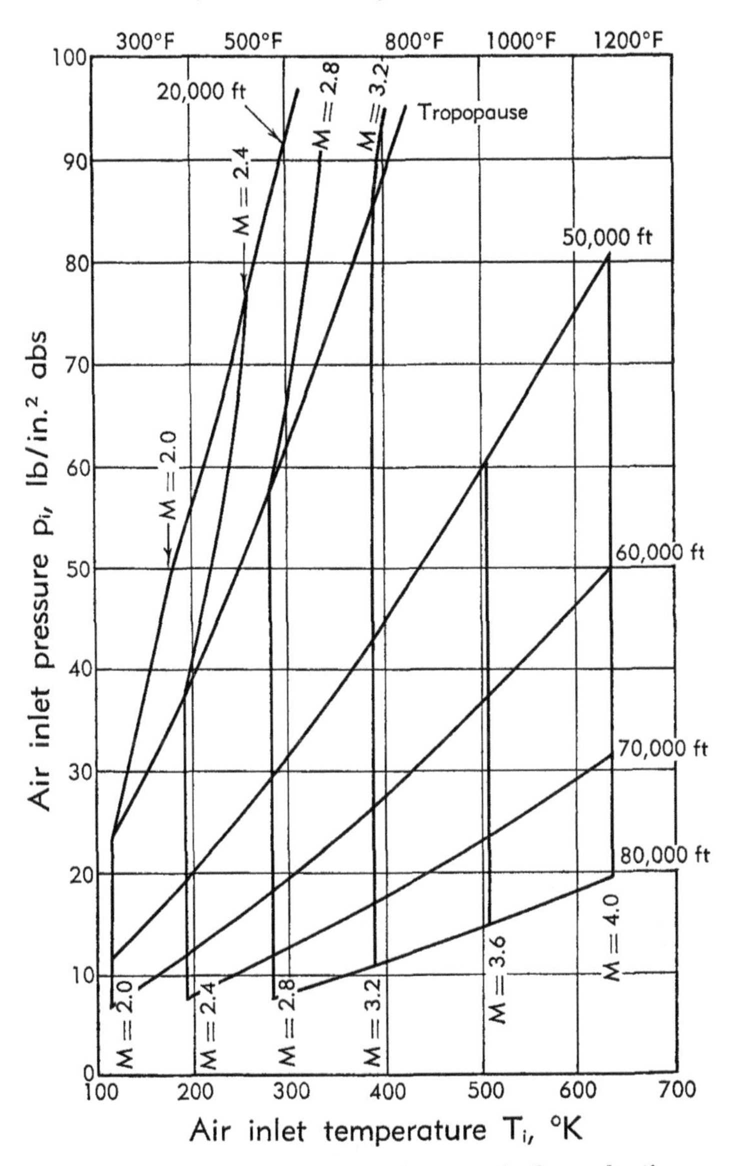


**Fig. B,1c.** Total temperature and pressure in the combustion chamber of a ramjet with two-shock intake (two-dimensional).

speeds between Mach 2 and 4. In spite of the absence of low temperature conditions, low pressures giving conditions unfavorable for combustion can still be encountered at high altitude.

The outstanding contrast with turbojet requirements is in the need for higher aerodynamic loading coupled with lower aerodynamic loss which arises in the lower supersonic speeds as a result of the limited pressure ratio that is available. Inlet velocities to the combustion system can be in the region of 250 feet per second, compared with 50–150 feet per second; pressure loss factors defined as the total head loss divided by the dynamic head based on total mass flow, inlet density, and maximum cross-sectional area must then be in the region of 2–6 velocity heads compared with 10–40. With normal tailpipe lengths, available residence time is in the region of 5–8 milliseconds. As a compensation for these fairly arduous requirements some of the secondary factors of the gas turbine case, such as the need for uniformity in the outlet temperature, are of less importance. The main differences between ramjet and afterburner combustion are that, with afterburning, inlet oxygen concentration is depleted and the inlet temperature is correspondingly increased; velocity is accordingly higher.

In general, combustion requirements, like other engine factors, are dominated by the flight plan of the aircraft or missile so that the distinction between the combustion problems of subsonic and supersonic flight is becoming increasingly important. In the subsonic cases the most important factors are probably the wide range of operating conditions and the special problems of maintaining stability and efficiency at low pressure and temperature. In supersonic flight, subatmospheric temperatures do not occur and combustion pressure levels are also higher so that the most arduous conditions for chemical reaction may be avoided. However, other requirements, such as that of wall cooling, become more severe.

## B,2. Techniques.

General Discussion. When gas turbine development started there were no ready-made processes or techniques capable of meeting all these requirements. Reciprocating engines had shown that high rates of heat release per unit volume could be achieved, but their operating conditions were different and their methods irrelevant. Furnace and boiler techniques were, and still are, limited to relatively low velocities and heat release rates and to a narrow range of operating conditions. It followed that new techniques had to be developed. A fortiori there was no body of theory which would have permitted designing from first principles. The process of combustion had attracted the attention of chemists ever since the days of Lavoisier, but inevitably they had concentrated their attention on the more easily observed slow combustion reactions at low temperatures and on the mechanisms whereby a complex hydrocarbon molecule is degraded and oxidized to simple products. In both the spark ignition engine and the diesel these low temperature reactions are significant, but in the turbojet and the ramjet they are unimportant. Even now, after ten years of intensive research, not only on the chemical but also on the physical and aerodynamic aspects of combustion, the basic theory, as presented, for instance, in Vol. II of this series, provides no quantitative treatment of high temperature reactions and no basis for a design procedure. The study of chain mechanisms and laminar flame processes remains an academic exercise having no real relation to practice.

The development of combustion techniques has therefore followed inevitably along empirical lines, and since the over-all process of changing fuel and air into hot uniform combustion products could be tackled in various ways, development effort has had the choice of a number of methods.

This point is fundamental. There is no unique solution to the combustion problem, and the essential processes of (1) mixing fuel, air, and piloting gases, (2) bringing about chemical reaction, and (3) diluting with excess air can be combined or separated in a variety of ways. Scientific treatment of technical combustion depends on the identification and analysis of the basic physical and chemical processes which are rate-determining or which in some way impose limits on the system, and these processes differ according to the practical system which is being used. Droplet burning would be irrelevant in a system using external fuel vaporization and the reaction rate would not impose any limitation in a design based on mechanical control of air-fuel ratio in the main combustion zone, but in other systems either of these can be the dominant factor. Examples could easily be multiplied. In view of this background, and since it is essential to simplify matters by reducing the number of elementary processes that are to be analyzed, it is necessary to start by considering in general terms the techniques which for various reasons, partly engineering and partly historical, have come to achieve practical importance.

Whittle in his first attempt at a combustion system turned to the analogy of the Primus stove, using the heat of the main reaction zone to vaporize the fuel. At its design condition this system worked well enough; its failure was due to a lack of flexibility. With a wide range of conditions to be covered this was, in fact, a major fault and the technique has never been used by itself since. In other ways Whittle's early design, Fig. B,2a, [*1*] had features the merit of which has been confirmed by experience and which has become almost axiomatic: the air-cooled flame tube, the separate primary zone to which only a part of the air is admitted, flame stabilization in a region of high turbulence, and secondary air injection through holes in the flame tube. The next step in the evolution of turbojet combustion was the system based on pressure jet fuel injection, swirling primary air, and stub tubes which provided the solution for the first successful Whittle engines of 1940–1941. This technique, which was worked out by Hawthorne and Lubbock [*1*], was in effect a development of boiler and furnace techniques, its most important features being the flame-stabilizing swirl flow produced in the primary zone and the use of a fuel spray produced by a pressure jet. Many current designs are lineal descendants of this scheme (Fig. B,2b). A variant of this type which has proved equally important is that introduced in the United States shortly afterwards, in which the primary air swirler was replaced by a blind end with radial air injection, giving a significantly different flow pattern.

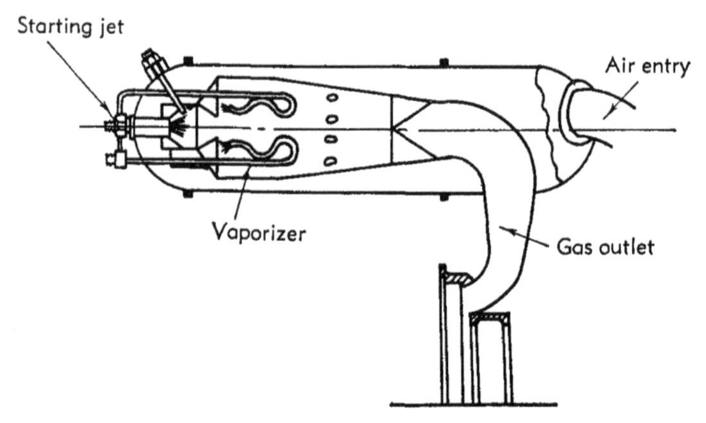


Fig. B,2a. Early Whittle combustion chamber with internal fuel vaporizer.

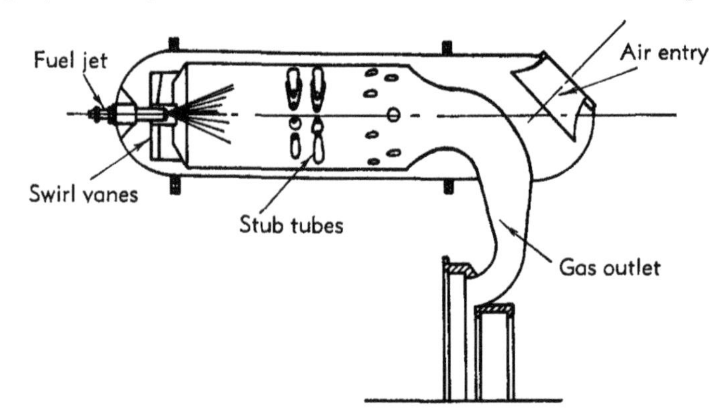


Fig. B,2b. First Whittle combustion chamber with fuel spray injection.

One other type which must be mentioned is that first investigated by Hottel and Williams and now of general application to ramjets and turbojet reheat systems [*2,3*], the method of precarburation followed by flame stabilization in the wake of a bluff body. In its original form this system used a simple V-shaped gutter as the flame holder, but in later developments the flame holder has evolved into more complex forms such as multiple gutters or perforated cones. There are two important features here, first the separation of the carburation process from the burning zone and next the stabilization of flame in only a small part of the mixture, followed by flame spreading from the pilot to the main charge.

This list is by no means comprehensive, but the systems mentioned are at least representative and illustrate the main practical techniques which contribute to gas turbine and reheat combustion and which require some discussion before we can go on to define the basic physical and chemical processes, such as evaporation of fuel droplets and reaction in homogeneous systems. These practical techniques are fuel injection, flame stabilization, and the continuance and completion of combustion, mixing, and mechanical design. It will be convenient to consider them in order.

*Fuel injection.* The first and most obvious requirement in fuel injection is that, if the fuel is not prevaporized, it must be finely divided so that the time required for evaporation or burning of the spray can be kept short. Less obvious, but equally important, is the part that fuel injection has been found to play in assisting the flow pattern in the stabilizing zone and in obtaining the required distribution of fuel in the reaction zone. Not only is the spray cone angle from a pressure jet significant, but the spray momentum is also important; in fact, investigations of combustion efficiency in turbojet chambers at low density conditions have shown that performance may sometimes be more effectively correlated with fuel spray momentum than with droplet size or any other yardstick. The direction and momentum of the fuel spray are therefore other factors that must be borne in mind.

In precarbureted systems fuel injection is again a crucial factor, and, in spite of the difference of mechanism, fuel distribution and droplet size are still important although both ramjet and reheat systems have sufficient air velocity and temperature to make the atomization process almost automatic. Fuel distribution in these systems is determined, as Probert [*4*] has pointed out, by the combined effects of the fuel-air momentum ratio which determines the penetration of the fuel jet and the position of the virtual source, and of the turbulent diffusion process by which the fuel dispersion is brought about.

*Flame stabilization.* Flame stabilization has been extensively studied both in the typical turbojet case in which the mixing of fuel spray and air is combined with the flame-holding flow pattern, and in the ramjet case where a precarburetted stream, which may be idealized as a uniform fuel-air mixture, forms a stable flame in the wake of a baffle. The former case has been studied especially by Clarke [*5*] and his coworkers; Fig. B,2c gives a pictorial representation of the flow pattern which is created in a representative configuration. The essential feature, so far as flame stabilization is concerned, is the toroidal flow reversal, driven in the main by the first row of radial air jets but assisted by the swirl of the axial air stream and the momentum of the conical fuel spray, which carries burning gases back to the region where fresh fuel and air are entering. The basic flow in the wake of a typical ramjet or afterburner baffle is not really dissimilar. As Hottel, Williams, and Scurlock [*6*, pp. 21–40] first pointed out, the shedding of eddies in the wake of a transverse rod or channel is suppressed when burning takes place and a stable flame anchoring reversal is formed. In the simpler case of the ramjet-type of baffle the essential "aerodynamic" factors are the size of the flow reversal region, the fraction of flow that enters it, and the heat loss. In the turbojet case not all of the air which passes through the radial holes enters the flow reversal and therefore the flow rate is no easier to define than before. In addition we have to allow for the heterogeneity of the fuel-air mixture and for severe concentration gradients. Clarke has analyzed this case in some detail.

*Combustion.* In the typical combustion system the distinction between the processes of flame stabilization and efficient burning is usually clear. Bluff-body pilots may provide a good flame anchor over the whole

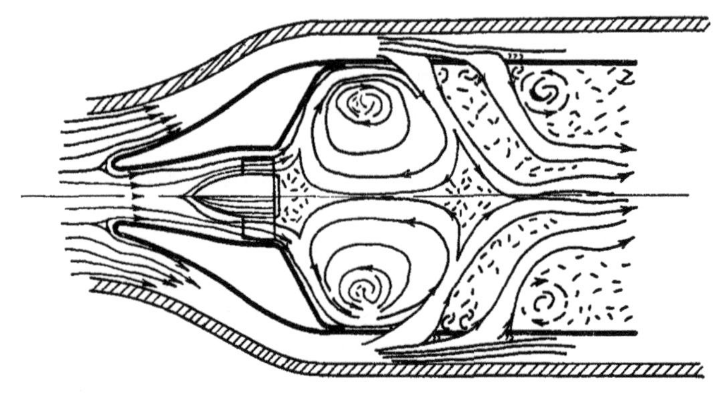

Fig. B,2c. Flow pattern in a spray injection chamber (after Clarke [*5*]).

of the operating range without necessarily igniting the whole of the charge. In a typical turbojet system, under arduous conditions, we can also get a stable flame in a small part of the primary zone, but with the major part of the charge passing through unburned. What is the nature of this combustion inefficiency and what are the factors that determine it? We can say on first principles that combustion loss can arise in a number of different ways:

1. In the first place a part of the fuel may pass through the system without being ignited at all. This can come about either through bad atomization leading to large drops which lack the time to burn through, or through bad mixing in the piloting or main reaction zone.
2. In the second place the reacting mixture may be quenched by the sudden admixture of excess air. In this case the unburned constituents are mainly carbon monoxide and hydrogen, although with some other "fragments" such as methane and formaldehyde. Spectroscopically this condition is characterized by the emission of the hydrocarbon

flame bands, first identified by Hawthorne and Gaydon [7] in 1941. Similar quenching may also occur in small or cluttered flame tubes by contact with cold surfaces.

3. Reaction may still be taking place at the outlet plane in over-rich eddies or local streams due to incomplete mixing at some stage of the process.
4. Reaction may be incomplete in a homogeneous charge simply because the residence time has been less than the time required for reaction.

In practice each and all of these mechanisms are operative and the following elementary processes must be taken into account: atomization, burning of droplets, reactions leading to ignition, mixing processes, the reaction rate subsequent to ignition, and flame-quenching processes.

*Mixing.* Of the secondary mixing process little need be said at this stage. The function here is, of course, the dilution of the hot, burned or burning, gases of the primary zone with secondary air to complete burning or with dilution air to reduce temperature to the turbine cycle temperature. Mixing by injection through holes in the flame tube wall is a clearly definable aerodynamic process in which the geometry of the system is the design variable, while flow velocity, pressure loss, and uniformity of outlet temperature are the operating variables. The stub tube of the early Whittle engine is, however, the prototype of a whole variety of possible mixing systems in which the dilution air, instead of being forced by virtue of its momentum to penetrate the flame, is ducted to the place where it is wanted. Here the aerodynamic process of mixing is complicated by evident mechanical and metallurgical factors, and with practical materials such ducted mixing systems also require the solution of a cooling problem.

*Mechanical design.* The combustion chamber is a lightly stressed component and it might well be thought that its mechanical design would be of trifling importance, but this is not necessarily the case, and under certain conditions mechanical failure can come about very quickly. The stresses developed in combustion chambers must evidently be a combination of alternating stresses induced in the light structure by pressure pulsations in the flow, alternating stresses caused by rapid heating and cooling, i.e. thermal shock effects, and steady thermal stresses caused by uneven heating. As to the relative importance of these three terms no generalization is possible and the subject awaits analysis. Whichever is the predominant stress pattern, metal temperature must be the determinant factor and wall cooling the main defense.

**B,3. Processes.** The object of the somewhat discursive previous discussion has been to underline the practical techniques which have found effective application in air-breathing jet engines and so to obtain a first indication of the elementary processes that are likely to be most important. The conclusions are as follows:

1. The inescapable process of air flow through the chamber in which compressibility effects and natural turbulence are not necessarily important but must obviously be taken into account; also the accompanying process of convectional heat transfer.
2. The equally basic process of chemical reaction in the (gaseous) burning zone or zones.
3. The various processes of mixing: fuel atomization, vaporization of fuel in air, mixing by turbulent diffusion, secondary mixing or dilution. Ideally none of these need impose limits on performance since they could all be dealt with by providing mechanical solutions; but in practice the perfect mechanisms are not available.
4. The secondary processes which may nevertheless be of the greatest importance: pressure pulsation in continuous burning, heat transfer to flame tube, etc.

The list is not an exhaustive one; for a comprehensive analysis of all the component processes which may contribute to turbojet and ramjet combustion the reader is referred to the work of Stewart [*8*, pp. 384–413].

*Air flow and heat transfer processes.* The important aerodynamic processes in combustion are those concerned with the various mixing processes which are discussed in a later section. All that need be said here is that we are concerned with fully turbulent flow (Reynolds numbers based on a compressor annulus of the order of $10^5$) and that while the possibility of Reynolds number effects must be borne in mind there is no reason to suppose that these are important. Compressibility effects are of marginal importance in the main combustion systems of turbojets but are more significant in ramjet and reheat systems. In general, therefore, both Reynolds number and Mach number should be taken into account.

*Fuel characteristics and combustion reactions.* In this fairly complex situation there is one big simplification to be made. In the continuous burning, constant pressure cycle of turbojet and ramjet engines we are almost exclusively concerned with combustion reactions taking place at high temperatures, say from 1000°K upwards. Under these conditions the idiosyncrasies and anomalies of low temperature reaction, the big differences between various hydrocarbons, the unpredictable combustion limits, and the uncertain effects of pressure and surface have largely disappeared. Chemically the process by which a hydrocarbon molecule is oxidized and broken down must still be complex, but the reaction as a whole behaves as a simple bimolecular one and the differences between hydrocarbons of different structure—straight chain and branched paraffins, olefins, aromatics, and so forth—are small. Add to this the fact that in practice we are concerned with fuels which are always blends of various hydrocarbon types and it follows that in characterizing fuels we need not concern ourselves with hydrocarbon composition as such. We can regard our basic fuels simply in terms of their physical properties and their elementary analysis. Fig. B,3a, for instance, shows the calorific value plotted against the carbon: hydrogen ratio [*9*]. These quantities, together with the boiling range and the viscosity, tell us almost all we need to know. Even the secondary characteristic of carbon-forming

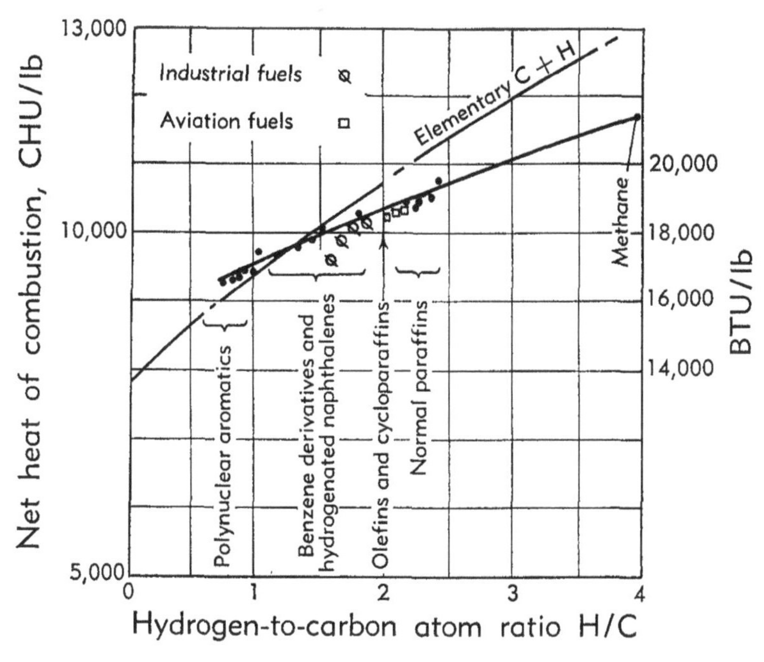


Fig. B,3a. The net heats of combustion and the hydrogen-to-carbon (atomic) ratios of some hydrocarbon fuels.

proclivity is quite closely defined in terms of the C : H ratio and the boiling point.

Evidence for these generalizations has gradually accumulated over the last ten years. In 1946 Lloyd [*10*] pointed out that the inflammability limits of hydrocarbon, alcohols, ketones, and other organic compounds could be quite well correlated on a thermal basis, the weak limits for instance being defined by a mixture strength giving a constant (theoretical) combustion temperature of 1300°C. This was confirmed by Egerton and Powling [*11*], who also showed that small additions of combustion promoters had negligible effect on either weak limits or flame speeds. From 1948 to 1952, Mullins [*12*] carried out a series of measurements of ignition delay by injecting a variety of combustibles into a stream of air at 1000–1300°K and showed that the differences between individual hydrocarbons, though measurable, were markedly less than those recorded at lower temperatures. His work also showed that the effect of pressure on the kerosene-air reaction was to reduce the ignition delay in rough proportion to the pressure increase, as would be expected in the case of a bimolecular reaction. Finally, Longwell [*13*] devised a combustion model which is even more directly relevant to jet engine combustion conditions, a spherical reaction vessel, into the center of which carburetted air can be injected at a high velocity, thus giving a high turbulence level with very rapid mixing of the fresh charge with burning mixture. The object here was to optimize all the mixing processes and in this way to measure the limits imposed by the homogeneous reaction process itself. Longwell's work showed (*inter alia*) that in his system the mass reaction rate was proportional to $p^{1.8}$.

This evidence led to the idea, worked out by Longwell and his coworkers [*13*], by Avery and Hart [*14*], and by Bragg [*15*], that the Arrhenius equation for a homogeneous bimolecular reaction might be applied to the combustion of complex hydrocarbons at high temperature.[2] The main assumption was that, whatever the mechanism may be, the process is controlled by some bimolecular reaction and that the characteristics of the higher hydrocarbons are sufficiently similar to make it legitimate to apply to mixed hydrocarbons the best-established kinetic data for a pure compound. The qualitative relationships which derive from the theoretical treatment have been very clearly expounded by Saunders and Spalding [*17*], but the quantitative treatment has been taken furthest by Herbert [*18*] and the very brief summary that follows is based on his work.

The textbook equation (see, for instance, II,D) gives the reaction rate per unit volume as $= PZe^{-E/RT}$, where $Z$ is the collision number which for given molecular species is proportional to $n_1 n_2 \sqrt{T}$, with $n_1$ and $n_2$ being the numbers of the reacting species per unit volume. The exponential term gives the fraction of the collisions having the requisite energy for reaction and $P$ is a steric factor.

For the idealized and simplified reaction which is assumed, $P$ is taken as constant irrespective of mixture strength. The reaction rate per unit volume is written as $KC_0 C_f \sqrt{T}\ e^{-E/RT}$ where $K$ is the over-all rate constant and $C_0$ and $C_f$ are concentrations of oxygen and fuel.

This equation at once shows the way in which the reaction rate varies in a given mixture as it progresses to completion. In the early stages the concentration of reactants is ample and the temperature is the limiting factor. In the final stages the temperature approaches its peak but the concentration of one or both reactants begins to approach zero and so

[2] The first examination of a combustion process controlled by a simple bimolecular reaction was, however, that of Childs [*16*], who in 1950 derived a correlation between combustion efficiency and the physical variables, pressure, air temperature, and mean velocity, based on inlet conditions and maximum cross-sectional area.

the rate falls. The maximum reaction rate therefore occurs at an intermediate condition which, as it turns out, corresponds to about 80-per cent "reactedness."

By equating the quantity of mixture reacting to an efficiency term times the quantity of mixture entering, a general expression may then be derived relating the engineering quantities: mass flow rate, volume, and pressure with the concentration terms, the temperature, and the reaction constants. For the particular case of weak mixtures of kerosene ($\mathfrak{M} = 168$) with air, for instance, the relationship is

$$\frac{w_a}{Vp^2} = 42.7 \frac{K}{R^2} \frac{e^{-E/RT_r}}{T_r^{\frac{3}{2}}} \frac{(1 - \epsilon)(1 - \epsilon\lambda)}{\epsilon}$$

where $w_a$ = air mass flow, lb/sec
$V$ = reaction volume, $ft^3$
$p$ = pressure, atm
$K$ = over-all rate const, $ft^3$/sec-lb-°$K^{\frac{3}{2}}$
$R$ = gas const
$E$ = activation energy
$T_r$ = reaction temperature
$\epsilon$ = oxygen consumption efficiency, i.e. oxygen consumed ÷ oxygen equivalent of fuel flow. (This definition of efficiency implies the simplifying assumption that inefficiency arises from unreacted fuel.)
$\lambda$ = mixture strength as a fraction of the stoichiometric proportion.

A similar assumption as to the nature of combustion loss in rich mixtures ($\lambda > 1$) gives

$$\frac{w_a}{Vp^2} = 42.7 \frac{K}{R^2} \frac{e^{-E/RT_r}}{T_r^{\frac{3}{2}}} \frac{\lambda(1 - \epsilon)^2}{\epsilon}$$

The next step is to assign values to the reaction kinetic constants $K$ and $E$ and to make use of the relationship (for a given initial temperature) between $\epsilon$ and $T_r$. The best available kinetic data are those given by Longwell's experiment referred to above. By using these, with the further assumption that the difference between Longwell's pressure exponent 1.8 and the whole number 2 is due to less than perfect homogeneity of the reaction conditions in the spherical bomb, a quantitative solution may be derived. This gives us, for the idealized case considered and within the limitations imposed by the assumptions made, the quantitative relationship between the air loading $w_a/Vp^2$, the equivalence ratio $\lambda$, inlet temperature $T_i$, and the efficiency. For the case of a constant inlet temperature (500°K, for instance) the resulting solution is given in the curves of Fig. B,3b (after Herbert [*18*]).

A further generalization (first suggested by Bragg) has been to find a unique relationship between the air-loading parameter and the inlet temperature. Herbert shows that the available data are well matched by the relationship

$$\text{maximum air loading} \sim e^{T_i/n}$$

where $n$ is a function of $\lambda$ best described by

$$n = 220\left(\sqrt{2} \pm \ln \frac{\lambda}{1.03}\right)$$

where the + refers to rich mixtures, $\lambda > 1.03$, and − refers to weak mixtures, $\lambda < 1.03$.

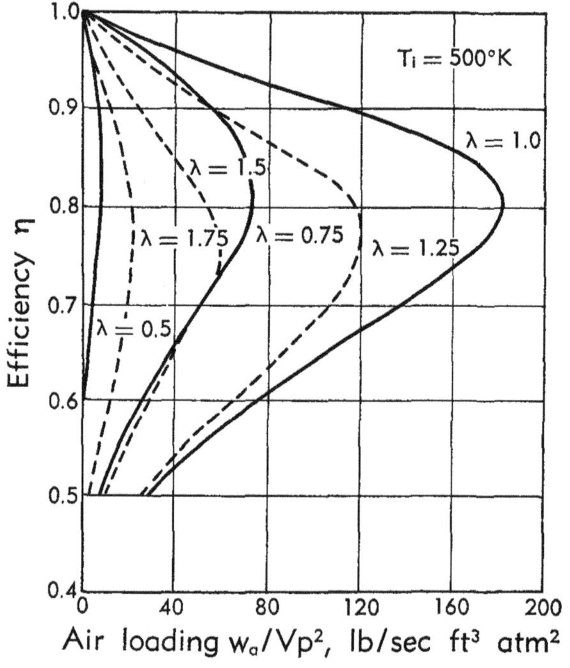


Fig. B,3b. Limiting reaction rates for a homogeneous adiabatic system (kerosene fuel) at various equivalence ratios and constant inlet temperature.

By this step the parameters have been reduced to three: the corrected air loading $w_a/Vp^2e^{T_i/n}$, the equivalence ratio $\lambda$, and the efficiency $\eta$.

This theoretical treatment incorporates a number of assumptions and some extrapolation and is based on somewhat tenuous reaction kinetic data. But imperfect though it may be, it does give a fair indication of the limits which the chemical reaction rate will set to the loading of combustion systems and of the behavior which must characterize systems in which this is the rate-controlling process. As to the effect of fuel characteristics, later work [*19*] has shown that in the stirred reactor the effects of hydrocarbon type are quite small. The reaction rates of *n*-heptane and iso-octane, for instance, were within 20 per cent of one another, half of the divergence being due to a difference in flame temperature.

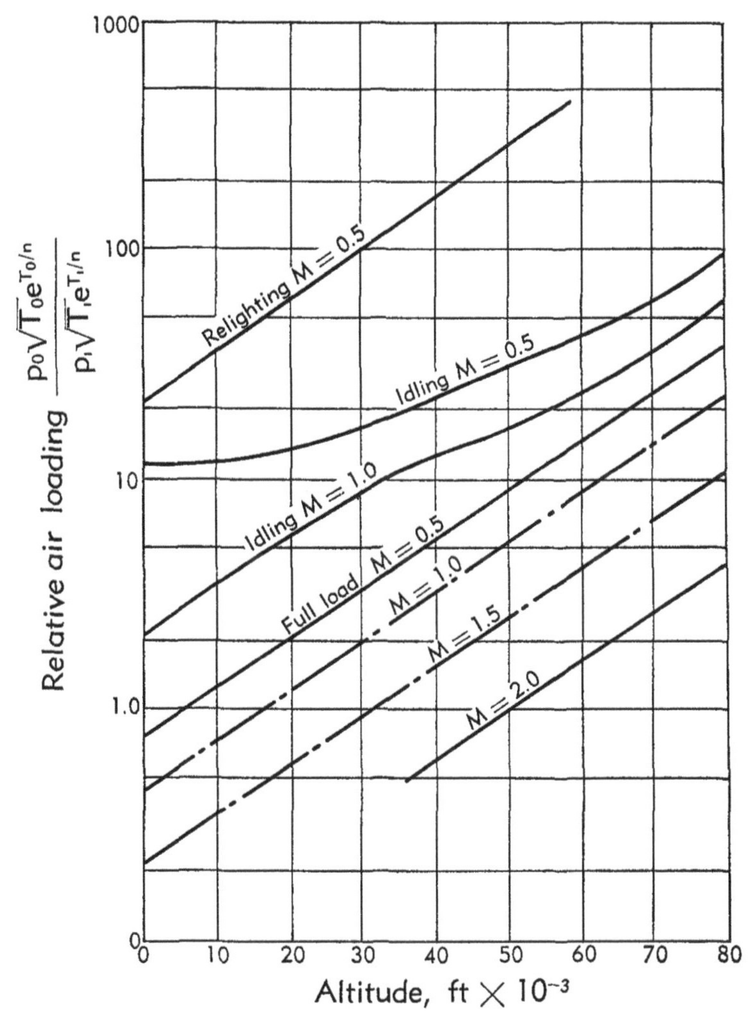


Fig. B,3c. Relative air loading as a function of flight speed and altitude, based on the "conventional" engine of Fig. B,1a. (Referred to sea level static condition.)

The air loading of a turbojet combustion system is, of course, a function of operating conditions. Since the nondimensional flow $w_a \sqrt{T_i}/p$ is substantially constant over the operating range, and since $V$ must also be constant, it follows that for a given engine the corrected air loading is proportional to $1/p \sqrt{T_i}\, e^{T_i/n}$.

This quantity is plotted in Fig. B,3c for the datum engine of Fig. B,1a to show the range of variation. It is clear that the effect of the operating range can outweigh that of the design factors. In absolute terms the corrected air loading of typical turbojet primary combustion zones at ground level static conditions is in the region 0.02–0.1 lb/sec-ft$^3$-atm$^2$ and the extreme operating conditions at high altitude may therefore imply corrected air loadings of 10 lb/sec-ft$^3$-atm$^2$ and over. Since the maximum corrected air loading (corresponding to blowout at optimum equivalence ratio) is about 35 lb/sec-ft$^3$-atm$^2$ and since a finite range of stability is obviously necessary, it is clear that the demand for operation at extreme conditions such as relighting at great altitudes can bring us very close to the limits imposed by the homogeneous adiabatic reaction process.

*Fuel injection.* Consider first the process of fuel injection by pressure (swirl) jet. This is a process which has been the subject of extensive research by Taylor [*20*], who has dealt theoretically with the nature of the flow in the jet, and by Fraser [*21*], Longwell [*22*], Joyce [*23*], Radcliffe [*24*], Carlisle [*25*], and others who have studied flow and atomization experimentally. As a result we know as much about this as about any of the component processes of combustion. Concerning the flow through swirl jets, for instance, Radcliffe has studied the performance of a family of injectors based on common design rules, using fluids covering a range of densities and viscosities. He showed that for this given type of jet there was a unique relation between $w_f/D^2\sqrt{\rho\Delta p}$ and $w_f/\mu D$

where $w_f$ = fuel flow rate, lb/hr
$D$ = orifice diameter, in.
$\Delta p$ = pressure across injector, psi
$\rho$ = fluid density, lb/ft$^3$
$\mu$ = absolute viscosity, centipoise.

$w_f/D^2\sqrt{\rho\Delta p}$ is a form of the discharge coefficient and $w_f/\mu D$ a form of the Reynolds number.

At $w_f/\mu D$ greater than 1600 the flow in the jet was found to establish itself with the development of the hollow conical spray, and above a value of 3000 (corresponding to $Re = 4000$) the discharge coefficient remained nearly constant. At low Reynolds numbers the effect of viscous forces is to thicken the fluid film in the final orifice and so to increase the discharge coefficient which for a fully developed spray is about 0.3. These critical values apply only to the particular design tested, but the behavior of other designs is qualitatively similar.

In practical applications, however, and with aircraft fuels of low viscosity, the convention is to regard the condition of low Reynolds number as of minor importance and to assume a constant discharge coefficient characterizing a given injector by its flow factor defined as

$$\frac{\text{volumetric flow rate}}{\sqrt{\text{fuel pressure drop}}}$$

the range of flow factors with which we are concerned in practice being from, say, 0.2 to 7. The flow rate is in units of imperial gallons/hr and the pressure drop in psig.

The mechanism of atomization by pressure jet was first elucidated by Fraser [26], who showed by flash photography how the expanding film of fuel is perforated before breaking down first into filaments and then into individual droplets. Quantitative studies have shown that it is difficult to generalize about the kind of droplet size distribution that results in the spray, but Joyce's work indicated that the sprays conformed approximately to the Rosin-Rammler distribution given by

$$\frac{R}{100} = e^{-(x/\bar{x})^n}$$

where $R$ is the percentage residue above size $x$, $\bar{x}$ the size factor, and $n$ the distribution factor. The distribution factor $n$ does not vary widely and is in general in the range $2.5 \pm 0.2$, but $\bar{x}$ varies widely, and the effects on atomization of such factors as injector size, fuel pressure, and viscosity have been examined in great detail. Conclusions may be summarized in the form

$$\text{Sauter mean diameter } (SMD) = \text{const} \times \frac{w_f^a \nu^b}{\Delta p^c}$$

the Sauter mean being the droplet size having the same specific surface as the spray while $a$, $b$, and $c$ are empirical exponents and $\nu$ is the kinematic viscosity. The value of $b$ is generally found to be about 0.2 and for the case of kerosene ($\nu = 2.5$ centistokes at 10°C) a fair approximation is given by

$$SMD = 182 w_f^{0.25}/\Delta p^{0.4} \text{ microns}$$

$w_f$ being in lb/hr and $\Delta p$ in psig.

These data all relate to the idealized case of atmospheric pressure and still air, and the effect of ambient pressure has yet to be determined.

How does this performance relate to the requirements? The obvious difficulty is, of course, that associated with operation over a wide range of fuel flows, for, long before the 100:1 range has been achieved with a simple jet having a reasonable top pressure, the swirl jet flow will have broken down altogether. The problem is illustrated by Fig. B,3d which shows the variation of fuel pressure, mean particle size, and cone angle over the operating range of flows. The same figure also illustrates the performance of two of the devices which have been used to overcome these troubles: the double injector which consists of a small jet tucked inside a large one with a change-over valve to bring the large one into action above a certain flow, and the rather more subtle spill control system in which the flow through the swirl chamber is maintained as net flow is reduced, by spilling fuel from the back of the swirl chamber and recirculating it. Both are effective although the changes in the cone angle are significant. Other devices for the maintenance of atomization at low flow

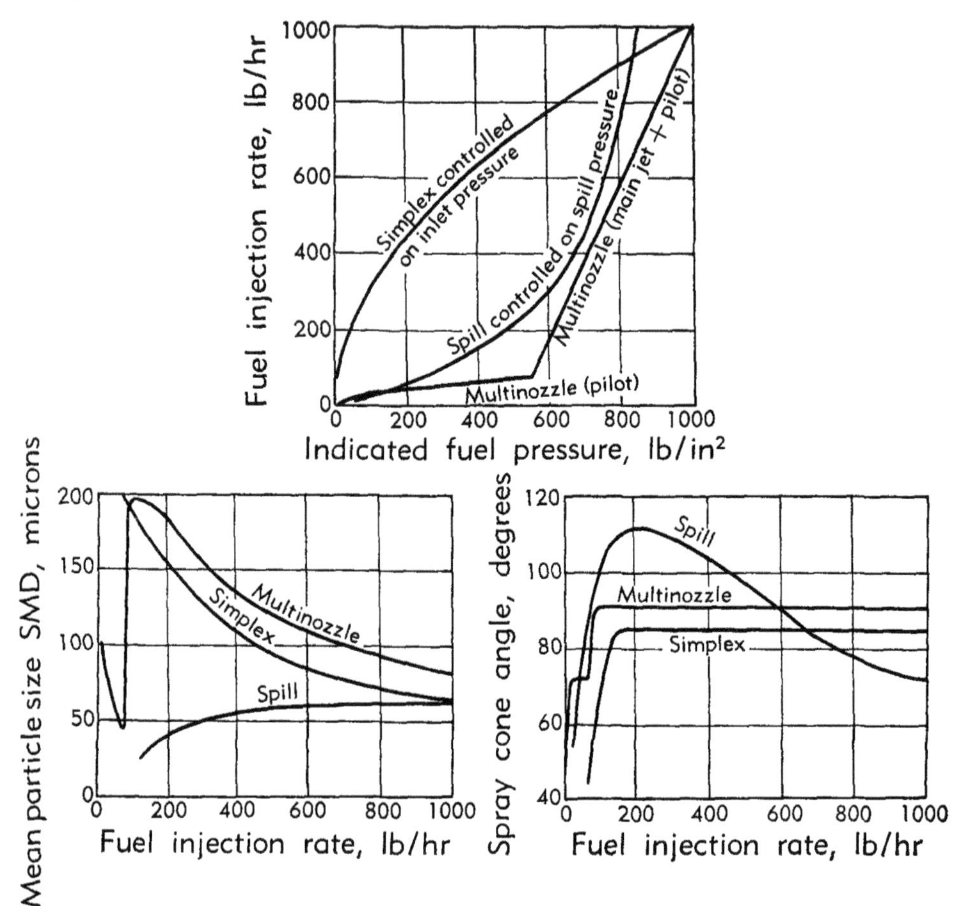


Fig. B,3d. The estimated performance characteristics of three types of fuel injection: simplex, spill, and multinozzle. (Top) fuel flow rate vs. fuel pressure, (bottom left) mean droplet size vs. fuel flow rate, (bottom right) spray cone angle vs. fuel flow rate.

are also available, e.g. the use of the energy of the air stream over the injector, the so-called air-boost system.

But the question remains whether the level of atomization obtained in this way is sufficient to provide the high rates of burning required. A partial answer may be obtained by combining a theoretical study by Probert [27] of the burning rates of a spray conforming to the Rosin-Rammler distribution with practical determinations by Godsave [28, pp. 818–830] of the evaporation constant for burning droplets. This gives the percentage of fuel in the spray remaining unburned as a function of time and the result is plotted for typical mean diameters (of the original spray) in Fig. B,3e. The comparison which this graph gives between the times required for spray burning and the residence times available is, however, no more than an approximation, for both the atomization data and the evaporation constant are taken from measurements made under idealized still-air conditions. In practice, air blast effects tend to improve

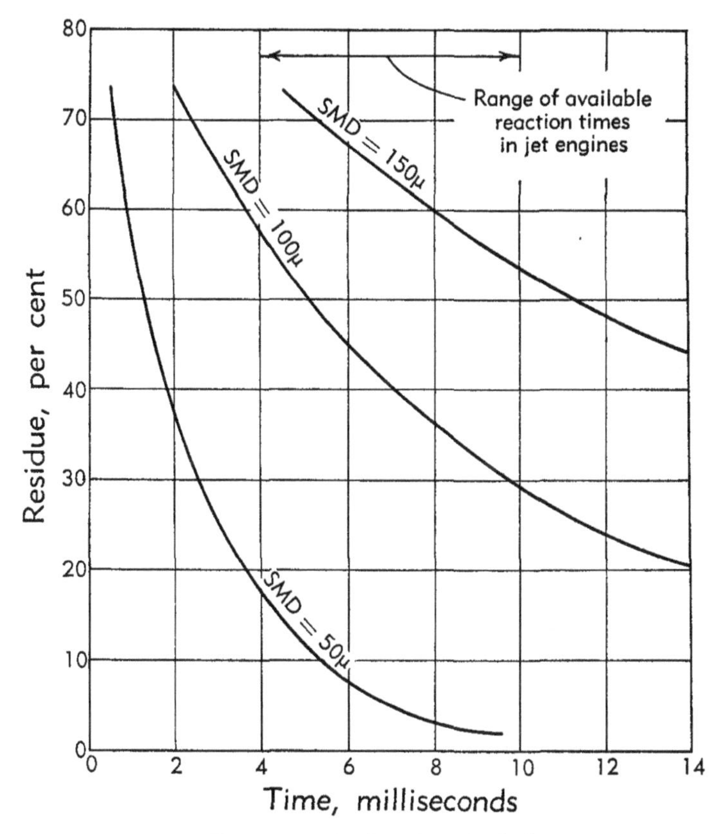


Fig. B,3e. The progressive burning of a fuel spray conforming to the Rosin-Rammler distribution law.

atomization, and turbulence increases the evaporation constant. Although a certain allowance must be made for these factors, nevertheless the figure gives a first indication of the course of events to be expected in a system controlled by droplet burning.

The second important technique of fuel injection is atomization by air blast. Here fuel metering is carried out either at a central distributor or by a simple orifice, and the energy of an air stream is used to break up the fuel into droplets. The physics of this process has also been investigated and the classical work is that of Nukiyama and Tanasawa which established the dependence of droplet size on the surface tension, density, and absolute viscosity of the liquid; on the relative flow rates of air and liquid; and on the relative velocity of air and liquid. Their work shows that for kerosene fuel the expression reduces to

$$SMD, \text{ microns} = \frac{11930}{v} + 58.5\left(\frac{w_f}{w_a}\right)^{1.5}$$

In practice the effect of $v$, the relative velocity between air and fuel (ft/sec), predominates over the effect of the relative flow rate $w_f/w_a$. Table B,3 shows the atomization obtained over a range of conditions. It is clear that with the air quantities and velocities available in air-breathing engines there is no difficulty in obtaining finely divided fuel sprays by this technique, hence the wide application of the method of natural air blast in precarburetted systems. It must, however, be noted that, as in the case of pressure jet injectors, the effect of ambient air pressure on the atomization process has not been effectively measured.

*Table B,3.* Mean droplet sizes (microns) obtained by air blast atomization over a range of conditions.

| Relative flow rates $w_f/w_a$ | Relative velocity $v$, ft/sec | | | |
|---|---|---|---|---|
| | 100 | 200 | 300 | 400 |
| 0.5 | 138 | 79 | 59 | 49 |
| 0.1 | 120 | 61 | 41 | 31 |
| 0.05 | 119 | 60 | 40 | 30 |

The process of fuel distribution by turbulent diffusion in the wake of the injector, which is, in practice, a more important variable than atomization has been studied by Longwell (II,J).

The third way of preparing a liquid fuel for combustion is, of course, by prevaporization, and this can be done in various ways:

a. By means of a fuel "boiler" heated either by the primary combustion gases (as in Whittle's early design) or by exhaust heat from the jet pipe.
b. By carburation of a part of the main air flow either in or immediately after the compressor, without heat addition.
c. By combinations of (a) and (b), i.e. by carburetting some part of the main air flow with the help of heat withdrawn from a region of hot gas.

The vaporization of hydrocarbon fuels in the presence of air has been investigated in some detail by Mullins [*29*] with the result that, for a fuel, the true boiling point-distillation curve of which has been determined, it is possible to calculate the various equilibrium conditions with fair accuracy. The more important of these equilibrium conditions are:

1. The bubble point, at which an infinitesimal part of the fuel vapor-air mixture, in proportions $1:R$, is in equilibrium with the bulk of the liquid fuel. (At the "normal" bubble point $R = 0$, and air is absent.)
2. The dew point, at which an infinitesimal part of the liquid fuel is in equilibrium with the bulk fuel vapor-air mixture in proportion $1:R$ (and at the "normal" dew point, $R$ is again 0).

Apart from these equilibrium conditions it is also possible to calculate the air temperature required (at a given air-fuel ratio, pressure, and fuel

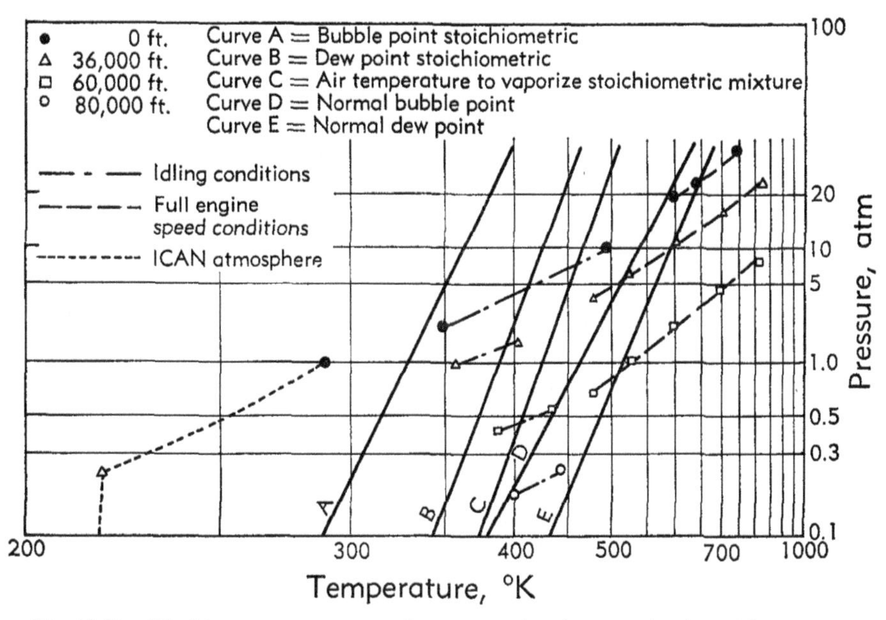


Fig. B,3f. Limiting temperatures and pressures for the vaporization of kerosene-air mixtures, compared with operating conditions for a turbojet engine.

temperature) to bring the system to an equilibrium state, such as the dew point. Fig. B,3f shows, as an example, the limiting temperatures and pressures required for the vaporization of kerosene-air mixtures (the full lines of the graph) and compares these with some of the operating conditions at the combustion chamber inlet in the datum engine of Fig. B,1a at various flight speeds. Vaporization is complete at conditions to the right of the dew point lines and it follows that prevaporization is possible with this fuel in the compressor delivery air of the datum engine at its top speed conditions, but not at the lower engine speeds.

*Aerodynamic mixing.* The various mixing processes occurring in high speed combustion systems have been studied in some detail. The mixing of fuel by turbulent diffusion into a fast air stream, for instance, has been studied by Longwell and his coworkers and is dealt with in Vol. II (p. 415ff.). Mixing of the fresh fuel-air charge with recirculated burning gases in piloting zones is, however, inevitably complicated by the concurrent chemical reaction from which it cannot readily be isolated. Tests in the cold, for instance, which can be made to give a very clear picture of the flow pattern, are not entirely relevant because of changes that may be brought about by the density differences present in the burning gas. The most comprehensive work on mixing in the piloting zone of a gas turbine (spray injection system) is that of Clarke [*5*], who made an effective semiquantitative analysis of this case and of the combined effects of the aerodynamic flow pattern, fuel injection, and droplet evaporation. The general picture is that flame is stabilized by the mixing of hot burning gases with the entering fuel and a small part of the entering air, the return flow of piloting gas being provided by a stable toroidal vortex driven in the main by air jets injected radially just downstream of the baffle (see Fig. B,2c). This vortex may, in some cases, be stabilized by applying swirl to a small part of the primary air which enters through an annulus surrounding the fuel jet. The conditions are too complex, however, to conform to any simple generalization.

A simpler case is that of the single phase system in which a precarburetted air stream burns in the wake of a flame holder or in a can. The flow pattern and the over-all behavior of such systems have been studied by Scurlock [*6*, pp. 21–40], Nicholson [*6*, pp. 44–68], Clarke, Longwell, and others. Here again the essential mixing process is that between the entering charge and the recirculated burning gas, but no generalization can yet be made as to what fraction of the charge enters the piloting region. A point of general interest made by Bragg and Holliday [*8*, pp. 270–295] is that a distinction must be drawn between continuous recirculation of the same element of hot gas, which implies wasted space, and a single cycle of flow reversal with a finite flow into and out of the piloting region. It is also important to remember that the piloting flow patterns in present use are not the only possible ones and are not necessarily the most effective.

The third of the basic mixing processes is that of dilution, i.e. the mixing of burning gases from the primary zone with secondary air to complete combustion and with excess air to bring about the final mixing to the level of the cycle temperature. Like the related process of flame stabilization, this must of course be done with the minimum of pressure loss and in a restricted space. In an idealized form the mixing of hot and cold air streams is more easily studied than any other of the component processes. Several such investigations have been made of the fundamental aerodynamics involved and for the trial of alternative mixing techniques. The main conclusions of this work follow:

1. The rate of mixing by molecular diffusion across the main interface bounding the two streams is quite negligible.
2. Mixing of two parallel streams by random turbulence is also so slow a process as to be useless in relation to the main mixing interface.
3. Devices for producing ordered turbulence are attractive at first sight, the turbulent wake of a cylinder being an obvious example. However, such devices are much less effective with highly turbulent streams flowing at different velocities than under idealized conditions, and they have never found practical application.
4. Rapid mixing requires the interleaving of the hot and cold streams either by cross stream injection or by some form of cross ducting, and with this technique is associated the turbulent mixing in the region of shear between the two streams. Every effective mixing system makes use of this principle.
5. The penetration of one stream across another and the pressure loss incurred in the process are functions of the momenta of the two streams and may be correlated by means of the momentum equations. The pressure loss of such a system can be treated theoretically.
6. Similarity between mixing systems of different dimensions requires geometrical similarity, constancy of Reynolds number, and constancy of the ratio (momentum of hot stream)/(momentum of cold stream).

In practice, mixing by cross stream penetration of air jets metered through circular holes has so far proved to be a most effective and useful technique. Not the least point in its favor is the fact that the cross section of a circular hole in a cylindrical shell is but little affected by distortion. While research has given useful guidance, the perfection of mixing arrangements in a new combustion chamber remains a matter of empirical experimentation and correction, mainly because of nonuniformity in both primary and secondary flows. In the long run, however, the aerodynamic advantages of ducted mixing may come to outweigh these factors, since the additional cooling problem which they introduce cannot be judged insoluble.

The work referred to above has all been concerned with "temperature mixing," the temperature being in fact the mean of successive eddies as measured by a thermocouple, but in the practical case the mixing of primary and secondary air must be on a molecular scale to bring the oxygen where it is wanted. Mixing by molecular diffusion across the individual eddies must therefore be taken into account as the only mechanism whereby the process can be completed. This mixing by diffusion may become increasingly important as working pressures rise, since the rate of mass transfer across a given concentration gradient remains roughly constant and independent of the absolute pressure level.

**B,4. Performance of the Complete Combustion System.** The questions now arise: How do the elementary processes, which we have just considered, interact with one another? How is the performance of practical combustion systems affected by operating conditions and how is performance best presented? What are the rate-controlling factors? The performance of a compressor can be largely summarized by the familiar plot of pressure ratio against nondimensional mass flow—what is the corresponding statement of combustion chamber performance?

Let it be said at once that there can be no such statement as simple as that which deals with purely aerodynamic events because of the real complexity of the process combining aerodynamic with physicochemical factors. And while the effect of pressure (Reynolds number) on compressor performance is secondary, its effect on combustion reactions is of the first order; similarly as to temperature. Since turbojet engines operate at very nearly constant (internal) Mach number, the effect of temperature variation on aerodynamic components is easily disposed of; but in combustion the reaction rate is an exponential function of temperature which therefore exerts a major influence. These difficulties are inherent in the nature of the combustion process and apply even if the matter is not further complicated by other factors, such as limitations imposed by fuel injection efficiency.

*Aerodynamic performance.* The factors which describe the aerodynamic performance of a combustion system are as follows:

1. Mach number, generally stated in notational form based on inlet total pressure and temperature and on a reference area which is taken as the maximum internal cross-sectional area of the chamber.

$$\frac{u}{\sqrt{\gamma \mathcal{R} T}} \text{ reduces to } \frac{w_a \sqrt{T_i}}{p_i A_{max}} \sqrt{\frac{\mathcal{R}}{\gamma}}$$

   or in the simplified form $w_a \sqrt{T_i}/p_i A_{max}$
2. Nondimensional pressure loss $\Delta p/p_i$
3. Length-diameter ratio, or length-annulus width ratio.
4. The temperature distribution factor

$$\frac{T_{max} - T_{mean}}{T_{mean} - T_i}$$

As to 1 and 2 we know that the combustion chamber Mach number is very nearly constant over the whole operating range of a given engine; $\Delta p/p_i$ will therefore also be constant. We can also combine 1 and 2 to give a pressure loss characteristic for the chamber by writing

$$\frac{\Delta p}{p_i M^2} = \frac{\Delta p}{p_i} \frac{\gamma p_i}{\rho u^2} = \frac{\gamma}{2} \phi$$

where $\phi$ is the pressure loss factor $\Delta p/\frac{1}{2}\rho u^2$. $\phi$ is not entirely determined by the geometry of the system—it is also influenced by the temperature rise, for a part of the pressure loss is associated with heat addition to the stream and with losses at the discharge end. It can be further analyzed as follows:

$$\phi = \phi_0\left[1 + K\left(\frac{\rho_i}{\rho_e} - 1\right)\right]$$

where $K$ is an additional constant and $\phi_0$ is the cold pressure loss.

These factors can be computed from the geometry of the system with an appropriate allowance for entry losses and for discharge coefficients, but they are usually determined experimentally.

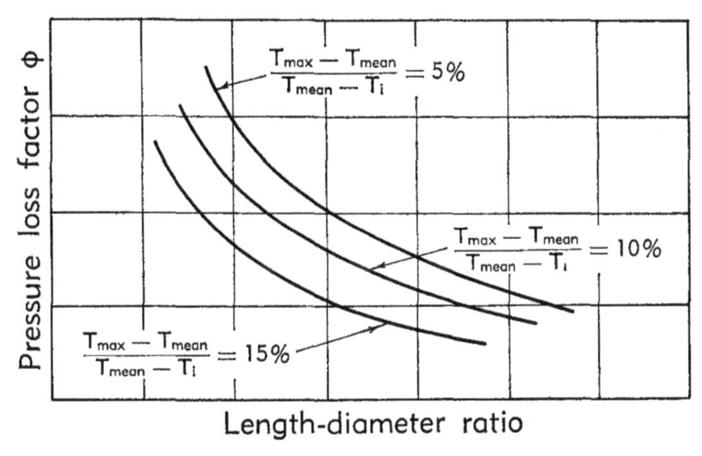


Fig. B,4a. The relationship between pressure loss factor and length-diameter ratio for constant temperature distribution factor.

There is also a relationship, for a given design, between the pressure loss factor, the length-diameter ratio, and the outlet temperature distribution. This cannot, in the present state of the art, be given exact mathematical expression, even if we simplify matters by considering only the mixing section of the chamber, but the type of relationship found is illustrated by Fig. B,4a.

It follows that, to define the aerodynamic characteristics of a combustion chamber, we must state the three factors: length/diameter ratio, pressure loss factor, and temperature distribution factor; but since there is no firm relationship among the three (except empirical correlations based on particular design procedures) no general comparison can be made between the aerodynamic performance of chambers of dissimilar shape.

*Thermodynamic efficiency.* The various possible forms of combustion loss have already been mentioned:

1. The reaction that has not had time to go to completion, in spite of homogeneous conditions, before the outlet plane or the plane in which chilling excess air is injected.
2. The fuel that has been unable to react completely for lack of oxygen.
3. The fuel that has failed to react because bad mixing, of one kind or another, has allowed it to pass through the reaction zone without reaching its ignition temperature.

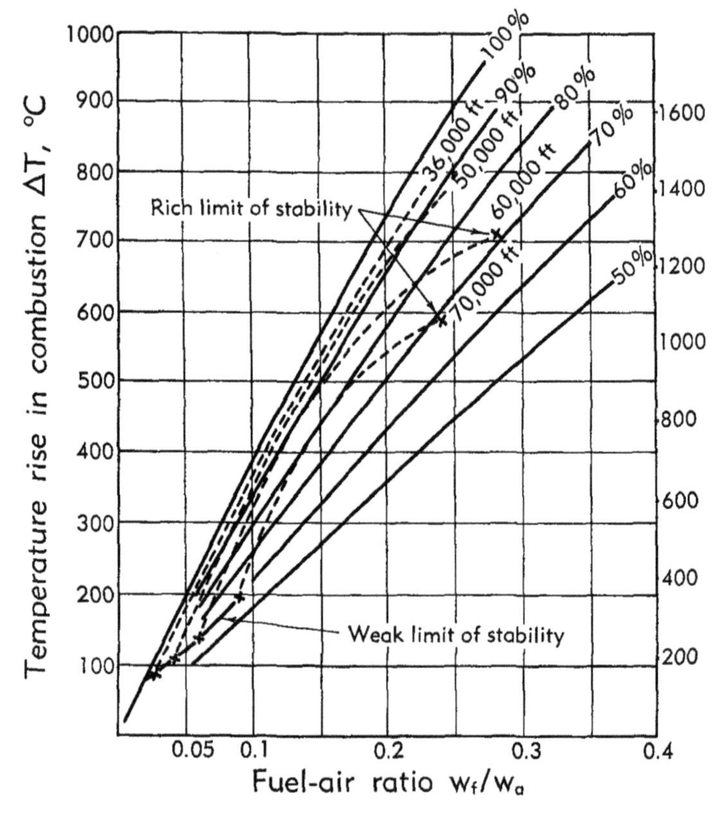


Fig. B,4b. Thermodynamic performance of a combustion chamber at cruising conditions for various altitudes.

Practical combustion systems exhibit all of these effects, and often enough one effect predominates over part of the operating range while another takes over at the end of the scale (e.g. [*30*]). In a ramjet, for instance, maximum thrust at low altitude implies low reaction loading and negligible losses as 1 and 3, but significant loss as 2; cruise conditions at altitude tend to reverse this situation. In considering the combustion efficiency of a chamber over its whole operating range one may therefore be forced to consider mixing factors as well as reaction rate.

The first requirement is for a straightforward statement of combustion efficiency and stability over the required operating range, and this is generally given in the form of Fig. B,4b which shows, at various altitudes, efficiency at the design point and operating range both in terms of fuel-air ratio and of temperature rise. For a given set of operating conditions, when the appropriate tests have been done, graphs of this kind give immediate answers to the obvious questions: efficiency at cruising conditions, stability range at idling, etc. But such a system of presentation allows of no extrapolation or interpolation to other test conditions and is merely a collection of ad hoc results. Hence there is clearly a need for the use of one or more combustion parameters with which the combustion efficiency (and stability) of practical systems may be correlated.

For systems that approximate to homogeneous adiabatic conditions there is no doubt that the air-loading parameter of Art. 3, together with the equivalence ratios, should provide such a correlation. The piloting zone of a gutter stabilizer, burning precarburetted fuel, is a case which should fulfill these conditions, and Herbert [*18*] has shown that the available stability data for a wide range of baffle types does correlate well against these parameters.[3] (If the fraction of flow entering the reversal zone remains constant, then $w_a/Vp^2$ reduces, for a simple baffle, to the frequently used expression $u/pl$ and this may equally be corrected by the temperature term $e^{T_i/n}$.) However, it is only in the piloting region that the homogeneous regime is approached; the efficiency of burning downstream of the baffle is a different story. In spite of the obvious inhomogeneity of the spray injection chamber, Bragg and Holliday have shown that such cases too can give good correlation between combustion efficiency and a reaction rate parameter at constant $\lambda$. The matter requires further study. It must be emphasized, however, that conditions in spray injec-

[3] The stability of practical flame holders, whether of the gutter or can type, must however be influenced also by aerodynamic factors, for even in the idealized case of the spherical reactor mixing must precede chemical reaction. This effect of mixing processes on reacting systems is not yet fully understood but it seems that the aerodynamic term can be given effective expression as follows: If air loading at blowout $\propto$ (Reynolds number)$^{-x}$, then at constant temperature and equivalence ratio, and assuming geometrical similarity it follows that

$$\frac{Q}{l^3p^2} = \text{const} \qquad \text{is modified to}$$

$$\frac{Q}{l^3p^2}(lp)^{\frac{2x}{1-x}} = \text{const}$$

where $l$ is a characteristic dimension.

The interpretation of Longwell's reactor experiments given in Part 3 implies that the exponent $\frac{2x}{1-x} = y = 0.2$ and other experiments with precarburetted air have given values of up to 0.5, the higher figure implying a less effective mixing of the entering stream with the reacting charge.

tion chambers are vastly different from those postulated in reaction rate theory and equally effective correlations have been demonstrated with quite different parameters. Thus Lloyd and Mullins [*31*, pp. 405–425] quote a result of Nicholson's work, showing an excellent correlation covering a range of equivalence ratios between efficiency and a fuel flow factor, implying a system dominated by fuel-mixing processes. Similarly, Lloyd and Hartley have shown that all available data on certain current designs, over a wide range of pressure, velocity, equivalence ratio, etc., correlates with the difference between a reaction temperature and the ignition temperature at the available pressure and reaction time. This result is shown in Fig. B,4c and is consistent with the predominance of combustion loss in form 3.

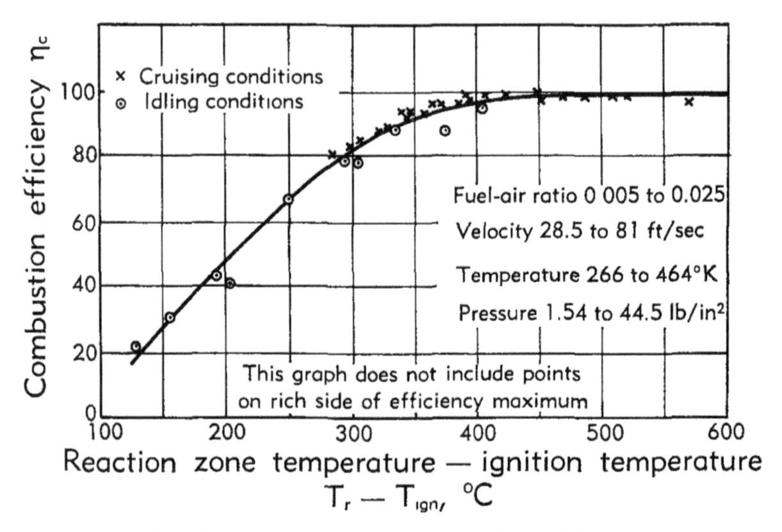


Fig. B,4c. Correlation of combustion efficiency with a "spontaneous" ignition parameter.

In general one may expect that improvement in combustion chamber design will tend, in spite of increased aerodynamic loading, to give conditions corresponding more and more closely to the homogeneous ideal, so that the air-loading factor will become increasingly applicable as a general measure of reaction zone duty and as a correlating factor.

To match the aerodynamic parameters (pressure loss factor, length-diameter ratio, and temperature distribution factor) we therefore have the combustion parameters which apply strictly only to the reaction zone proper: corrected air-loading rate and combustion efficiency (equivalence ratio being taken at the design condition to be 1). If for a given design we know both the variation of combustion efficiency with air loading (assumed predominant) and the relationship between the three aerodynamic parameters, then the dimensions of a chamber for a defined duty can, as Bragg and Holliday have shown [8, pp. 270–295], be worked out.

Even in cases where no single factor is rate-determining, an understanding of the constituent processes can help in many ways towards a rational handling of test and development problems. A particular example of this is the interesting question of combustion chamber scaling. Stewart [8, pp. 384–413] has shown in a careful analysis of the problem that if temperature level and fuel-air ratio are kept constant then conditions of pressure and velocity can be established which, in a model, should give a performance similar to that of a full-size unit, in respect to the majority of the component processes. Fig. B,4d, for instance, illustrates

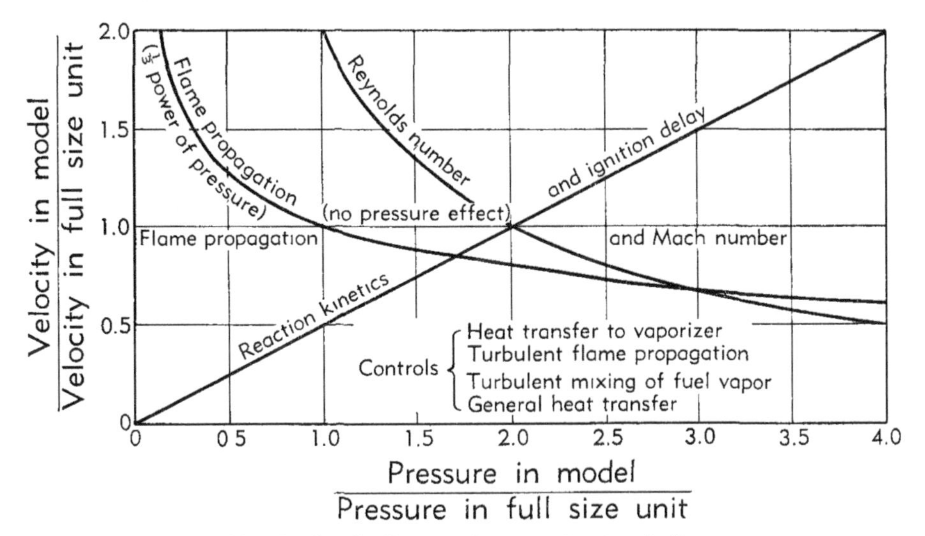


Fig. B,4d. Scaling requirements to give similar conditions for a half-size vaporizing chamber.

the pressure and velocity relationships required to give similarity between a given vaporizing system and its half-size model in respect to the turbulence level and heat transfer reaction rate compressibility, spontaneous ignition, etc. The condition for similarity is seen to be

$$u = \text{const}$$
$$pl = \text{const}$$

A necessary proviso with spray injection systems is that the fuel spray should also be appropriately scaled to maintain the constancy of spray momentum/air momentum, and because of the Reynolds number effects this condition may have to be established empirically.

*Loading factors and over-all performance.* The foregoing argument has demolished combustion intensity as a unique parameter for the loading of a complete combustion system without putting anything else in its place; indeed it has shown that no such "omnibus" parameter can be representative of both the aerodynamic and the chemical processes. These must therefore be treated separately, the aerodynamic loading factor for a chamber of given shape being simply Mach number, while the chemical factor is the corrected air loading which for this purpose may legitimately be based on the total flame zone volume. The performance of a given combustion chamber may therefore be effectively summarized as follows:

Aerodynamic performance: Mach number, outlet temperature distribution, and pressure loss factor.
"Chemical" performance: Combustion efficiency at a number of fuel-air ratios as a function of air loading.

Incidentally, a summary of this kind is of course appreciably more comprehensive than the statement of compressor performance in terms of the usual characteristic on nondimensional air mass flow and pressure ratio coordinates. To make this latter comparable, information would also be needed on outlet velocity distribution and on Reynolds number effects.

To make a general comparison of two different combustion techniques, however, more information is needed about aerodynamic performance. It then becomes necessary to know, for each case, the relationship between two of the aerodynamic parameters with the other held constant, e.g. one of the curves of Fig. B,4a.

## B,5. Cited References.

1. Whittle, F. The early history of the Whittle jet propulsion gas turbine. *Proc. Inst. Mech. Engrs. 152*, 419–435 (1945).
2. Longwell, J. P. Combustion problems in ram jet design. *J. Aeronaut. Sci. 16*, 707–713 (1949).
3. Avery, W. H. 25 years of ram jet development. *Jet Propulsion 25*, 604–614 (1955).
4. Probert, R. P. Application of research to gas turbine combustion problems. *Inst. Mech. Engrs. and Am. Soc. Mech. Engrs. Joint Conference on Combustion, Sec. 5*, 70–76 (1955).
5. Clarke, J. S. The relation of specific heat release to pressure drop in aero-gas-turbine combustion chambers. *Inst. Mech. Engrs. and Am. Soc. Mech. Engrs. Joint Conference on Combustion, Sec. 5*, 24–31 (1955).
6. *Third Symposium on Combustion, Flame and Explosion Phenomena.* Williams & Wilkins, 1949.
7. Gaydon, A. G. The spectra of chilled hydrocarbon flames. *Proc. Roy. Soc. London A179*, 439 (1942).
8. *AGARD Selected Combustion Problems II.* Butterworths, London, 1956.
9. Lloyd, P. Lectures on internal combustion turbines: The fuel problem in gas turbines. *Inst. Mech. Engrs. Symposium*, 220–229 (1948).
10. Lloyd, P. The spontaneous ignition of liquid fuel in a hot gas stream. *Proc Sixth Intern. Congress Appl. Mech.*, Paris, 1946.
11. Egerton, A., and Powling, J. The limits of flame propagation at atmospheric pressure. I: The influence of 'promoters.' *Proc. Roy. Soc. London A193*, 172 (1948).

12. Mullins, B. P. Studies on the spontaneous ignition of fuels injected into a hot air stream. *Fuel 32*, 211–252, 327–329, 451–492 (1953).
13. Longwell, J. P., and Weiss, M. A. High temperature reaction rates in hydrocarbon combustion. *Ind. Eng. Chem. 47*, 1634–1643 (1955).
14. Avery, W. H., and Hart, R. W. Combustor performance with instantaneous mixing. *Ind. Eng. Chem. 45*, 1634–1637 (1953).
15. Bragg, S. L. Unpublished work at Rolls Royce Ltd.
16. Childs, J. H. Preliminary correlation of efficiency of aircraft gas-turbine combustors for different operating conditions. *NACA Research Mem. E50F15*, 1950.
17. Saunders, O. A., and Spalding, D. B. The chemistry and physics of combustion. Part II: Chemical and physical factors controlling the rate of combustion. *Inst. Mech. Engrs. and Am. Soc. Mech. Engrs. Joint Conference on Combustion, Sec. 10*, 12–22 (1955).
18. Herbert, M. V. *AGARD Combustion Researches and Reviews 1957. A Theoretical Analysis of Reaction Rate Controlled Systems.* Agardograph No. 15. Butterworths, London.
19. Weiss, N. A., Lang, R. J., and Longwell, J. P. Combustion rates in spherical reactors. *Ind. Eng. Chem. 50*, 257 (1958).
20. Taylor, G. I. The boundary layer in the converging nozzle of a swirl atomiser. *Quart. J. Mech. and Appl. Math. 3*, 129–139 (1950).
21. Fraser, R. P. Liquid fuel firing. *Inst. Mech. Engrs. and Am. Soc. Mech. Engrs. Joint Conference on Combustion, Sec. 2*, 32 (1955).
22. Longwell, J. P. *Chem. Eng. Sc. D. Thesis*, Mass. Inst. Technol., 1943.
23. Joyce, J. R. The atomisation of liquid fuels for combustion. *J. Inst. Fuel 22*, 150 (1949).
24. Radcliffe, A. The performance of a type of swirl atomiser. *Proc. Inst. Mech. Engrs. 169*, 93–100 (1955).
25. Carlisle, D. R. Communication on the performance of a type of swirl atomiser by A. Radcliffe. *Proc. Inst. Mech. Engrs. 169*, 101–103 (1955).
26. Dombrowski, N., and Fraser, R. P. A photographic investigation into the disintegration of liquid sheets. *Phil. Trans. Roy. Soc. London A247*, 101 (1954).
27. Probert, R. P. The influence of spray particle size and distribution in the combustion of oil droplets. *Phil. Mag. 37*, 94 (1946).
28. *Fourth Symposium on Combustion.* Williams & Wilkins, 1953.
29. Mullins, B. P. The vaporization of fuels for gas turbines. *J. Inst. Petroleum 32*, 703–737 (1946); *33*, 44–70 (1947).
30. Childs, J. H., and Graves, C. C. Relation of turbine-engine combustion efficiency to second-order reaction kinetics and fundamental flame speed. *NACA Research Mem. E54G23*, 1954.
31. *AGARD Selected Combustion Problems.* Butterworths, London, 1954.

# SECTION C

## *EXPERIMENTAL TECHNIQUES*

H. C. HOTTEL
G. C. WILLIAMS

**C,1. General Discussion.** The function of the ideal gas turbine combustion chamber is to combine streams of fuel and air, convert completely the chemically bound energy of the fuel into thermal energy and produce with minimum loss of total pressure a stream of burned products having a desired spacewise distribution of temperature and velocity. The effect of the combined aerodynamic and combustion losses in a given burner on the performance of a given engine is most simply determined by running the engine with the burner. This represents at once the grossest scale of measurement, often the most expensive one to perform, and one which usually gives the least information pertinent to improving the burner performance. The measurements required to determine progress of the combustion and of the flow processes increase in number and complexity with the fineness of the measurement scale. The streams may at many points in the chamber be in nonequilibrium states, and stream properties may be expected to vary with time at a given point as well as with position in the flow cross section.

*Principal measurements.* As noted in Sec. B, the quantities of interest in evaluating the quality of performance of a combustion chamber are: burner efficiency $\eta_b$ and stagnation pressure loss across the burner $p_3^0 - p_4^0$. Neglecting minor heat losses, the gas turbine burner is operated as a constant total enthalpy process. The energy balance between stations 3 and 4 may be written

$$w_a h_{a_3}^0 + w_{f_3} h_{f_3}^0 = (w_a + w_f) h_4^0 \tag{1-1}$$

Burner efficiency has been loosely referred to as fractional chemical change. However, when one goes to evaluate this numerically as the enthalpy change due to actual chemical reaction divided by the enthalpy change associated with complete reaction, it becomes apparent that, with the possibility of phase change and with two different feed-stream temperatures, that of the fuel and that of the air, there is no unambiguous chemical-bond energy associated with the process. One can use as a base the temperature of the stream carrying the most "sensible" energy, namely the air; or one can ignore the difference between this base and some arbitrary and convenient base used in the tabulation of thermodynamic properties. Let the burner inefficiency be defined as the enthalpy of the exhaust stream, of chemical composition corresponding to actual exhaust conditions but in the vapor state, divided by the enthalpy of the fuel stream in the vapor state, both evaluated *at* (not above) the base temperature and pressure (a state where the enthalpies of $N_2$, $O_2$, $CO_2$, $H_2O$ (vapor), and the complete oxidation products of any other elements present are assigned zero values). Let these enthalpies, per unit mass, be called the chemical enthalpies or enthalpies of combustion, $h_{c_4}$ for the exhaust gases and $h_{cf_3}$ for the feed. To complement these quantities, a physical enthalpy may be defined as the enthalpy change resulting from changing the stream from $p$ and $T$, without chemical change, to vapor at the base temperature and pressure. With these definitions, the total enthalpies of the fuel and of the products are now expressible as the sum of the chemical and physical (sensible, latent, and kinetic) quantities, the latter identified by the subscripts.

$$h^0_{f_3} = h^0_{sf_3} + h_{cf_3} \tag{1-2}$$

$$h^0_4 = h^0_{s_4} + h_{c_4} \tag{1-3}$$

where

$$h^0_{sf_3} = h_{sf_3} + \frac{V^2_{f_3}}{2} \tag{1-4}$$

$$h^0_{s_4} = h_{s_4} + \frac{V^2_4}{2} \tag{1-5}$$

The velocity to be used in Eq. 1-4 and 1-5 is the maximum resultant velocity $V$, which is not necessarily equal to $u$, the velocity parallel to the duct axis. Note that the resultant value $V$ may be known without knowledge of its direction. Burner efficiency may then be defined in chemical terms as

$$\eta_b = 1 - \left(\frac{w_a + w_f}{w_f}\right)\left(\frac{h_{c_4}}{h_{cf_3}}\right) \tag{1-6a}$$

or, by use of the energy balance equation (Eq. 1-1) in sensible energy terms as

$$\eta_b = \frac{(w_a + w_f)h^0_{s_4} - w_a h^0_{sa_3} - w_f h^0_{sf_3}}{w_f h_{cf_3}} \tag{1-6b}$$

If the fuel is all in the vapor state and the specific heats are independent of pressure, $\eta_b$ may be expressed in terms of total temperatures as follows:

$$\eta_b = \frac{w_g \bar{c}_{p_g}(T^0_4 - T_b) - w_a \bar{c}_{p_a}(T^0_{a_3} - T_b) - w_f \bar{c}_{p_f}(T^0_{f_3} - T_b)}{w_f h_{cf_3}} \tag{1-6c}$$

wherein $w_g$ is the total gas flow rate ($\equiv w_a + w_f$) and $\bar{c}_p$ is the mean constant-pressure specific heat between a base temperature $T_b$ and the temperature $T$.

As an approximation, combustion efficiency is sometimes computed on the basis of the temperature rise above the inlet air temperature $T^0_{a_3}$; i.e. the ideal value of $T^0_4$, $T^0_{4_1}$ for $\eta_b$ of unity, is computed from Eq. 1-6c, and the efficiency is expressed as a ratio of actual temperature rise to the ideal

$$\eta_b \text{ (based on temperature rise)} = \frac{T^0_4 - T^0_3}{T^0_{4_1} - T^0_3} \tag{1-6d}$$

If the total temperature $T^0_4$ in the above relation is expressed in terms of stream temperature $T_4$ by the usual expression, derivable from Eq. 1-5 by substitution of $c_{p_m}$ for a mean specific heat of the burned gases between $T_4$ and $T^0_4$, one obtains

$$h^0_{s_4} - h_{s_4} = c_{p_m}(T^0_4 - T_4) \tag{1-7}$$

It is clear that the stagnation temperature rise in this equation is assumed not to include the effect of any phase change or chemical energy transformation. Consequently a stagnation temperature measurement on a stream carrying a significant amount of liquid fuel is not by itself susceptible to rigorous interpretation. Since the gas turbine is usually operated with quite lean fuel-air ratios and with relatively low turbine inlet temperatures, in the absence of liquid fuel at the burner exhaust there would be substantially no change in gas composition on stagnation from $p_4$ and $T_4$ to $p^0_4$ and $T^0_4$; i.e. $h_{c_4}$ should have the same value whether evaluated from a gas analysis of a stream sample taken at station 4 or from a knowledge of the fuel-air ratio and a stagnation temperature measurement made on the same stream at that station.

The measurement of aerodynamic loss is straightforward in the rare case where the variations in stream temperature and velocity over the flow cross section are small, or where the stream velocities are low enough to make the resultant total pressure substantially independent of the method of mixing the streams passing a cross section of interest. The total pressure loss is usually expressed as the sum of (1) the unavoidable loss due to uniform change in stream velocity on heat addition and (2) skin friction, flame stabilization (velocity inequalities, drag, and turbulence), and mixing losses required by the particular chamber. The unavoidable loss, of the order of 2 "cold" velocity heads, usually amounts to less than one fourth the total loss in low-loss chambers and less than one tenth in high pressure drop burners.

To determine the over-all performance or combustion and aerodynamic efficiencies of a specific chamber under given operating conditions, the following information is required:

1. Weight rates of flow of air and fuel at burner inlet ($w_{a_3}$ and $w_{f_3}$) and of combustion gases at burner exit $w_{g_4}$ ($= w_{a_3} + w_{f_3}$).
2. Mass mean values of stagnation pressure, stagnation temperature, and composition of fuel and air streams at the inlet, and of burned products at the exit. In the usual case where the streams are nonuniform at the flow cross sections designated as inlet and outlet for the chamber, point-to-point traversing is required to determine the mass mean quantities.

Specification of stagnation pressure, stagnation temperature, and chemical state is sufficient to determine the work-producing capacity, above any arbitrary base, of a uniform system for which an equation of state is known. If the system is nonuniform and in motion, knowledge of the relative mass flow rates of its components is also required.

Thus, at a given cross section of flow it is necessary to establish, at each point, the stagnation pressure, the stagnation temperature, the mass velocity normal to the traverse plane, and the composition. If any one of the first three of these four is not measured directly, two additional measurements are required. If the resultant velocity $V$ and its direction are determined, only three other measurements need be taken. Possible combinations appear in Table C,1.

*Table C,1.* Measurement combinations required for traverse of burner streams.

| | Combination numbers | | | | | | | | |
|---|---|---|---|---|---|---|---|---|---|
| | 1 | 2 | 3 | 4 | 5 | 6 | 7 | 8 | 9 |
| Stagnation pressure $p^0$ | √ | √ | | | | | √ | √ | √ |
| Stagnation temperature $T^0$ | √ | √ | √ | √ | | | | √ | |
| Static pressure $p$ | | | √ | √ | √ | √ | √ | √ | |
| Static temperature $T$ | | √ | √ | √ | √ | √ | √ | | √ |
| Resultant velocity $V$ | | | | | √ | √ | | | √ |
| Chemical composition $y$ | √ | √ | √ | √ | √ | √ | √ | √ | √ |
| Mass velocity $G$, normal to traverse plane | √ | | √ | | √ | | | | |
| Linear velocity $u$, normal to traverse plane | | √ | | √ | | √ | √ | √ | √ |

*Additional measurements.* While the above measurements may suffice to determine loss characteristics of a burner, additional measurements may be required to indicate sources of loss, to follow the combustion and mixing processes, to indicate directions for improvement or to show corrective measures needed.

The additional measurements can include those of:

1. The scale and intensity of turbulence and of local vortex flow (IX,F).
2. The spacewise distribution of fuel, and for liquid fuels the distribution of fuel particle size (II,J).

3. The progress of vaporization of the fuel (thermal cracking of the fuel and partial combustion may accompany the vaporization process).
4. "Unmixedness"—timewise variation in composition or temperature at a point (see Art. 9).
5. Carbon formation and deposition.
6. The extent of the reaction zone, primarily visible flame length (timewise variation is again a factor).
7. The temperatures of the burner walls (IX,D,1) and the pressures exerted on the walls. (Wall temperatures are of interest not only because of temperature limitations on service life, but because of the role played by the walls in heat transfer, both by convection and by radiation. Wall pressure measurements can yield significant information about gas flow patterns near the walls.)

Experimental techniques are not developed for precisely determining all of these directly; some must be evaluated only semiquantitatively from crude direct measurements or be inferred from other measurements.

**C,2. Temperature Measurements.** To the extent that burning gases are not in thermodynamic equilibrium, they cannot be said to have an unambiguous value of temperature (See I,H and II,A). Despite marked departure from equilibrium of certain types of energy levels in the molecules of a burning gas, however, the translational and in some cases the rotational energy levels do conform, with substantially no time lag, to the Maxwellian distribution which gives meaning to the term temperature; and temperature-sensing devices, which depend on their attainment of thermal equilibrium with the random translational energy of the gas, may be expected to yield temperatures with an ambiguity that is small even when combustion is in progress.

THERMOCOUPLES. For probing a flow cross section in a gas turbine chamber, there is no present tool as generally useful as a thermocouple, but the user must take precautions to minimize the introduction of each of six major sources of error:

*Radiation error.* This must be held low by one or more of the accepted methods: shielding [*1*], treating the junction surface to give it a low emissivity, or making convection dominate radiation by use of small junction size or high gas velocity (see IX,D,2 and IX,I,5). Tight coverage of the junction with a flattened tube section of silver (to 1600°F), gold (to 1800°F), or platinum (to the limits of a base metal couple) has been found one of the most effective modifications of a bare couple [*2,3*]. Platinum, markedly poorer than silver or gold but necessary when the temperature is high, has an emissivity when clean, $\epsilon_c$, which varies linearly with temperature from 0.12 ± 0.01 at 1000°F to 0.22 ± 0.01 at 2600°F, and which departs exponentially from its "clean" value, to a "dirty" value $\epsilon_D$ which depends primarily on gas cleanliness (mostly measured by suspended unburned residue or luminosity), with a time constant $k$, dependent primarily on temperature [*4*].

$$\epsilon = \epsilon_C + (\epsilon_D - \epsilon_C)(1 - e^{-kt}) \tag{2-1}$$

*Table C,2a.* Typical values of $\epsilon$ and $k$.

| | $\epsilon_C$ | $\epsilon_D - \epsilon_C$ | $k$, hr |
|---|---|---|---|
| Brick-lined combustor: slightly luminous exhaust gas, fuel-air = 0.08, 2475°F, $p$ = 1 atm, $M$ = 0.2. | 0.21 | 0.34 | 1.7 |
| Inconel combustor: lean exhaust gas, small amounts of incandescent particles; fuel-air = 0.03, 1670°F, $p$ = 1 atm, $M$ = 0.3. | 0.17 | 0.28 | 3 |
| Inconel combustor: slightly lean, highly luminous, with exhaust residue; fuel-air = 0.055; 2230F°. | 0.20 | 0.20 | 2.8 |
| Jet engine exhaust gas: | | | |
| combustor outlet, 1600°F | 0.16 | 0.24 | 2.8 |
| tailpipe, 1200°F | 0.13 | 0.17 | 2.8 |
| ramjet engine, 3000°F | 0.25 | 0.15 | 1 |

Clean gases at around 2300°F, lean for 20 hours or cycling between rich and lean for 200 hours, had no effect on the emissivity of the platinum couple used.

*Stagnation error.* Cognizance must be taken of an interest in two kinds of temperature, the moving-stream or enthalpy temperature $T_e$ and the total temperature $T^0$ corresponding to adiabatic stagnation of the gas ($T^0 - T_e = V^2/2c_p$). The couple junction in the absence of other sources of error will read $T_e$ plus a temperature increase due to heating consequent on junction friction in the moving air stream. This additional temperature rise is for gases a fraction $r$ ($<1$) of the temperature rise $T^0 - T_e$ due to adiabatic stagnation of the gas stream (see IX,D,1 and 2). To convert the couple reading $T_c$ ($\equiv T_e + r(T^0 - T_e)$) to the desired $T^0$ or $T_e$, the velocity $V$ and fraction $r$ must both be known. The latter, in general a function of Mach, Prandtl, and Reynolds numbers, is for air flowing over bare butt-weld couples in a subsonic stream about 0.66, 0.72, 0.78, and 0.86 for transverse flow across smooth wire, transverse flow over a round-beaded junction, axial flow over a beaded junction, and axial flow along smooth wire, respectively [*5*]. Suitable stagnation probe designs can raise the value of $r$ to a substantially constant value near 1 (see IX,D,2) [*5,6*].

*Lead wire losses.* These losses from the couple junction are due to the immersion of too short a run of junction leads in an isothermal environment. If a couple is immersed in gas at $T_{g1}$ to a depth $L$ and beyond that point is immersed in gas at $T_{g2}$ for an infinite distance, and the couple reads $T_c$, it may be shown that if radiation and stagnation effects are absent, the fractional departure $(T_c - T_{g1})/(T_{g2} - T_{g1})$ from the desired reading $T_{g1}$ is given by

$$\frac{T_c - T_{g1}}{T_{g2} - T_{g1}} = \frac{1}{\cosh C + B \sinh C} \tag{2-2}$$

where

$$C = L\sqrt{\frac{h_1P_1}{k_1A_1}} \quad \text{and} \quad B = \sqrt{\frac{k_1A_1h_1P_1}{k_2A_2h_2P_2}}$$

$P$ and $A$ are the total perimeter for heat transfer from the gas to the couple leads and the total cross section for axial heat transfer, respectively, of the lead wires or protection tube; $h$ and $k$ are the gas-lead heat

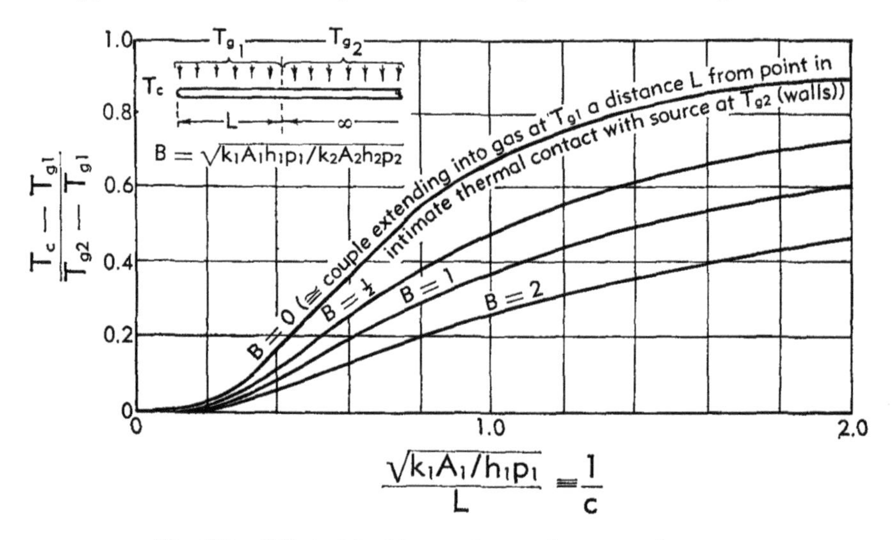


Fig. C,2. Effect of lead immersion on thermocouple error.

transfer coefficient and lead thermal conductivity; subscript $_1$ refers to properties in that part of the system embedded a distance $L$; and subscript $_2$ to the remainder of the system. The above equation includes the special case of a couple lead extending from a point in perfect thermal contact with a wall at $T_{g2}$ and immersed in gas at $T_{g1}$ for a depth $L$, for which case $B = 0$.

Fig. C,2, representing Eq. 2-2, shows the magnitude of the lead-loss error to be expected. For example, the use of a $\frac{1}{4}$-in. O.D. stainless protection tube with 0.01 in. wall thickness, with $h_1 = h_2 = 10$ and $k_1 = k_2 = 10$ (English hour-foot units) and a 1-inch immersion $L$ in gas at $T_{g1}$ causes a fine wire couple attached to its end to read gas temperature $T_{g1} + 5.5$ per cent of the difference $T_{g2} - T_{g1}$, where $T_{g2}$ is the temperature of gas bathing all but one inch of the long tube; or to read gas temperature $+11.1$ per cent of the difference $T_w - T_{g1}$, where $T_w$ is the temperature of the wall from which the protection tube protrudes one inch. Doubling the isothermal zone length at $T_{g1}$ reduces these errors to 0.3 and 0.6 per cent respectively. In the case of an unprotected couple, the two lead wires of which have markedly different thermal properties, the solution is considerably more complex than that described by Eq. 2-2; guidance can usually be obtained from the knowledge that the couple reading will lie between the values calculated by applying Eq. 2-2 to each of the two lead wires separately.

*Time lag error.* In a measurement of change of temperature due to rapid changes in operating conditions, a high heat capacity of the couple and low heat transfer coefficient at its surface combine to cause its temperature change to lag that of the stream in which it is immersed. A couple subjected to sudden change in environment undergoes all but $1/e$th of the change in the time $Vc_p\rho/hS$, where $Vc_p$ is the heat capacity of the sensing element of effective surface $S$ at which the transfer coefficient is $h$; or it registers 90 per cent of the change in the time $2.3Vc_p\rho/hS$. (See, in addition, IX,D,2.)

*Catalysis of combustion.* Thermocouples, resistance thermometer wires, or protection tubes of platinum or its alloys may yield abnormally high temperatures due to surface catalysis of incompletely burned products, but the temperature rise is not predictable. The phenomenon is complicated by the existence of both high and low temperature effects, which are different. Fiock, et al [*7*] measured a temperature rise on platinum of 12–20°F in turbine combustion chamber gases at 800°F. Adding one half per cent propane increased the temperature rise about 50 per cent; adding hydrogen increased it still more. The effect decreased with a rise in gas temperature, disappearing at 1100–1200°F. This low temperature phenomenon was absent on surfaces of silver, gold, stainless steel, nickel, and ceramic coatings.

Van Langen and Hubbard [*8*] report a temperature difference of about 100°C between bare and protected Pt-Pt:Rh high velocity couples placed in a coke-oven gas flame at 1300–1700°C. They claim identification of the rise with the presence of hydrogen in the flame, and say there was no rise in other flames tested when the analyses showed no hydrogen. The variation in temperature rise, however, does not at all parallel the variation in hydrogen content of the flame.

Gilbert and Lobdell [*9*, p. 285], using platinum at 1300–1750°C as a resistance thermometer in a rich acetylene flame, found no evidence of catalytic action in the maximum temperature region at 2230°C, probably because the system had reached chemical equilibrium; but they did find a strong catalytic action in cooler parts of the gases on the upstream side of the maximum temperature zone.

Leah and Carpenter [*9*, p. 274], comparing the effect of flame gases on fine rods—either 0.0005-in. bare Pt-Rh or 0.0002-in. Pt-Rh covered with quartz to a diameter of 0.0005-in.—found differences up to 250°C between the bare and coated wires at the hottest parts of the flames of premixed CO-$O_2$, CO-air, propane-air, ethylene-air, and benzene-air. To permit use of the wires without burnout they always studied rich or lean flames; and the lower reaction velocity associated with such mixtures may account for the effect not found by Gilbert and Lobdell in the acetylene flame. At a point 2 cm or so above the flame front (gas temperature 1100–1200°C) the difference of the bare and covered wires had dropped to about 25°C. The working conclusion that can be drawn from these several investigations is that, in combustion gases not in thermodynamic equilibrium, platinum may act as a catalyst; and the intermediate temperature range in which there is no trouble, if such a range exists, has not been adequately established.

*Loss of calibration.* The choice of suitable couple materials and protection tubes for the temperature and gas composition involved, the life of the couple wire as affected by size and environment, and the elimination of cold-junction errors and of immersion errors due to metal inhomogeneity—these are all matters of prime importance, but adequately covered in standard reference books on temperature measurement [*10*]. National Bureau of Standards tests indicate that chromel-alumel couples withstand attack by sulfur compounds in gases at 1200–1400°F for 60 hours, provided the shield material around the couple junction is silver, gold, stainless steel, nickel, or ceramic coatings; and extended use at 2000°F is feasible. The presence of silver, however, causes rapid sulfur corrosion of the alumel wire, even at 1200°F [*11,12*]. NACA studies indicate that, for gas turbine measurements, Pt-Pt:Rh couples show no significant change in calibration during test periods of 20–200 hours [*4*].

In the design of a couple for a given probing operation the minimizing of all six types of errors is not favored by the same changes in design. For example, when a steep temperature gradient exists transverse to the main gas flow direction and a temperature traverse of the flow cross section is to be made, it may be necessary to accept a moderate radiation error due to omission of several concentric shields around a couple junction in order to minimize the error due to inadequate immersion of the junction in an isothermal environment (see Art. 8). For a thermocouple system of any design, construction on the smallest possible scale should favor the suppression of the radiation error, lead wire losses, and the time lag error; should be substantially without effect on the stagnation error; and should be pursued to the limit established by the loss of calibration or by construction difficulties. (For more details on the use of thermocouples, see IX,D,1 and D,2, and IX,I,5.)

Radiation Methods. Although radiation methods of temperature measurement find much less frequent application in conventional gas turbine performance evaluation than thermocouples, the identification

*Table C,2b.* Radiation methods of high temperature measurement.

- I. Dependent on the use or measurement of absorption and emission.
  - A. With background radiator.
    - 1. *Use of Kirchhoff's law.*
      - a. *Null methods.*
        - (1) *Monochromatic*
        - (2) *Total radiation*
      - b. *Non-null methods.*
        - (1) *Monochromatic*
        - (2) *Total radiation*
    - 2. Measurement of relative absorption of two spectral lines.
  - B. *With background mirror*—monochromatic.
- II. Dependent on emission measurement only.
  - A. Based on relative intensities of spectral lines.
  - B. Based on radiating properties of suspended carbon.

of the source of difficulties in combustion chamber performance may well necessitate examination of the progress of combustion at points where diluent air has not yet brought the temperature down. Radiation methods have special value when the gas temperature is too high or the gas is too active chemically for the use of thermocouples, or when the temperature difference of gas and wall is so great that thermocouple radiation errors are difficult to suppress; but, with a few exceptions to be mentioned, there is the disadvantage that the temperature measured is a value averaged over the optical path through the gas. The most useful and reliable radiation methods depend on the laws of equilibrium radiation —laws independent of the chemical or physical nature of the radiator. These are Planck's law giving the intensity of monochromatic radiation from a black body,

$$E_{B,\lambda} = \frac{c_1\lambda^{-5}}{(e^{c_2/\lambda T} - 1)} \tag{2-3}$$

and Kirchhoff's law stating that the ratio of emission from a body to that from a black body at the same temperature, called the emissivity $\epsilon$, is equal to the absorptivity $\alpha$ which the body exhibits for incident radiation *with which it is in equilibrium.* Table C,2b classifies radiation temperature-measuring methods, with an italicizing of those methods not requiring special knowledge of the source of radiation within the gas. All methods listed under *use of Kirchhoff's law* in the table make use of the same principle of measuring intensity $E_g$ from the gas in question (at desired temperature $T_g$), intensity of the background surface radiator $E_s$, and intensity $E_{g+s}$ of the background surface viewed through the gas. Two relations may be written:

$$E_{g+s} = E_g + E_s\tau_g \tag{2-4}$$

in which $\tau_g$ is the transmittance of the gas for radiation from the background; and a definition,

$$\epsilon_g \equiv \frac{E_g}{E_{b,g}} \tag{2-5}$$

in which $E_{b,g}$ is the intensity of black body radiation at gas temperature and $\epsilon_g$ is the emissivity of the gas, defined by the equation. These relations are used in a variety of ways, with corresponding variations in exact meaning of the terms; their use will be discussed in relation to the items of Table C,2b.

*Monochromatic null method.* The monochromatic transmissivity of the gas is the complement of its monochromatic absorptivity (ignoring reflection and scatter which are very small), and the latter in turn equals the monochromatic emissivity if the absorptivity is measured with radiation with which the gas can maintain equilibrium. Let the background $S$ (generally an electrically heated strip-filament lamp) be varied in temperature until the target intensity is independent of whether $S$ is viewed directly or through the gas (i.e. until $E_{g+s} = E_s$). Then from Eq. 2-4, $E_s(1 - \tau_g) = E_g$; from Kirchhoff's law, $1 - \tau_g = \epsilon_g$; and combining these with Eq. 2-5 gives $E_{g,b} = E_s$. In words, the background monochromatic intensity, at match, equals that of a black body at gas temperature. Experimentally the measurements may be made in succession, plotting $E_s$ against current through $S$, then $E_{g+s}$ against current through $S$, and taking the intersection of the two curves; or a system of mirrors may be set up to permit looking at $S$ alternately through and around the gas, adjusting the current until there is no difference. High spectral resolution is not necessary, though it improves the precision of the method when a spectral line is being used. The radiation instrument may be an optical pyrometer if the gas exhibits soot luminosity, an infrared spectrometer using a wavelength in the water vapor or carbon dioxide spectrum, an ultraviolet monochromator adjusted to the bands of neutral OH [*13*] or a sodium D line spectrometer. (If the latter is used, the technique is of course simplified to making the D line or a part of it disappear against the background radiation from $S$). Note that the background $S$ need not be a black body, but that means must be provided for determining its black body brightness temperature at the wavelength used, such as by an optical pyrometer together with sufficient knowledge of the emitting characteristics of the background to convert from intensity in the red to intensity at the wavelength used in the matching. Only in the calibrating step must allowance be made for the presence of a window or lens between the background and the pyrometer. The method depends for its validity on the choice of a wavelength at which the emitter maintains equilibrium between its energy states responsible for the emission and its translational energy states. Carbon particles at $0.65\mu$ (red screen of an optical pyrometer), sodium atoms emitting at the NaD line, OH bands except near the blue-cone tip of a flame [*13,14*], $H_2O$ at $2.7\mu$, and $CO_2$ at $4.3\mu$ satisfy this condition in most flames. (Leah and Carpenter [*9*, p. 274], however, present evidence that within the burning zone of slow-burning (rich or lean) premix flames the D line method may yield temperatures 250°C higher than the translational temperature given by compensated quartz-covered resistance wires, but their results may have been affected by the steep temperature and concentration gradients existing in the flames used.) If reversal of the NaD line is being attempted with soot present in the gas, there will be some loss of sensitivity. This will be less, if the gas is uniform in temperature, the higher the sodium concentration of the gas. If the gas is nonuniform in temperature, however, a high sodium concentration is undesirable. In the temperature-averaging process along an optical path, the greatest weight is given to the highest temperature and to the end of the path nearest the spectrometer, and the nonuniformity of the weighting process along the path is greater the higher the sodium concentration. Since temperature gradients are usually steepest at the edges of a gas mass, minimum weight is given the cold mantle by minimizing the sodium. This must of course be balanced against the attendant loss of sensitivity as the sodium is decreased. An example is illuminating.

Consider a cylindrical gas mass consisting of a core at 1600°K surrounded by a mantle 10 per cent (160°) cooler, and occupying one-tenth of the radius of the hotter core. As sodium is added to make the NaD line emissivity successively 0, 0.25, 0.50, 0.75, and 1.0, match will occur at background brightness temperatures 15, 15, 15.9, 18.3, and 160° below the gas-core temperature. One concludes that the minimum sodium concentration consistent with retention of sensitivity is to be used. From the fact that the absorption coefficient and therefore the gas emissivity varies throughout the D line, the above calculation indicates that the centers of the D line doublet will disappear at a lower temperature than the edges if the gas viewed is cooler on its spectrometer side. The substantial elimination of cool-mantle error—in fact, the conversion of the method to one capable of local temperature determination—is achieved by adding sodium in a small stream locally within the gas just upstream of the plane of measurement.

Transmission measurements have been adapted to the determination of temperatures of rapidly fluctuating flames by Curcio, Stewart, and Petty [*13*], who obtained a response time of 0.1 millisecond with apparatus designed to use the ultraviolet band of OH in the flame, together with a hydrogen arc background, a beam chopper, and a photocell pickup.

*Total radiation null method.* The use of total rather than substantially monochromatic radiation does not change the equations except for making all the terms refer to total radiation. The instrument used is a nonselective total radiation pyrometer, i.e. one with a gold or aluminized mirror rather than a quartz lens. The background radiator at $T_s$ must now be a true black body since reliance is being put on the applicability of Kirchhoff's law throughout the spectrum. Windows in the optical path can still be used, provided that their transmittance for total radiation from $S$ is known as a function of $T_s$ and that they transmit sufficiently well in the portion of the spectrum in which the main gas emission lies. This method has been tested on gas containing $CO_2$ as the only radiator, under conditions in which the gas was isothermal throughout the zone containing $CO_2$. When the product, partial pressure of $CO_2$ times the path length through the gas was as low as 0.017 ft atm and the gas temperature was 1166°R, the error in the method was 6°F; at 2350°R the error was 12–20°F for $P_{CO_2}L$ of 0.017–1.68 [*15*]. The main objection to the method is the cool-mantle error. Local addition of $CO_2$ or water vapor, analogous to the addition of sodium, is not feasible, both because it would unduly change the main stream and because the gas of interest is most generally combustion products already containing $CO_2$ and $H_2O$.

*Monochromatic non-null method.* In the simpler-to-use, non-null method the background temperature is not under control and cannot therefore be adjusted to make $E_{g+s} = E_s$. However, the assumption that the monochromatic absorptivity of the gas $1 - \tau_g$ equals its emissivity is a good one, despite the fact that the intensity used in the measurement is not that corresponding to a black body at gas temperature. Measurements are made of $E_g$, $E_s$, and $E_{g+s}$. Eq. 2-4 and the assumption of Kirchhoff's law now yield

$$\epsilon_g = \frac{E_s + E_g - E_{g+s}}{E_s} \tag{2-6}$$

and Eq. 2-5 then yields $E_{g,b}$ from which the gas temperature is known.

*Total radiation non-null method.* The non-null method applied to total radiation measurements does not have the validity of the three methods discussed above. The assumption that the total absorptivity of the gas for radiation from a black source at other than gas temperature equals the total gas emissivity is unwarranted, and only true for a "gray" gas. The error introduced by nongrayness is less the nearer the background temperature $T_s$ is to the unknown gas temperature. The method has been used with some success on luminous flames, with a chromic-oxide-blackened refractory wall serving as the background surface at $T_s$ [*16,17*].

*Relative absorption of two spectral lines.* This method applied to two or more lines, for example, of the OH, CH, or NO spectrum, is discussed elsewhere (IX,M).

*Background mirror.* Simpler than a controlled-temperature background is the use of a mirror behind the hot gas [*18,19*]. A single thickness of the gas has a monochromatic intensity $E_1$ given approximately[1] by $\epsilon_g K_1 e^{-c_2/\lambda T}$. Using $1 - \epsilon_g$ for the transmittance of the gas, a double thickness of gas achieved by the use of a mirror of reflectivity $R$ has an inten-

[1] Using Wien's instead of Planck's law, ignoring the 1 relative to $e^{c_2/\lambda T}$. The error in $E$ is $<1$ per cent when $\lambda T < 3100\mu°K$.

sity $E_2^y$ given by $[\epsilon_g + \epsilon_g(1 - \epsilon_g)R]K_1e^{-c_2/\lambda T_g}$. Eliminating $K_1$ by determining the intensity $E_b$ of a calibrating source at known brightness temperature $T_b$ (for which $E_b = K_1e^{-c_2/\lambda T_b}$), one obtains for gas emissivity

$$\epsilon_g = \frac{1 + R - E_2/E_1}{R} \tag{2-7}$$

and for gas temperature,

$$\frac{1}{T_g} - \frac{1}{T_b} = \frac{\lambda}{c_2} \ln\left(\epsilon_g \frac{E_b}{E_1}\right) \tag{2-8}$$

where $c_2$ is the second Planck constant, 14,387$\mu$°K. $\lambda/c_2$, if not known, may be determined experimentally by sighting on a calibrating source at a succession of temperatures.

With an optical pyrometer, a spectrometer or a photomultiplier tube and interference filter [*19*] may be used for the background mirror method, with the wavelength chosen subject to the limitations discussed under *Monochromatic null method.* The flame need have a luminous emissivity of only a few per cent to make the method workable with an optical pyrometer. The method cannot be used on total flame radiation since the transmittance of a gas for its own radiation varies throughout the spectrum.

Since all the radiation methods so far discussed yield a weighted-mean temperature, limited as discussed under *Monochromatic null method,* the methods do not agree when the gas temperature is nonuniform, and those methods operating at a wavelength where the flame emissivity is greatest give most weight to the side of the gas facing the receiving instrument.

*Relative emission intensities of spectral lines.* Measurements of the relative intensities of spectral lines, such as the rotational bands of diatomic molecules like CH and OH, yield less reproducible values of temperature than the methods italicized in Table C,2b; but they may be more informative if interest is in the temperature levels associated with nonequilibrium processes (see IX, Fig. I,3b and Plate M,3b).

*Radiating properties of suspended carbon.* Knowledge of the radiating and absorbing characteristics of the suspended soot in gas flames may be used as a basis for temperature determination, by measuring monochromatic intensity at each of two wavelengths [*18*]. The instrument used may be an optical pyrometer provided with two color screens such as red and green [*18,20*], or for a rapid response system a spectrometer with phototube receivers operating at 0.69$\mu$ and 0.90$\mu$ [*21,22*], or a beam splitter, two interference filters, and phototube receivers [*23*] (see also IX, Fig. I,3a).

Other methods of temperature measurement, including the use of temperature-dependent flow devices [*24*] and the Schmidt hot wire method [*25*], are discussed in IX,I.

**C,3. Pressure Measurements.** As noted in Table C,1, the minimum measurements required for evaluating the burner performance include in any case either static or stagnation pressure in the plane of a traverse. The determination of velocity components may require, in addition, measurements of directional values of total pressure.

Standard devices, wall taps, and impact or Pitot static tubes of standard construction in traversing rigs or in rake array are usually adequate for measurements at the burner entrance and exit planes [*26*].

For the study of internal flow, the use of wall taps for measuring static pressure is usually impracticable. Velocity components normal to the walls due to vortex flow and the bridging by unburned or coked fuel of the small tap holes required [*27*] can introduce large and variable errors in measured static pressure. Warping by thermal stress of the usual thin burner walls can often produce erroneous results. Probes are therefore recommended for the measurement of static pressure as well as of total pressures. Probe positioning can be made independent of wall warping; probes can be oriented to determine static pressure and directional values of total pressure; they can be withdrawn and examined for freedom from deposits.

In flow cross sections large enough to permit its use, the airfoil Pitot meter of First and Silverman [*28*] can be used to determine static and total pressures and flow direction. For flow passages too small to permit the insertion of a Pitot static tube without undue flow disturbance, the data on pressure distribution on cylinders normal to flow have been used together with measurements of stagnation pressure and pressure at 90 degrees from the stagnation point on tiny rotatable cylinders placed in the stream of interest. The stream static pressure and velocity for a stream of known composition and temperature are then computed by successive approximations for the available correlations [*29*] of pressure coefficients with Reynolds number for cylinders.

Gas turbine burners present the experimental difficulties of measuring pressures in small flow passages and in gas flow zones of high gradients in velocity and temperature. Both of these require the use of small probes for accurate work. The high temperatures encountered in some parts of the burner volume, however, necessitate cooling of the probes or their fabrication from thermally stable materials such as ceramics. Either of these alternatives involves compromises on smallness of size or in precision of fabrication. According to Gracy, et al [*30*], the Kiel tube, which is bulky even when uncooled, and simple total pressure tubes with sharp leading edges which are impossible to cool adequately are least sensitive to misalignment with flow. Probes fabricated from high-melting metals may fail, like tungsten on being exposed to a high temperature oxidizing zone, like molybdenum due to thermal embrittling, or like platinum due to catalyzing combustion on its surface.

The combustion process may generate some rather large pulsations in static pressure which it may be desirable both to measure and to eliminate the effects of on measurement of mean pressures. Zaback [*31*] discusses the response of measuring systems to oscillating pressures, and Bradfield and Yale [*32*] describe small Pitot tubes with fast pressure-response times. Draper and Li [*33*] describe their high speed response straingauge, which must be insulated from high temperature or cooled.

The combined effects of the combustion and mixing processes can produce fluctuations in velocity, temperature, and composition comparable with turbulence of quite high intensity. These fluctuations may be far from isotropic, but some indication of their effects on the measurement of static and impact pressures is given by a comparison with the effects of isotropic fluctuations as evaluated in a theoretical approach by Goldstein [*34*] and determined experimentally by Fage [*35*]. At isotropic turbulence levels as high as 25 per cent, the error in stagnation or static pressure is estimated to be high by 6 per cent of the velocity head. Turbulence levels in the wake of bluff objects and in confined flames have been estimated [*36*, pp. 21–40] to rise as high as 20 per cent. Taylor and Comings [*37*] present the results of experiment on the effect of stream turbulence on impact pressure measured in isothermal air jets.

**C,4. Velocity Measurements.** The resultant velocity of a gas stream of known composition can be computed from a knowledge of stream pressures and temperatures in combination as shown in columns 2, 3, 4, 7, and 8 of Table C,1. Measurements of pressure or temperature may be inconvenient or difficult to accomplish without disturbing the flow, especially within a burning zone. Velocity determination by pulsed addition or pulsed tracking of tracers such as dust particles or ions, by the measurement of leakage current, and by acoustic measurements are discussed in IX,C,1,2,3,4. Because of the ionization effects in flames and the probable presence of large but unknown gradients in temperature and composition, ionization, corona discharge, and acoustic measurements are not readily applicable in close proximity to flames or in high temperature flame gases. The use of illuminated particles in the study of flows in flame zones is discussed in IX,J, and experiments with self-luminous particles in flames are described by Nicholson and Field [*36*, pp. 44–68].

The regularity of eddy-shedding from bluff bodies may be exploited for indirect velocity determination by determining the frequency of vibration of fine reeds immersed in the flow. Relations are available [*38*] correlating the ratio of the product of eddy-shedding frequency $f$ by the reed diameter $d$ to the stream velocity $V$ normal to the reed with Reynolds number of the flow based on reed diameter. Since the Strouhal number $(fd/V)$ is but a weak function of Reynolds number over some ranges, and is approximately equal to 0.2, proper choice of reed stiffness-length ratio and reed diameter should be possible [*39*, pp. 170, 351] to permit direct evaluation of the stream velocity from measured reed vibration frequency in a given range of velocities of interest. If such combination of reed characteristics and Reynolds number is not available, reed frequency can be calibrated with stream velocity. Reed vibration can obviously be measured by other than optical means; thus reeds may be used where optical tracking of particles is impossible. Velocity determination by reed frequency measurement is, however, subject to error due to velocity, temperature, or density fluctuations in the stream or to exposure of the reed or vibration-sensing device to high temperature gases. The accuracy of which the method is capable has not been established.

**C,5. Mass Velocity Measurements.** The local flow rate of mass per unit area is usually determined from $\rho u$, the product of stream density and stream velocity at a point, the density and velocity each being based on the measurement of one or more stream properties. Density determination generally depends on a calculation involving an equation of state for the stream (thus implying a knowledge of the molecular constitution of the stream in addition to the obviously needed values of static temperature and pressure). The determination of gas density by the measurement of the refractive index of the gas stream appears attractive because of its simplicity: The refractive index of most combustion gas mixtures is almost linear in gas density. The density of a mixture of known composition may thus be determined from a single measurement of the refractive index $\eta$ of the mixture and a knowledge of the contribution to refraction due to each component.

$$\eta_\lambda - 1 = \rho \sum \alpha_{i\lambda} y_i \tag{5-1}$$

where $\eta_\lambda$ is the refractive index of the gas mixture of mass density $\rho$ for light of wavelength $\lambda$; $\alpha_{i\lambda}$ is $n_{i\lambda} - 1$, the refractivity of pure component $i$ at unit mass density; and $y_i$ is the mass fraction of component $i$ in the gas mixture. If the gas composition is unknown, an additional refraction measurement is required at a known value of the gas density. Unfortunately, the refractometer does not lend itself readily to use in high temperature zones or to measurements in systems having large temperature gradients.

The heat loss from a stationary hot wire immersed in the stream can be used to determine $\rho u$ (see IX,F,2) if the temperature and physical properties such as specific heat, thermal conductivity, and viscosity are known for the stream.

Aerodynamic phenomena susceptible of reasonably precise measurement and dependent only on the product $\rho u$ are unknown. Direct determination of mass velocity by stationary instruments, therefore, seems improbable of attainment. If the duct carrying a stream to be measured can be provided with a U-bend and be rotated about its axis, the measurement of bending stress (due to the Coriolis force) on the radial arms of the U-bend may be used, according to Li and Lee [*40*] to determine mass flow rate. Although this method apparently gives good time-mean flow rate measurements even in rapidly pulsating flow, it is obviously not applicable to the determination of the local mass velocity in a stream having a gradient in mass velocity across its flow area.

Other measuring techniques which suggest themselves are the use of rotating total head tubes and the measurement of lift on a cylinder rotating about its axis perpendicular to the stream flow. Both of these possible methods pose obvious problems of minimum size of flow area measurable and difficulty of interpretation of data taken near solid surfaces. In addition, the accurate measurement of lift on a small cylinder presents a difficult experimental task.

As a simple illustration of the general method of adding to the significance of readings from a stationary instrument by imposing a velocity of known magnitude and direction on the device, consider two total head tubes placed with their openings at the same radius from an axis of rotation parallel to the stream flow, but with one tube having a yaw angle in motion of $\theta$ to the direction of mass flow and the other having a yaw in motion of $\pi - \theta$. If the two-tube array is rotated with a constant speed $V$ of the tube openings in a stream of density $\rho$ and velocity parallel to the axis of rotation of $U$, for incompressible flow the total pressure $p_u^0$ on the tube pointing upstream will be

$$p_u^0 = \frac{\rho}{2}(V \sin\theta + U \cos\theta)^2 + p$$

and the total pressure $p_d^0$ on the tube pointing downstream will be

$$p_d^0 = \frac{\rho}{2}(V \sin\theta - U \cos\theta)|(V \sin\theta - U \cos\theta)| + p$$

The measured difference, for $V \sin\theta > U \cos\theta$,

$$p_u^0 - p_d^0 = 2\rho\,(V \sin\theta)\,(U \cos\theta)$$

depends on the known quantities $V$ and $\theta$ and is linear in $\rho U$, the desired quantity. If $V \sin\theta$ is markedly larger than $U \cos\theta$ the implicit assumption of equal pressure coefficients near unity for both total head tubes is probably a reasonable one. It should be noted that the precision of the suggested method depends on the use of total head tubes having high sensitivity to yaw angle.

The method of measuring lift for a rotating cylinder requires lift measurements at two or more values of the ratio $V/U$ of rotational velocity to stream velocity. If the Reynolds number range for the rotated cylinder is such that the lift coefficient $C_L$ is linear in $V/U$, as indicated by Reid [41] and Betz [42], the lift coefficient can be expressed as $C_L = a + b(V/U)$. With values of lift $L_1$ and $L_2$ measured at $V_1$ and $V_2$,

$$L_1 = \left(a + b\frac{V_1}{U}\right)\frac{\rho U^2}{2}, \quad L_2 = \left(a + b\frac{V_2}{U}\right)\frac{\rho U^2}{2}$$

$$L_1 - L_2 = b(V_1 - V_2)\frac{\rho U}{2}$$

It would be necessary to make an independent calibrating run at known stream velocity $U$ to determine the constant $b$. The available data do not indicate strong dependence of $b$ on Reynolds number. No experiments are known, however, for cylinders as small as would be required for use in gas turbines.

It has been suggested that the mass velocity of a stream of unknown composition and physical properties could be determined by measuring the heat loss from a hot wire maintained at each of several different known temperatures or given each of several different known velocities by rotation. The reader is cautioned against attempting this since it can readily be shown that the effects of the physical properties, including stream density, are combined in such a manner in the heat loss equation as to be inseparable and such an experiment will permit determination of (1) the stream velocity if the wire is moved and (2) the stream temperature if the wire temperature is changed, but not $\rho U$.

## C,6. Measurements of Gas Composition.

Sampling a Fluctuating Stream. When the composition, temperature, and velocity of a stream are varying rapidly and in random fashion, the withdrawal of a gas sample representative of the space-average composition of the gas stream at the sampling point is a matter of considerable difficulty. If the velocity of gas withdrawal into a Pitot-type sampling probe is held equal to the velocity of the approach stream and the gas flow through the sampling tube is isothermal, a proper sample will be collected. Since these conditions are difficult to meet, it is desirable to have some knowledge of errors to be expected, or criteria for probe design to minimize errors. Sampling velocities varying from one quarter to three times the stream velocity have caused collection, from gas turbine chambers, of gas samples having computed flame temperatures which vary as much as 500°F [43]. In other gas turbine chamber studies, varying the sampling velocity has produced a two-fold variation in indicated fuel-air ratio or oxygen content [44].

Gas sampling may be accomplished with a small-diameter Pitot-type tube facing the stream and provided, close to its mouth, with a small chamber (Helmholtz resonator type); a long tube in which damping effects are accentuated; or a lateral sampling probe, with axis aligned with the stream and gas withdrawal through side-inlet holes. Leeper [*45*] has analyzed the performance of the first of these types in some detail, for the special case of isothermal sampling of a stream of constant velocity and varying density. Even for this simplest of experimental conditions the analysis is quite complex, and incomplete. Sampling at a velocity $V_s$ in a stream of velocity $V_0$ produces pressure fluctuations at the probe mouth, caused by the deceleration (or acceleration) of parcels of different density; and with these pressure fluctuations are associated fluctuations in flow rate through the sampling tube. Leeper's theory leads to the conclusion that the sampling error $E$, measured by the difference in mean density of the sample $\bar{\rho}_s$ and that of the stream $\bar{\rho}$, should vary linearly with $(V_s/V_0 - V_0/V_s)$; and that the proportionality constant of the variation should be proportional to the stream velocity $V_0$ (in the group $\omega l/V_0$), dependent on the wave form of the density variation (he examined square waves), and a complicated function of three other groups, $\zeta_1$ the damping ratio, $\omega/\omega_n$ the frequency ratio, and $(\rho_u - \bar{\rho})/\bar{\rho}$ the ratio of density wave amplitude to the mean space density.

$$\zeta_1 = \frac{\mu Z_1^2}{2\bar{\rho}\omega_n r_0^2}, \quad \omega_n = \sqrt{\frac{c^2(\pi r_0^2)}{v_2 l}}$$

where $\mu/\bar{\rho}$ is the kinematic viscosity, $r_0$ is the sampling tube radius, $Z_1$ is a constant (first root of zero-order Bessel function), $\omega_n$ is the natural frequency (radians/sec) of the sampling tube-gas chamber combination, $c$ is the velocity of sound, $v_2$ is the surge chamber volume, $l$ is the probe length from mouth to surge chamber, and $\omega$ is the frequency (radians/sec) of the density variation of the main stream.

The validity of Leeper's analysis has received only partial testing. Wohl [*46*] used a conventional probe to sample a simulated stream of unmixed $CO_2$-air of 16 density variations per sec ($\omega$ = 100 rad/sec) and at simulated stream velocities of 25, 50, and 95 ft/sec. Whether or not a probe has the internal geometry characteristic of a Helmholtz resonator, it should nevertheless have an error proportional to $[(V_s/V_0) - (V_0/V_s)]$ and to $V_0$ so long as the frequency ratio is kept constant. The error in per cent $CO_2$ was in fact found proportional to the quantity $[(V_s/V_0) - (V_0/V_s)]$, and the proportionality constant divided by the stream velocity $V_0$ was substantially constant, about $-0.019$ per cent $CO_2$ per unit of $[(V_s/V_0) - (V_0/V_s)]$ and per ft/sec stream velocity. An absolute comparison of theory and experiment could not be made because the probe was not of the Helmholtz type. Leeper obtained some data on the sampling of an unmixed $CO_2$-air stream of 1100 cycle density frequency

($\omega = 6970$ rad/sec), using a Helmholtz resonator probe of 0.023-in. radius, 1.08-in. length, and $7 \times 10^{-3}$-in.$^3$ volume. He found that the experimentally determined error in per cent $CO_2$ in the sample was $-0.0066$ per unit of $[(V_s/V_0) - (V_0/V_s)]$ and per ft/sec stream velocity, whereas theory predicted the somewhat lower value $-0.0035$. Analysis of the expected performance of a smaller Helmholtz probe ($r_0 = 0.01$ in.; $l = 0.30$ in.; $v_2 = 0.002$ in.$^3$) indicated that, for conditions in the stream for which it was designed, it should cause a somewhat lower sampling error than the one just discussed.

The presence of fluctuations in temperature as well as composition in the stream to be sampled adds to the difficulty of obtaining a representative sample since heat transferred to and from the probe walls to the gas being sampled causes oscillatory forces to be set up in impact probes, thereby causing erroneous samples to be taken even if the sampling velocity equals the stream velocity. Consequently, it is desirable to minimize this heat transfer by using probes which mix the sampled

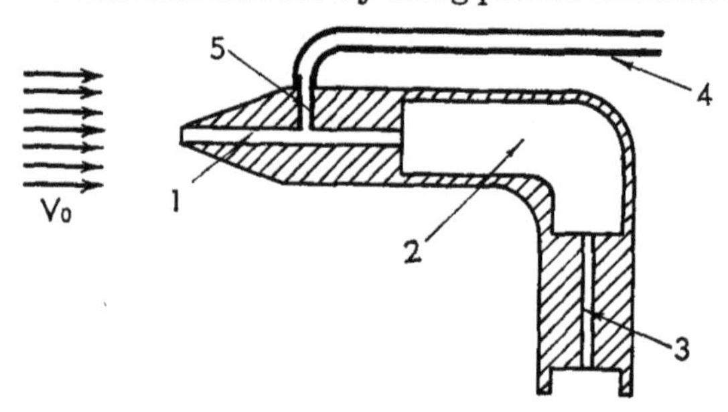


Fig. C,6. Helmholtz-type metering probe.

gases as soon as possible after withdrawal from the stream. A Helmholtz-type probe, with its short tube and close-coupled mixing chamber, accomplishes this action. Since the quantitative performance of such a probe in an isothermal variable-density stream compares favorably with the more conventional type, and since the Helmholtz type is superior for variable-temperature streams, its use is recommended. The gas chamber should be large enough to contain the gas sampled from several wavelengths of main-stream density variation to make the chamber gas properties remain relatively constant; this is particularly important in the sampling of variable-temperature streams.

A Helmholtz-type impact probe design has been recommended [*43*] which eliminates the need for metering the sampled gases to match $V_s$ to $V_0$. Fig. C,6 illustrates the probe which, in addition to the normal components, is provided with a side-outlet tube 5 from a pressure tap on the side of the main sampling tube 1. Ideally, if the pressure at 6 equals the static pressure in the free stream the velocity in the tube will equal stream velocity. In practice, Kapstein [*47*] found that even with a very much larger probe than Leeper recommends, correction must be made for pressure drop from the probe mouth to the pressure tap. With a tap at 14 diameters downstream in a 0.085-in. diameter probe in a stream at atmospheric pressure and velocity $V_0 = 55$–$110$ ft/sec, he found a sampling velocity ratio $V_s/V_0$ of 0.65–0.74 when static pressures were matched, vs. 1.0 for frictionless flow. This was in good agreement with the application of boundary layer theory. Kapstein's tap could profitably have been placed two or three probe diameters from the mouth.

The usual precautions must be taken to obtain the correct static pressure of the main stream, necessitating consideration of the flow pattern near the probe.

Two other probe applications merit brief comment. (1) The lateral sampling probe, with orientation parallel to the main stream and side holes for induction of the sample, introduces an error, in the sampling of a variable-density stream, which is proportional to the sampling velocity. Consequently, a velocity as near zero as is practicable should be used, with perhaps an extrapolation of results to zero velocity. (2) In the sampling of a supersonic stream, if the flow in the sampling tube is above Mach 1 so that a swallowed-shock condition exists, the sample is representative. It may be desirable, however, to verify the absence of a bow shock wave by a spark or schlieren photograph.

Representative sampling of a stream of randomly varying velocity is not possible unless the variation is slow enough to permit the assumption that the flow in the probe has time to adjust itself to main-stream conditions. In that case a sampling-error analysis such as Leeper's can be used, in combination with separate probe information concerning the velocity fluctuation spectrum, to carry out an integration of the error due to the sampling velocity's not matching the stream velocity. Much work would be necessary to establish the validity of the procedure.

The sampling of a stream containing suspended liquid droplets may be desirable to determine, for example, the fraction of the fuel which exists locally in droplet form. This is feasible if other disturbing factors such as temperature or gas-phase inhomogeneity or velocity fluctuations are small enough to be neglected. The method depends on the principle that the "collection efficiency" [*48*] of a small sharp-edged impact tube for droplets of greater than 10-micron diameter in streams of velocity greater than 100 ft/sec is substantially 1, from which it follows that the rate of entry of droplets into the probe mouth is uninfluenced by gas sampling velocity $V_s$ in the probe, over the range zero to values exceeding the free stream velocity $V_0$. Let $f$ be the original fuel-air ratio necessary to produce the sample obtained, based on analysis by some convenient methods such as completion of the combustion over heated copper oxide or platinum, followed by carbon dioxide analysis. Let $f_{liq}$ and $f_{vap}$ be the ratios to be determined, of liquid fuel to air and of vaporized or burned fuel to air, respectively, in the main-stream gas at the sampling point. When a sample is withdrawn from the probe at velocity $V_s$ less than the stream velocity $V_0$, its composition $f$ will be higher in fuel content than the main stream and be given by

$$f = f_{vap} + \frac{V_0}{V_s} f_{liq} \tag{6-1}$$

Kapstein [47] has used this relation, determining $f$ for different sampling rates, plotting $f$ vs. $V_0/V_s$ and putting a straight line through the data to determine $f_{liq}$ as the slope and $f_{tot} = f_{liq} + f_{vap}$ as the intercept on the line, $V_0/V_s = 1$. The probe was made from hypodermic tubing 0.109-in. O.D. and 0.085-in. I.D., with sharpened nose and a pressure tap along its side. The tap permits direct use of the probe as a gas meter by connecting through a manometer to a suitable static pressure measuring device in the main stream. For the application made, a study of fuel distribution and fractional vaporization, the probe gas was led into a preheater coil of $\frac{3}{8}$-in. copper tubing surrounding a rather large-capacity vaporizer consisting of 2.5 feet of 1-in. steel tubing packed with copper turnings and heated with 25 feet of #20 chromel-A heating wire. The gases were then led to a combustion train.

The accuracy of the above method is dependent primarily on the quality of the calibration of the probe as a gas meter. A study by Dussourd and Shapiro [49] of how to measure the velocity and total pressure of an aerosol with an impact probe supports the assumption made above that a 0.1-in. probe is small enough for $50\mu$ particles at velocities as low as 50 ft/sec. However, the influence of the pressure rise $\Delta p$ due to particle deceleration within the probe may become significant. From Dussourd and Shapiro's study it may be shown that below a value of $\rho_g D/\rho_{liq} d$ of 0.25, this pressure rise is given, for distances within a few probe diameters of the mouth, by

$$\Delta p = 6 \frac{x}{d} \left( \frac{V_0 - V_s}{V_0} \right)^2 \frac{w_{liq} \rho_g}{w_g \rho_{liq}} \left( \frac{\rho_g V_0^2}{2} \right) \left( \frac{d(V_0 - V_s)\rho_g}{\mu_g} \right)^{-0.3} \tag{6-2}$$

where $x$ is the distance from probe mouth to side tap, $w_{liq}/w_g$ is the mass ratio of liquid to gas in the free stream, $\rho_g$ and $\mu_g$ are the density and viscosity of the liquid-free gas, $d$ is the mean particle size of liquid of density $\rho_{liq}$, and $D$ is the probe mouth diameter.

Gas Analysis. If the performance of a gas turbine is to be evaluated, some knowledge of composition of the combustion products is necessary, regardless of what other measurements are to be made (see Table C,1); but the extent of the information necessary varies with the details of the objective. Although the minimum information needed is the molecular weight and the heat of combustion of the products, interest in element balances to check on the internal consistency of the data, in the progress of combustion, and in identification of the cause of inadequate combustion performance dictate a need for a fairly complete chemical analysis both of the fuel and of the combustion products.

Many methods of gas analysis are available for application to gas turbine combustion problems; most of them require for success more specialized training than is apparent to one having an elementary knowledge of their principles of operation. The following should be included for consideration: (1) conventional chemical volumetric analysis by successive absorption in reagents, followed by quantitatively measured combustion; (2) mass spectrometer examination of molecules and molecular fragments obtained by ionization of the gas; (3) infrared absorption; (4) condensation and fractional distillation; (5) gas chromatography; (6) thermal conductivity; (7) other miscellaneous instrumental methods mostly restricted to rapid examination of two-component or effectively two-component systems; and (8) gas analysis trains. For effective application of these methods a detailed exposition is necessary, and the reader is referred to such comprehensive works as those of Mullen [*50*] and Altieri [*51*]. Space is available here only to indicate some of the problems, limitations, and recent developments.

*Absorption and combustion.* This is the conventional or standard method of analysis. Its advantages are that it requires relatively inexpensive equipment, can be readily adapted to changes in the general objective of the analysis, and permits an unambiguous interpretation of the data. Its disadvantages are its slowness relative to some of the instrumental methods and the considerable number of sources of error, including absorption by most reagents of *some* of those compounds they are not intended to absorb (leading to techniques of equilibration of the reagents with the gas to be analyzed), desorption of previously absorbed gases from the reagents, incomplete combustion (as influenced by the temperature of the filament or the rate of flow of gases over it), manifold-volume errors, deterioration of reagents, stopcock-grease absorption, and stopcock leaks. In consequence, the results of analyses of the same gas sample by different cooperating laboratories, as reported by Shepherd [*52,53*], show a surprisingly large spread in results reported.

The generally held view that a rigid sequence of operations can be applied, regardless of sample composition, is a mistaken one; and it is necessary to follow procedures recommended on the basis of analysis of samples similar in composition to those of interest. As a single example consider some of the recommended solutions for the absorption of carbon monoxide. Acid cuprous chloride is widely used, but because its equilibrium back-pressure of carbon monoxide builds up rapidly with use, two pipettes should be used in series; it also absorbs oxygen and alkaline gases. Ammoniacal cuprous chloride is widely used, but it absorbs, in addition to oxygen and acid gases, olefines and acetylenes to a certain extent. Cuprous sulfate-beta naphthol is used, but it is slower than the cuprous chlorides and tends to absorb ethylene and acetylene. Except when CO is present in small amounts its determination is best made as a part of the combustion analysis. The latter, following absorption analysis for $CO_2$, $O_2$, unsaturated hydrocarbons, and perhaps CO, uses either a heated copper oxide tube or a heated platinum wire with oxygen or air. Copper oxide is used at 285–295°C to burn $H_2$ and CO in the presence of hydrocarbons. At 310°C ethane starts to burn, but 700°C is required for completion, and operation above 500°C is not practicable. A platinum spiral which is faintly red will burn $H_2$ (mixed with oxygen) completely; CO partially or completely, depending on the intimacy of contact and the number of repeated passes; and $CH_4$ not at all. Raising the temperature to a white heat assures combustion of methane and any other hydrocarbons or remaining CO. Complete combustion data include the determination of contraction due to burning and to condensation of water formed (TC), carbon dioxide formed ($CO_2$), and oxygen consumed by combustion ($O_2$). Although some gas mixtures require for analysis in principle only the first two of these three quantities, it is a mistake not to include oxygen consumption as part of the data obtained. Where two quantities suffice, ($TC + CO_2$) may then be considered a single piece of data, and the accuracy of analysis is thereby markedly improved [*52*] (the method of course requires knowledge of the purity of the oxygen used). It should perhaps be pointed out that although a combustion analysis on a mixture of gaseous combustibles and nitrogen does not yield enough information to identify the composition in detail unless there are known to be but two or three gases present besides nitrogen, it does permit determination of the amounts of carbon, net hydrogen, and nitrogen in the sample. This information is often sufficient, indicating as it does the air requirement, molecular weight, and approximate heating value of the sample. The reader is referred to the literature for material on precision determination of oxygen [*54*], constant volume analysis, to be preferred when samples are small [*55,56*], microanalysis [*57,58*], the determination of small concentrations of CO [*59*], catalytic fractional combustion of $H_2$, CO, and $CH_4$ over Pt on a ceramic carrier [*60,61*], the description of Orsat-type equipment [*62,63*], and summaries and reviews of available methods and recent developments [*64,65,66,67*].

*Mass spectrometer.* This instrument uses a small gas sample, the molecules of which are ionized, producing molecular fragments of different "appearance potentials," and then accelerated and brought onto the exit slit in turn by variation of a magnetic or electrical field which determines the mass-charge ratio of the fragments hitting the slit. An ion collector measures the amount associated with each mass-charge ratio. Because of the identity of the mass-charge ratio of fragments from different gases, interpretation often involves the solution of simultaneous linear equations; and for some gas combinations the solution is of low accuracy. The reduced velocity of bombardment of the gases being analyzed, to cut down the severe fragmentation conventionally obtained, can contribute to better accuracy of analysis [*68*]. Shepherd [*53*], on the basis of comparison of analyses of a carburetted water gas by 51 cooperating laboratories using both absorptiometric and mass spectrometer methods, concludes that the mass spectrometer is somewhat inferior for $H_2$ and CO both in accuracy and in spread of the results, but superior for light hydrocarbons and for oxygen in small concentration. The over-all accuracy of the two methods is indicated by calculating the density and heating value from the reported gas analyses and comparing the results with direct accurate measurements. The true heating value 549 was more closely approached by the calculations based on mass spectrometer analyses, $550 \pm 7$, than by those based on absorption and combustion, $545 \pm 3$; but the spread of the data by the latter method was markedly less. The true density of 0.6475 may be compared with $0.648 \pm 0.002$ based on chemical analysis versus $0.650 \pm 0.007$ based on the mass spectrometer. The marked spread in results from different laboratories using the latter instrument may be expected to decrease as the use of this relatively new tool for gas analysis becomes better understood. For an exposition of the instrument see [*69*, pp. 1991–2058] and also [*70,71,72,73,74,75*].

*Infrared absorption.* The method depends on the fact that, except for $H_2$, $O_2$, and $N_2$, gases encountered in gas turbine operation have infrared absorption bands at wavelengths indicative of their molecular structure and of strengths varying with their concentration. The complications in interpretation are due to the sharing of parts of the spectrum by more than one type of molecule, the "broadening" of absorption bands of one kind of molecule by the presence of others, and the complex character of the gas concentration-absorption relation especially when the spectral resolution is low. For analyses of combustion gases, the components of which have not all been identified, the infrared spectrometer is a powerful tool in the hands of a specialist. When the components to be analyzed are all known, the measurements of absorption may be made at as many wavelengths as there are components, the wavelengths being chosen as far as possible to make each unique to a single species. The composition is then obtainable directly, from a previous calibration. If more than one component is responsible for absorption at a particular wavelength, which is often the case, a system of simultaneous equations must be solved; and under some circumstances the resulting accuracy of analysis is low. Despite the overlapping spectra of the components present in a gas, infrared analysis is possible without using a dispersion-type instrument if all the components which may be present are known in advance. To analyze for a particular component the radiation from a high temperature source such as a Globar is passed through a so-called interference cell containing those other gases, both known to be present in the gas sample and to share spectral regions with the component being analyzed, and a beam of radiation is thereby obtained which is absorbable in substantial amount only by the component in question. This beam is then passed through the sample cell, split into two beams and, after being sent in parallel through two cells, one of which (the "filter" cell) contains a high concentration of the component being analyzed, falls on the two junctions of a differential thermocouple. The ability of the filter cell, containing component $x$, to change the intensity of one of the beams depends on how much $x$-absorbing radiation has already been removed in passage through the sample cell. The presence of unsuspected components can produce serious error. The reader is referred to the literature for detailed descriptions of commercially available instruments [*65*,*76*,*77*], special instruments [*78*,*79*,*80*,*81*], techniques [*82*,*83*], and accuracy to be expected [*53*,*82*].

*Condensation and fractional distillation.* This procedure is used to analyze a hydrocarbon mixture too complex to be identified by combustion methods [*84*,*85*,*86*,*87*]. It is rapidly being replaced by mass-spectrometric, infrared, and chromatographic analyses where the number of analyses to be made justifies the investment in apparatus for the newer techniques.

*Gas chromatography.* This relatively new method depends on the difference in the phase equilibrium constants of the different components of the gas sample when placed in contact with a solid phase or a liquid phase supported on a solid absorbent. Its chief interest in the present application is in the analysis of partially burned medium hydrocarbons or their oxidation products; but it is in the process of development to include analysis of the usual components of engine exhaust gases. Two quite distinct applications of the principles of gas chromatography are used, referred to as the elution-partition and the adsorption-displacement methods. In the former a carrier gas, preferably hydrogen or helium, flows at a carefully controlled constant rate of 50–400 $cm^3$/min. through a thermal conductivity cell (or past it, with diffusion connection) into a "column" or tube packed with an adsorbent (such as silica gel or charcoal), or with a support (such as acid-washed diatomaceous earth) which holds a liquid absorbent (such as petroleum oils, phthalates, silicones, waxes, glycols), thence to another thermal conductivity cell and finally through a flowmeter. A gas sample of 0.3–20 $cm^3$ is introduced into the flowing stream just before the column and is adsorbed or absorbed by the column packing, but moves along the packing with the carrier gas, the least firmly held component moving fastest and finally desorbing into the carrier gas as it comes to the end of the column. Its presence in the carrier gas is recorded by the conductivity cell as an error-function-type curve superimposed on the steady conductivity representing the carrier gas. The area under the curve is proportional to the volume of the component in the sample, and the proportionality constant is almost independent of the nature of the component if hydrogen or helium is used as the carrier. By the proper choice of sample size, flow rate, operating temperature, and particularly the column contents, the various component peaks will show no overlap. An analysis requires about 30 minutes, and an accuracy of 0.2–2 per cent of the amount of any adsorbable component present is claimed. Considerable experimentation will be needed before $CO_2$, $O_2$, $N_2$, and CO can be separated satisfactorily with sufficient accuracy to compete with older methods.

In the absorption-displacement method a much larger sample is used, about 350 $cm^3$. This is displaced with mercury into the base of the adsorbent section containing charcoal, alumina, or silica gel, the most readily adsorbed components remaining near the column inlet. A displacement gas which is adsorbed preferentially to any of the components of the gas sample, such as acetone, ethyl acetate, or chloroform, is now introduced continuously into the base of the adsorber, thereby driving off the least strongly held component first. The thermal conductivity record obtained is entirely different from that of the elution partition method, and consists of a step curve, the height of the steps representing the conductivity of the components and the length their volume. The last step is the displacing gas. Carbon dioxide and methane are readily separated.

Enormous progress is certain to come in the development of chromatography for gas analysis, chiefly because of its ability to exploit the different phase equilibria of different adsorbents. An extensive literature exists, which is growing rapidly, including [*88,89,90,91,92,93,94,95,96*].

*Thermal conductivity.* Heat conduction from an electrically heated wire on the axis of a cell carrying the gas to be analyzed causes the wire temperature to differ from that in a comparison cell by an amount proportional to the difference in conductivity of the gases. The wire temperature difference is measurable in a Wheatstone bridge circuit with such accuracy that $CO_2$ in air has been determined to 0.005 per cent, and 1 per cent benzol in air has been determined with an accuracy of one part in 500, or 0.002 per cent on the total sample [*97*]. The method as described is limited to the analysis of a two-component system or its equivalent, i.e. one in which either the ratio of all the components but one is fixed or their conductivities differ much less from each other than from that of the component being determined. It is possible, however, to analyze a three-component mixture by making use of one of the following principles: (1) the larger a conductivity cell, the more the heat loss from the wire depends on convection rather than conduction; and convection depends on molecular weight and specific heat as well as conductivity. Thus by using cells of two different sizes, two different components in a third gas may be determined [*98*]. (2) Pressure affects heat loss by convection but not by conduction (unless the pressure is very low). Consequently the difference in the unbalance of two cell pairs operating at different pressures offers a basis for analyzing a three-component mixture [*99,100*]. (3) The temperature coefficient of conduction as well as the conductivity may be utilized to analyze a three-component mixture, by operating the cell pair at each of two wire currents. Combinations of these methods could in principle permit analysis of a four-component mixture, but minor variations of the base-gas or fourth-component composition could cause trouble. The merits and limitations of thermal conductivity analysis are discussed by Weaver [*101*].

*Other miscellaneous instrumental methods.* Discussion is restricted to a rapid examination of predominantly two-component systems. (1) Pauling's suggestion [*102*] that the paramagnetism of oxygen could be used to produce a simple device for oxygen analysis has led to the development of highly reliable equipment capable of determining the partial pressure of oxygen to 1 part in 500 over almost any range. The magnetic susceptibility of gases commonly encountered is between 0 and 0.6, except for oxygen, nitric oxide, and nitrogen dioxide, for which the values are 100, 45, and 4. Consequently, except for a very minor correction for background gas composition, the device reads partial pressure of oxygen directly and is of great value in combustion studies. (2) The velocity of sound in a perfect gas is proportional to $\sqrt{\gamma RT/\mathfrak{M}}$; consequently its applicability to two-component gas analysis is measured by the spread in $\gamma/\mathfrak{M}$ for the two gases, where $\gamma$ is the specific heat ratio and $\mathfrak{M}$ is the molecular weight. The method has not had much use in gas analysis; Mullen [*50*] summarizes the status of development. (3) Gas interferometry permits accurate determination of the refractivity of a mixture. Since this is the weighted mean of the refractivity of the components weighted in proportion to their partial pressures, the method is applicable to an effectively two-component system. With good equipment, 1 per cent $CO_2$ in air can be determined to about 1 part in 150. For details see Tilton and Taylor's "Refractive Index Measurement" in [*103*].

*Gas analysis trains.* The most generally required knowledge of turbine-gas composition is the air-fuel feed ratio, the unburnedness as measured by the oxygen required to complete the combustion or heat generated in completing it, and the fraction of the fuel fed which appears as heavy unburned hydrocarbons. Such information can be obtained without a complete gas analysis, by setting up a train of apparatus in which several measurements are made continuously. For example, the partially burned sample may be passed through a furnace tube where its combustion is completed, thence through a drier to a thermal conductivity meter to determine $CO_2$. The fuel-air ratio is then obtainable directly by stoichiometry. If part of the sample is run first through a condensate trap to drop out any unburned heavy fuel, the $CO_2$ content of the finally burned sample indicates the fuel-air ratio exclusive of the fuel condensed. Inclusion of Pauling oxygen meters before and after the furnace tube permits determination of the oxygen consumed in completing the combustion. Alternatively, a catalytic combustion device may be used which, by measurement of the temperature rise around a heated wire on which the combustion is completed, gives a reading proportional to the heat of combustion of the residual unburned components in the gas. Calibration of such a device is necessary. An example of a gas analysis train is presented by Gerrish and Meens [*105*]. Various gas analysis trains are available on the market, combining measurements of absorption, thermal conductivity (sometimes at more than one point in the train), and stream-property modifications associated with combustion. The Gooderham Analyser [*106,107*] is an example of a continuous absorption device for determining the same components which are analyzable by conventional Orsat-type volumetric equipment.

This discussion of gas analysis procedures has been general in character rather than specific in its application to gas turbines. The reason is that the objective in making a gas analysis can vary so greatly in gas turbine research, development, and performance studies that any rigid recommendation of a preferred technique to be followed is made substantially impossible.

## C,7. Traverse Measurements.

*Temperature.* In addition to the requirement for temperature measurement of burner exhaust gases to determine the progress of combustion, it is often necessary, because of thermal limitations on the strength of turbine nozzles and blades, to know the exhaust gas temperature distribution. Thus, knowledge that the mean exhaust temperature lies below the thermal limits for downstream elements of the engine may be misleading in the presence of large gradients in exhaust gas temperature. These temperature gradients can be so steep that the use of temperature-sensing probes which average the gas temperature over an area as small as a sixteenth of a square inch may produce erroneously smooth traverses. If the size of the power plant elements downstream of the traverse plane is larger than that of the area over which the probe averages the gas temperature, a probe-measured temperature below the thermal limit for the downstream elements usually assures safe operating temperatures. The mean exhaust temperature determined by traversing the exhaust stream with such a large probe, however, generally gives a false measure of combustion efficiency.

The four thermocouple probes sketched in Fig. C,7a were used to make temperature traverses on a radius in the exhaust plane of an annular burner. The measured temperatures are shown in Fig. C,7b as the hollow symbols in a plot of temperature vs. radial distance from the axis of the annulus. The solid circles are corrected gas temperatures for probe 1, based on the measured thermocouple junction temperatures and calculated corrections (see below) for radiant heat transfer and stagnation

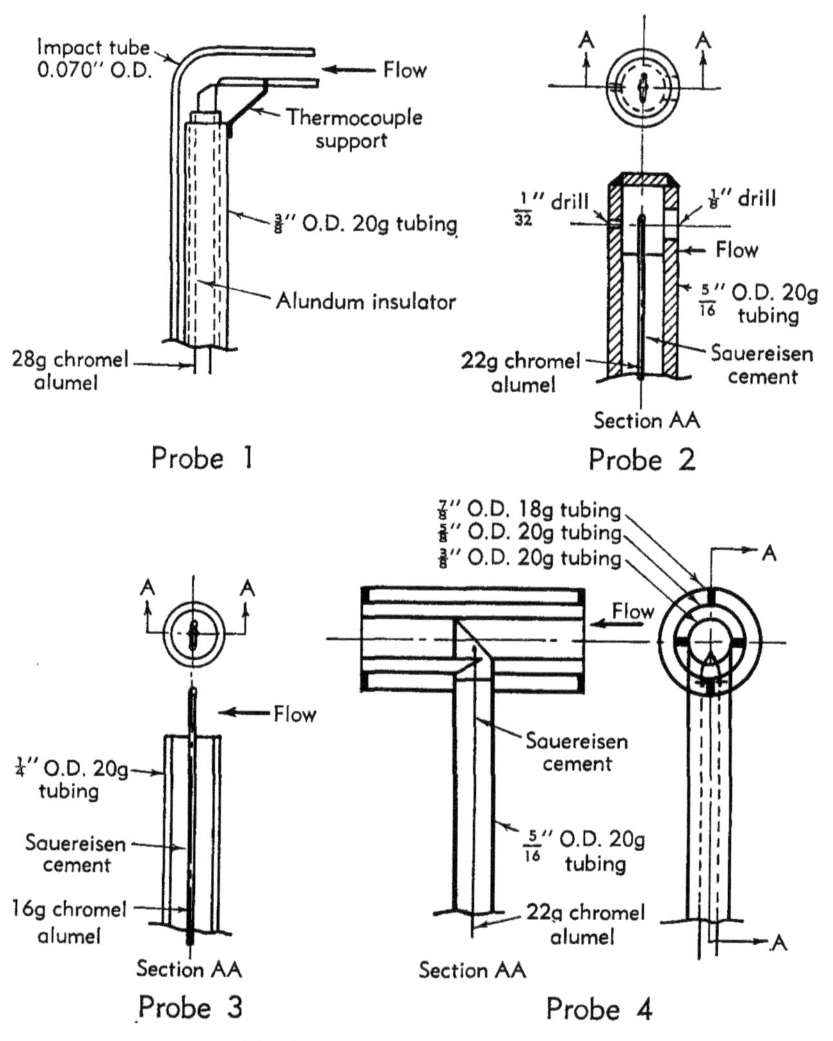


Fig. C,7a. Thermocouple probes.

recovery. Static pressure at the exhaust plane was one atmosphere; the average exhaust gas velocity was about 500 ft/sec.

On comparison of the measured temperatures for probes 1 and 3, it is seen that the temperature of bare 16-gauge wire (probe 3), commonly used for its ruggedness, is as much as 1000°R below that of the more fragile 28-gauge wire of probe 1. Probe 2, a modification of the Massachusetts Institute of Technology stagnation probe [*5*], because of radiation from the relatively large shield surrounding the thermocouple junction and the low heat transfer coefficient from gas to junction due to the

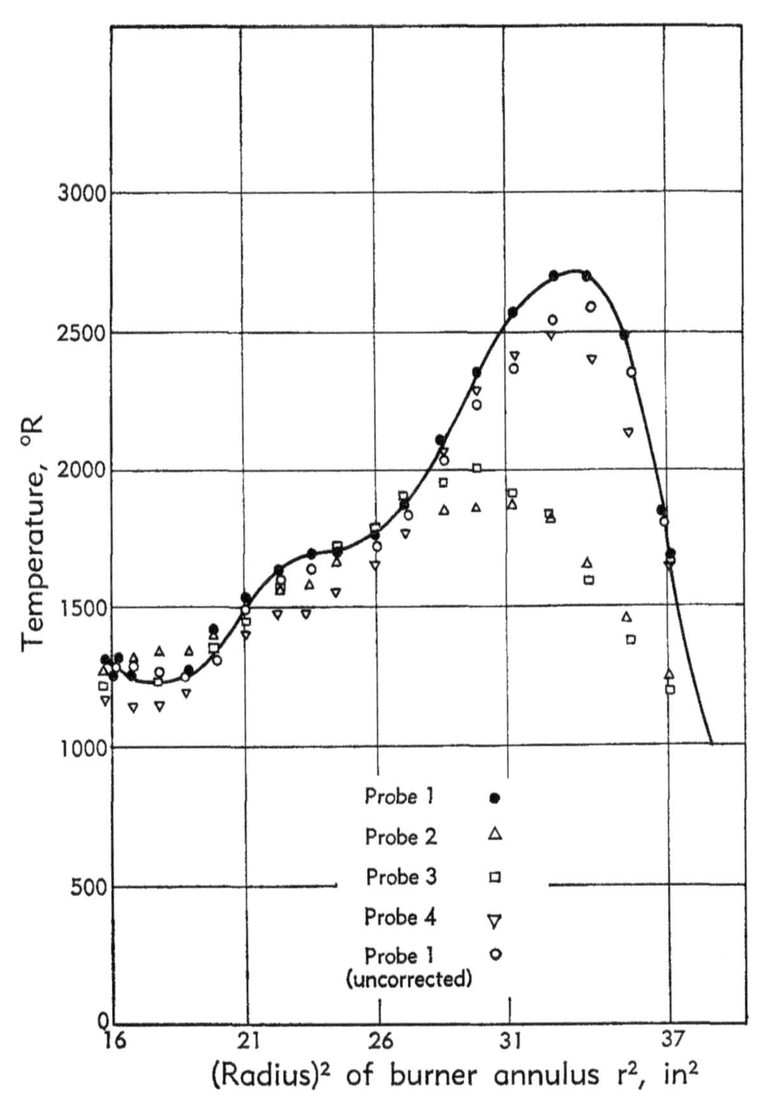


Fig. C,7b. Typical traverses made with probes of Fig. C,7a.

low gas rate through the shield, gives readings close to that of probe 3, in spite of the smaller 22-gauge wire junction. The three-shield design of probe 4 represents an attempt to reduce the radiation error and the shield spacing, based on recommendations of King [*1*]. A comparison of the temperature measurements for probe 4 with those for probe 1 shows that for gradients in gas temperature less than about 3000°F per inch, the junction temperatures of probe 4 are closer to the true gas temperature than are those of probe 1. At the larger radii, where the temperature gradient within the exhaust gas is about 9000°F per inch, probe 4 readings are over 200°R below those of probe 1 and over 300°R below the corrected measurements for probe 1. It is probable that a multishielded couple could be made sufficiently small to give good local measurements in a region of high temperature gradient. It seems likely, however, that in addition to being more costly to fabricate in the very small sizes required, the life of such a small device would be less than that of a bare wire of even smaller effective size.

The comparisons already made between the results obtained with the several types of thermocouple probes shown indicate the difficulty of developing a probe design with adequate provision for reducing measurement errors and yet having sufficient strength for long life or of sufficiently simple construction that short life is of little concern. A reasonable solution appears to be the employment of the probe of simplest design—a bare wire of the smallest diameter which will give a life in use commensurate with the effort required to fit it into a traversing rig with the desired accuracy of positioning and with the time required for traversing. A bare, butt-welded junction of the same diameter as the fine couple wires and orientated so that the flow is normal or parallel to the wire axis presents a device simple to construct and, by comparison with geometrically more complex probe designs, straightforward to correct for junction-temperature errors due to radiation, conduction, and convection. The general correction procedure is illustrated here by the method used to obtain the corrected temperature points shown in Fig. C,7b for probe 1.

A typical traverse, such as that shown in Fig. C,7b, is made along a line joining the centers of the upstream extremities of the impact tube and the couple junction. To obtain the points plotted in Fig. C,7b, the probe entered from the outer radius and was moved in one-eighth-inch steps from the outer radius to the inner radius of the burner exhaust plane. The couple readings were thus one station behind the impact tube reading since the spacing between the tube and the couple, believed to be near the minimum without causing appreciable flow interaction, was one-eighth inch. The procedure to correct the measured junction temperature is then one of successive but rapidly convergent approximations. From knowledge, for a given station in the traverse, of the impact pressure and temperature and static pressure (assumed uniformly one atmosphere for the corrections in Fig. C,7b), a trial value of stream velocity, stream density, and stream temperature is determined for an assumed molecular composition of the stream (assumed here to correspond to that of a completely burned mixture having an adiabatic flame temperature equal to the measured probe temperature—a close approximation for normal gas turbine air-fuel ratios and for combustion efficiencies high enough to insure absence of liquid fuel in the gas stream). A steady state heat balance on the thermocouple junction at $T_c$ equates the rate of heat input from the gas by convection $q_c$ (gas radiation is normally negligible in comparison with convection) to the rate of heat output from the couple junction by radiation $q_r$ to its surroundings, and by conduction $q_k$ along the couple lead wires.

$$q_c = q_r + q_k \tag{7-1}$$

where

$$q_c = A_c h_c (T_{aw} - T_c) \tag{7-2}$$

$$q_r = \sigma \epsilon_c A_c (T_c^4 - T_s^4) \tag{7-3}$$

and $q_k$ can be estimated from the curves of Fig. C,2.

The contribution of $q_k$ can usually be neglected for (1) gas flow parallel to the axis of the couple and its leads, (2) flow normal to the couple axis and its leads, with small gas temperature gradients along the axis of the wires, and (3) very fine wire couples and leads (see Fig. C,2).

The convection heat transfer coefficient $h_c$ in the above expression for $q_c$ is based on the "adiabatic wall temperature" $T_{aw}$ ("wall" is here that of the measuring instrument) corresponding to the temperature of the flowing stream $T_g$ plus the fractional attainment $r$, of the stagnation temperature rise,

$$r = \frac{T_{aw} - T_g}{T_g^0 - T_g} \tag{7-4}$$

(See *108*, pp. 261–267, 312] or [*5*] for values of $r$ and $h_c$ for various couple configurations and stream conditions.) The effective temperature $T_s$ of the surroundings for back-radiation to the couple is estimated from the fraction $F_i$ of the surroundings at each temperature level $T_i$,

$$T_s^4 = \sum F_i T_i^4 \tag{7-5}$$

This relation is applicable when the surroundings are black. Rigorous allowance for the effect of different emissivities of the different parts of the surroundings can be made if warranted [*107*, Chap. 4].

Eq. 7-1 through 7-4 are combined with Eq. 1-7 to give an expression for the stagnation temperature $T_g^0$,

$$T_g^0 = T_c + \frac{\sigma \epsilon_c}{h_c}(T_c^4 - T_s^4) + (1 - r)\frac{V^2}{2c_p} \tag{7-6}$$

With known values for $T_c$, $T_s$, and $\epsilon_c$, the solution of Eq. 7-6 for $T_g^0$ still requires the evaluation of $V$, $h_c$ and $c_p$. The evaluation of $V$ from measurements of $p^0$ and $p$ requires a method of determining $\rho$, and prediction of $h_c$ requires a knowledge of $\rho V$.

In general $\rho$ and $V$ are determined by simultaneous solution of the following:

$$p^0 - p = \int_0^V \rho V dV, \quad \text{where } \rho = \frac{p}{\mathcal{R}T} \tag{7-7}$$

$$T^0 - T = \int_0^V \frac{VdV}{c_p}, \quad \text{where } c_p = c_p(T,p) \tag{7-8}$$

For subsonic flow at temperature levels involving only slight amounts of dissociation, Eq. 7-7 and 7-8 can be combined and integrated with constant values for $c_p$ and $\mathcal{R}$.

$$V = \left\{2c_pT^0\left[1 - \left(\frac{p}{p^0}\right)^{\mathcal{R}/c_p}\right]\right\}^{\frac{1}{2}} \tag{7-9}$$

Further, if the gas stream can be considered incompressible, Eq. 7-7 can be integrated separately.

$$V = \left[2\left(\frac{p^0}{p} - 1\right)\mathcal{R}T\right]^{\frac{1}{2}} \tag{7-9a}$$

To obtain a trial value of $V$, the measured thermocouple value of $T_c$ is inserted in Eq. 7-9 or 7-9a for $T^0$ or $T$, together with the measured or known values of $p$, $p^0$, $\mathcal{R}$, and $c_p$. The corresponding trial value of gas density is calculated from the perfect gas law, using the static pressure for $p$ and $T_c$ for $T$.

The trial values of $\rho$ and $V$ are then used to determine $h_c$ and to solve Eq. 7-6 for a trial value of $T_g^0$. With this value of $T^0$ new values of $V$ and $\rho$ are computed as above to determine a new value of $T_g^0$ from Eq. 7-6. Unless the stream velocity is near sonic, or the temperature is high enough for dissociation to be important, the computation of two values of $T_g^0$ is sufficient.

The largest calculated correction shown for probe 1 in Fig. C,7b, where the inlet air temperature is 700°R, is 180°R. This correction from a measured local value of 2560–2740°R corresponds to about a ten per cent increase in local value of combustion efficiency. At the same point, the 2400°R temperature measured by the three-shielded probe 4 corresponds to a ten per cent lower local combustion efficiency.

*Gas sampling.* As noted in Art. 6, for streams varying timewise in density the minimum sampling error is obtained when samples are withdrawn at stream velocity. When the gas to be sampled contains raw fuel or incompletely burned products, it is necessary to chill the sample rapidly to maintain its composition. The process of rapidly cooling a stream flowing at high velocity causes pressure changes in the stream and makes difficult a determination, by conventional pressure measurements, of whether or not the sampling velocity matches the stream velocity. In addition, the flow pulsations originating from the chilling of a stream of fluctuating temperature introduce the possibility of increasing the sampling error over that of an isothermal system.

A modified version of probe 1 to permit gas sampling is shown in Fig. C,7c. In use the probe is advanced in $\frac{1}{8}$-inch steps so as to read first temperature, then impact pressure. From these two readings the stream

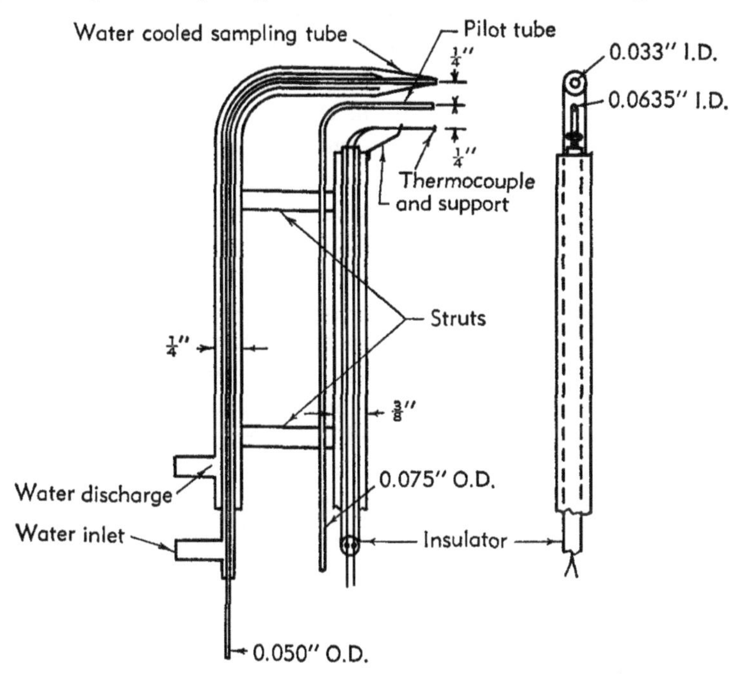


Fig. C,7c. Combination thermocouple, impact pressure, and gas-sampling probe.

velocity is approximated, then the gas-sampling probe is advanced $\frac{1}{8}$ inch and the gas sample is taken so that the sampling velocity into the sampling tube is close to that of the gas stream at that point.

An example of the effect of sampling velocity on gas composition is shown in Table C,7.

If it is reasonably assured that the stream to be sampled does not

*Table C,7*

Stream velocity, 435 ft/sec
Burner air-fuel ratio, 71.5 lb air per lb fuel
Corrected measured gas temperature, 1972°R

| | | | |
|---|---|---|---|
| Sampling velocity, ft/sec | 26 | 270 | 435 |
| Gas composition: per cent $CO_2$ | 2.15 | 2.60 | 2.83 |
| per cent $O_2$ | 17.85 | 17.07 | 16.70 |
| per cent $N_2$ | 80.01 | 79.94 | 80.08 |
| per cent combustibles | 0.26 | 0.39 | 0.39 |
| Local air-fuel ratio, from analysis | 74.4 | 65.5 | 60.9 |
| Combustion efficiency | 84.6 | 86.7 | 87.8 |
| Calculated temperature, °R | 1510 | 1635 | 1705 |

have a rapidly fluctuating temperature, the measurement of static pressure and stagnation pressure and temperature will, as shown above, permit determination of the stream velocity for use in sampling. It is, however, not recommended that reliance be placed on the absolute magnitudes of performance criteria determined from analysis of chilled gas samples withdrawn from highly turbulent, high velocity streams.

*Methods of averaging.* The precise definition of combustion efficiency in terms of the fraction of the theoretical conversion of chemical enthalpy of the fuel to sensible enthalpy of combustion products requires an evaluation of the sensible enthalpy of the exhaust gases. Generally this requires traversing the exhaust stream to determine the local specific enthalpy of the exhaust gases and the associated local mass velocity. The enthalpy of the entire exhaust stream is then determined by an integration, over the exhaust cross section, of the product of local specific enthalpy by local mass velocity. If interest lies solely in the mean value of exhaust gas enthalpy it is sometimes feasible to determine the enthalpy by a calorimetric measurement on the entire flowing stream. This may, for example, depend on measurement of the temperature of a mixed stream resulting from spraying water of known temperature at a known rate into the exhaust gases, the mixed temperature being measured at a plane downstream where the mixture is homogeneous and all vapor. The method lacks precision when the temperature just before quenching is so high that quenching is attended by chemical change, since a small error in the determination of the chemical change is equivalent to a large error in the determination of sensible enthalpy. Determination by the calorimetric method of the local value of stream enthalpy, as contrasted to determination of the enthalpy of the entire stream, can in principle be accomplished by local adiabatic sampling at the local stream velocity. However, the condition that the sampling shall be adiabatic is difficult to meet; and the condition that the sampling rate shall be right cannot be realized without knowledge of the local velocity which often requires knowledge of the temperature itself. Consequently, the technique has not been applied to any appreciable extent.

If the determination of a mass-average exhaust gas temperature is satisfactory for the purpose at hand, a simple plotting and graphical integration technique can be employed for most ranges of velocity and temperature encountered at turbojet exhausts. Assume that a probe such as No. 1 of Fig. C,7a is used to traverse a radius of an annular burner having inner and outer radii $r_1$ and $r_2$. Assume, further, that the probe temperature $T_c$ equals $T_g^0$, the local stagnation temperature of the exhaust gas; that the flow velocity is low enough for the flow to be incompressible; and that the average fuel-air ratio is lean and its variation across the exhaust stream is sufficiently small so that the exhaust gas composition can be considered uniform for the purpose of calculating the gas density from its pressure and temperature. In the general case the mass velocity and temperature of the gas vary circumferentially as well as along radii of the annulus, so that a double integration is required. As an illustration of the method of averaging, axial symmetry is assumed. The mass-average exhaust gas temperature $T^0_{g_m}$ is then defined in terms of the local mass velocity $\rho V$, and the local gas temperature $T^0_g$ in the differential annular flow area, $dA$.

$$T^0_{g_m} = \frac{\int_0^A T^0_g \rho V dA}{w_g}$$

where $dA = 2\pi r dr$ and $w_g = \int_0^A \rho V dA$.

From Eq. 7-7,

$$p^0 - p = q = \frac{\rho V^2}{2}, \quad \rho = \frac{p}{\mathcal{R} T_g}$$

$$\rho V = \left(\frac{2pq}{\mathcal{R} T_g}\right)^{\frac{1}{2}}$$

Since $T^0_g \cong T_g \cong T_c$ and $p \neq f(A)$,

$$T^0_{g_m} = \frac{\int_{r_1}^{r_2} (qT_c)^{\frac{1}{2}} d(r^2)}{\int_{r_1}^{r_2} (q/T_c)^{\frac{1}{2}} d(r^2)} \tag{7-10}$$

Plots of $(qT^0_g)^{\frac{1}{2}}$ and of $(q/T^0_g)^{\frac{1}{2}}$ vs. $r^2$ then provide the basis for performing the integrations required to determine $T^0_{g_m}$. Typical data obtained with probe 1 are plotted in Fig. C,7d. The average enthalpy corresponding to $T^0_{g_m}$ determined as above is strictly the mass-average enthalpy only when the specific heat of the gas is a constant. A more rigorous integration to obtain the mass-average enthalpy $h_{g_m}$ requires knowledge of $c_p$ for the gas as a function of temperature level, and performance of the integration with $c_p$ as a variable.

$$h_{g_m} = \frac{\int_{r_1}^{r_2} c_p (qT^0_w)^{\frac{1}{2}} d(r^2)}{\int_{r_1}^{r_2} (q/T_w)^{\frac{1}{2}} d(r^2)}$$

The mass-average temperature $T^0_{g_m}$ determined by dividing the area under the curve of $(qT^0_g)^{\frac{1}{2}}$ by that under the curve of $(q/T^0_g)^{\frac{1}{2}}$ in Fig. C,7d is 1760°R. If the mass velocity had been assumed to be uniform over the exhaust cross section the integration would have been performed with $\rho V$ outside the integral sign; and the area-average temperature thus obtained from the dotted curve of Fig. C,7d is 1860°R. Comparison of mass-average and area-average temperatures for other traverses of equal precision has shown differences of over 300°R in the two temperatures,

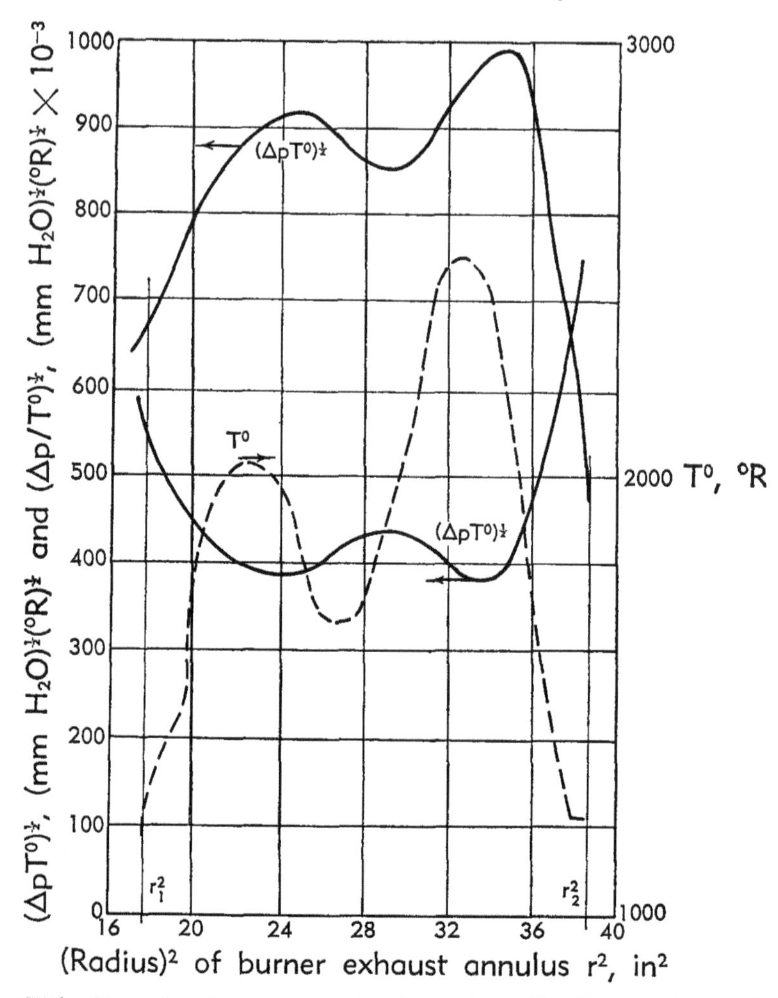


Fig. C,7d. Examples of traverses made with probe of Fig. C,7c showing total temperature, mass velocity, and total temperature-mass velocity product (approximately proportional to enthalpy flux).

demonstrating that failure to allow for variation in local mass velocity can produce quite misleading values of average exhaust temperature.

## C,8. Measurement of Time-Dependent Properties.

Composition. The restriction of primary air to a gas turbine combustion chamber, to favor combustion by keeping the temperature high, makes the progress of combustion depend largely on the attainment of microscale intimacy of mixing. The departure from uniformity is characterized by the mean amplitude of fluctuation in composition and the scale of the fluctuation or, in combination with stream velocity, the time factor. These quantities are the analogues to the intensity and scale which characterize turbulence and, similarly, require for their determination in general a probe rapidly responsive to changes in composition. In a chemically reacting gas system, however, a significant inference concerning the mean amplitude of fluctuation in composition can be made solely from the chemical analysis of a time-mean gas sample, provided one accepts certain assumptions [*36*, pp. 266–300]. The first of these is that the union of fuel and oxygen molecules is so rapid compared to the mixing of fuel-rich and oxygen-rich eddies that any instantaneous point-composition is restricted to one containing either excess oxygen or excess unburned gases (fuel vapor or its decomposition or partial combustion products), but not both. The presence of both oxygen

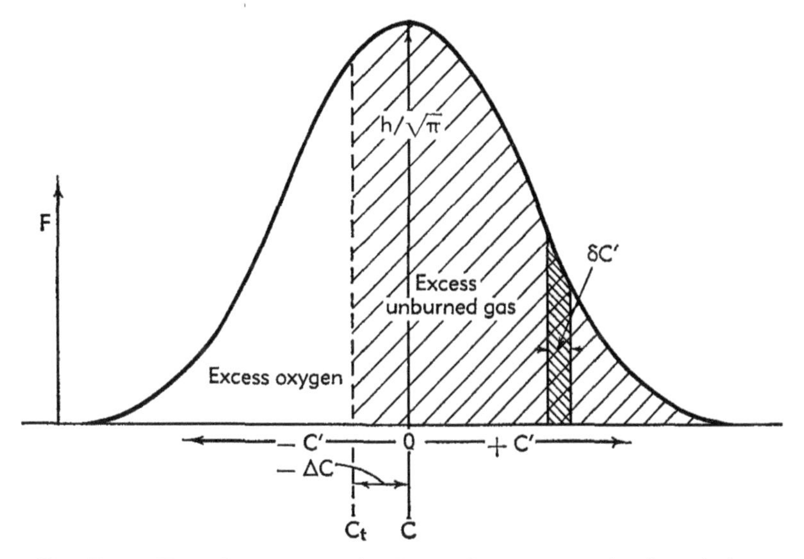


Fig. C,8a. Normal error curve for fluctuating concentration in turbulent flames formed from fuel and air streams mixing during combustion.

*and* unburned gases in a slowly withdrawn sample is then indicative of their successive existence at the top of the sampling probe, followed by a chilling which prevents reaction except at eddy "interfaces" as the sample is withdrawn. The second assumption is that the composition at a point shows no effect of selective molecular diffusion, i.e. the carbon-hydrogen ratio in the wet sample is that of fuel and the oxygen-nitrogen ratio that of air.

Let all gas compositions be expressed as mole fractions of fuel vapor calculated from the gas analysis back to the unreacted state; and let $C_i$ represent the instantaneous mole fraction of fuel vapor at a point. This may be resolved into a time-mean value $C$ and a fluctuating component $C'$, so that $C_i = C + C'$. Let it be assumed that the fluctuating concentration component $C'$ obeys the normal error laws. Fig. C,8a illustrates the relationship. The ordinate $F$ at a given value of $C'$ represents the relative frequency of occurrence of a fluctuation of that magnitude.

$$F = \frac{h}{\sqrt{\pi}} e^{-h^2C'^2} \tag{8-1}$$

Thus the fractional time that the concentration is between $C + C'$ and $C + C' + dC'$ is represented in Fig. C,8a by the ratio of the elementary cross-hatched area to the area under the whole curve; and since the latter is one, to $F \cdot dC'$.

The constant $h$ of the error curve is related to $\overline{C'^2}$, the mean square value of $C'$, by

$$\frac{1}{2h^2} = \overline{C'^2} \quad \text{or} \quad h = \frac{1}{\sqrt{2}\sqrt{\overline{C'^2}}} \tag{8-2}$$

Let the mole fraction of fuel vapor in a mixture of proportions for complete combustion be $C_t$; and let $\Delta C$ be the difference between the time-mean concentration $C$ and that of the stoichiometric mixture $C_t$, $\Delta C$ being positive at a point where the average composition is fuel-rich. Fig. C,8a illustrates this case ($-\Delta C$ is negative). When the instantaneous concentration $C_i$ ($= C + C'$) exceeds $C_t$, unburned gas will be entering the sampling tube at that instant (all such points lie in the shaded area to the right of the broken line of the figure), and the concentration of unreacted fuel vapor is proportional to $C_i - C_t$, or to $C' + \Delta C$. When $C_i$ is less than $C_t$, oxygen will be present in the gases entering the sampling tube at that instant (unshaded area under the curve), and $C' + \Delta C$ will be the fuel-vapor equivalent of oxygen in the gases. Referring to Fig. C,8a, the total amount of unburned gas present in the sample is proportional to the product of its concentration times the fractional time that such concentration prevails, or to $\int (C' + \Delta C)F dC'$. Since this integral is the moment of the shaded area about the broken line, the ratio of unburned gas to oxygen found in the time-mean sample is proportional to the ratio of the moment of the shaded area to the moment of the unshaded area about the broken line. It may be shown [*36*, pp. 266–300] that this ratio is a function of $h/\Delta C$, or of $\sqrt{\overline{C'^2}}/\Delta C$; the relation appears in Fig. C,8b. In using the figure the sign of $\Delta C$ may be neglected and the value of (oxygen equivalent of unburned gas)/(free oxygen) or its reciprocal used, whichever is greater than 1.

Since unburned oxygen, unburned fuel gas, and $\Delta C$ are all capable of evaluation from a gas analysis, Fig. C,8b yields the value of $\sqrt{\overline{C'^2}}$, the root-mean-square of the fluctuating mole fraction, a quantity which is seen to be an index of the intimacy of mixing. The larger the quantity the less intimate the mixing; and the name "unmixedness factor" has been suggested. Although the term is dimensionless, equal values of it in two different systems in which the stoichiometric composition $C_t$ is entirely different do not indicate equal departures from perfection. Better as a characterization is $\sqrt{\overline{C'^2}}/C$, the quotient of the rms fluctuating composition by the time-average value. This may more properly be designated the "unmixedness" at the sampling point. The terms $\sqrt{\overline{C'^2}}$ and $\sqrt{\overline{C'^2}}/C$ are the analogues of turbulent velocity $u'_{rms}$ and turbulence intensity $u'_{rms}/V$.

It is to be noted that although the derivation presented here for obtaining unmixedness from the analysis of a time-average gas sample may lack validity under some experimental conditions, the unmixedness

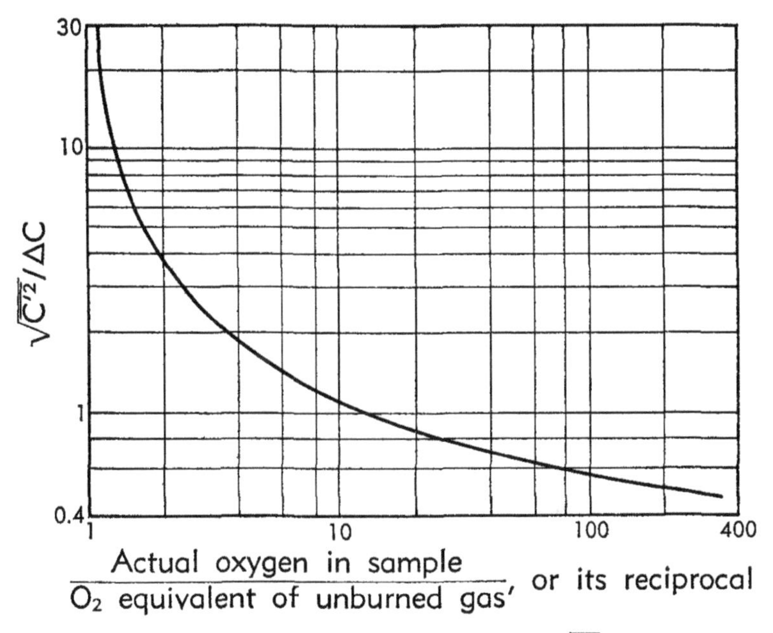


Fig. C,8b. The unmixedness factor $\sqrt{\overline{C'^2}}$.

as defined does exist; and it could in principle be determined independently of any assumptions as to how combustion proceeds if probes were available for quantitative sensing of rapid changes in composition. No such probe exists at present which responds rapidly only to fluctuating composition in the presence of fluctuations in temperature and velocity. Values of unmixedness obtained by the method described above, even though their absolute value may be in question, should serve to indicate semiquantitatively the progress of mixing along a combustion chamber. A somewhat more rigorous treatment of the problem, allowing for asymmetry in the distribution function for concentration, has been presented by Richardson, et al [*9*, pp. 814–817].

TIME-VARYING TEMPERATURE AND VELOCITY. Simultaneous measurement of instantaneous values of temperature and velocity in a gas stream in which these properties vary rapidly with time is not known to have been attempted in experimentation on gas turbine burners. The range of temperatures to be encompassed, the fluctuation frequencies encountered, and the difficulty of separating effects due to combined action of temperature and velocity variation give strong arguments against success in applying simple instrumentation to such measurements. Optical density measurement does not permit distinction between fluctuation of velocity (IX,F) and that of temperature, emission, or absorption spectroscopy does not permit evaluation of lower levels of temperature, and few of the optical methods are applicable to probing of flow cross sections in high temperature streams. The instruments most probably applicable to such measurements are the hot wire anemometer and the fine wire thermocouple.

*Short response time system.* If it is possible to use sufficiently fine wires and if the temperature levels to be measured are low enough so that the radiant heat transfer can be neglected, two fine wire anemometers carrying different currents or one anemometer and one fine wire thermocouple can be used to distinguish between the velocity and temperature fluctuation. The couple and one of the anemometers, if its current is small enough, will assume the stream temperature if the wires are small enough, and by simultaneous comparison with the other probe carrying a larger current the variation in velocity can also be determined. Unfortunately, wires with sufficiently rapid response time are so fragile that they present extreme experimental difficulty. A wire of 0.001-in. diameter in an atmospheric pressure stream at 600 ft/sec changes 65 per cent of the gas temperature change for a temperature pulse 2 feet long (300 cycles), only 8 per cent for a pulse 2 inches long (3600 cycles). The sensing wire can, within limits, be kept large enough for strength if at the same time the electronic circuitry for voltage measurement is modified to provide compensation for time lag of the wire by a graded attenuation of low frequency signals, or if it is desired to measure only scales or frequencies but not amplitudes of temperature and velocity fluctuations. As the temperatures to be measured increase, however, the increased wire gauge required for strength and stability introduces a temperature error due to thermal radiation which can no longer be neglected. Under such conditions the complexity of circuitry and of data analysis would be so increased that the measurement of fluctuating temperatures and velocities by probe wires seems of dubious utility.

*Long response time.* A wire of adequate sturdiness placed in a stream of fluctuating temperature will normally respond so slowly as to produce a steady reading corresponding, in the absence of radiation and conduction, to a temperature between the high and low extremes of stream temperature fluctuations. Illustrative of this effect are the curves of

Fig. C,8c which were calculated for a thermocouple immersed in a stream of steady velocity but having temperature fluctuations quite short in duration compared to the response time of the couple. In addition to the assumption of zero radiation and conduction it was also assumed that the gas flow pattern around the couple, and hence the heat transfer rate to the couple, came immediately to the steady flow state corresponding to the instantaneous gas temperature, i.e. pulse durations were assumed long compared to the ratio of wire width to stream velocity. The heat transfer coefficient between the gas and the couple was assumed to be

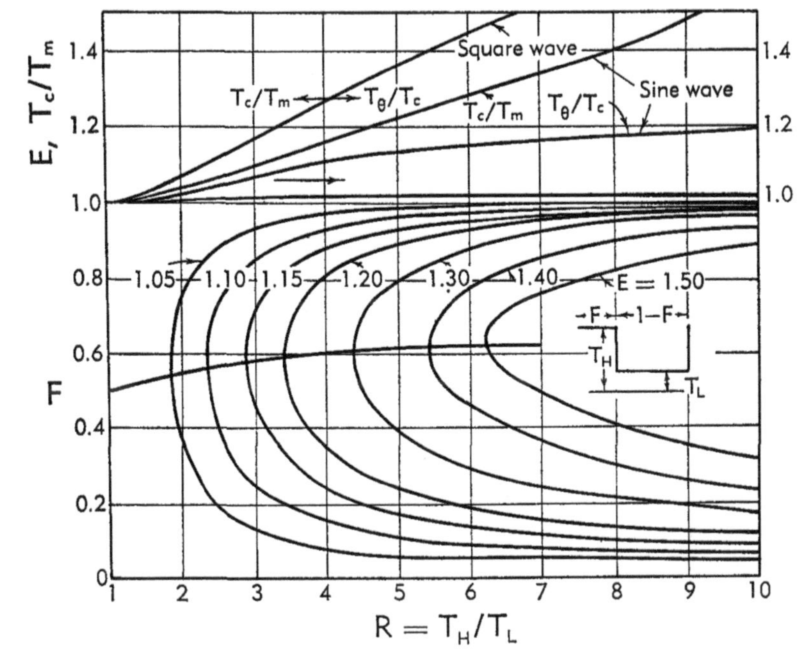


Fig. C,8c. Effect of time-varying temperature on error of a massive thermocouple.

inversely proportional to the square root of the absolute temperature of the gas. The stagnation temperature rise was assumed to be negligible. The abscissa $R = T_H/T_L$ of Fig. C,8c is the relative amplitude of the fluctuation from the highest gas temperature $T_H$ to the lowest gas temperature $T_L$. The ordinate of the lower portion of the figure represents the fraction $F$ of a cycle during which the gas temperature is at $T_H$, when the temperature fluctuation is of rectangular wave form. The mass-mean temperature $T_m$ of such a stream of perfect gas of constant specific heat is given by the following relation:

$$T_m = \frac{T_H T_L}{FT_L + (1 - F)T_H}$$

The parametric curves are for constant values of the ratio $E$ of the couple temperature $T_c$ to the mass-mean temperature $T_m$

$$E = \frac{T_c}{T_m} = \frac{\left(\frac{FR^{1/2}}{1-F} + 1\right)\left(\frac{F}{R} + 1 - F\right)}{1 + \frac{F}{(1-F)R^{1/2}}}$$

For example, the curve of $E = 1.30$ represents combinations of $R$ and $F$ for which the couple temperature is always 30 per cent above $T_m$. The couple fractional error, $E - 1$, is always less than 0.3 for values of $R < 4.4$, the abscissa at which the curve $(\partial R/\partial F)_E = 0$ intersects the $E = 1.30$ curve. The curves in the upper portion of Fig. C,8c illustrate the effect on $E$ on changing the wave form from square ($F = 0.5$) to symmetrically sinusoidal $T = T_L\left(\frac{R+1}{2}\right)\left[1 + \frac{(R-1)\sin\theta}{R+1}\right]$. The right-hand ordinate of the upper curve is the ratio of the time-average stream temperature $T_\theta$ to the couple temperature. For both square and symmetrically sinusoidal waves, $T_\theta/T_L = (R+1)/2$. It is noted that for sinusoidal fluctuation of stream temperature the couple temperature is closer to the time-mean value than to the mass-mean value, whereas for the square wave the reverse is true, i.e. $T_c - T_m < T_\theta - T_c$.

Clearly the curves of Fig. C,8c are at most of academic interest if sufficient data are available to permit evaluation of the quantities $F$, $T_H$, and $T_L$, or their counterparts for other wave forms. As discussed above, however, such measurements are extremely difficult, although it is possible that uncompensated measurements may give reasonable approximations for $F$ and some idea of the values of $R$. Inspection of Fig. C,8c shows that if $F$ has values near zero or unity or if $R$ is less than 2, the error in temperature measurement will be less than 5 per cent of $T_m$.

Somewhat similar calculations to those for Fig. C,8c can be made for streams of fluctuating velocity and for combined velocity and temperature fluctuations. Velocity fluctuation alone would produce a variation in error of velocity measurement with the ratio of velocities somewhat similar to that shown for temperature, although the velocity error for given $F$ and $R$ would be less than the temperature error. In addition there is little reason to expect velocity fluctuations as large in magnitude as those of temperature, which latter can in poorly operating systems vary from the temperature of air entering the combustion chamber to that of adiabatic combustion of the local fuel-air mixture. Curves for the combined effects of velocity and temperature fluctuations on, say, errors in measurement of mass-mean temperature would show effects on measurement error less than or greater than those of Fig. C,8c, depending on the correlation between the temperature and velocity fluctuations.

## C,9. Cited References.


1. King, W. J. *Trans. Am. Soc. Mech. Engrs. 65*, 421–428 (1943).
2. Dahl, A. I., and Fiock, E. F. *Trans. Am. Soc. Mech. Engrs. 71*, 60–61, 153–160 (1949).
3. Dahl, A. I. Extension of the calibration of thermocouples with pressed radiation shields. *Natl. Bur. Standards 7th Quart. Progress Rept. 3620*, Jan.–Mar. 1950.
4. Glawe, G. E , and Shepard, C. E. *NACA Tech. Note 3253*, 1954.
5. Hottel, H. C., and Kalitinsky, A. J. *J. Appl. Mech. 67*, A25 (1945).
6. Scadron, M. D., Gettleman, C. C., and Pack, G. J. *NACA Research Mem. E50I29*, 1950.
7. Fiock, E. F., et al. *Natl. Bur. Standards Project 3620, Rept. 13, 14*, 1948.
8. Van Langen, J. M., and Hubbard, E. H. Measurements in the flame with a suction pyrometer with a bare thermocouple. *International Flame Radiation Research Document F31/a/5*, Apr. 1955.
9. *Fourth Symposium on Combustion.* Williams & Wilkins, 1953.
10. Am. Inst. of Phys. *Temperature—Its Measurement and Control in Science and Industry.* Reinhold, 1941.
11. Dahl, A. I., and Freeze, P. D. Some effects of sulfur on silver-shielded chromel alumel couples. *Natl. Bur. Standards Rept.*, Oct. 1948.
12. Fiock, E. F., and Dahl, A. I. *Final Rept. on Natl. Bur. Standards Project 2616*, Jan. 1951.
13. Curcio, J. A., Stewart, H. S., and Petty, C. C. *J. Opt. Soc. Amer. 41*, 173–179 (1951).
14. Wolfhard, H. G., and Parker, W. G. Combustion processes in flames, Part VI. *Roy. Aircraft Establishment Rept. Chem. 457*, Mar. 1949.
15. Hottel, H. C., and Mangelsdorf, H. G. *Trans. Am. Inst. Chem. Eng. 31*, 517 (1935).
16. Mayorcas, R., and Riviere, M. *International Flame Radiation Research Document D3/b/8*, Ijmuiden, Holland, 1952.
17. Daws, L. F., and Thring, M. W. *J. Inst. Fuel 25*, 528 (1952).
18. Hottel, H. C., and Broughton, F. P. *Ind. Eng. Chem., Anal. ed. 4*, 166–175 (1932).
19. Sanderson, J. A., Curcio, J. A., and Estes, D. V. *Phys. Rev. 74*, 1221 (1948).
20. Cooper, M. A. *Sc. D. Thesis in Physics.* Univ. Witwatersrand, Union of South Africa, 1932.
21. Uyehara, O. A., Myers, P. S., Watson, K. M., and Wilson, L. A. *Trans. Am. Soc. Mech. Engrs. 68*, 17–28 (1946).
22. Myers, P. S., and Uyehara, O. A. *S.A.E. Quart. Trans. 1*, 592–601, 611 (1947).
23. Monnot, G. *Compt. rend. 228*, 906–908 (1949).
24. Fuller, L., and Marlow, B. *Natl. Gas Turbine Establishment Mem. M189*, June 1953.
25. Schmidt, H. *Deut. Phys. Ges. 11*, 87 (1909).
26. Huston, W. B. *NACA Tech. Note 1605*, 1948.
27. Rayle, R. E. *S. M. Thesis.* Dept. Mech. Eng., Mass. Inst. Technol., 1949.
28. First, M. W., and Silverman, L. *Ind. Eng. Chem. 42*, 301–308 (1950).
29. Goldstein, S. *Modern Developments in Fluid Mechanics*, Vol. II. Clarendon Press, Oxford, 1938.
30. Gracy, W., Letko, W., and Russell, W. R. *NACA Tech. Note 2331*, 1951.
31. Zaback, I. *NACA Tech. Note 1819*, 1949.
32. Bradfield, W. S., and Yale, G. E. *J. Aeronaut. Sci. 18*, 697–698 (1951).
33. Draper, C. S., and Li, Y. T. *J. Aeronaut. Sci. 16*, 593–610 (1949).
34. Goldstein, S. *Proc. Roy. Soc. London A155*, 571–575 (1936).
35. Fage, A. *Proc. Roy. Soc. London A155*, 576–596 (1936).
36. *Third Symposium on Combustion, Flame and Explosion Phenomena.* Williams & Wilkins, 1949.
37. Taylor, J. F., and Comings, E. W. *Proc. Midwest Conference on Fluid Dynamics*, 204–215 (1951).
38. Spivack, H. M. *J. Aeronaut. Sci. 13*, 289–301 (1946).

39. Den Hartog, J. P. *Mechanical Vibrations.* McGraw-Hill, 1940.
40. Li, Y. T., and Lee, S. *Trans. Am. Soc. Mech. Engrs. 75*, 835 (1953).
41. Reid, E. G. *NACA Tech. Note 209*, 1924.
42. Betz, A. *Z. Ver. deut. Ing. 69*, 11 (1925).
43. Leeper, C. K. *A Study of Jnmixed Gaseous Streams. S. M. Thesis in Mech. Eng.*, Mass. Inst. Technol., 1948.
44. Longwell, J. P. Private communication. *Esso Lab. Process Div.*, Jan. 1950.
45. Leeper, C. K. *Sampling of Unmixed Gaseous Streams. Sc. D. Thesis*, Mass. Inst. Technol., 1954.
46. Wohl, K. Private communication. *Univ. Delaware*, approx. 1950.
47. Kapstein, S. *Determination of Fuel Distribution in a Volatile Spray Cloud. S. M. Thesis*, Mass. Inst. Technol., 1951.
48. Langmuir, I., and Blodgett, K. B. Mathematical investigation of water droplet trajectories. *Army Air Force Tech. Rept. 5418*, Dec. 1944.
49. Dussourd, J. L., and Shapiro, A. H. *Proc. Heat Transfer and Fluid Mech. Inst.*, Univ. Calif., Los Angeles, 1955.
50. Mullen, P. W. *Modern Gas Analysis.* Interscience, 1955.
51. Altieri, V. J. Gas analysis and testing of gaseous materials. *Am. Gas Assoc.*, New York, 1945.
52. Shepherd, M. *Anal. Chem. 22*, 881–885 (1950).
53. Shepherd, M. *Anal. Chem. 22*, 885–889 (1950).
54. Stone, H. W., and Skavinski, E. H. *Ind. Eng. Chem., Anal. ed. 17*, 495–498 (1945).
55. Clifton, V. E. *Industrial Chemist 22*, 200–202 (1946).
56. Fréling, E., and Dugleux, P. *Rev. Inst. franç. pétrole et Ann. combustibles liquides 5*, 3–8 (1950).
57. Nash, L. K. *Anal. Chem. 18*, 505 (1946); *22*, 108 (1950).
58. Dobrinskaya, A. A., Neiman, M. B., and Adreev, E. A. *Zavodskaya Lab. 16*, 934–938 (1950).
59. Schnidman, L. *Am. Gas Assoc. Monthly 25*, 405–408, 428, 503–506, 522–523 (1943).
60. Knorre, G. F., et al. *Zavodskaya Lab. 10*, 102 (1941).
61. Knorre, G. F., et al. *Chem. Zentr. 22*, 2759 (1942).
62. Brooks, F. R., et al. *Anal. Chem. 21*, 1105 (1949).
63. Matuszak, M. P. *Fisher Gas-Analysis Manual for Use with Apparatus of the Orsat Type*, 3rd ed. Eimer and Amend, Pittsburgh, Pa.
64. Spence, R., and Smales, A. A. *Ann. Repts. Progress Chem. 46*, 291 (1949).
65. Mendoza, M. P. *Bull. Brit. Coal Utilisation Research Assoc. 14*, 409–423 (1950).
66. Darby, H. T. *Anal. Chem. 22*, 227–234 (1950); *24*, 244–241 (1952).
67. Nash, L. K. *Anal. Chem. 23*, 74–81 (1951).
68. Bennett, W. H. *Chem. Eng. News 29*, 552 (1951).
69. Weissberger, A., ed. *Physical Methods of Organic Chemistry*, 2nd ed., Vol. I, Part 2. Interscience, 1949.
70. Thode, H. G., and Shields, R. B. *Repts. Progress in Phys. 12*, 1–21 (1948–1949).
71. Robertson, A. J. B. *Proc. Roy. Soc. London A199*, 394 (1949).
72. Coggeshall, N. D., and Kerr, N. F. *J. Chem. Phys. 17*, 1016–1021 (1949).
73. Burmaster, K., and Evans, E. C. *Instruments 23*, 242–245 (1950).
74. Rock, S. M. *Anal. Chem. 23*, 261 (1951).
75. Bernandet, P., and Maydoff, R. *Abstr. Physik Ber. 30*, 168–169 (1951).
76. Happel, W. J., and Coggeshall, N. D. *Instruments 23*, 552–556 (1950).
77. Koppius, O. G. *Anal. Chem. 23*, 554 (1951).
78. Watkins, J. M., and Gemmil, C. L. *Anal. Chem. 24*, 591 (1952).
79. Vengerov, M. L. *Doklady Akad. Nauk. U.S.S.R. 46*, 200–205 (1945).
80. Vengerov, M. L. *Compt. rend. Acad. Sci. U.S.S.R. 46*, 182–187 (1945); *51*, 195–197 (1946).
81. Vengerov, M. L. *Nature 158*, 28–29 (1946).
82. Coggeshall, N. D., and Saier, E. L. *J. Appl. Phys. 17*, 450–456 (1946).
83. Pristera, F. *Appl. Spectroscopy 6*, 29–44 (1952).

84. Podbielniak, W. J. *Petroleum Refiner 30*, 145–155 (1951).
85. Smittenberg, J. *Rec. trav. chim. 67*, 703–719 (1948).
86. LeRoy, D. J. *Can. J. Research 28B*, 492–499 (1950).
87. *California Natl. Gas Assoc. Bull. TS-411*, 1941.
88. Staff Report. *Chem. Eng. News*, 1692–1696 (Apr. 1956).
89. Barrer, R. M., and Robins, A. B. *Trans. Faraday Soc. 49*, 807 (1953).
90. Griffiths, J., James, P., and Phillips, C. *Analyst 77*, 897 (1952).
91. Muller, R. H., ed. *Anal. Chem. 27*, 6 (1955).
92. James, A. T., and Martin, A. J. P. Gas-liquid chromatography: A technique for the analysis and identification of volatile materials. *Brit. Medical Bull.*, Dec. 1954.
93. Phillips, C. S. G. *Discussions Faraday Soc. 7*, 241–248 (1949).
94. Smet, W. M. *Discussions Faraday Soc. 7*, 248–255 (1949).
95. Cramer, E., and Muller, R. *Mikrochemie ver. Mikrochim. Acta 36/37*, 553–560 (1951).
96. Criddle, D. W., and LeTourneau, R. L. *Anal. Chem. 23*, 1620–1624 (1951).
97. Greep, R. C. *Mass. Inst. Technol., Dept. of Chem. Eng. Internal Rept.*, 1932.
98. Minter, C. C., and Burdy, L. M. J. *Anal. Chem. 23*, 74–81, 143–147 (1951).
99. Minter, C. C. *Ind. Eng. Chem., Anal. ed. 19*, 464 (1947).
100. Pritchard, F. W. *J. Sci. Instr. 29*, 116–117 (1952).
101. Weaver, E. R. *Physical Methods in Chemical Analysis*. Academic Press, 1951.
102. Pauling, L., Wood, R. E., and Sturdivant, J. H. *J. Am. Chem. Soc. 68*, 795–798 (1946).
103. Berl, W. G., ed. *Physical Methods in Chemical Analysis*. Academic Press, 1950.
104. Gerrish, H. C., and Meens, J. L., Jr. *NACA Wartime Rept. E128*, first issued as *Advance Restricted Rept. 3J07*, Oct. 1943.
105. Gooderham, W. J. *J. Soc. Chem. Ind. 59*, 1 (1940).
106. Gooderham, W. J. *Analyst 72*, 520 (1947).
107. McAdams, W. H. *Heat Transmission*. McGraw-Hill, 1954.

# SECTION D

## *FUEL INJECTION*

A. RADCLIFFE

**D,1. Introduction.** In this section attention is confined to the problem of injecting fuel into air or vitiated air. In so doing, mention is made of methods that are used in rocket motors, but this is only because it is possible to use the same techniques in both rockets and gas turbines. When a gas turbine is designed, the air mass flow, the temperature rise, and the combustion efficiency required are all defined so that the fuel flow requirements can be fairly readily calculated. Since most gas turbines consume fuel of organic origin which when burned in air gives a temperature rise of about 2000°C, and since the temperature that turbine materials can tolerate is about 1000°C, it is clear that the fuel to be injected is generally less than that for a stoichiometric mixture. But as explained in Sec. A, combustion can only be achieved at or about stoichiometric mixtures. The ideal mixture has the desired fuel-air ratio and is as homogeneous as a mixture of molecules of different kinds can ever be. Since it takes a finite time either for a flame to pass through even an ideal mixture, or for it to react completely, other less perfect mixtures may behave as though they were ideal. For example, small droplets of fuel may evaporate and disperse in the surrounding air so quickly that the ideal mixed state is achieved at a rate that is high compared to flame propagation or to reaction. Because of this and like effects it is possible—as Burgoyne of Imperial College, and Browning and Krall [*1*, pp. 159–163] have shown—for droplets of about 20-microns diameter dispersed in air to burn so that macroscopically the behavior is that of a vapor-air mixture. Thus the difference between droplet-air and vapor-air mixtures is not as great as might be thought likely and the decision about whether to inject liquid or vaporized fuel is not clear-cut. Also it is not always possible to be sure that any one of the methods of fuel injection to be described is always to be preferred to another. As far as it is practicable the advantages, disadvantages, and typical uses of each method are discussed as described.

**D,2. Liquid Injection through Simple Holes.** The simplest way of putting liquid fuel into a combustion chamber is to pass it through a small circular hole. If the velocity is low the liquid is omitted as a pencil of fuel. Omitting gravity there are two main reasons for the distortion of this pencil. At low velocities the instability is that discussed by Rayleigh [*2*]. Cylindrical filaments of liquid of radius $a$ and of length more than $2\pi a$ are unstable and tend to break into large droplets, each of which contains the liquid that once was in a length $2\pi a$ of the filament. When the liquid velocity is higher the interaction of gas and liquid results in surface instability and the formation of a wavy jet, secondary ligaments, and finally droplets. The droplets so formed are smaller than those due to surface tension forces. Larger droplets moving in a gas at a sufficient velocity break into smaller droplets. Photographic evidence [*3*] shows that when the relative liquid-air velocity increases slowly the front surface of a sufficiently large droplet is indented. The indentation increases until the liquid is in the approximate form of a hemispherical vessel with a thick rim. The rim contains some 70 per cent of the liquid. The liquid film tears and concentrates as ligaments. These and the rim break into droplets. When the liquid-air velocity increases suddenly a drop changes shape and approximates to a plano-convex lens with the plane surface meeting the air. Droplets are torn from the lens periphery and the lens disrupts into small droplets. Although the mechanism of breakdown depends greatly on the suddenness of velocity change, the size of the droplet that breaks at a given velocity is not greatly different. For slow changes the critical droplet size is given by

$$d(u_f - u_a) = 8000\sigma \tag{2-1}$$

where $d$ is in microns, $u_f$ and $u_a$ are the droplet and air velocities in meters per second, and $\sigma$ is the surface tension in dynes per cm.

A suddenly applied air velocity is equivalent to about 1.41 times a gradually applied velocity. Eq. 2-1 holds to velocities of as much as that of sound in air. It is interesting to notice that at a fuel injection pressure of 1000 lb/in.$^2$ the maximum velocity that can be imparted to kerosene is 130 meters per second, and at this velocity the critical size for droplets of kerosene is about 14 microns. Since kerosene injected through a circular hole into nominally stationary air is decelerated continuously so that some of the liquid is not effectively exposed to the full 130 meters per second of relative velocity, it is not surprising that many droplets larger than 14 microns are found.

Although it is thus possible to force liquid through a simple circular orifice so that it atomizes well in still air, the pressures necessary are high. A variation in fuel flow can be obtained in two ways. It is possible to vary the fuel pressure near the discharge hole which implies the use of a wide range of high pressures with the consequence that the highest pressure required may be beyond the ability of ordinary pumps. Alternatively arrangements may be made to vary the number or size of fuel discharge holes. In either case the mechanical difficulties are severe. For this reason there is a tendency when using a simple jet in gas turbine systems and ramjets to supply the necessary relative fuel-air velocity by high speed air rather than high speed liquid.

Nukiyama and Tanasawa [4] measured the drop sizes produced when a liquid is injected into moving air, a process which is often called natural air blast atomization. Their main interest was to discover what happens when fuel is injected through a small hole into the venturi of a carburetor such as is usually used to prepare the correct fuel air mixture for a spark ignition piston engine. A metered air flow was supplied to the venturi and fuel was injected in the same direction as, or across, the air stream. Measurements of droplet size were made by collecting the droplets on slides covered with Valvoline. The atomized liquid consisted of mixtures of water, alcohol, and glycerine. They showed that the direction of injection only affected the droplet size insofar as it affected the relative air-to-fuel velocity. The droplet size distribution was found to be represented reasonably well by the equation

$$\frac{dN}{dx} = Bx^2e^{-bx^\delta} \tag{2-2}$$

where $N$ is the number of droplets; $x$ the droplet diameter; and $B$, $b$, and $\delta$ are constants for a particular spray.

There are many possible mean diameters. As did Sauter [5], Nukiyama and Tanasawa calculated the surface mean diameter of the spray or, more exactly, the diameter of the droplet which has the same surface-to-volume ratio as the spray. They found that this diameter was related to the surface tension $\sigma$ (dynes/cm), the relative air-to-liquid velocity $\Delta u$ (meters/sec), the liquid density $\rho_f$ (grams/cm$^3$), the liquid viscosity $\mu_f$ (poises), the liquid flow rate $V_f$ (cm$^3$/sec), and the air flow rate, $V_a$ (cm$^3$/sec) by the equation

$$d_0 = \frac{585\sqrt{\sigma}}{\Delta u\sqrt{\rho_f}} + 597\left(\frac{\mu_f}{\sqrt{\sigma\rho_f}}\right)^{0.45}\left(\frac{1000V_f}{V_a}\right)^{1.5} \tag{2-3}$$

Inserting into this equation values of $\sigma$, $\rho$, and $\mu$ representative of kerosene (i.e. 30, 0.8, and 0.016) we have

$$d_0 = \frac{3580}{\Delta u} + 45.4\left(\frac{1000V_f}{V_a}\right)^{1.5} \tag{2-4}$$

or making the appropriate allowance for the density of kerosene and air

$$d_0 = \frac{3580}{\Delta u} + 86.8\left(\frac{m_f}{m_a}\right)^{1.5} \tag{2-5}$$

where $m_f$ and $m_a$ are liquid and air mass flows. Since $m_f/m_a$ will not normally exceed $\frac{1}{10}$ and since $\Delta u$ will not normally much exceed 100 meters per second it is clear that the contribution of the $m_f/m_a$ term to droplet size will usually be quite small and that for natural air blast at atmospheric pressure the surface mean diameter (*SMD*) is almost inversely proportional to the relative air-to-liquid velocity. Lewis, et al. [*6*] considered the effect of density on atomization using gases with temperatures and pressures with properties shown in the table.

| Gas | Density, g/liter | Viscosity, centipoise | Velocity, ft/sec |
|---|---|---|---|
| Nitrogen | 1.18 | 0.017 | 700 |
| Ethylene | 1.13 | 0.010 | 700 |
| Helium | 0.169 | 0.019 | 1100 |

Their measured droplet sizes for nitrogen were about one half of the sizes predicted using Eq. 2-4. The sizes of the droplets produced using ethylene and helium were, respectively, one half and twice the produced droplets using nitrogen. It should be noted that Nukiyama and Tanasawa did not include effects of gas density and gas viscosity in Eq. 2-3; neither did they include the effect of scale, or size.

The way in which a single spray such as the one described by Eq. 2-3, 2-4, and 2-5 mixes with air is discussed in II,J.

The natural air blast process is best used where large amounts of high velocity air are available as in a ramjet or afterburner. A typical application in a reheat system is described by Johnstone [*7*]. Kerosene is injected through 10 holes each of diameter 0.052 in. into high velocity gases before they are diffused. The arrangement is such that the liquid is injected upstream so that the air-fuel relative velocity is as high as possible and since there is a distance of 4 feet between fuel injectors and baffles the fuel and air is quite well mixed before entering the burning zone.

The flow out of a simple hole drilled into a pipe in which fuel flows is complicated by the pattern of flow in the pipe, the kind of edge the hole has, and the Reynolds number at the hole. For fuels such as kerosene the fuel flow is given to a first approximation by the equation

$$w_f = 9000d^2(\rho_f\Delta p)^{\frac{1}{2}} \tag{2-6}$$

where $w_f$ is the fuel flow in lb/hr, $d$ the hole diameter in inches, $\Delta p$ is the pressure drop in lb/in.$^2$, and $\rho_f$ is the fuel density (g/cm$^3$).

**D,3. Flat Sprays.** Flat sprays may be produced in a wide variety of ways. The liquid emitted from two equal holes may be allowed to impinge at an angle. A flat spray is then formed in the plane which is normal to the plane of the axes of the two holes and which bisects the angle between the axes. The two holes have to be quite accurately aligned and so such a method is not often used.

An alternative method is to use a slot in a plane surface. Consider an infinite horizontal slot in a vertical infinite surface. Let liquid be at a high pressure on one side of the plane surface. If we neglect the effect of gravity we can see (1) that the liquid approaching the slot has no components parallel to the slot and (2) that the vertical components of momentum of liquid approaching the slot from above and below it cancel. Thus the liquid issues as a flat sheet. Similarly liquid issues horizontally from a horizontal slot cut through a vertical cylindrical tube. From the slot in a flat sheet the lines of flow are parallel. From a cylinder the lines of flow are radial. Although there are end effects it is

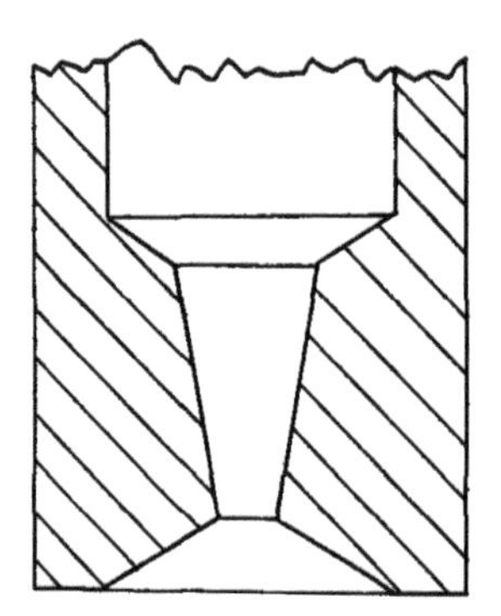

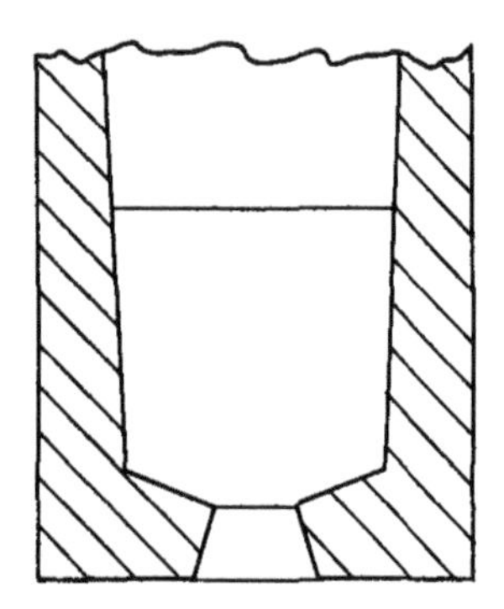

Fig. D,3. Sections through a nozzle for producing a flat spray. It is formed by the intersection of two slots at 90° to each other.

possible to achieve flat sprays by cutting relatively short slots in plane or cylindrical surfaces or to use any other arrangement which causes the flow into the nozzle to be substantially from two opposed directions.

Slots may be cut at right angles from either side of a flat plate. Liquid enters one slot and flows in towards the intersection, passes through the hole at the intersection, and spreads out as a flat spray in the second slot. Fig. D,3 shows two normal sections through such a nozzle which produces a flat spray that spreads out to take the approximate shape of a sector of a circle of angle about 75°.

The behavior of flat sprays has been investigated by Dombrowski and Fraser [*8*]. They used a nozzle formed by cutting two slots at right angles so that the hole was a rectangle 0.086 by 0.042 cm and the liquid spread in a plane parallel to the shortest edges. For ethyl alcohol of surface tension 24 dynes/cm the flat spray was well developed at a pressure of 5 lb/in.$^2$. A pressure of about 15 lb/in. was needed to spread water of surface tension 73 dynes/cm to the same extent. For flow such that the Reynolds number is less than 20,000 and the liquid velocity is less than 7 meters per second the flow is fairly smooth until waves produced by air friction lead to disintegration. At higher Reynolds numbers turbulence in the liquid causes perforations in the sheet. As with cylindrical jets, viscosity effects are such as to oppose movements which break up the liquid. Consequently the liquid spreads further before disintegrating.

Flat sprays are often used in agriculture and fire-fighting installations and may have applications in annular combustion systems but, as yet, little use of them has been seen in gas turbines or ramjets.

**D,4. Swirl Atomizers.** In 1902, Korting constructed an atomizer in which fuel passed between spiral vanes in an annular passage before

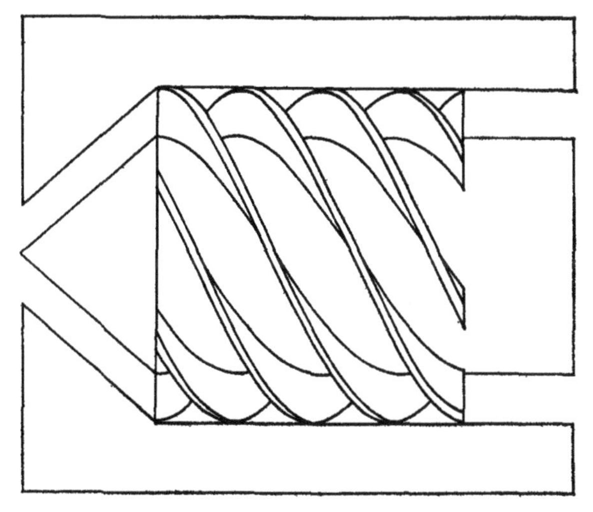

Fig. D,4a. Korting atomizer.

entering a space between two concentric conical surfaces (Fig. D,4a). The fuel swirled inwards and left the atomizer in the form of a hollow cone. Since that date many swirl atomizers have been made. Fig. D,4b, D,4c, and D,4d show some examples. The main difference between the modern atomizer and that of Korting is the absence of the solid central cone and the use of designs that make it relatively easy to ensure concentricity of the various parts. Fig. D,4b shows an atomizer in which the liquid is introduced to the swirl chamber via helical slots. Fig. D,4c shows the use of drilled ports and Fig. D,4d is a section of yet a further variation in which the entry ports are cut in a flat disk. There is no great difference in the performance of the various kinds of atomizer when they are well made, and the choice as to which is to be used is normally dictated by manufacturing and assembly considerations. Since tolerances

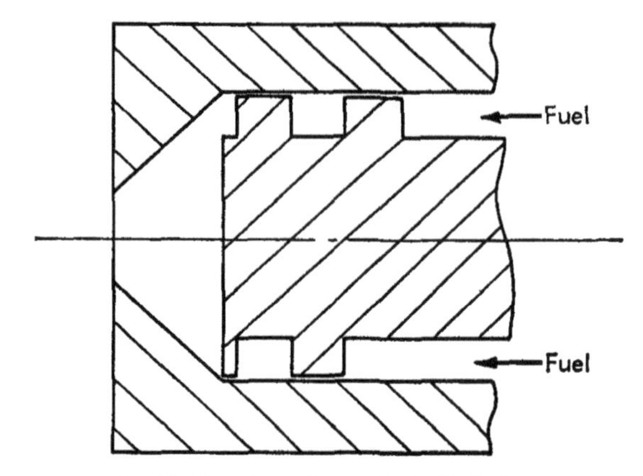


Fig. D,4b. Atomizer with helical slots.

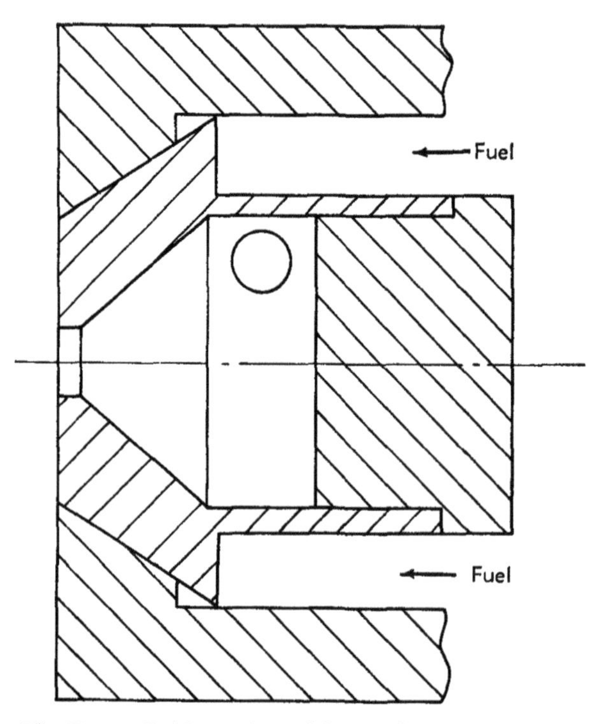


Fig. D,4c. Swirl atomizer with drilled tangential ports.

of the order of fractions of a thousandth of an inch may be needed on the dimensions and concentricity, the manufacture of atomizers demands good equipment and control. The essentials of a swirl atomizer are (1) arrangements for feeding fuel tangentially into a swirl chamber so that the inlet ports and chamber show symmetry, (2) enough ports or slots to avoid a "lobed," or unsymmetrical spray cone, and (3) a circular outlet orifice with its axis coincident with that of the swirl chamber.

In a design satisfying these considerations liquid entering through the swirl ports spirals towards the orifice where it has both axial and tangential velocity so that it spreads out as a film of roughly conical shape on leaving the orifice. The shape is not precisely conical for several reasons. For example the paths of the particles of fluid leaving the orifice if produced back do not meet at a point. Aerodynamic interaction of the spray and surrounding gases distorts the cone, especially at high chamber pressures and high fuel injection pressures [*9*]. The pull of surface tension may also distort the cone. The flow described only holds when the atomizer is sufficiently large and the fuel pressure is sufficiently high. At very

Fig. D,4d. Swirl plate showing six slots through which fuel enters swirl chamber.

low pressures the performance is different, and as the pressure is increased the outflow passes through the following regimes:

1. Liquid dribbles out of the orifice.
2. Liquid leaves the orifice as a pencil with ribs or other distortion from a circular cross section.
3. Immediately on leaving the orifice the liquid spreads out conically, but surface tension—and perhaps air forces—cause the liquid to be pulled into a tulip shape so that the film coalesces. A little beyond the point of coalescence the liquid may separate into several streams which break into large droplets.
4. The tulip opens and the liquid has the shape of the cup with a conical base; *ligaments and droplets come away from the rim of the cup.*

5. The liquid film is nearly conical. Unstable waves form in the surface and ligaments and droplets of fuel develop in turn.
6. The liquid moves away from the orifice in conical shape. The film breaks up quickly and is very short.

G. I. Taylor [*10*] has considered the flow of a real, i.e. viscous liquid, in a swirl atomizer. He points out that there will be a boundary layer on the conical surface leading to the final orifice and that this boundary layer

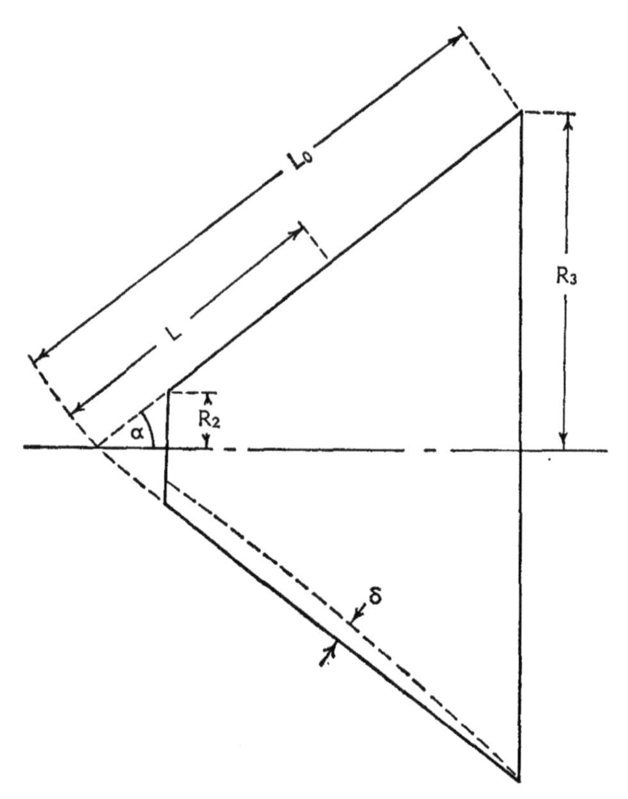


Fig. D,4e. The boundary layer (thickness $\delta$) in a conical liquid sheet from a swirl atomizer cone.

will move under the influence of two forces. The swirling fluid, which is outside the boundary layer in the sense that it is further from the conical surface but which is inside the layer in that it is nearer to the axis of the atomizer, tends to make the boundary layer rotate with it. The pressure drop between the inlet and final orifice tends to move the boundary layer in the direction of the generators of the cone. A boundary layer may move down the back wall of the swirl chamber and part of it may pass along the liquid surface at the air core and out through the orifice. Fig. D,4e represents the cone and boundary layer where $R_2$ is the radius of the outlet orifice, $R_3$ is the greatest radius of the cone, $\alpha$ is the semi-angle of the cone, and $L$ and $L_0$ are respectively the distance along the generators to a point in the conical surface and to the base of the cone. The thickness of the boundary layer is $\delta$ and the angle which the flow at the surface of the cone makes with the generators is $\chi$.

Taylor used the Polhausen method [*11*] to investigate the flow in the boundary layer and obtained the relation between $(\delta/L_0)\sqrt{\Omega/\nu \sin\alpha}$ and $\delta/L_0$ shown in Fig. D,4f where $\nu$ is the kinematic viscosity of the liquid, and $\Omega$ is the vortex circulation divided by $2\pi$, which will be constant everywhere outside the boundary layer, i.e. in the potential flow

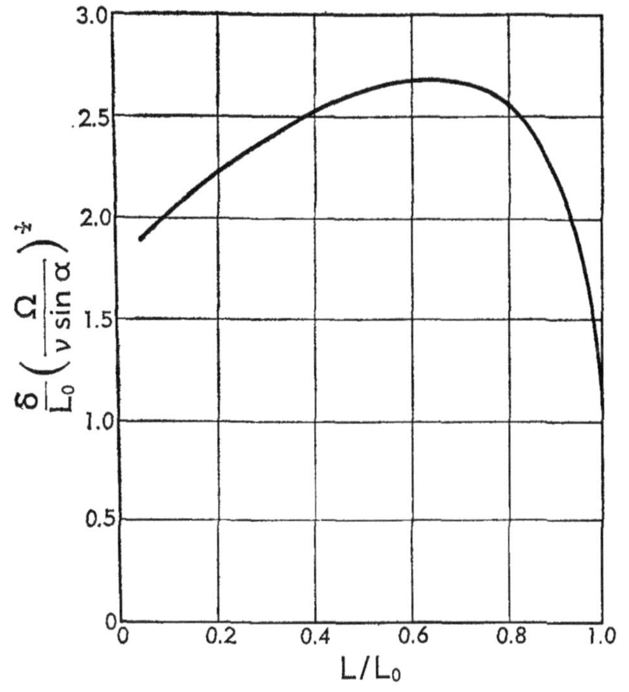


Fig. D,4f. The relation between $\left(\frac{\delta}{L_0}\right)\left(\frac{\Omega}{\nu \sin\alpha}\right)^{\frac{1}{2}}$ and $\frac{L}{L_0}$.

region. It would be the product of the radius $R_2$ of the orifice and the tangential velocity $v_t$ at that radius if there were no wall friction. From Bernoulli's equation $v_t$ cannot be greater than $\sqrt{2\Delta p/\rho_f}$ where $\Delta p$ is the pressure drop and $\rho_f$ is the fluid density, so that an upper limit to $\Omega$ and a lower limit to $\delta$ at the outlet can be found from Fig. D,4f. The value of the boundary layer thickness at the orifice which is thus found is quite close to those observed. Taylor also calculated the angle $\chi$ which the flow infinitesimally close to the surface of the cone makes with the generators (Fig. D,4g). The angle at the orifice when $R_2/R_3$, and hence $L/L_0$ is

0.2, is 19.3°. This implies that liquid in contact with the surface of the cone has an inward motion as it approaches the final orifice.

It is to be noted that $\delta$ increases as $\Omega/\nu$ decreases so that the final orifice tends to run full at low pressures and higher viscosities. These effects are confirmed by observation.

The motion of the liquid film and the antisymmetric oscillations excited by the relative motion between the liquid film and air have been examined theoretically by Squire [*12*]. Taking surface tension $\sigma$, liquid velocity $u_f$, air density $\rho_a$, film thickness $2h$, and wavelength $\lambda$ it is shown that instability occurs when $\sigma/\rho_f u_f^2 h$ is less than unity and that when it is

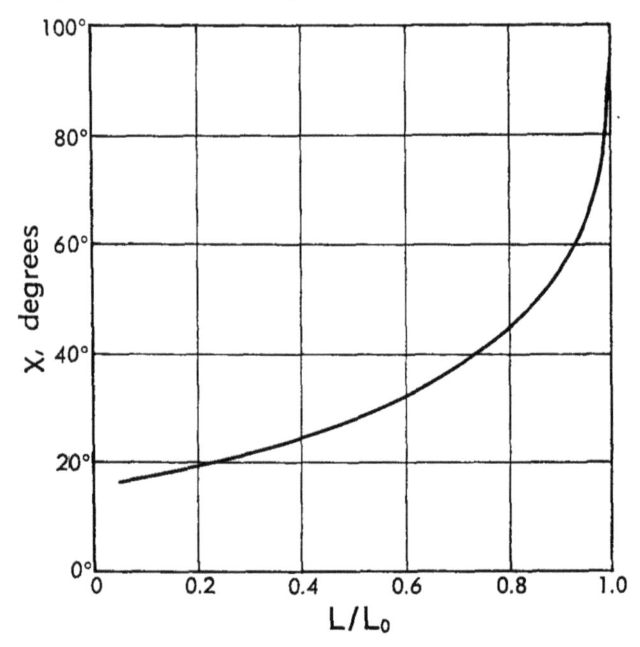


Fig. D,4g. The angle $\chi$ which surface drag makes with generators of cone along the length of the generators.

much less than unity the wavelength which is most unstable and grows most quickly is given by $\lambda = 4\pi\sigma/p_a u_f^2$. The measured and calculated wavelength for a series of sprays photographed at the National Gas Turbine Establishment, Farnborough, were in reasonable agreement allowing for difficulty in estimating liquid velocities. The breakup of the liquid film is associated with the generation of unstable waves when the liquid velocity and air pressure are not too low. The crests and troughs of the unstable waves tend to separate into ligaments of fluid which in turn break into droplets.

The mass flow per second $m_f$, through a swirl atomizer, may reasonably be expected to depend upon the pressure drop $\Delta p$, the fluid density $\rho_f$, the fluid viscosity $\mu$, and the shape and size of an atomizer. If we consider only atomizers of a fixed geometrical shape the flow may be expected to depend upon a length dimension which we may define as $L$.

A dimensional analysis of the problem shows that there are two important groups: discharge coefficient $(m_f^2/L^4\rho_f\Delta p)^{\frac{1}{2}}$ and Reynolds number $m_f/\mu L$, and that one should be a function of the other, say

$$\frac{m_f^2}{L^4\rho_f\Delta p} = f\left(\frac{m_f}{\mu L}\right) \tag{4-1}$$

Measurements are reported on an atomizer of a shape as shown in [*13*, Fig. 19a]. Tests were carried out on a series of atomizers scaled so that the final orifices varied from 0.020 to 0.070 in., with liquids of viscosities 0.5 to 20 centistokes and densities 0.75 to 1.6. Taking the final orifice

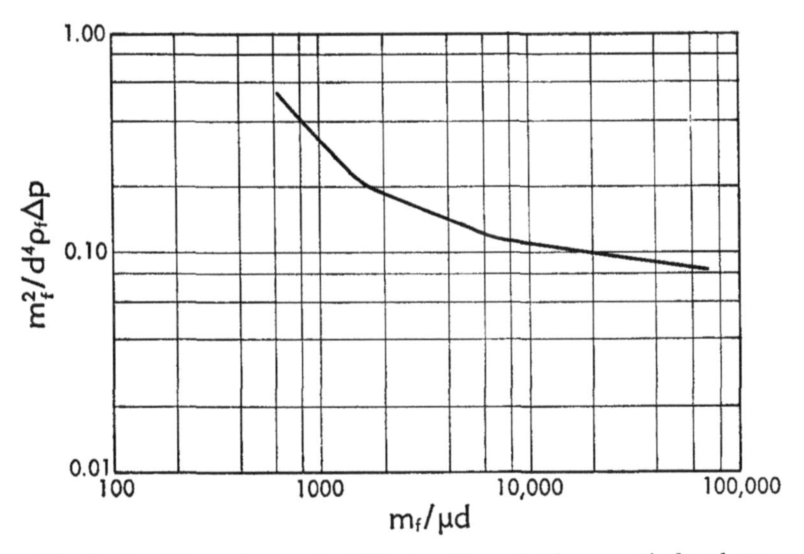


Fig. D,4h. The variation of $m_f^2/d^4\rho_f\Delta p$ with $m_f/\mu d$ for a particular shape of atomizer (consistent units).

diameter as $d$, the diameter of the two inlet ports was $d$, the diameter of the swirl chamber was $7.5d$, the length of the cylindrical part of the swirl chamber was $2d$, the radius of offset of the tangential parts was $3d$, the length of the final orifice was $2d/3$, and the angle of the swirl cone leading to the orifice was 90 degrees. The measurements were all made with the bubble fully opened. The relation between $m_f^2/d^4\rho_f\Delta p$ and $m_f/\mu d$ is shown in Fig. D,4h. In general the bubble opens when $m_f/\mu d$ exceeds 1600. If $m_f/\mu d$ is less than 1600 and $\mu$ is low the cone may not be properly formed. Points to the extreme left of the curve can only be obtained with a satisfactory conical spray when $\mu$ is as high as 20 centipoises. In other words, reducing Reynolds number by increasing $\mu$ has a different effect on the spray than reducing Reynolds number by reducing $m_f/d$. A suggestion of the reason for these effects can be found by reference to the Weber number. If the Weber number is defined as $\sqrt{du_f^2\rho_f/\sigma}$, where $u_f$ is the average fuel velocity at the orifice, $d$ the orifice diameter, $\rho_f$ the fuel density, and $\sigma$ the surface tension then $We^2$ is proportional to $m_f^2/\sigma\rho_f d^d$. A reduction in Weber number means that this group is reduced. A reduction in Weber number, as defined above, means that the inertia forces of the emerging liquid are made smaller in comparison to the forces arising from surface tension. The latter eventually predominate and cause the liquid sheet to converge. These events are also, of course, influenced by the density and viscosity of the air (or gas) into which the fuel is being sprayed.

[13] also showed the flow through, and cone angle from, atomizers of the same construction but of different shape. Commenting on this

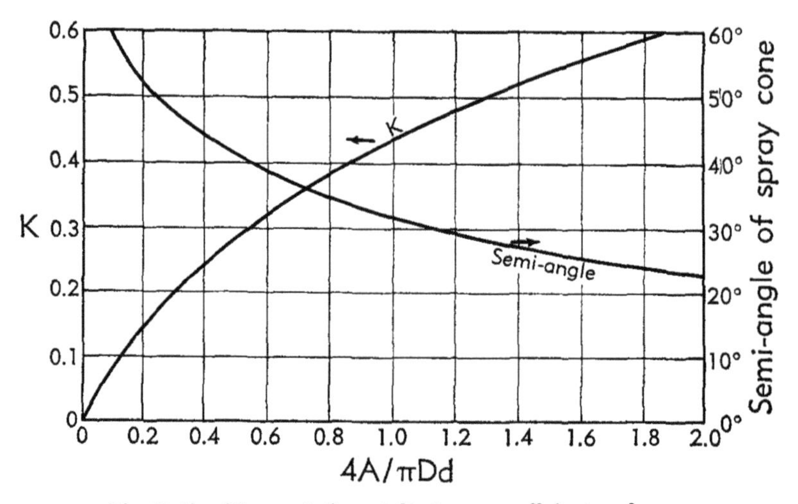


Fig. D,4i. The variation of discharge coefficient and cone angle with swirl chamber geometry; inviscid theory.

Carlisle pointed out that G. I. Taylor had suggested an inviscid flow theory for atomizers from which a relation between the discharge coefficient $(m_f^2/d^4\rho_f\Delta p)^{\frac{1}{2}}$ and $4A/\pi Dd$ might be deduced. Here $A$ is the inlet port area and $D$ the swirl chamber diameter. Carlisle has shown experimentally that for kerosene the measured discharge coefficients depart from the theoretical ideal as either $D/d$ and the over-all length-to-diameter ratio of the swirl chamber is varied. Fig. D,4i, D,4j, and D,4k show the ideal curve and corrections. Theory [14] suggests that the relation between the cone angle and the discharge coefficient should be as shown in Fig. D,4i. In fact the spray angle deviates from the theoretical angle as the product of discharge coefficient $K$ and orifice diameter $d$ changes. Taking the latter to be measured in thousandths of an inch the deviation is about $-10°$ when $Kd$ is 0 and varies almost linearly with $Kd$ to be $+16°$ when $Kd$ is 30.

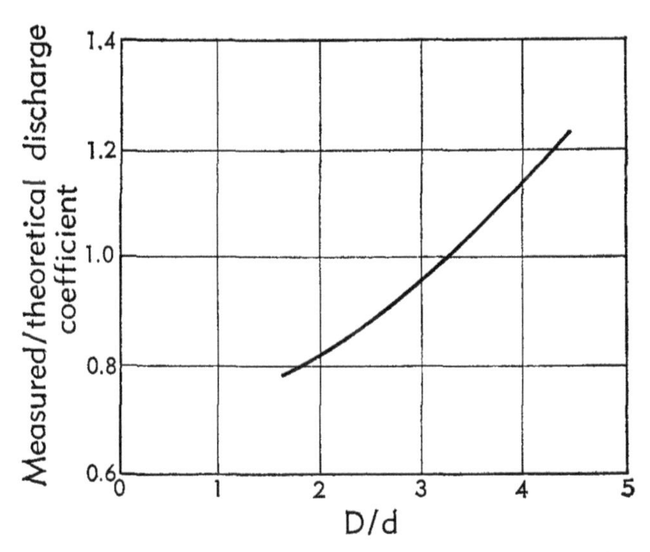


Fig. D,4j. The dependence of the ratio of actual to theoretical discharge coefficient upon the ratio chamber diameter to orifice diameter.

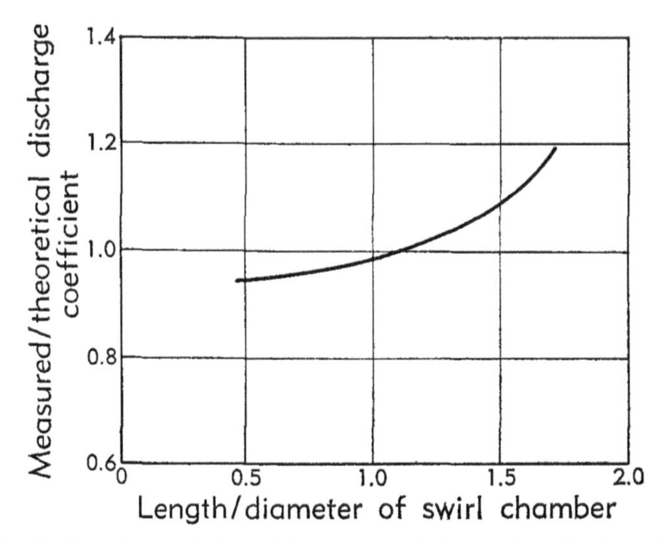


Fig. D,4k. The variation of discharge coefficient ratio with the ratio of the total length of swirl chamber to diameter of swirl chamber.

Fig. D,4h shows that flow through an atomizer is a function of viscosity. But clearly the effect of viscosity is small for parts of the curve where $m_f/\mu d$ is greater than 2000 and $m_f^2/d^4\rho_f\Delta p$ is almost constant. Thus the flow through an atomizer is approximately proportional to the square root of the fluid pressure drop. For this reason the flow number—usually taken to be the flow in gallons per hour divided by the square root of the pressure drop in psi—is a convenient measure of an atomizer size even though the flow number is not a number and is not entirely independent of fuel pressure, density, or viscosity.

The droplets produced by a swirl spray operating at fixed conditions have a fairly broad distribution which is often defined in terms of the Rosin-Rammler [*15*] distribution first used to describe the size spectrum of coal particles. It is

$$V = V_0 e^{-(x/\bar{x})^n} \tag{4-2}$$

where

$x$ = droplet diameter
$\bar{x}$ = a parameter of the dimension of droplet diameter
$n$ = a parameter
$V_0$ = total volume of droplets
$V$ = volume of droplets of greater diameter than $x$

The parameter $n$ which varies between 2 and 4 fixes the tightness of the distribution about the size $\bar{x}$. The parameters $\bar{x}$ and $n$ vary from spray

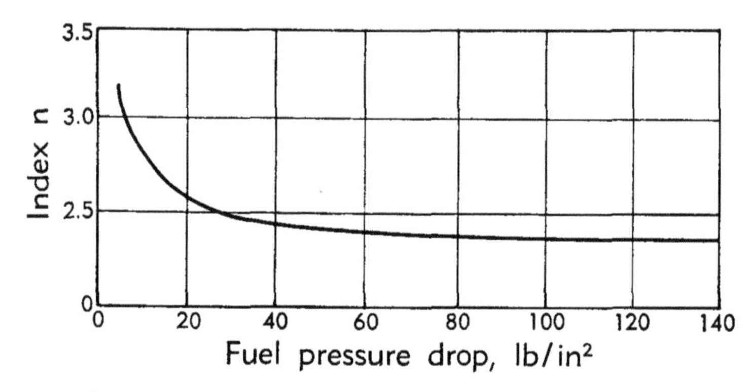


Fig. D,41. Variation of fuel distribution index $n$ with fuel pressure drop.

to spray. Needham [*16*] examined data on sprays of wax which were used at a suitable temperature to simulate kerosene; he was able to show that for atomizers of flow number 1 to 4.5 there was an approximate correlation of $n$ with injection pressure (Fig. D,41) and that for the same sprays

$$\bar{x} = 270 \frac{w_f^{0.25}}{p^{0.4}} \tag{4-3}$$

where $\bar{x}$ is measured in microns, the flow $w_f$ in lb/hr and the pressure $p$ in psi. The surface mean diameter, often called Sauter mean diameter [*5,17*], is given by

$$SMD = 180 \frac{w_f^{0.25}}{p^{0.4}} \tag{4-4}$$

Similar results were published by Joyce [*18*]. Since the flow number ($FN$) of an atomizer is taken to be proportional to $w_f/p^{0.5}$ it is clear from Eq. 4-4 that the droplet size from an atomizer is proportional to $(FN)^{0.25}p^{-0.275}$. Although there is room for some argument about the exact value of these indices, only Longwell (II,J,1) obtains results which depart greatly from Eq. 4-4. Since the velocity at which liquid leaves a swirl atomizer is proportional to the square root of the pressure drop (neglecting losses due to friction) it can be seen that for a fixed liquid flow the *SMD* is proportional to $u^{0.8}$. This compares with a natural air blast system where the *SMD* is proportional to velocity (Eq. 2-5). For both types of atomization, fuel flow also has an effect on the mean particle size.

The effect of varying fuel viscosity on atomization is relatively small. Data in [*16*] suggest that particle size increases as the one-fifth power of fuel kinematic viscosity. The effect of combustion chamber pressure on the atomization from swirl atomizers is not known. For Diesel injection systems [*19,20*] the effect of changing air pressure is small and it is probable that the effect of pressure is small for swirl atomizers. Turner and Moulton [*21*] discuss the sprays from two kinds of nozzle and show that the droplet size is proportional to the surface tension raised to powers of 0.713 and 0.594. Thus we can summarize and say that when fuel of viscosity $\mu$ centistokes and surface tension $\sigma$ dynes/cm is sprayed into air at NTP the surface mean diameter is given by

$$SMD = \frac{22.8\mu^{0.2}\sigma^{0.6}w_f^{0.25}}{p^{0.4}} \tag{4-5}$$

The effect of changing the atmosphere receiving the spray may well be small.

The fuel required by a jet engine is a function of forward speed, altitude, and engine revolutions and may be increased or decreased during accelerations or decelerations.

At 60,000 feet the air pressure and mass flows in an engine for, say, cruise conditions are about one fifteenth of the pressure and mass flow at ground level. Hence an engine that demands fuel flows varying by a factor of 7 at each altitude may well demand flow variations of more than 100 to 1. Since the flow through a swirl atomizer is proportional to the root of the pressure drop in the atomizer this implies that if a simple system is used the fuel pressure range varies by 10,000 to 1 and calls for very high maximum or very low minimum pressures. Since pressures of 1000 lb/in.$^2$ are high for kerosene and pressures of much under 10 lb/in.$^2$ give poor atomizer performance, methods of avoiding excessive pressure variations have had to be found.

The first solution is to use more than one swirl atomizer and to increase the number used as the fuel flow is increased. Arrays of seven atomizers, each of small cone angle arranged so that either the central atomizer alone or the central atomizer plus the six outer atomizers can be used, have been tried. Unfortunately the sprays tend to coalesce with severe deterioration in atomization. This and the general complexity of the atomizer arrangement make the system practically unusable.

An alternative system was used with some success in the National Gas Turbine Establishment "T" scheme development. Two atomizers were arranged back to back so that a small atomizer injected fuel upstream and a larger atomizer injected fuel downstream. One or both atomizers could be used for small and large fuel flows.

An ingenious arrangement of two atomizers is used satisfactorily on, for example, Rolls Royce gas turbines. A small flow number pilot atomizer fits inside a larger main atomizer. The fuel flows are arranged to pass through separate supply lines or through pressure-controlled valves so that the small atomizer operates until the fuel demand reaches a desired level. Thereafter increasing quantities of fuel are supplied first at low pressures and later at higher pressures to the main atomizer. The cone angles of the atomizer are chosen so that the fuel from the two atomizers mixes and even when the supply pressure to the main atomizer is small the momentum of the fuel from the small atomizer is sufficient to spread and atomize the fuel from both atomizers.

Fig. D,4i shows that the discharge coefficient of swirl atomizers increases as the inlet port area is increased, and many atomizers have been made which exploit this fact. Pillard and similar atomizers arrange that the back face of the swirl chamber is movable. When fully forward the area of inlet ports giving access to the swirl chamber is small; as the back face is withdrawn a greater area of inlet port, or perhaps more inlet ports, are exposed. Although the movement of the back face can be controlled manually it is usual for a spring to oppose the force due to fuel pressure on the back wall of the atomizer; then as the fuel pressure increases the atomizer port area increases so that the flow increases partly because the discharge coefficient is increased and partly because the applied fuel pressure drop across the atomizer is greater. It is essential for satisfactory operation that the piston move smoothly and be stable at all positions, otherwise the fuel flow may vary intermittently or cyclically and give rise to fluctuation in the combustion chamber heat release.

The Lucas Duplex [*22*] is typical of a variation in this kind of atomizer. It is arranged that the fuel supply to different sets of inlet ports to the same chamber be controlled by a valve in the fuel line. When a low flow is required the fuel pressure is low and a pressure-controlled valve permits fuel to enter the chamber by two ports. At higher pressures the fuel gains access to the chamber by a further set of four inlet ports.

Peabody and Fisher patented a spill type of swirl atomizer in 1921. They appreciated that the flow through the normal outlet of a swirl atomizer could be reduced by taking fuel out of holes in the back plate of the atomizer; i.e. by spilling the fuel. If the hole in the back plate of the swirl chamber is large enough it can be arranged so that all the fuel is spilled (Fig. D,4m). The fuel that does go through the normal atomizer outlet can have quite a high velocity so that it atomizes well. When most of the fuel is spilled, that which passes through the atomizer into the combustion chamber has a small axial velocity. Therefore the cone angle is high. This variation of cone angle with spill can be a handicap in designing a combustion chamber. The air core may extend into the spill line so that air is mixed with the fuel. Pockets of air in the fuel lines are a nuisance. If the spilled fuel is fed into the pump intake the pump performance may be impaired. For this reason a spill arrangement that cannot collect air has advantages. Dowty spill atomizers collect the spilled fuel in an annulus leading into the back face of the swirl chamber.

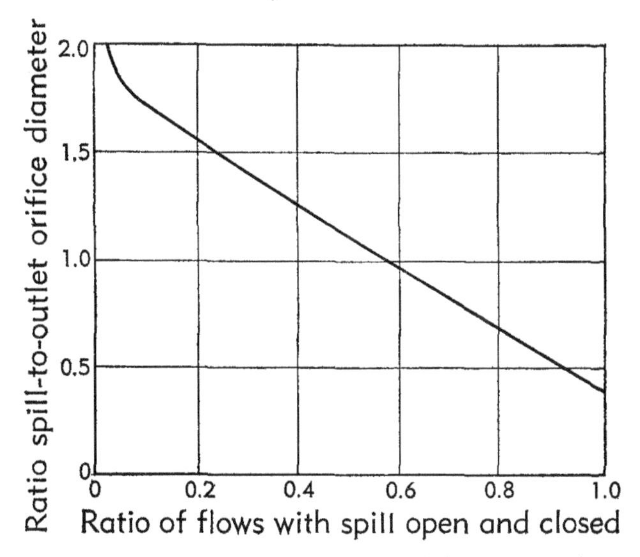


Fig. D,4m. The dependence of the ratio of minimum to maximum flow (at a given inlet pressure) upon the ratio of spill to orifice diameter.

Alternatively it is possible to take the fuel from a slot near the final orifice [*22*]. In Fig. D,4n the flow versus pressure performance of a spill atomizer is represented graphically.

Since droplet size has been shown to be inversely proportional to the relative fuel-to-air velocity for simple swirl atomizers it is to be expected that the same will apply to spill atomizers. So the atomization might well be predicted if it were known what part of the theoretically possible velocity appeared in the spray. Although this is not known explicitly it is reasonable to assume that losses are greater in a spill atomizer than in a simple atomizer and that this explains why the atomization at a given inlet pressure and flow are worse than for a simple swirl atomizer delivering the same outlet flow at the same pressure. In [*12*] it is noted that there are few data available about droplet size from spill atomizers,

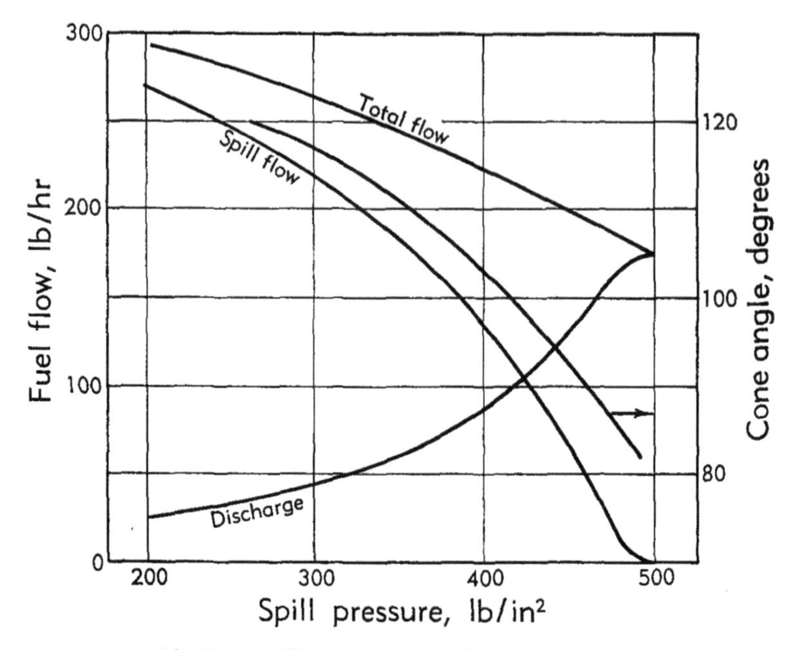


Fig. D,4n. Kerosene flows and cone angle for a spill atomizer. Inlet pressure 600 psi.

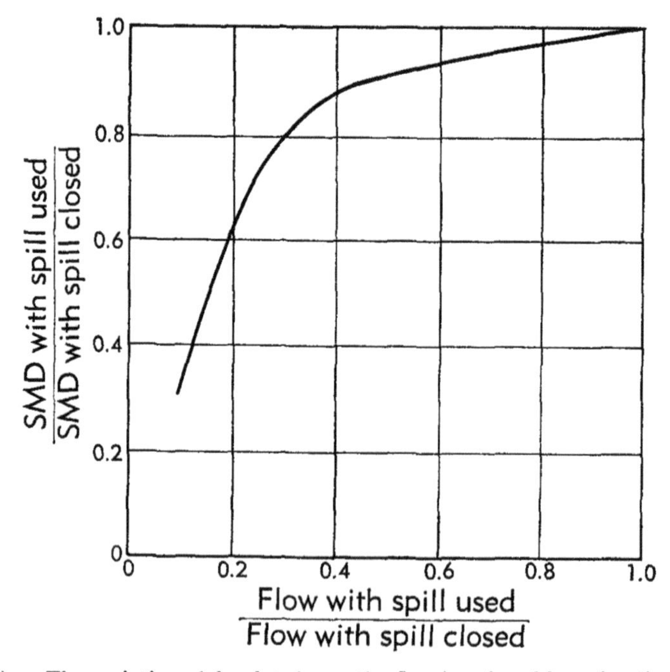


Fig. D,4o. The variation of droplet size as the flow is reduced by using the spill line.

but that the available evidence indicates that the surface mean diameter for spill closed is $325w_f^{0.318}/p^{0.530}$ microns ($w_f$ is the fuel flow in lb/hr and $p$ pressure drop in lb/in.$^2$) and that the effect of reducing the flow by opening of the spill line is as shown in Fig. D,4o.

The spill atomizer and other devices can be compounded. Lubbock [*18*] designed an atomizer with a piston sensitive to the pressure as the back plate of a swirl atomizer. As fuel pressure increases the piston uncovers a greater inlet port area and throttles the spill line. Fletcher and Ashwood [*23*] used a twin atomizer. The main swirl jet faced downstream in the combustion chamber and was spill-controlled. A pilot jet supplied from the spill line faced upstream. Darling [*24*] records the use of a spill atomizer which is arranged so that a piston can pass through the spill orifice and close the outlet orifice. This prevents drips from the orifice and permits the constant circulation of fuel through the supply and spill system, so that the fuel can be brought to the desired temperature and viscosity at the nozzle before spraying begins.

**D,5. Air Blast Atomization.** A swirl atomizer cannot operate so as to form a satisfactory conical film if the oil viscosity is too thick. The tolerable viscosity depends on atomizer size and fuel pressure as discussed in Art. 4. If the fuel viscosity is in excess of 20 centistokes at the atomizer it may well be necessary to use air blast atomization. When the air that is to be burned has a high velocity, a simple injection system based on single holes is frequently sufficient. In conventional gas turbine combustion chambers the gas velocities are too low for natural air blast atomization.

In some circumstances air assistance may be given to swirl atomizers. If the main trouble with the swirl atomizer is that at low flows a bubble rather than a cone is formed, a little air supplied via the air shroud may be sufficient. Such an atomizer has been used by the Westinghouse Electric Corporation to atomize a heated heavy fuel (bunker C). A swirl atomizer is surrounded by a concentric cap. Air supplied at a pressure of several pounds per square inch to the annulus between atomizer and cap passes through an air swirler and spirals inward to an orifice in the cap through which both fuel and air pass. It has been shown by H. Clare in unpublished work that such a system also improves the atomization of kerosene at low fuel pressures. For kerosene an applied air pressure of 0.25 lb/in.$^2$ is enough to decrease the mean particle size by 60 per cent.

When fuel is too viscous for a swirl atomizer, more drastic air atomization is necessary. Romp [*25*] gives diagrams of a wide variety of steam and air blast atomizers designed for use in oil-fired furnaces. Broadly the general principles followed are to put the oil and gas in concentric tubes so that they mingle where the gas velocity is at a maximum. The Laidlaw

Drew atomizer is one such atomizer. The fuel is in an inner pipe and meets swirling air in a converging annulus. The applied air pressure for use in a furnace is about 15 to 40 lb/in.². The performance of an atomizer in which fuel flows radially inward to a swirling air stream (Fig. D,5a) is described

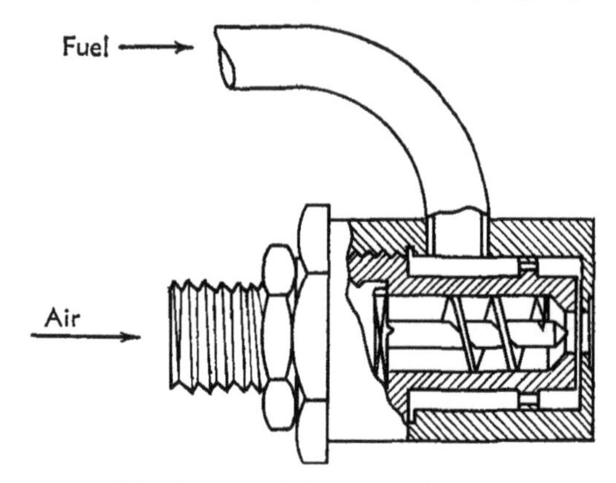


Fig. D,5a. Air blast atomizer.

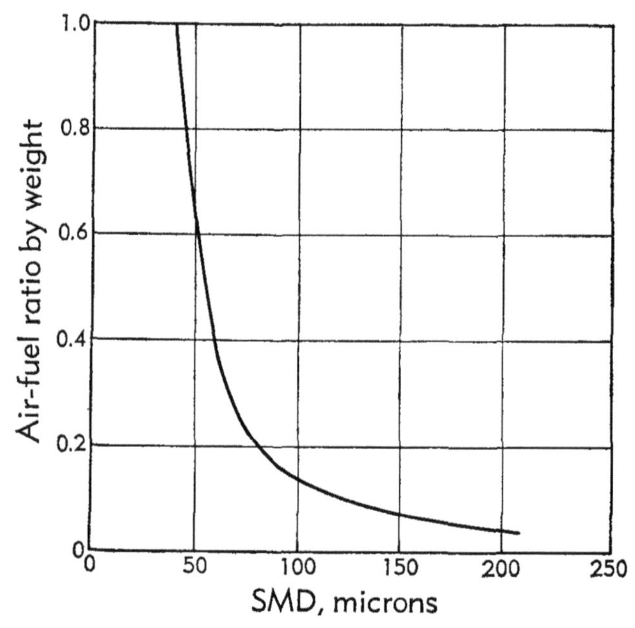


Fig. D,5b. Dependence of droplet size on air fuel ratio for an air blast atomizer. Fuel viscosity 40 centistokes.

in [*26*]. The atomizer is normally used at an air pressure sufficient to choke the air orifice. The droplet size distribution conforms quite closely to a form represented by

$$\frac{dN}{dx} = mNe^{-mx}$$

where $N$ is the number of droplets, $x$ is the droplet diameter, and $m$ is a parameter. The mean droplet size is $1/m$ and the diameter of the droplet having the same surface-to-volume ratio as the spray (i.e. *SMD*) is $3/m$. Although somewhat affected by fuel pressure the main factor controlling droplet size is the air-fuel ratio. Fig. D,5b shows the relationship for an atomizer with an orifice of 0.25 inches and for a fuel of viscosity 40 centistokes as determined by use of the wax simulation technique. The droplet size is approximately proportional to the square root of the orifice dimensions. This is in line with observations on swirl atomizers.

**D,6. Disk or Cup Atomizers.** Mechanical atomizers which consist of rotating disks or cups have been in use in oil-fired, steam-raising plants since about 1900. As yet such atomizers have been little used in gas turbines for no very obvious reasons. Fig. D,6a shows a typical arrangement. Fuel dribbles on to the inside surface of a rotating cylindrical tube. Centrifugal forces spread the film fairly uniformly over the surface. The tube is joined to an expanding section. Centrifugal forces drive the fuel

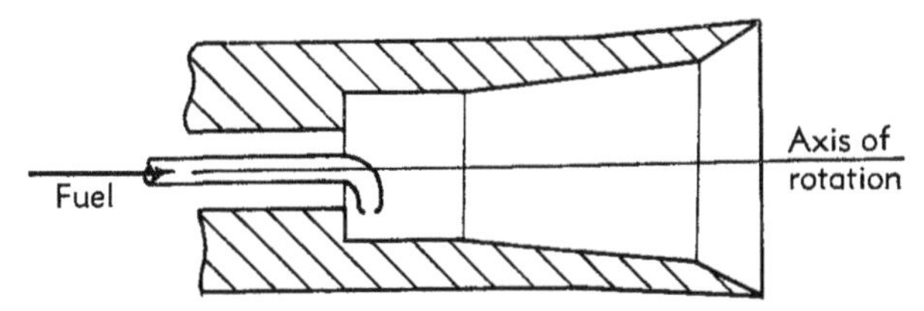


Fig. D,6a. Diagram of rotating cup atomizer.

along the expanding walls of the cup to a lip from which the liquid departs as the mass of liquid becomes sufficient, so that the constraint of surface tension becomes too small to prevent momentum forces which are very nearly tangential to the lip of the cup from moving the liquid in a straight line. Hinze and Milborn [27] have described the performance of cup atomizers fairly completely.

If it is arranged that fuel arrives steadily on, and spreads smoothly over, the cylindrical part of the cup and if the cup rotates without vibration the liquid spreads uniformly; otherwise it is possible for the fuel to spread unevenly so that the amount of fuel arriving at different parts of the lip varies. In general the path of the fuel on the surface of the cup follows the line of generators of the inside surface of the cup, and friction keeps the velocity relative to the cup low so that on arriving at the cup lip the main component of velocity is tangential. As the liquid flow is increased from zero the liquid pattern leaving the lip has three fairly distinct stages. In the first stage small quantities of liquid accumulate at points on the lip and the centrifugal forces on the gradually increasing bulge increase until they overcome surface tension and tear a drop of liquid from the lip. This droplet formation is repeated at intervals of time and distance around the lip of the cup. The same phenomenon can be seen when a liquid is allowed to drip under the influence of gravitational forces from the lower horizontal edge of a flat plate. In the second stage the rate of supply of liquid to the neck of what might otherwise be a droplet is sufficient to maintain flow into a ligament of fluid. Since each part of the fluid tends to move tangentially at disk lip velocity from the lip, it is a simple exercise in geometry to draw the shape of a typical ligament. As the fuel flow is increased the number of ligaments increases until finally they begin to touch and the final stage is reached. At these

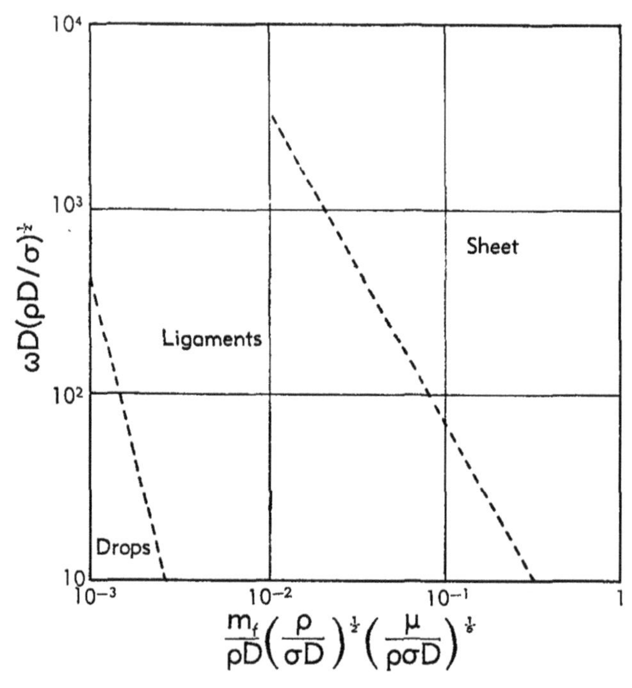


Fig. D,6b. Criteria for formation of droplets, ligaments, and sheet by a spinning cup atomizer.

higher flows the liquid leaves the disk as a flat film which breaks up into ligaments and droplets as do conical films. Using atomizers of diameters 1–10 cm, Hinze and Milborn showed that the three stages of flow were separated into regions by lines on graphs of $\omega D(\rho D/\sigma)^{\frac{1}{2}}$ and $(m_f/\rho D)(\rho/\sigma D)^{\frac{1}{2}}(N/\rho\sigma D)^{\frac{1}{6}}$ as in Fig. D,6b. At low lip velocities the film breaks into droplets which have a narrow distribution curve with an index in the Rosin-Rammler equation of about 8. This contrasts with a value of about 2.5 for swirl atomizers. Such an atomizer has potential advantages when droplets of more uniform size are needed. At flows which are too low to be of much importance to gas turbines very uniform sprays can be produced [*28,29*].

The rotating cup atomizer can be combined with air blast atomization by blowing air parallel to the axis of rotation and across the lip of the atomizer [*30*].

Atomization of a similar kind results when liquid is dribbled down radial pipes attached to a disk. If the exit from the pipe is small, centrifugal forces develop a pressure so that rotational speed and liquid pressure both play a part in the atomization process.

**D,7. Vaporization.** Although injection of fuel as liquid into a gas turbine combustion chamber has certain advantages in that the liquid drops mix well with the air it does not burn until vaporized. Unburned liquid can collect in the combustion chamber and be a nuisance in many ways. Therefore the injection of vapor or liquid which immediately vaporizes has value.

It is interesting to consider the conditions that have to be satisfied to obtain a vapor-air mixture. Generally the fuel-air ratio and combustion inlet pressure are settled by design conditions. The vapor pressure and density of the fuel dew point conditions are either known or have to be calculated or measured for a range of temperatures. Assuming the vapor to be at the dew point the pressure and density of the air can be determined at each temperature. Hence the maximum possible vaporized fuel-air ratio at each temperature can be calculated. The temperature of the mixture which will just hold the required fuel air ratio can be determined. If the specific heat of air and liquid fuel and the latent heat of vaporization are known, the heat content of the desired vapor and air mixture is calculable and if this exceeds the heat content of the unmixed fuel and air the heat to be added to either fuel, air, or both can be determined. Mullins [*31,32*] discusses this problem and concludes by presenting nomograms from which the necessary temperature increase of either fuel or air can be determined for a 100-octane fuel and a kerosene (burning oil).

If the vaporizing of the fuel is tackled by arranging that the air is heated and cold liquid fuel is injected into the warmed air the system reverts to the injection of liquid fuel into a gas and the arrangements may naturally take the form discussed in previous articles. Alternative methods allow for the warm air to evaporate the fuel from a porous surface as a wick, but such methods are not well adapted to cope with the large flow rates of a gas turbine.

If the fuel is heated the methods used can be divided into two main categories. The fuel may be heated to the necessary temperature on its way to the combustion zone and kept under a high pressure, so that it does not vaporize until after it has passed through a nozzle and the pressure falls to the combustion chamber pressure. The fuel flow is controlled partly by fuel pressure but mainly by varying the area of the nozzle—

which takes the form of a valve. It is necessary in such a system to avoid cracking the fuel. Most hydrocarbon fuels tend to crack if held at 350°C for a long period, so that to avoid cracking at high temperatures the fuel must be very rapidly heated and vaporized as soon as possible after it is hot enough. Because of this complication and the relatively inflexible fuel control this kind of vaporization, which is sometimes called flash vaporization, is seldom used except for specialized combustion experiments.

The simpler method is to put the fuel at combustion chamber pressure—or thereabouts—into pipes or vessels where it can be heated and so vaporized at the chamber pressure. In principle this is what is done with most gas turbine vaporizing systems. The fact that air is passed through the same pipes and so is heated at the same time assists vaporization and ensures that even if not properly vaporized the fuel is certain to be kept moving into the combustion space. Examples of the use of such systems are to be seen in the engines of Armstrong Siddeley, England, and the Wright J65, U.S.A. In the Mamba [*33*], primary air and fuel are supplied to the longer leg of a tube bent in the form of a J. The air sweeps the fuel through this and the mixture emerges from the shorter limb of the J so that it is moving against the general direction of air flow. In burning, the gas heats the J-tube (sometimes called a walking stick) and vaporizes the incoming fuel. A similar system is used in the Viper [*34*] and J65 [*35*]. The walking stick in the Viper and J65 is slightly modified. It is made of three straight sections joined by right angle bends, and the long arm contains a swirling pin for better mixing.

**D,8. Gaseous Injection.** The use of gaseous fuels has not presented very severe problems to the gas turbine designer. The chief difficulty is to arrange a suitable distribution of the fuel within the combustion chamber. Hawthorne, Weddel, and Hottel [*36*, p. 266] discuss the mixing of gas jets entering a gaseous medium and point out that the process is controlled by the momentum and buoyancy of the jet. For small density differentials and for all high velocity, small diameter jets the buoyancy effects are unimportant. For a fully turbulent gaseous jet the fuel concentration at a point depends only on the axial and radial distances from the orifice when the distances are measured in terms of the diameter of the orifice. In a gas turbine combustion chamber the amount of air that is to be burned with the fuel and the amount that is to be kept for cooling is fixed by design considerations. Hence the ratio of the diameter of the fuel orifices to the chamber diameter has to be adjusted to give the desired fuel-concentration distribution pattern. Successful gas-burning gas turbines have been built by both Brown Boveri [*37,38*] and Ruston and Hornsby. In the Brown Boveri 10,000-kw gas turbine to be installed at

Bucharest, the tests in Switzerland were carried out using oil fuel at a pressure of 35 atm. before the conversion to gaseous injection at a pressure of about 11 atm which is near the final compressor delivery pressure. It is reasonable to assume that the combustion chambers were similar for both fuels and that the chief objective of the conversion was to arrange a similar distribution of the gaseous fuel as had been achieved with the liquid fuel. The Ruston and Hornsby gas turbine is sold in similar versions for liquid and gaseous fuels and much the same combustion system is used for both. Again the main objective is to arrange similar distribution. The gaseous fuel control is obtained partly by changing pressure and partly by the use of valves. Flow in such conditions is discussed in III,A and III,B.

**D,9. Solid Fuels.** The injection of solid fuels into a gas turbine is a much more complex matter than the injection of liquids or gases and is in general only accomplished by transport on a moving grate or belt or as a suspension in some gas or liquid. Except when the burning is carried out at atmospheric pressure and a heat exchanger is used to heat compressed air indirectly, each system involves passing the coal through air locks on an equivalent system from a low to a high pressure.

If the coal is to be carried by compressed air it is necessary to dry the coal to discourage coagulation. A further essential step which can be carried out concurrently or after the drying is pulverization. Over fifty papers were presented to a Conference on Pulverized Fuel arranged by the Institute of Fuel in June 1947 concerning the preparation and use of pulverized fuel. Fitton [*39*] divides the coal-pulverizing systems into two main classes: (1) mechanical and (2) pneumatic. Mechanical systems are subdivided to cover a wide speed range; but in all of them it is arranged that the coal be crushed by balls, rollers, or wheels, or in some analogous fashion. In some, the drying air sweeping through the mill selectively carries away particles of sufficiently small size. Pneumatic systems are of two kinds. In one the air blows coal so that it impacts on either hard surfaces or other pieces of coal. In a second kind, due to Yellott and Kottcamp of Bituminous Coal Research Incorporated, the coal is crushed to particles which pass a $\frac{1}{16}$-in. sieve, and later a rapid decrease in pressure results in the small particles of coal exploding to be even finer.

Hurley [*40*] described methods of injecting a suspension of powdered fuel in air into a furnace. The suspension passes through long narrow venturi-shaped slots or through venturi-shaped pipes into the combustion space. The velocity at the venturi throat is designed to prevent the blowback of flames. The equipment required to prepare such fuels is of course cumbersome and weighty and not at all likely to be carried by aircraft.

## D,10. Cited References and Bibliography.

*Cited References*


1. *Fifth Symposium on Combustion.* Reinhold, 1955.
2. Rayleigh. *Proc. London Math. Soc. 10 (4)*, 1878–1879.
3. Lane, W. R. *Ind. Eng. Chem. 43*, 1312–1317 (1951).
4. Nukiyama, S., and Tanasawa, Y. *Trans. Soc. Mech. Eng. Japan 4*, 86–93, 138–143 (1938); *5*, 62–75 (1939); *6*, II-7, II-15 (1939); *6*, II-18, II-28 (1940). Transl. in *Can. Defence Research Board* paper by E. Hope, Mar. 1950.
5. Sauter, J. *Forsch. Gebiete Ingenieurw. 312*, 1926.
6. Lewis, H. C., et al. *Ind. Eng. Chem. 40*, 67–74 (1948).
7. Johnstone, A. P. *Flight 56*, 285–287, 363–365 (1949).
8. Dombrowski, N., and Fraser, R. P. *Phil. Trans. Roy. Soc. London A247*, 101–130 (1954).
9. DeCorso, S. M., and Kemeny, G. A. *Trans. Am. Soc. Mech. Engrs. 79*, 607–615 (1957).
10. Taylor, G. I. *Quart. J. Appl. Math. 3*, 129–139 (1950).
11. Pohlhausen, K. *Z. angew. Math. u. Mech. 1*, 257 (1921).
12. Squire, H. B. *Brit. J. Appl. Phys. 4*, 167–169 (1953).
13. Radcliffe, A. *Proc. Inst. Mech. Engrs. London 169*, 93–106 (1955).
14. Taylor, G. I. *Proc. Seventh Intern. Congress Appl. Math. 2, Part 1*, 280–285 (1948).
15. Rosin, P., and Rammler, E. *J. Inst. Fuel London 7*, 29 (1933).
16. Needham, H. C. *Power Jets (Research and Development) Rept. R1209*, 1946.
17. Sauter, J. *NACA Tech. Mem. 390*, 1926.
18. Joyce, J. R. *J. Inst. Fuel London 22*, 150–156 (1949).
19. Lee, D. W. *NACA Rept. 425*, 1932.
20. Sass, F. *Kompressorlose Dieselmaschinen.* Springer, Berlin, 1929.
21. Turner, G. M., and Moulton, R. W. *Chem. Eng. Progr. 49*, 185 (1953).
22. Watson, E. A. *Proc. Inst. Mech. Engrs. 158*, 187–201 (1948).
23. Ashwood, P. A., and Fletcher, W. G. *Brit. Patent Specification 647, 344*, 1950.
24. Darling, R. F. *Proc. Inst. Mech. Engrs. 168*, 159–165 (1954).
25. Romp, H. A. *Oil Burning.* Martinus Nijhoff, The Hague, 1937.
26. Clare, H., and Radcliffe, A. *J. Inst. Fuel London 27*, 510–515 (1954).
27. Hinze, J. O., and Milborn, H. *J. Appl. Mech. 17*, 145 (1950).
28. Walton, W. H., and Prewitt, W. C. *Proc. Phys. Soc. London B62*, 341–350 (1949).
29. May, K. R. *J. Appl. Phys. 20*, 933–938 (1949).
30. van de Putte, W. L., and van den Bussche, H. K. J. *World Power Conference*, The Hague, Sept. 1947.
31. Mullins, B. P. *J. Inst. Petroleum 32*, 703–737 (1946).
32. Mullins, B. P. *J. Inst. Petroleum 33*, 44–70 (1947).
33. *Engineering 165*, 583–588 (1948).
34. *Aeroplane 85*, 139–144 (1953).
35. *Flight 65*, 787, 788 (1954).
36. *Third Symposium on Combustion, Flame and Explosion Phenomena.* Williams & Wilkins, 1949.
37. Seippel, C., and Weber, H. *Brown Boveri Rev. 33*, 263–269 (1946).
38. Pfenninger, H. *Brown Boveri Rev. 40*, 144–166 (1953).
39. Fitton, A. *Inst. Petroleum Rev. 2*, 18–26 (1949).
40. Hurley, T. F., and Cook, R. *J. Inst. Fuel London 11*, 195–202 (1938).


*Bibliography*


Giffen, E., and Muraszew, A. *The Atomisation of Liquid Fuels.* Chapman and Hall, 1953.

# SECTION E

# *FLAME STABILIZATION*[1]

J. HOWARD CHILDS

## *CHAPTER 1. FLAME STABILIZATION BY BLUFF BODIES IN GASEOUS FUEL-AIR MIXTURES*

**E,1. Introduction.** After obtaining the proper fuel-air mixture by the techniques discussed in Sec. B and D, the next requirement for gas turbine primary combustors and afterburners is the stabilization of the flame. The mean flow velocity in these combustors greatly exceeds the flame speed. Consequently, the flame would be extinguished unless some device were used to create a local low velocity region in which the flame is anchored. A flame stabilized in this manner can then propagate into the main stream at a rate determined by the flame speed and the local flow velocities. In turbojet primary combustors the flame is stabilized inside perforated liners; these devices will be considered in Chap. 3. In this chapter the discussion will be restricted to bluff-body flame holders of the type used in afterburners.

A system of flame holders such as those shown in Plate E,1 is employed in most current afterburners. Similar flame holders are also used in many ramjets. The flame holders must be properly located with respect to the fuel injectors so that the correct fuel-air mixture passes over the flame holders; the fuel-air ratios in the burning zone must be confined within narrow limits near the stoichiometric. The flame holders shown in Plate E,1 consist of V-shaped gutters with the apex of the gutter pointing upstream. The V-gutter flame holder is the type most commonly used. However, any bluff body immersed in the stream can serve as a flame holder, since the purpose of the flame holder is to create a low velocity region of recirculatory flow in its wake.

For a given flame holder and fuel, and approach-stream conditions of pressure and temperature, there exist definite velocity and composition limits within which a flame can be stabilized on the flame holder. Fig. E,1 shows typical stability limits for flames burning from circular rods arranged normal to the flow [*1*]. The flame can be maintained to higher

[1] Manuscript received June 1956.

flow velocities for mixture compositions near the stoichiometric. For velocities and fuel-air ratios lying outside the curves of Fig. E,1 the flame "blows off" and is extinguished.

Many studies of flame stabilization have been conducted with bluff-body flame holders of various shapes. These studies show the effect of the various flow variables and design variables and also provide insight

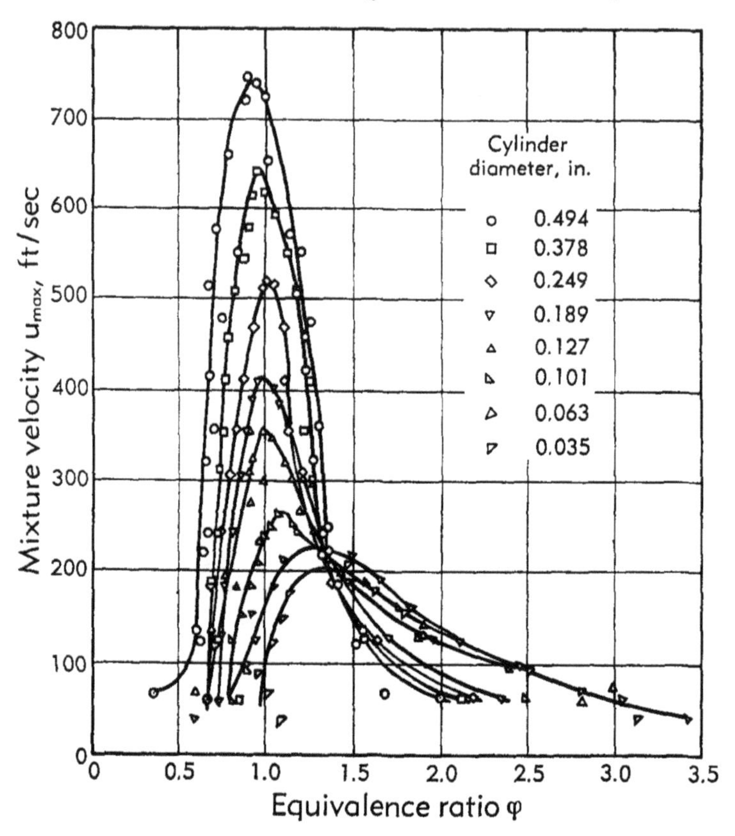


**Fig. E,1.** Stability limits of flames burning from cylindrical rods of various sizes [*1*]. Fuel, hydrocarbon blend (commercial paint thinner); inlet temperature, 150°F; static pressure, approximately 1 atmosphere.

into the mechanism of flame stabilization. Such data are of obvious importance to the combustor designer. Before considering these flame stabilization data, it is advantageous to first examine the flow patterns about bluff bodies.

## E,2. Flow about Bluff Bodies.

*Isothermal flow.* Consider first the case of isothermal flow about a cylinder with its main axis normal to the flow [*2,3*].

As the fluid diverges to pass the cylinder, the pressure on the cylinder surface decreases, and, as the fluid converges on the downstream side, the pressure rises to its original stagnation value. The boundary layer formed on the upstream side of the cylinder flows around the cylinder until a rising pressure is encountered. At this point, because of the dissipation of its directed energy by viscosity, the boundary layer separates from the cylinder. A few cylinder diameters downstream of the separation point, the wake fluid begins to curve back to form a pair of vortices. At Reynolds numbers between 1.0 and about 30, a stable vortex pair exists in the wake of the cylinder. The flow is completely laminar, and the low velocity wake is quite small. For Reynolds numbers between 30 and 150, the boundary layer and the separated boundary layer (or vortex layers) are still laminar. However, the vortex layers roll up into individual vortices which shed from the cylinder and pass downstream in two parallel, stable rows (a Kármán vortex street). The region directly behind the cylinder is essentially unstable and consists of a recirculation zone induced by vorticity leaving the cylinder before the rolling-up process is complete.

As the Reynolds number is increased above 150, vortices continue to shed from the cylinder, but a transition to turbulent flow in the vortex layers now occurs at some distance downstream of the cylinder. With further increase in the Reynolds number, this transition region moves upstream toward the cylinder. For Reynolds numbers near $10^4$, the transition to turbulent flow begins at the point where the flow separates from the surface of the cylinder. For Reynolds numbers greater than $2 \times 10^5$, the transition to turbulence precedes the boundary layer separation.

In general, this picture of the flow about circular cylinders also applies to isothermal flow about other bluff bodies, such as spheres and disks perpendicular to the air stream. However, the values of Reynolds number at which the various transitions occur are affected by the geometry of the body. For example, the transition to turbulence in the vortex layers begins at a Reynolds number of about $3 \times 10^4$ for spheres, as contrasted to $1.5 \times 10^4$ for cylinders [*3*].

A typical isothermal flow pattern around a V-gutter flame holder is shown in Plate E,2. The photograph was obtained using flash photography and balsa dust as flow tracers [*4*]. The low velocity recirculation zone is apparent in the photograph. The Kármán vortex street is also visible.

A more detailed discussion of isothermal flow about bluff bodies is available in IV,B.

*Flow with flame present.* The point of boundary layer separation from a cylinder does not appreciably shift due to heat addition, as shown by pressure measurements about the cylinder surface with and without a flame present [*5*, p. 733].

Laminar flame fronts have been observed [*6*] in flames burning from bluff-body flame holders for Reynolds numbers well above 100. Evidently, the combustion process is responsible for delaying the transition to turbulent flow that occurs in the vortex layers with isothermal flow. Due to conduction and mixing between the hot gases in the recirculation zone and the gases in the vortex layers, the temperature of the vortex layer will increase with distance from the separation point [*2*]. This increase in temperature can be sufficient to reduce the local Reynolds number by a factor of 10 to 20, since $Re \sim T^{-1.7}$ due to density and viscosity effects. Consequently, if heating of the vortex layers is sufficiently rapid to appreciably increase their temperature before transition occurs, the vortex layers may be stabilized for large distances downstream of the flame holder.

As the Reynolds number is increased, however, the transition region in the vortex layers moves toward the flame holder despite the effects of heating. Furthermore, because of the quenching action of the flame holder, heating of the gases close to the flow separation point will be small. Finally, therefore, when the Reynolds number is sufficiently high to insure transition to turbulence near the separation point with isothermal flow, the flame surface might be expected to become turbulent regardless of heating effects. From this, Zukoski [*2*] concludes that transition from a laminar to a turbulent flame front would be expected at Reynolds numbers of about $10^4$, since this is the value where flow transition occurs at the separation point with isothermal flow.

At low Reynolds numbers a stable vortex pair exists in the wake of a cylinder on which a flame is seated [*6,7*]. At higher Reynolds numbers, however, the flow in the flame holder wake becomes unstable, as shown by the work of Nicholson and Field [*8*, pp. 44–68]: small particles of sodium acetate were placed in the air stream. In the flame zone these particles emitted light which was recorded by high speed motion pictures. Many of the particles in the flame zone were observed to travel upstream toward the flame holder at velocities as high as one third of the approach stream velocity. At 200-ft/sec stream velocity with a $\frac{3}{4}$-in. plate flame holder, this flow reversal was observed not only in the center of the flame holder wake, but also near its edges. It was therefore concluded that a stable pattern of eddies did not exist in the wake of the baffle. It appeared that large eddies of varying direction of rotation existed and that eddy-shedding probably occurs.

During combustion, flow recirculation has been observed to take place in the flame holder wake up to a distance of 3 to 10 times the flame holder width [*6*; *8*, pp. 44–68; *9*]. The recirculation-zone length increases with the fuel-air ratio and varies considerably with the flame holder shape [*9*].

The flow patterns about a typical flame holder are shown in Fig. E,2. The length of each arrow indicates the magnitude of velocity at that point, while the direction of the arrow indicates the flow direction. The heavy line in Fig. E,2 indicates the location of the flame front, which is drawn similar to those shown by photographs of flames burning from various flame holders [*6*; *8*, pp. 44–68; *10*]. Velocities are low in the region of recirculatory flow behind the flame holder. Velocity gradients exist within the recirculatory region and very large gradients occur at the interface between the recirculatory region and the main stream [*6*]. Such velocity gradients will generate a high level of turbulence in the flame zone. This turbulence provides a highly wrinkled flame front and increased flame-spreading rate. Farther downstream from the flame holder the flow recirculation ceases and the high temperature gases generated by the flame accelerate until their velocity exceeds that of the surrounding unburned mixture. In the region where the velocity of the burned gases is about the same as that of the main stream, no additional turbulence will be generated; however, the turbulence generated upstream

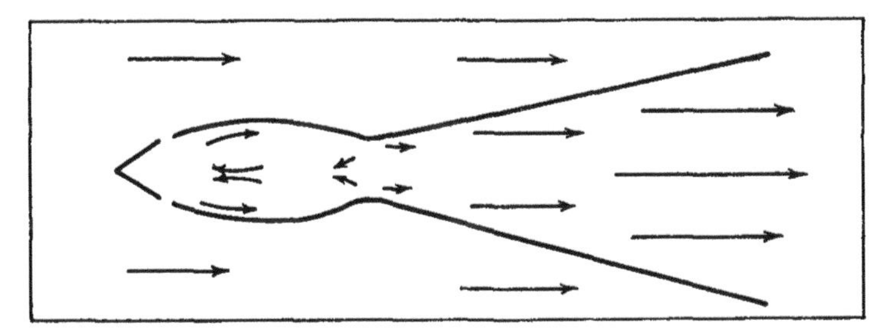

Fig. E,2. Diagram of flow about V-gutter with flame present.

may persist and provide a high flame-spreading rate in this region. In the region where the velocity of the burned gases exceeds that of the surrounding gases, a second region of high velocity gradients is created. In this region additional turbulence is generated and the flame-spreading rate is increased.

A detailed discussion of the mechanism of flame propagation into the main stream is beyond the scope of this section, and the discussion that follows will therefore be restricted to the phenomena occurring in the sheltered wake of the flame holder.

## E,3. Effect of Flow and System Variables on Flame Stabilization.

*Flame holder size.* As shown in Fig. E,1, an increase in flame holder size results in wider flame stability limits. Several investigators have found that the data for various flame holder sizes can be correlated by plotting $u_{max}/D^n$ as a function of the fuel-air mixture composition (where $u_{max}$ is the blowoff velocity and $D$ is the characteristic dimension of the flame holder). For circular rod flame holders with their main axis normal to the stream, Scurlock [6] found that a value of 0.45 for $n$ correlated the data, as shown in Fig. E,3a.

In the relation $u_{max} \sim D^n$, most of the other studies (e.g. [1,2,11]) have yielded a value of approximately 0.5 for $n$. However, three exceptions have been noted; one exception is that of [12], where a value of $n = 0.85$ was obtained. Zukoski [2] has satisfactorily explained this discrepancy by showing that a distinct change in the relation between blow-off velocity and Reynolds number occurs near $Re = 10^4$. Fig. E,3b from

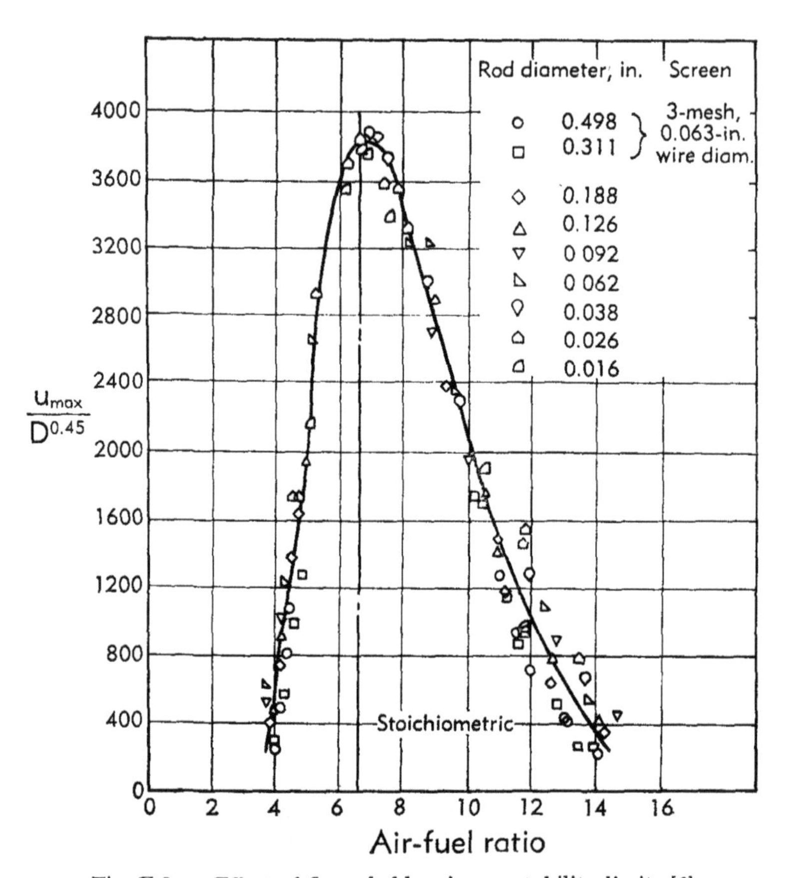


Fig. E,3a. Effect of flame holder size on stability limits [6].

[2] shows the data of several investigators plotted so as to demonstrate this transition. The high value of $n$ was obtained in [12] because $n$ was determined from a single curve faired through all the data; for those data above the transition Reynolds number a value of $n \cong 0.5$ is obtained.

The transition in Fig. E,3b occurs in the Reynolds number regime where the transition to turbulence occurs at the flow separation point with isothermal flow (see Art. 2). The transition in Fig. E,3b may therefore be attributed to a change from a laminar to a turbulent flame front; this has been verified by schlieren photographs of the flame [*2*].

The photographs of [*2*] clearly show a transition from laminar to turbulent flames as flame holder size is increased. For cylindrical flame holders having diameters less than 0.1 in., the flame front is laminar for at least 30 diameters downstream at all Reynolds numbers at which the flame holders will operate. For very high Reynolds numbers with these small flame holders, the flame is residual; that is, a flame exists in the sheltered wake of the flame holder but does not propagate into the free stream gases. Under such conditions, the nonburning flow downstream from the residual flame forms a turbulent wake, which quickly develops into a typical Kármán vortex street.

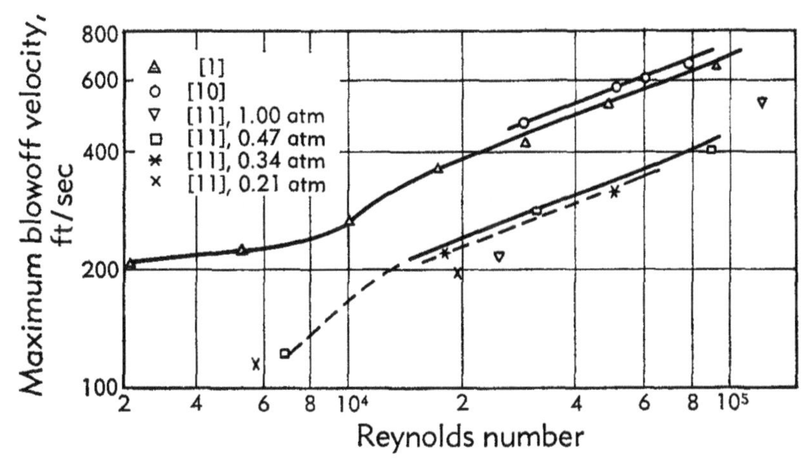


Fig. E,3b. Variation of maximum blowoff velocity with Reynolds number [*2*].

For flame holder diameters about $\frac{1}{8}$-in., the flame fronts are laminar at only the lowest Reynolds numbers [*2*]. As the Reynolds number increases the flame structure becomes progressively more granular. At the higher Reynolds numbers a complete transition has occurred to give a condition reasonably interpreted to be a turbulent flame front.

Other exceptions to the relation $u_{max} \sim D^{0.5}$ are found in [*8*, pp. 40–44; *13*]. In both of these cases $u_{max} \sim D^{1.0}$. With these two exceptions, all available blowoff data at Reynolds numbers above the transition regime ($\text{Re} \cong 10^4$) of Fig. E,3b show $u_{max} \sim D^{0.5}$ approximately. $\text{Re} < 10^4$ at blowoff is only encountered with a flame holder dimension below about 0.1 in., which is far below the sizes used in practical combustors.

The flame holders used in the investigation of [*8*, pp. 40–44] were long cylinders mounted with the main axis parallel to the flow, thus providing an appreciable dimension in the direction of flow. The Reynolds numbers based on the distance from the leading edge of the flame holders therefore became quite large before reaching the flow separation point.

Consequently, a transition to turbulence in the flame holder boundary layer probably occurs upstream of the flow separation point in this case. Zukoski and Marble [*14,15*] suggest that this phenomenon accounts for the different dependence of $u_{max}$ on $D$ in [*8*, pp. 40–44]. If this explanation is correct, then it would be expected that $u_{max} \sim D^{1.0}$ for all cases where the Reynolds number is large enough to produce transition to turbulence ahead of the point of flow separation. This would include small size flame holders having an appreciable dimension in the direction of flow and all very large flame holders regardless of the shape of their leading edges. By the same token, if the Zukoski and Marble thesis holds, then an extension of the curve of Fig. E,3b to Reynolds numbers well above $10^5$ should show another transition and a steeper slope to the curve at very high Reynolds numbers. In Art. 4, some additional data will be presented which support Zukoski and Marble's arguments.

The peaks in the blowoff velocity curves occur at mixtures near stoichiometric for the larger flame holders of Fig. E,1b. For the small-diameter rods, however, the peaks in the curves shift to fuel-air ratios greater than stoichiometric. This same trend has also been observed in [*2*] where it was also demonstrated that for methane, having a lower molecular weight than oxygen, the maximum blowoff velocity shifts toward mixtures leaner than stoichiometric for small flame holder sizes. These trends parallel those observed by Lewis and von Elbe [*7*], who showed that the minimum ignition energy was obtained at richer-than-stoichiometric mixtures for heavy fuels, and at leaner-than-stoichiometric mixtures for fuels of low molecular weight.

Lewis and von Elbe suggested that the difference in diffusion rates for the light and heavy molecules accounts for the observed effects on ignition. This same explanation also applies to the blowoff of flames from flame holders [*1,2*]. Combustible material enters the recirculation zone by transfer across the interface between the free stream gases and the recirculation zone. For low Reynolds numbers corresponding to flame holder diameters below about 0.1 in., it has been shown that this interface is laminar. Consequently, for small flame holders most of the mass will be transported into the recirculation zone by molecular diffusion. For the hydrocarbon blend used to obtain the data of Fig. E,1, the oxygen diffuses much more rapidly than the hydrocarbons because the molecular weight of oxygen is lower than the effective molecular weight of the hydrocarbon blend. This phenomenon would account for the shift of the peak velocities to mixture compositions richer than stoichiometric for the small-diameter flame holders in Fig. E,1b, since a fuel-rich mixture in the free stream would be required to produce a stoichiometric mixture in the flame holder wake in this case. This diffusion mechanism has been substantiated by the analysis of gas samples drawn from the burning zone behind the flame holder [*2*].

For the larger flame holders the flow is turbulent in the interface between the hot and cold gases. Here the principal mass transfer is accomplished by turbulent diffusion. Fuel and oxygen are therefore diffused at equal rates and the peak blowoff velocity occurs near stoichiometric.

*Approach stream turbulence.* Scurlock [*6*] has studied the effects of approach stream turbulence by placing screens upstream of the flame holder. Fig. E,3c shows the effect of changing turbulence from 0.4 per cent and a scale of 0.013 in. to an intensity of 2.7 per cent and a scale of 0.061 in. For the small flame holder the effect of increased scale and intensity was large but for the larger flame holder the effect was very small.

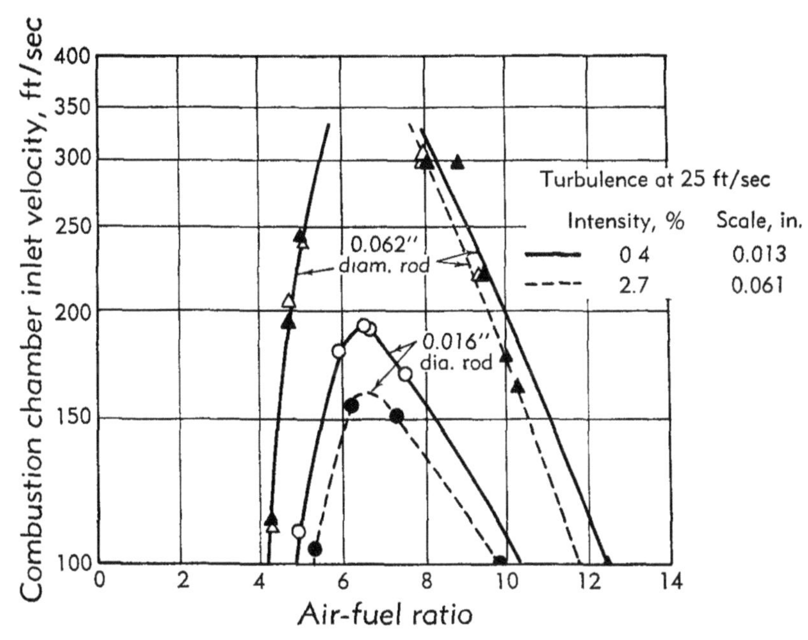


Fig. E,3c. Effect of approach stream turbulence on stability limits [*6*]. Fuel, Cambridge (Mass.) city gas; inlet temperature, 80–125°F; pressure, 1 atmosphere.

For flame holders of the size used in most afterburners ($D = 1.5$ in., or larger) there is probably no measurable effect of approach stream turbulence, although large swirls or nonuniformities in the flow profile would be expected to have an effect.

*Pressure.* The investigations of DeZubay [*12*] and of Weir, Rogers, and Cullen [*16*] have both shown that flame stability is improved by an increase in pressure. Fig. E,3d presents a correlation [*12*] showing the effect of baffle diameter, pressure, and mixture velocity using propane as fuel. Here, the fuel-air ratio at blowout is plotted as a function of the dimensional group $u_{max}/D^{0.85}p^{0.95}$. The flame was stabilized on circular flat disks. As previously noted, an exponent of 0.5 on $D$ provides a better correlation of the data for Reynolds numbers above $10^4$. It also appears that an exponent of about 0.9 on pressure provides a better correlation if only the data for Re > $10^4$ are used. In [*16*] no single exponent on pressure correlated all the data; the required exponent varied with $D$ and $f$. Therefore, considerable uncertainty exists regarding the proper relation between pressure and blowoff velocity.

Two different theories have been advanced to account for the effect of pressure on flame stability limits. One theory [*13*] suggests that the effect of pressure is due to the increased ignition lag accompanying a decrease in pressure. The other theory [*17*] suggests that the observed pressure effect is the result of its influence on the chemical reaction rate

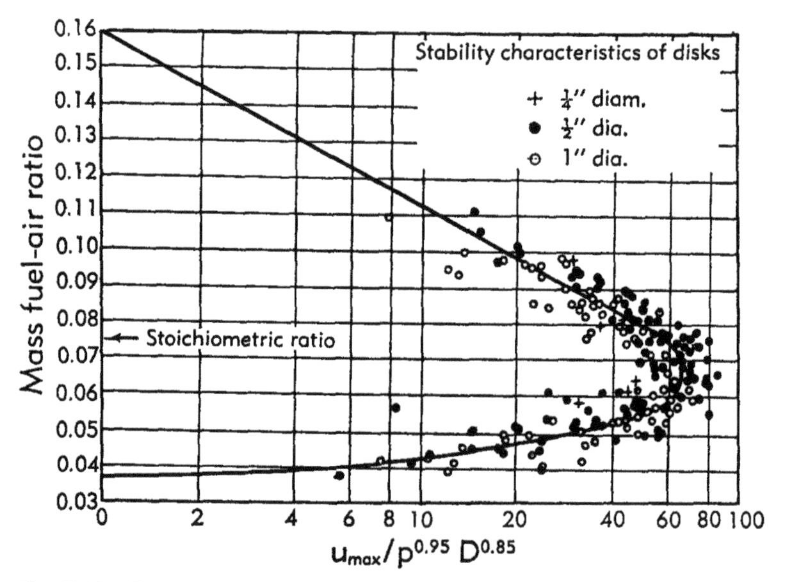


**Fig. E,3d. Effect of pressure on stability limits [*12*]. Disk flame holders. Fuel, commercial propane. Units: u, ft/sec; p, lb/in.²; D, in. Inlet temperature, 80°F.**

in the hot gases in the flame holder wake. Both of these theories are discussed in more detail in Art. 4.

*Temperature.* Several investigations have been made of the effect of inlet temperature on flame stability limits [*1*; *8*, pp. 40–44; *11*]. Haddock [*1*] showed that the maximum velocity for stable burning was proportional to the inlet temperature to the 1.2 power. This same relation also correlated the data of [*11*]. Typical data from Haddock's investigation are shown in Fig. E,3e.

The beneficial effects on flame stability accompanying an increase in inlet temperature are not surprising. For one thing, the increased initial temperature brings the incoming mixture closer to its ignition temperature, thereby reducing the required heat and mass transfer from the recirculation zone. Also, if the flame speed is important in the stabilization process, then the increased flame speed [*18*] could also account, in part, for the observed effect.

*Flame holder shape.* As previously mentioned, flame holders having an appreciable dimension in the flow direction will produce transition to turbulence in the boundary layer upstream of the flow separation point. For such flame holders, $u_{max} \sim D$, approximately. For flame holders not having a long dimension in the flow direction, such as rods and disks normal to the stream, $u_{max} \sim D^{0.5}$, approximately. A comparison of the work of Longwell [*8*, pp. 40–44] with that of DeZubay [*12*] shows that DeZubay could maintain stable flames to velocities considerably higher for comparable operation conditions and flame holder sizes. DeZubay used disks as flame holders. Longwell used flame holders that might be

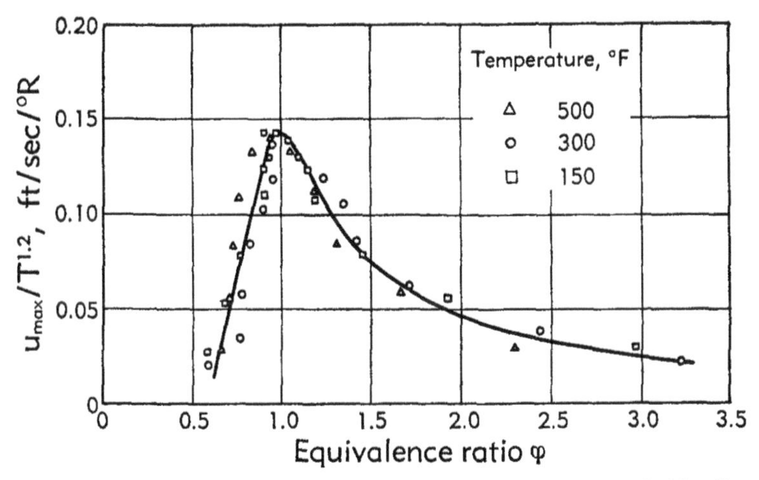


Fig. E,3e. Effect of inlet temperature on stability limits [*1*]. Flame holder diameter, 0.127 in.; fuel, hydrocarbon blend (commercial paint thinner); pressure, 1 atmosphere.

expected to induce boundary layer turbulence (cylinders with main axis parallel to flow). The lower velocity limit observed by Longwell might therefore be attributed to the change to turbulence in the flame holder boundary layer.

Flame holders having only a small dimension in the flow direction (closed- and open-edged gutters, flat plates, and cylindrical rods) all provide about the same stability limits [*6*]. In another study [*8*, pp. 40–44] little difference was found for another series of flame holder, all of which had an appreciable dimension in the flow direction, except when the trailing edge was streamlined. Streamlining reduced the flame stability limits, as would be anticipated because of the reduced size of the recirculation zone.

The stability of a two-dimensional flame holder (long cylinder normal to flow) appears to be somewhat greater than for a three-dimensional flame holder (sphere) of the same characteristic dimension [*19,20*].

Obstacles of the same size with different shapes are known to produce vortices of different sizes. It would therefore be expected that some parameter better than $D^n$ might exist for indicating recirculation zone size. Mestre [*13*] shows that an improved correlation of the stability limits for different flame holder shapes is obtained by using the product

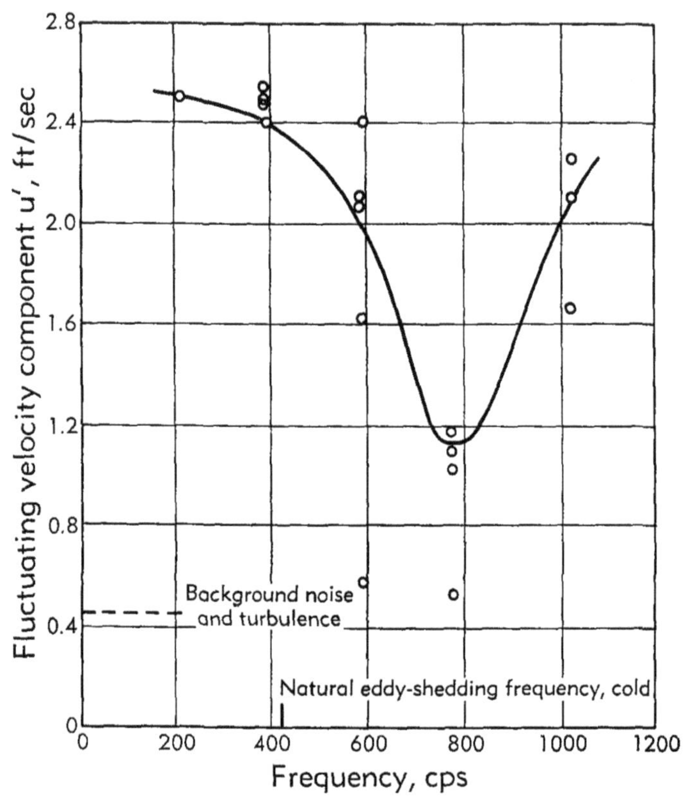


Fig. E,3f. Disturbance velocities required to produce blowoff at a fixed velocity and fuel-air ratio [*22*]. $u_0$, 50 ft/sec; 0.306-in. flame holder; fuel, propane.

of the characteristic dimension and the flame holder drag coefficient, rather than the characteristic dimension alone. Hence

$$\frac{u_{\max}}{(DC_d)^n} = \varphi(f) \tag{3-1}$$

*Flame holder temperature.* An increase in flame holder temperature produces an increase in the flame stability limits [*5*, p. 743; *6*; *8*, pp. 21–40; *21*]. This is to be expected, since the increased heat transfer to the gases in the recirculation zone produces a higher wake-gas temperature. However, the magnitude of the change in stability limits with flame holder temperature depends on the flame holder size and is not large in some cases [*1*,*21*].

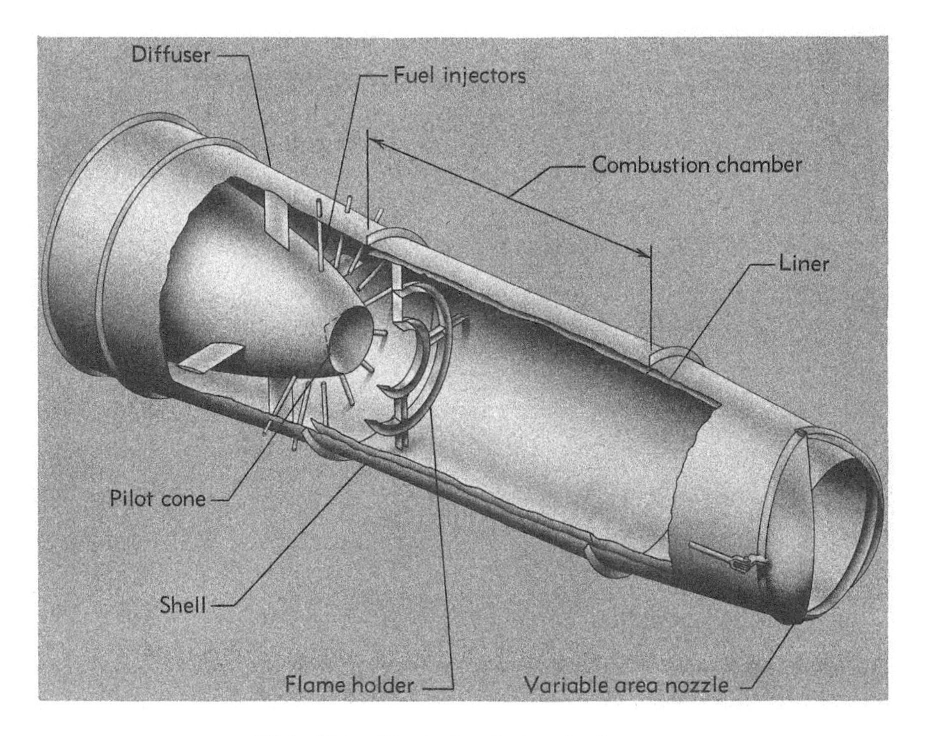


Plate E,1. Typical turbojet afterburner.

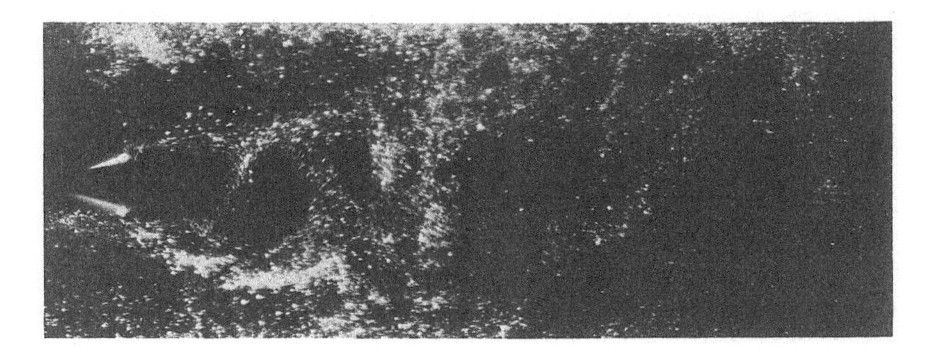

Flow →

Plate E,2. Balsa dust flow picture of isothermal flow about V-gutter.

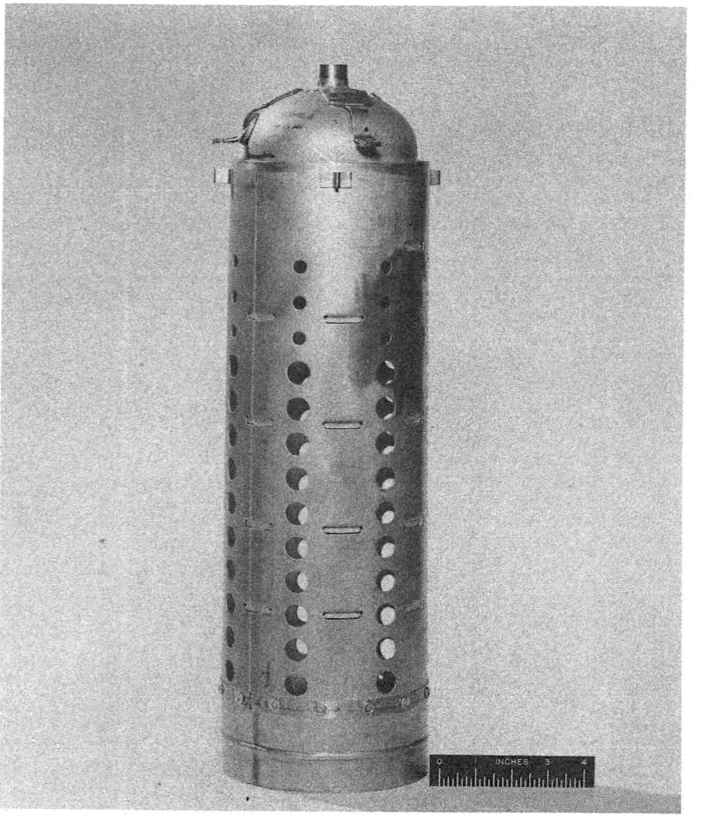

Plate E,9. J33 combustor liner.

A higher flame holder temperature causes the flame front to remain laminar for a greater distance downstream of the flame holder [*2*]. This trend is to be expected because the increased flame holder temperature serves to increase the heat transfer into the vortex layer. The delay in the transition to turbulence in the vortex layer is then due to the same effects as those previously discussed for adiabatic flow as contrasted to isothermal flow about the flame holder. However, for flame holder temperatures as high as 2000°F, the transition Reynolds number is increased by only 20 per cent.

*Acoustic disturbances.* Blackshear [*22*] studied the effect of acoustic disturbances on the flame-spreading rate and flame stability. A 60-degree, V-gutter flame holder was attached to the downstream end of a long, rectangular-sectioned chamber which was connected through a tube to a piston-type speaker. The sound generated by the speaker issued from a slot just upstream of one edge of the flame holder. The imposed disturbances were found to narrow the stability range slightly. The reduction in the stability range varied with the frequency of the imposed disturbance. Fig. E,3f shows the disturbance amplitude required to cause flame blowoff for various disturbance frequencies. Minimum disturbance amplitude was required at a frequency of 780 cps. The data of Fig. E,3f are for a velocity of 50 ft/sec and a fuel-air ratio that is but slightly above the lean blowoff limit (0.042) at 50 ft/sec. At a slightly higher fuel-air ratio of 0.047, blowoff was not induced with speaker inputs as much as 10 times the values employed to produce the amplitudes shown in Fig. E,3f.

In practically all cases investigated [*22*], imposed disturbances resulted in an increase in the heat release rate in the combustor. When the flame holder was located a considerable distance from the open exhaust end of the duct, the flame would become noisy even without disturbances being introduced from the speaker. In general, these "rough-burning" flames would show an increased heat release rate similar to that for imposed disturbances. An exception was noted, however, near the lean blowoff limit. Here the flame began to emit a low frequency noise at about 85 cps, and the heat release rate dropped abruptly. Shadowgraph pictures of the flame at these conditions showed that the flame was intermittently extinguished in the downstream portion of the duct.

The frequency of disturbances such as these would be expected to depend on the duct length downstream of the flame holder. It has been observed [*6*; *8*, pp. 44–68; *23*] that one of the dominant frequencies of pressure pulsation can usually be correlated with the calculated frequency, assuming the duct length to be one half the wavelength.

The range of operating conditions over which rough burning occurs is also affected by the duct length [*5*, p. 733; *22*]. Such phenomena undoubtedly account for some of the differences in blowoff limits obtained by different investigators.

*Area blockage by flame holder.* The velocity past the edge of the flame holder is the significant velocity that determines flame stability. The duct width is therefore important because the fraction of the duct area that is blocked by the flame holder determines the velocity past the edge of the flame holder. Duct width has been varied experimentally [*5*, p. 733] and flame blowoff demonstrated to occur at the same value of $u$ regardless of the differences in $u_0$.

The stability parameter $u/D^n$ must be low enough to lie within the stable operating range of the flame holder, and the lower the value of $u/D^n$, the wider the stable operating range of the fuel-air ratios. This parameter first decreases and then increases as the flame holder width $D$ is increased progressively for a fixed duct size. The increase in $u/D^n$ at high values of $D$ occurs because the area blockage becomes sufficient to cause a large increase in velocity past the flame holder for a fixed duct size and fixed approach stream velocity. For one-dimensional, incompressible flow

$$u = u_0\left(\frac{A_0}{A_0 - A_b}\right) = u_0\left(\frac{1}{1 - A_b/A_0}\right) \tag{3-2}$$

Then for any given inlet pressure the stability parameter $u/D^n$ can be expressed

$$\frac{u}{D^n} = \frac{u_0}{D^n}\left(\frac{1}{1 - A_b/A_0}\right) \tag{3-3}$$

To obtain the flame holder blockage ratio $A_b/A_0$ giving the optimum (minimum) value for the stability parameter, solve for

$$\frac{d(u/D^n)}{d(A_b/A_0)} = 0$$

Two solutions are obtained depending on the general shape of the flame holder elements. For flame holders giving $A_b/A_0 \sim D$, such as V-gutters, long cylinders, or similar shapes arranged with their main axis normal to the flow,

$$\text{optimum}\,\frac{A_b}{A_0} = \frac{n}{n+1} \tag{3-4}$$

For this case, optimum $A_b/A_0 = \frac{1}{3}$ for $n = \frac{1}{2}$ and optimum $A_b/A_0 = \frac{1}{2}$ for $n = 1$.

For flame holders giving $A_b/A_0 \sim D^2$, such as cones, spheres, disks, etc,

$$\text{optimum}\,\frac{A_b}{A_0} = \frac{n}{n+2} \tag{3-5}$$

For this case, optimum $A_b/A_0 = \frac{1}{5}$ for $n = \frac{1}{2}$ and optimum $A_b/A_0 = \frac{1}{3}$ for $n = 1$.

Fig. E,3g shows relative values of the flame stability parameter for various blockage ratios and different flame holder shapes. Since one-dimensional, incompressible flow equations were used in the analysis, the curves of Fig. E,3g are not exact. However, the location of the minimum points on the curves would be changed very slightly by allowing for compressibility.

The curves of Fig. E,3g make it obvious that the afterburner designer must give due consideration to the flame holder area blockage as well as flame holder width. Since V-gutter flame holders are used in most current

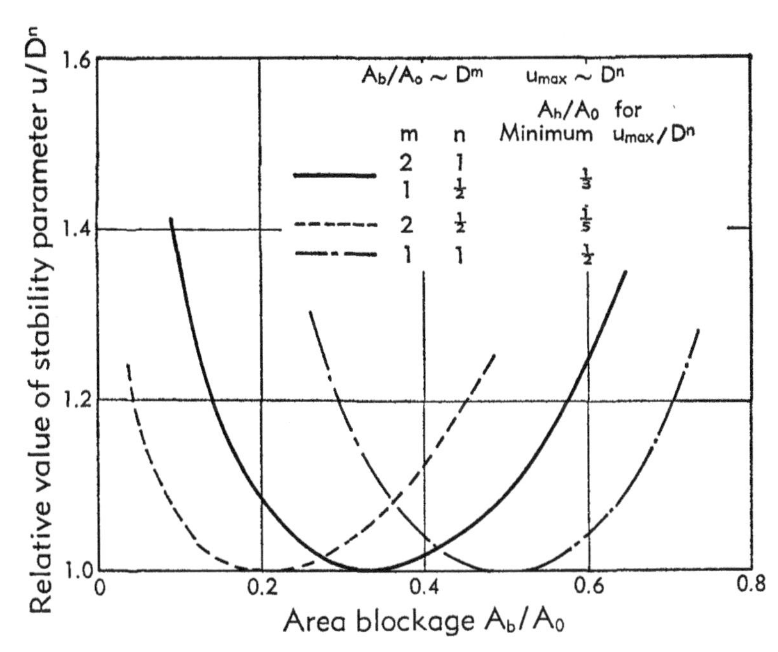


Fig. E,3g. Effect of flame holder area blockage on flame stability parameter. Constant duct width and approach stream velocity.

afterburners, the curves for $m = 1$ in Fig. E,3g are the ones that apply to these burners. The factors that must be considered in the design of these afterburners are as follows:

1. The spacing between the flame holder elements determines the burner length required; the greater the spacing, the greater the burner length required to obtain closure of the flames propagating into the unburned mixture from adjacent flame holders. Or, conversely, the greater the spacing of the flame holders, the lower the combustion efficiency for a fixed burner length.
2. For any given spacing of the flame holder elements, the width $D$ of these elements determines the maximum velocity and minimum pressure at which stable flames can be maintained. Knowing the most

severe velocity and pressure conditions required of the burner the designer must select $D$ so as to obtain values of $u/p^{0.95}D^n$ sufficiently low to be within the stable operating range of the flame holders. If flame stability were the only consideration, the designer would always select $D$ to give the area blockage corresponding to the minimum point on the curve of Fig. E,3g.

3. The flame holder area blockage determines the pressure losses across the flame holder. To keep pressure losses low, in some cases it may be desirable to either increase the flame holder spacing (which reduces the combustion efficiency) or to reduce the flame holder dimension below the value corresponding to the minimum point for the appropriate curve in Fig. E,3g (which reduces the operating range).

*Flame stability equation.* Combining the effects of the various flow and system variables leads to the following expression for flame stability:

$$\frac{u_{\max}}{(DC_d)^n p^{0.95} T^{1.2}} = \phi(f) \tag{3-6}$$

where $n \cong 0.18$ for $Re < 10^4$; $n \cong 0.5$ for $Re > 10^4$ and $Re <$ value necessary to cause transition ahead of the separation point; $n \cong 1.0$ for $Re$ sufficiently high to produce boundary layer transition prior to flow separation. This expression is of obvious utility to the combustor designer; it shows the increase in flame holder width that will be necessary for any increase in the severity of the operating conditions. The relation between $u_{\max}$ and the pressure is somewhat uncertain. Also the effect of inlet temperature has not been investigated at the high temperatures encountered in afterburners and in high Mach number ramjets. Additional research is therefore needed to verify the flame stability relation.

## E,4. Theories of Flame Stabilization.

*Stabilization controlled by rate of heat flow from recirculation zone into fresh gas.* A number of investigators have described the mechanism by which a flame is stabilized on a bluff body; among the earliest were Williams, et al. [*6*; *8*, pp. 21–40; *24*], who suggested a mechanism as follows: On a thermal basis alone, the quantity of energy required from the flame holder would be increased by (1) increasing the velocity $u$ past the flame holder, (2) increasing the difference between the ignition temperature and the approach stream temperature $(T_{ig} - T_0)$, (3) increasing the thickness of the mixing (preignition) zone, or (4) increasing the heat capacity per unit volume. If, as in [*6*], the thickness of the mixing zone is assumed proportional to the thermal diffusivity $k/\rho C_p$ and inversely proportional to the laminar flame velocity $S$, then the heat supply rate required to raise the mixing zone gas to its ignition temperature can be written as

$$H_{req} \sim \frac{uk}{\rho C_p S} \rho C_p (T_{ig} - T_0) \tag{4-1}$$

The heat supplied from the recirculation zone is given by

$$H_{avail} \sim \left[ \left( \frac{Du}{\nu} \right)^c \frac{k}{D} \right] D(T_f - T_0) \tag{4-2}$$

where $c$ is a constant and where the bracketed term includes the factors affecting the heat transfer coefficient from heated cylinders. In a two-dimensional duct, the area for heat transfer is proportional to the rod diameter $D$.

At blowoff, the heat required is infinitesimally greater than the heat supplied from the recirculation zone; or $H_{req} = H_{avail}$ when $u = u_{max}$. This leads to the relation,

$$\frac{u_{max}}{D^n} = \phi(f) \tag{4-3}$$

where the new exponent $n$ on the rod diameter replaces $c/(1 - c)$. This relation between $u_{max}$ and $D$ agrees with the experimental data correlations of Art. 3.

*Stabilization controlled by dwell time of fuel-air mixture in layer adjacent to recirculation zone.* Zukoski and Marble [*15*] suggest that the dwell time of the fuel-air mixture in the mixing zone is the factor that determines flame stability. They present the following model for the flame stabilization process: The incoming mixture flows past the flame holder and enters the mixing zone that lies between the unburned gases and the burned gas in the recirculation zone. If the velocity is low enough and the temperature high enough in the mixing zone, then the incoming mixture will be ignited before flowing past the end of the recirculation zone. This burning material then forms a propagating flame as it is swept downstream. However, if the velocity past the flame holder is increased, a condition will be reached where the material in the mixing zone will not be ignited before reaching the end of the recirculation zone. For such a condition, as the gas flows downstream from the end of the recirculation zone, further mixing with cool gas will quench any chemical reactions started in the mixing zone, and no propagating flame will be formed. Any further increase in velocity will result in much unignited material entering the recirculation zone and will cause blowoff to occur abruptly. At the blowoff condition, the time required for ignition of the mixing zone gases is equal to the dwell time of the gases in this zone; this time, given by $\tau = L/u_{max}$, is the critical time for the flame stabilization process. With the postulated blowoff mechanism, this time would be expected to be constant for a given fuel and fuel-air mixture, regardless of the flame holder shape, etc.

In support of this theory, it has been shown [*15*] that for a given fuel and fuel-air ratio the value of $\tau$ is approximately constant for a range of flame holder sizes and shapes; these data are tabulated in Fig. E,4a. In addition, measurements of recirculation zone temperature [*15*] show

| Flame holder geometry | D, in. | $\tau$, sec $\times 10^4$ |
|---|---|---|
| 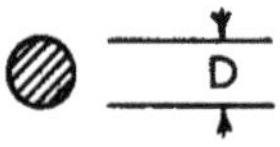 D | $\frac{1}{8}$<br>$\frac{3}{16}$<br>$\frac{1}{4}$ | 3.09<br>2.85<br>2.80 |
|  D<br>14-mesh screen | $\frac{1}{4}$ | 3.00 |
|  D | $\frac{1}{4}$<br>$\frac{3}{8}$<br>$\frac{1}{2}$<br>$\frac{3}{4}$ | 2.38<br>2.70<br>2.65<br>2.58 |
| D | $\frac{1}{4}$<br>$\frac{3}{8}$<br>$\frac{1}{2}$<br>$\frac{3}{4}$ | 3.46<br>3.12<br>3.05<br>3.03 |
| 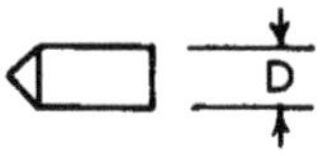 D | $\frac{3}{4}$ | 3.05 |
| 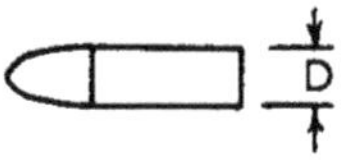 D | $\frac{3}{4}$ | 2.70 |

Fig. E,4a. Values of characteristic time, $\tau = L/u_{max}$ [*15*]. Stoichiometric fuel-air ratio.

that this temperature does not decrease by a large amount as the velocity is increased approaching blowoff. This would indicate that the recirculation zone is filled with almost completely burned material and its effectiveness as an ignition source does not decrease sufficiently to induce the blowoff.

Measurements of recirculation zone length $L$ [*15*] show that, with a flame anchored on the flame holder, $L/D^n$ is constant for Reynolds numbers above some minimum value. For long cylinders with their main axis normal to the flow, $n = \frac{1}{2}$; for cylinders with conical upstream ends and with the main axis parallel to the flow, $n = 1$. The larger value of $n$ for the latter case was attributed to transition to turbulence in the flame holder boundary layer upstream of the flow separation point. This was verified by inducing boundary layer transition with the flame holders giving $n = \frac{1}{2}$ for undisturbed flow; with induced transition a value of $n = 1$ was obtained.

Since $L \sim D^n$,

$$\frac{u_{\max}}{L} \sim \frac{u_{\max}}{D^n} \sim \frac{1}{\tau} = \phi(f) \tag{4-4}$$

for a given approach stream pressure and temperature. This is identical with Eq. 4-3 and with the correlations of the experimental blowoff data for various flame holder sizes.

The time $\tau$ is the ignition lag of the mixing zone gases. Work on the autoignition of kerosene [*25*] has shown that the ignition lag varies inversely with pressure. As suggested by Mestre [*13*], this effect of pressure on ignition lag can account for the observed dependence of flame stability on pressure:

$$\frac{u_{\max}}{D^n} \sim \frac{1}{\tau} \sim P$$

for a given fuel and fuel-air ratio. Or, with $f$ as a variable,

$$\frac{u_{\max}}{pD^n} = \phi(f) \tag{4-5}$$

This equation is almost identical with the empirical equation for correlating flame blowoff data at various pressures (Art. 3).

A theoretical analysis of ignition and combustion in a laminar mixing zone has been presented by Marble and Adamson [*26*].

*Stabilization controlled by chemical reaction rate in flame holder wake gases.* The recirculation zone behind a bluff body is assumed to be a definite volume $V$ in which a homogeneous chemical reaction occurs at steady state in the analysis made by Longwell, Frost, and Weiss [*17*]. Unburned mass feeds into the zone at a constant rate and is assumed to be instantaneously mixed with all the other material within the zone. Burned material leaves the zone at constant rate with a temperature and composition identical to that within the zone. Heat losses to the stabilizer or to the main stream are neglected. If $w_a$ represents the mass rate at which air enters the recirculation zone, $f$ represents the fuel-air ratio of the incoming mixture, and $\eta$ represents the fraction of fuel burned within the recirculation zone, then the fuel burned is $w_a f \eta$. This quantity must be equal to the fuel consumed by chemical reaction. It is assumed that combustion within the recirculation zone occurs according to a second order reaction rate equation. A second order equation is required to account for the observed effect of pressure. Since the over-all chemical reaction occurs by some unknown chain mechanism, the foregoing assumption is tantamount to assuming that some bimolecular reaction step in the over-all chain takes place at a much slower rate than all the other reaction steps. The actual reactants in this rate-determining step are unknown; however, their concentrations can be assumed proportional to the air and fuel concentrations $C_a$ and $C_f$, respectively. The collision factor $Z$ combines the collision frequency, the steric factor, and the concentration proportionality constants. Thus the material balance can be written

$$w_a f\eta = VC_aC_fZe^{-E/RT} \tag{4-6}$$

Now $C_a$ and $C_f$ can be related to the initial concentration and the initial temperature through $f$ and $\eta$. With the appropriate substitutions, Eq. 4-6 can be written as

$$\frac{w_a}{Vp^2} = \frac{Ze^{-E/RT}}{T^2}\frac{(1-\eta)(1-\eta y)}{\eta\alpha} \tag{4-7}$$

where $\mathfrak{M}$ = molecular weight of fuel; $T$ = adiabatic flame temperature corresponding to fraction burned; $y$ = fraction of stoichiometric fuel concentration if mixture is lean; $y$ = unity if mixture is rich; and

$$\alpha = \left(\frac{Rf}{\mathfrak{M}} + \frac{R}{29}\right)^2$$

If $w_a/Vp^2$ from Eq. 4-7 is plotted against $\eta$, curves such as those shown in Fig. E,4b are obtained for various concentrations, using a fixed activation energy $E$. The equation is actually triple-valued in part, but only the values indicated by the solid curves are of practical interest. A blowoff curve of $w_{a,max}/Vp^2$ against concentration can be obtained by cross-plotting the maxima of such curves against concentration. The maximum can be obtained directly by setting the derivative of Eq. 4-7 equal to zero, solving for $\eta$ by trial and error, and substituting values of $\eta$ and $T$ into Eq. 4-7 to obtain values of $w_{a,max}/Vp^2$. Blowoff curves obtained from Eq. 4-7 are presented in Fig. E,4c.

The air flow $w_a$ into the recirculation zone is proportional to the velocity $u$, to the stream density, which in turn is proportional to the static pressure $p$, and to some function of the diameter $D$. The zone volume $V$ depends almost entirely on some function of $D$. Hence,

$$\frac{w_{a,max}}{Vp^2} \sim \frac{u_{max}}{D^n p} \tag{4-8}$$

which has the same form as the correlating factor used by DeZubay [*12*]. Longwell [*17*] has plotted the data of DeZubay in accordance with Eq. 4-7 and 4-8; this is presented in Fig. E,4d.

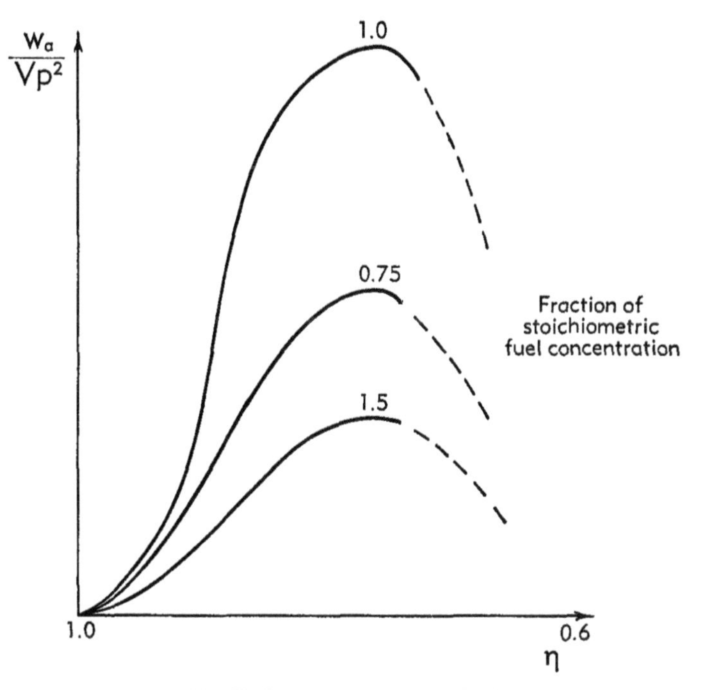


Fig. E,4b. Plot of Eq. 4-7 [*17*].

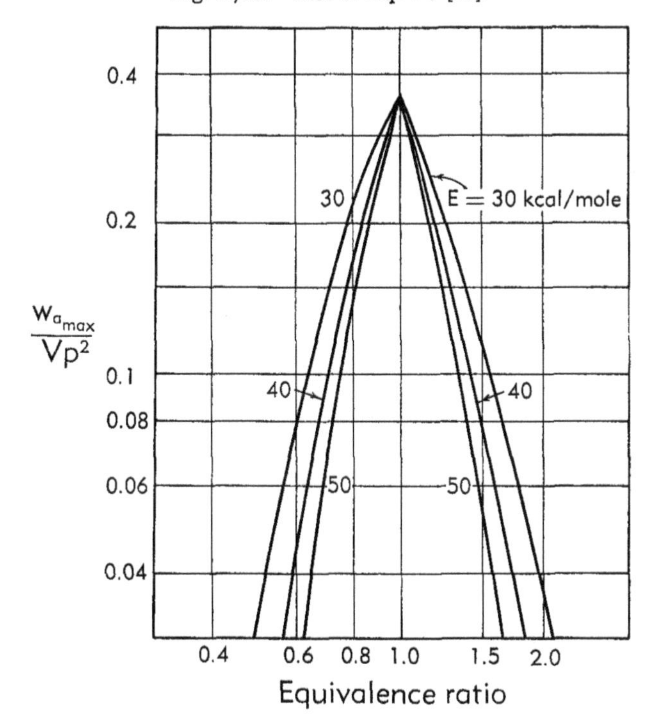


Fig. E,4c. Blowoff curves from Eq. 4-7 [*17*].

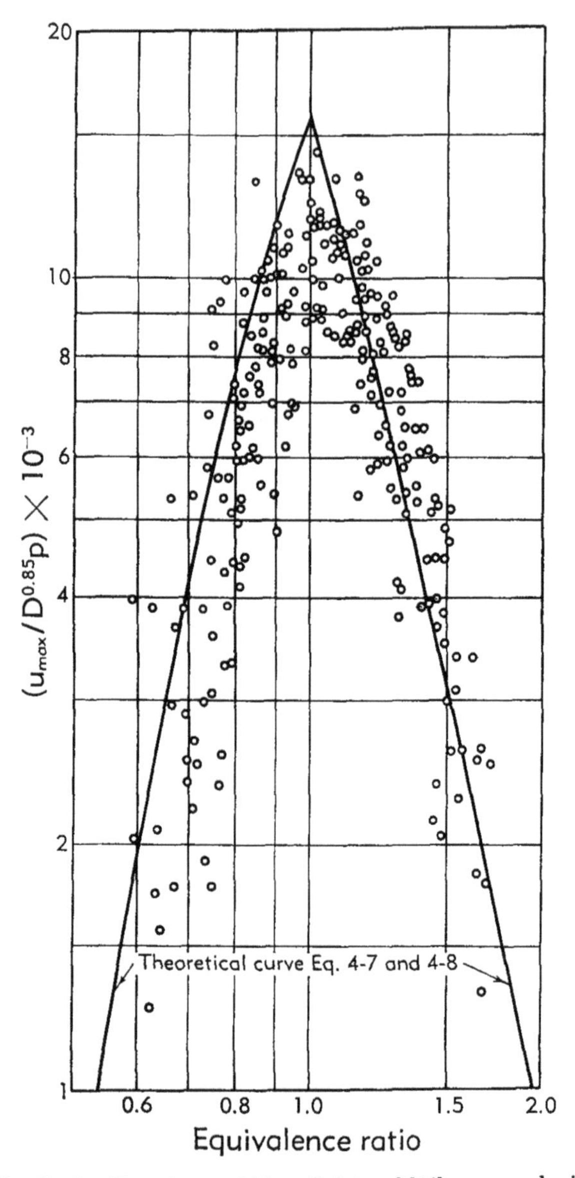


Fig. E,4d. Experimental blowoff data of [*12*] compared with theoretical curve for $E$ = 40,000 cal/mole [*17*].

The curve in Fig. E,4d was calculated from Eq. 4-7 and 4-8; however, the collision factor $Z$ and the energy of activation $E$ were selected to provide the best fit of the experimental data. The required value of $Z$ was $4 \times 10^{12}$ $cm^3$/gram mol and the required value of $E$ was 40 kcal/gram mol. These values are reasonable when compared with values obtained from classical kinetics.

*Discussion of the various theories.* The three different theories are similar in their gross aspects. All agree that the hot gases in the recirculation zone serve as the heat source to initiate combustion in the incoming charge. The first theory attributes blowoff to insufficient heat (and mass) transfer to raise the mass of gas in the mixing zone to its ignition temperature; the role of velocity is accounted for by its effect on the quantity of fresh gas entering the mixing zone and on the heat (and mass) transfer rate between the recirculation zone and the mixing zone. The second theory (Zukoski and Marble) changes this picture somewhat and attributes the velocity role entirely to its effect on the dwell time of the gases in the mixing zone. The third theory (Longwell, et al.) suggests that the rate of chemical reaction (hence, heat generation) in the recirculation zone becomes insufficient; the role of velocity is attributed to its effect on the rate of mass flow into the recirculation zone.

The mechanisms assumed for the chemical and heat transfer processes in the flame holder wake in the second and third of these theories appear markedly different upon first consideration. The second theory assumes that considerable time (ignition lag) elapses while the gases are heated up to their ignition temperature, after which reaction is essentially instantaneous. The third theory assumes instantaneous mixing, followed by reaction at a finite rate. However, these two mechanisms are essentially governed by the same basic laws, the laws of chemical kinetics, if one considers that the ignition lag is a reaction rate phenomenon.

Williams and Shipman [*5*, p. 733] point out that flame blowoff has been observed [*5*, p. 733; *6*; *8*, pp. 44–68] to first take place at some point downstream from the flame holder; the point of flame termination then moves toward the flame holder until no flame remains. In addition, flash pictures [*5*, p. 733] show that, near the blowoff limit, flames exhibit striations in the center of the flame near the downstream end of the recirculation zone. These striations are believed to be due to unburned or incompletely burned gases. This evidence would indicate that decay of the propagating flame occurs first and this is followed by a depletion in the supply of hot combustion gases that recirculate from the downstream region into the flame holder wake. Flame blowoff would then follow the decay of the propagating flame. This lends support to the mechanism proposed in the second theory.

The strongest evidence in support of the second theory, however, lies in the data of Zukoski and Marble [*15*]. As previously noted: (1) $\tau = L/u_{max}$ is nearly constant for a variety of flame holder shapes and sizes: (2) $L \sim D^{1/2}$ for those flame holders giving $u_{max} \sim D^{1/2}$; (3) $L \sim D$ for those flame holders giving $u_{max} \sim D$; and (4) less change occurs in recirculation zone temperature near blowoff than might be expected with the third theory. (This latter point is also verified by Westenberg, et al. [*9*].)

# CHAPTER 2. FLAME STABILIZATION BY BLUFF BODIES IN AIR-FUEL SPRAY MIXTURES

**E,5. Introduction.** The preceding chapters have dealt exclusively with flame stabilization in premixed vapor fuel-air mixtures. This corresponds to the actual case in most turbojet afterburners, such as the one depicted in Fig. E,1a. Afterburner inlet temperatures range from 1050 to 1250°F in most current turbojets, and these temperatures are well above the final boiling point of jet fuel. In most afterburners the fuel injectors are located 15 inches or more upstream of the flame holders. This distance is sufficient to insure complete vaporization of the fuel before it reaches the flame holders.

However, in some combustion systems, considerable quantities of liquid fuel may persist in the mixture approaching the flame holder. This would be the case for afterburners having a very short distance between fuel injectors and flame holders and for some ramjets where the inlet temperature may be below the final boiling point of the fuel at some flight conditions.

Where substantial amounts of liquid fuel are found in the approach stream, the mechanism of flame stabilization may be quite different than that previously discussed for gaseous mixtures. The only extensive study of flame stabilization in air-fuel spray mixtures was made by Hottel, Williams, May, and Maddocks [*5*, p. 715], and practically all of the ensuing discussion is taken from their paper.

**E,6. Mechanism of Flame Stabilization.** With all or most of the fuel in the liquid state, the recirculating eddy in the downstream side of a flame holder is the source of ignition on which flame stability depends, just as for the case of gaseous fuel. With liquid fuel, however, the fuel-air ratio in the recirculation zone may differ from that of the approach stream by a greater amount than for the case of gaseous fuel.

Stability limits for Diesel fuel sprays uniformly dispersed in air are presented in Fig. E,6a [*5*, p. 715]. The hollow tube flame holder could be electrically heated. Fig. E,6b shows a cross plot of these same data on the coordinates used in preceding figures. Fig. E,6a and E,6b show that a stable flame can exist at far lower fuel-air ratios than are possible with homogeneous gaseous mixtures. It is therefore obvious that the fuel-air ratio in the flame holder wake must be considerably higher than in the approach stream for at least some of the operating conditions.

The mechanism of flame stabilization is illustrated in Fig. E,6c. Liquid fuel droplets in the approach stream collect on the flame holder as indicated in the diagram. The larger the droplets, the greater their tendency to collect on the flame holder; very small droplets are inclined to follow the flow streamlines around the flame holder and hence give a low value of "collection efficiency." During combustion, the liquid accumulation on the flame holder has been observed as a narrow band along each side of the rod at the point of flow separation [*5*, p. 715]. Heat from the recirculation zone is transferred to the accumulated liquid fuel on the flame

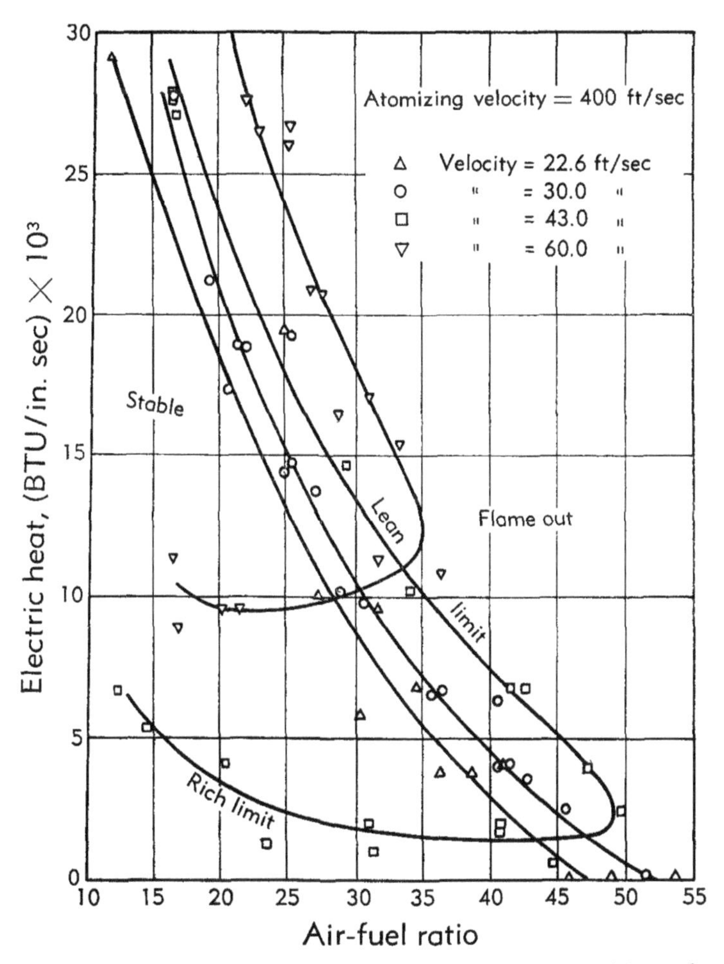


Fig. E,6a. Stability limits for Diesel fuel at various velocities and electric heat inputs [*5*, p. 715]. Atomization velocity 400 ft/sec.

holder, both directly and indirectly through the flame holder. By this means, some of the accumulated fuel is vaporized, and it then mixes with air diffusing from the main stream. For a stable flame to persist, a sufficient quantity of this fuel-air mixture must enter into the recirculation zone to release the heat necessary to make up for the heat losses from the recirculation zone.

High speed movies [5, p. 715] show that some of the liquid is sloughed off the flame holder. Any heat that is transferred to this portion of the fuel while it is on the flame holder is wasted as far as the flame stabilization process is concerned. As the fuel-air ratio is increased, the amount of fuel being sloughed off the flame holder also increases until the heat available for fuel vaporization is reduced to the point where the flame is extinguished. This phenomenon accounts for the rich limits shown by some of the curves of Fig. E,6a. This same limit was also reached at the

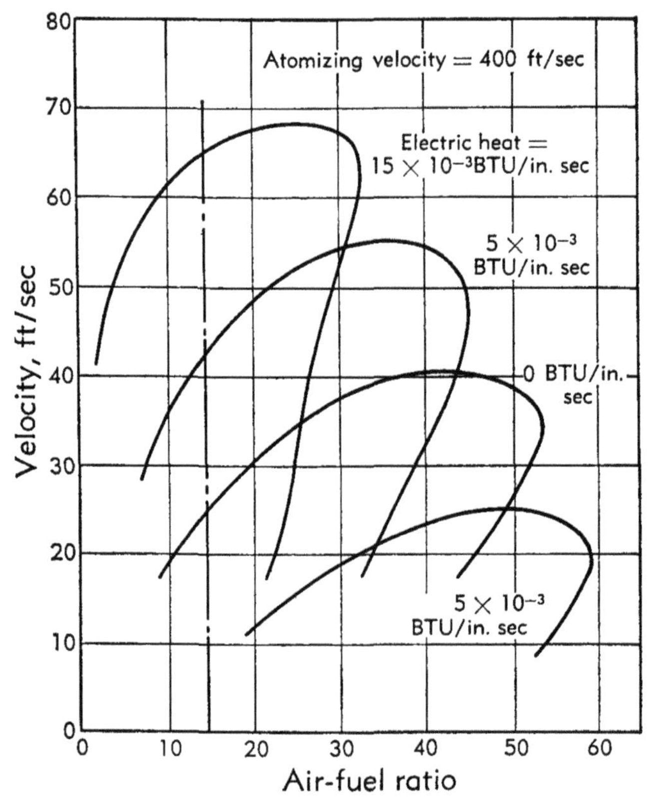


Fig. E,6b. Stability limits for Diesel fuel at various velocities and electric heat inputs [5, p. 715]. Cross plot of data of Fig. E,6a.

higher flow velocities by maintaining constant fuel-air ratio and reducing the electrical heat input to the flame holder. The end result is the same: insufficient heat flow into the accumulated liquid on the flame holder to produce the necessary vapor fuel concentrations in the recirculation zone. While this stability limit is reached due to an increase in fuel-air ratio and is therefore referred to as a rich limit, the flame extinction is seen to be actually the result of fuel-lean mixtures in the recirculation zone.

As the fuel-air ratio is reduced, the heat loss due to liquid sloughing is also reduced. The flame holder temperature increases gradually as the fuel-air ratio is reduced until the lean limit of Fig. E,6a is reached. At this fuel concentration, the flame holder temperature rises rapidly, the liquid accumulation is observed to disappear from the flame holder, and the flame is extinguished. The observed rapid increase in flame holder temperature means that the liquid is no longer collecting and vaporizing on the flame holder. This is attributed to vapor binding, a phenomenon often encountered in evaporation [*27*]. Fuel drops striking the hot flame

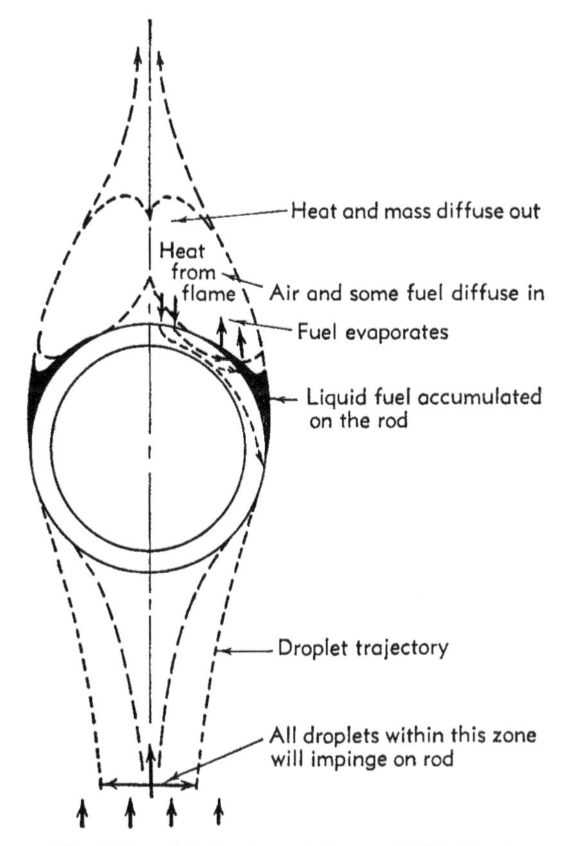


Fig. E,6c. Mechanism of flame stabilization in sprays of fuel impinging on rods [*5*, p. 715].

holder flash-boil and form a film of vapor that insulates them from further heat transfer. The heat flow into the droplets thus drops abruptly at this point and the flame holder temperature rises rapidly, increasing the tendency toward vapor binding. Obviously, the vapor fuel concentration in the recirculation zone decreases markedly, and this accounts for the flame blowoff. This same stability limit can also be reached by increasing the external heat supplied to the flame holder (see Fig. E,6a).

This mechanism of flame stabilization can be further described with the use of Fig. E,6d. Curve A shows the heat flow required to vaporize the quantity of liquid from the flame holder necessary to produce the eddy zone temperatures shown as the abscissa. Since the air flow into the eddy zone will be practically constant for a fixed flow velocity, the required heat flow is directly proportional to the fuel-air ratio in the eddy zone. Hence, curve A of Fig. E,6d has the familiar shape of the curves of fuel-air ratio against combustion temperature. Curves *B* and *C* show values of latent heat actually transferred to two cases: *B* corresponds to zero external heat input, and *C* represents a case where the

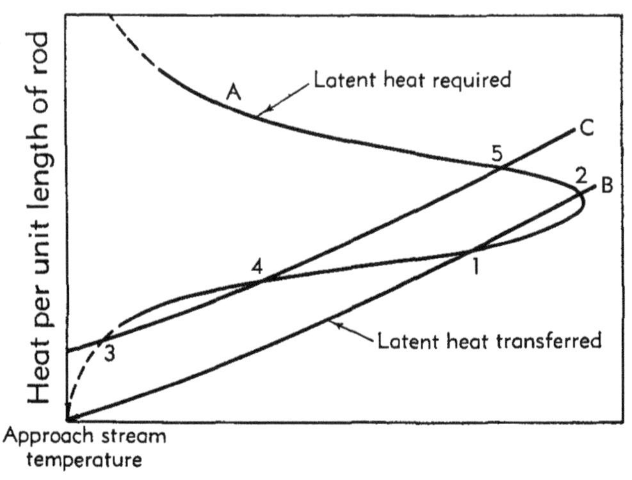


Fig. E,6d. Schematic presentation of equilibrium conditions on the flame holder [*5*, p. 715].

flame holder is electrically heated. While the exact values are not known, these curves must have the general shapes indicated in the figure, and this is all that is required for the discussion that follows:

Considering first curve *B* for no electric heat, there are two equilibrium points, 1 and 2. By considering the effect of a minor perturbation in eddy zone temperature, it becomes apparent that point 2 represents a stable equilibrium, while point 1 is a metastable case. Curve *C*, for external flame holder heating, has three points of intersection, designated 3, 4, and 5. Points 4 and 5 are similar to points 1 and 2 discussed above, with point 5 representing the stable equilibrium. Point 3 was believed to lie outside the combustible mixture range in [*5*, p. 715]. Thus the stable flames obtained with liquid fuel sprays correspond to conditions such as points 2 and 5 on the fuel-rich side of stoichiometric, or very near the stoichiometric mixture.

Fig. E,6e summarizes the general trends observed for flames produced from liquid sprays and lists the causes of flame blowoff in the various

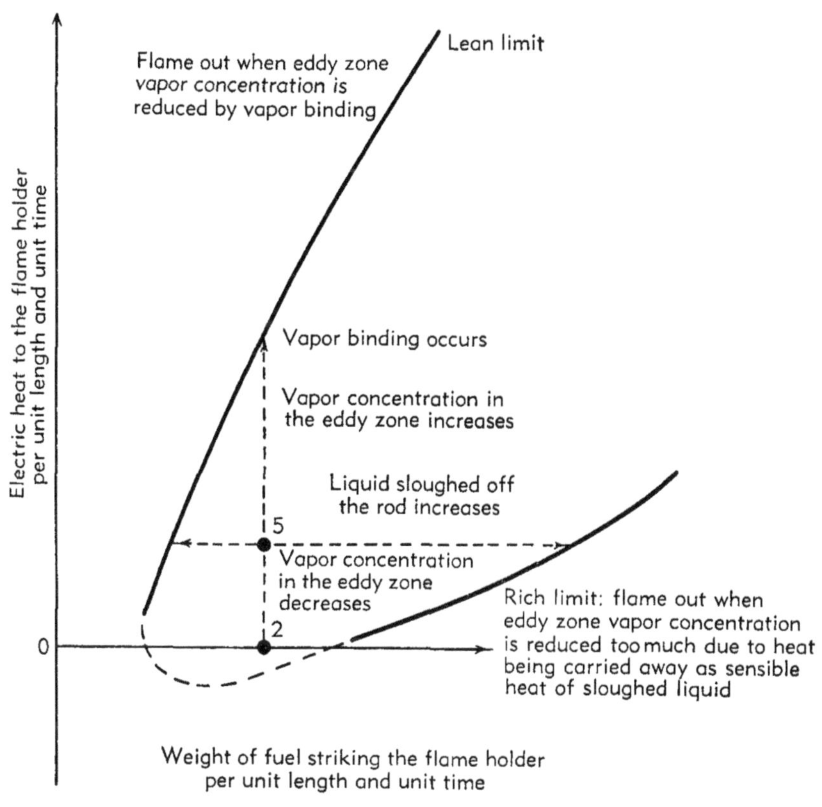


Fig. E,6e. Summary of stability limits and causes of blowoff for flames stabilized in fuel sprays [*5*, p. 715].

regions of operation. The points labeled 2 and 5 in Fig. E,6e represent the corresponding points in Fig. E,6d.

**E,7. Effect of Fuel Atomization and Fuel Volatility.** The effect of fuel atomization on flame stability limits is shown in Fig. E,7a [*5*, p. 715]. The fuel was atomized by injecting the fuel axially into high velocity air jets. The higher the atomization velocity (the relative velocity between the air and fuel jets) the smaller the mean drop size. For the smaller drop sizes the stability limits are shifted toward richer fuel-air mixtures. As previously mentioned, the fraction of the total liquid droplets that collect on the flame holder diminish with decrease in the mean drop size of the spray. Calculations [*5*, p. 715] show that the shift in stability limits noted in Fig. E,7a is probably due entirely to this change in droplet collection efficiency. The curves for the three different atomization velocities closely approach a common curve when the data are plotted against the computed rate of fuel collection by the rod.

The effect of fuel volatility was investigated by measuring the stability limits for various mixtures of Diesel fuel and propane [*5*, p. 715]. The data for a single approach stream velocity are shown in Fig. E,7b. Since the oxygen equivalent of a unit weight of propane is not the same as for a unit weight of Diesel fuel, the data are plotted against the reduced air fraction, rather than fuel-air ratio. The reduced air fraction is defined as the ratio of the air mass flow to the sum of the air mass flow and the air flow required for stoichiometric proportions with the fuel. Different abscissas are used in the two parts of Fig. E,7b. In the first part of the

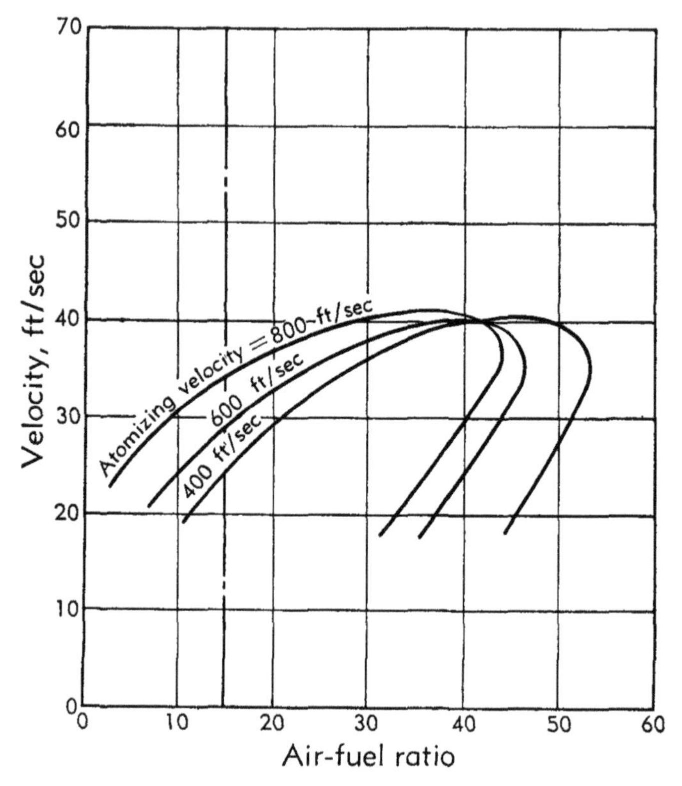


Fig. E,7a. Effect of atomizing velocity on stability limits [*5*, p. 715]. Diesel fuel.

figure, the reduced air fraction is based on the Diesel fuel alone; in the second part of the figure the reduced air fraction is based on the total fuel. Fig. E,7b shows that propane has practically no effect on the Diesel fuel concentration required at the lean stability limit until very high propane concentrations are reached. This might be expected, since at the lean limit where the vapor binding occurs, the rate of Diesel fuel vaporization will suddenly drop almost to zero; if the propane-air concentration is appreciably below the lean stability limit of the propane-air flame, then flame blowoff will occur just as though no propane were present.

The addition of propane to the Diesel fuel made it possible to burn to richer mixtures and to stabilize the flames to higher approach stream velocities as shown in Fig. E,7c [*5*, p. 715]. The effect of propane in extending the rich limits can be explained in terms of the same physical picture previously described: At the rich limit with Diesel fuel, the liquid sloughing off of the rod causes sufficient heat loss to reduce the Diesel fuel vaporization to too low a rate. Hence, adequate fuel is not supplied

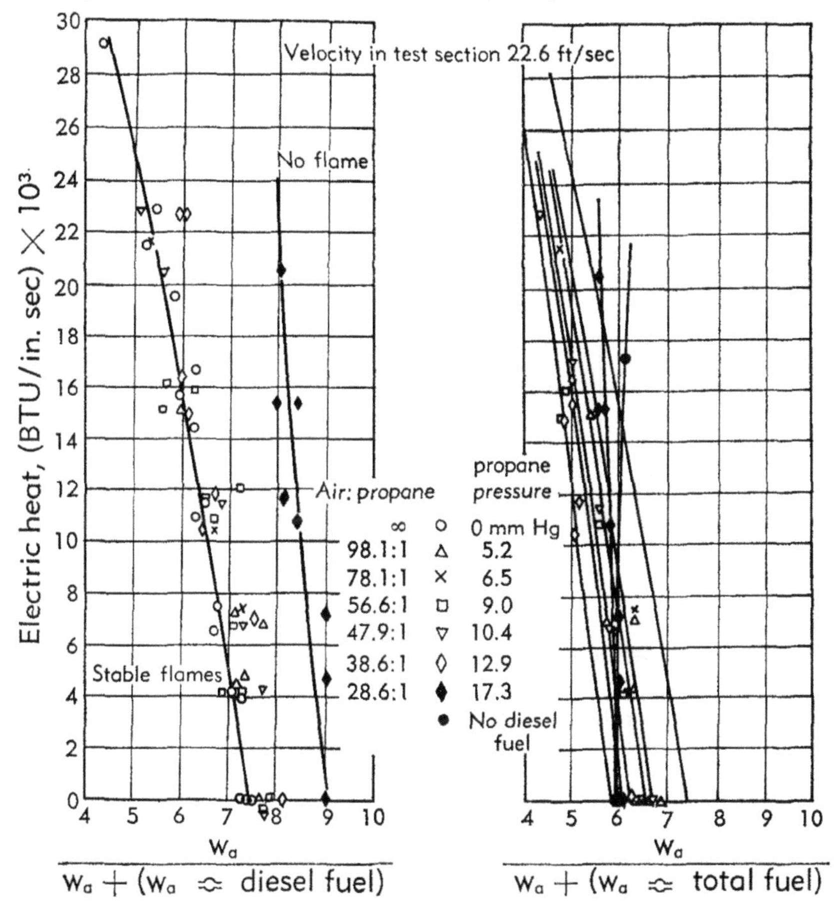


Fig. E,7b. Effect of fuel volatility on lean stability limits [*5*, p. 715]. ($w_a$ ≏ fuel) indicates the air flow required for stoichiometric proportions with the fuel. Diesel fuel-propane mixtures.

to the recirculation zone. The addition of propane obviously will increase the fuel concentration in the recirculation zone and thus extend the rich limits.

In [*5*, p. 715] the effects of increasing the fuel volatility are summarized as follows:

1. Due to evaporation, the mean drop size of the spray and thus the collection efficiency will be reduced.

2. Due to the reduced boiling point, the sensible heat required to bring the liquid to its boiling point should be reduced.
3. Due to the reduced boiling point, the heat flux at which vapor binding occurs will probably be lowered slightly.

As a result of these changes, both the lean and the rich stability limits shift to richer fuel-air mixtures with increased fuel volatility, just as for the case of finer fuel atomization.

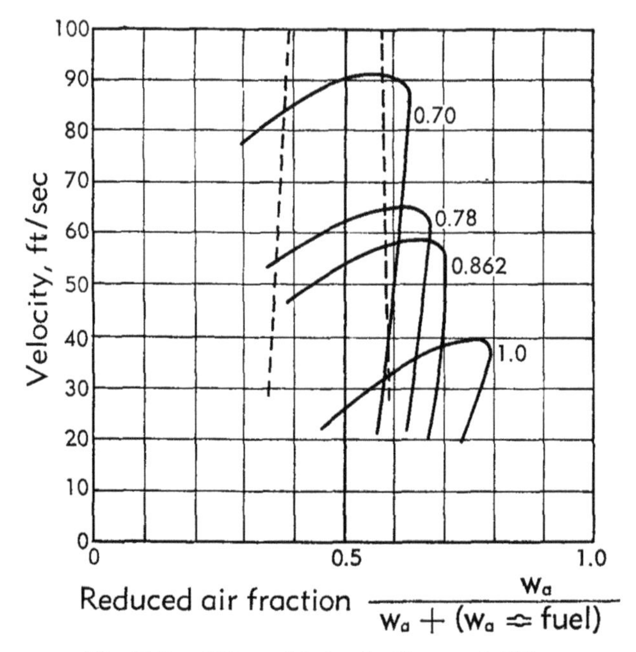


Fig. E,7c. Effect of fuel volatility on stability limits [*5*, p. 715]. Diesel fuel-propane mixtures.

**E,8. Effect of Flame Holder Size and Shape.** Fig. E,8a presents flame stability limits for Diesel fuel sprays and cylindrical rod flame holders of varying diameter [*5*, p. 715]. An increase in flame holder diameter results in a reduction in the maximum velocity and also shifts both the lean and rich limits to leaner mixtures. These trends are in marked contrast to those previously discussed for gaseous fuel-air mixtures. Similar data at higher atomization velocities do not show the progressive curtailment of the lean limits with reduction in flame holder size.

The effect of flame holder size on flame stability cannot be explained on the basis of the change in droplet collection efficiency: An increase in flame holder size will decrease the collection efficiency. This should produce a higher flame holder temperature for given operating conditions. The net result of this would be an extension of the rich and maximum velocity limits and a narrowing of the lean limit with increasing flame holder size; the exact opposite of this actually occurs. Obviously, then, other factors must account for the observed effects. One factor that could explain the observed phenomena is the reduction in the heat transfer coefficient that accompanies an increase in flame holder size (for a rod, $h \sim d^{-0.5}$, approximately). The reduced heat transfer per unit area accompanying an increase in flame holder size would increase the tendency toward flame extinction due to excessive liquid cooling [*5*, p. 715].

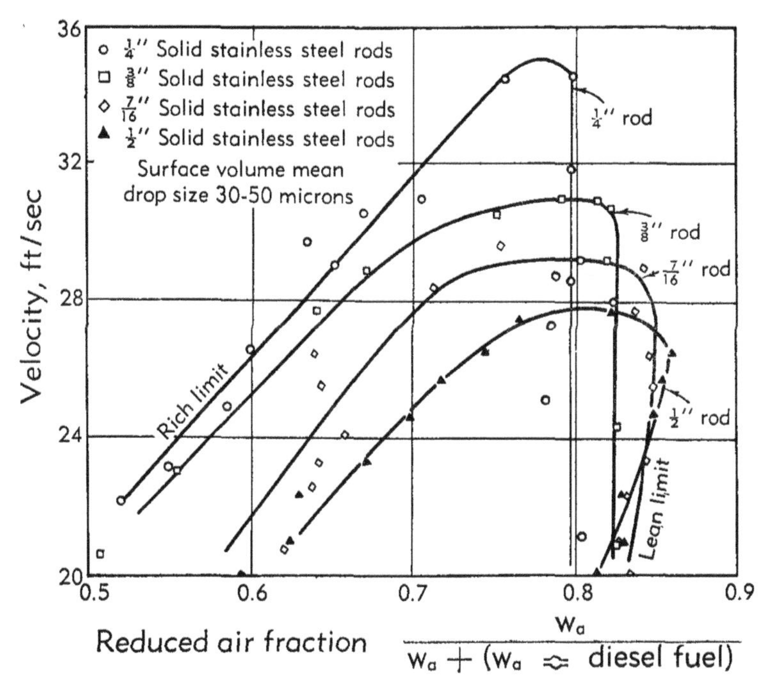


Fig. E,8a. Effect of flame holder size on flame stability in air-fuel spray mixtures [*5*, p. 715]. Diesel fuel.

Flame holder design can have a large effect on flame stability limits obtained with liquid fuel sprays. Some of these effects are shown in Fig. E,8b, which compares the stability limits of a solid rod, a rod with triangular-shaped fins extending downstream, the same rod rotated 180° to place the fins upstream, and a kaolin-filled rod [*5*, p. 715]. All the rods were ¼ in. in diameter. With the fins extended downstream, the rich limits and the maximum velocity limit were considerably extended over those for the solid rod. This would be expected because of the increased rate of heat flow from the flame zone into the rod.

The rich limits and the peak velocity for stable burning were considerably reduced with the fins pointed upstream. This is due to the cooling of the flame holder by the fins.

From the foregoing discussion it is evident that flame holder design features that affect the heat transfer to the flame holder can have a pronounced influence on the flame stability limits with a liquid fuel spray. It would also be expected that flame holder shape would be important

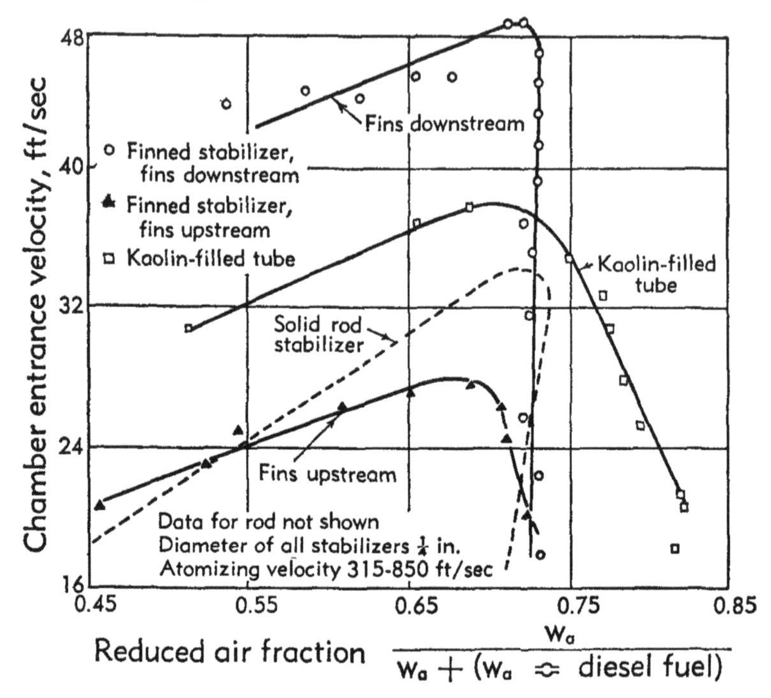


Fig. E,8b. Effect of flame holder design on flame stability in air-fuel spray mixtures [*5*, p. 715]. Diesel fuel.

because of its effects on the droplet collection efficiency and on the liquid sloughing tendency.

# *CHAPTER 3. FLAME STABILIZATION INSIDE PERFORATED LINERS*

**E,9. Introduction.** In turbojet primary combustors and in some ramjets the flame is stabilized inside a perforated liner, or "can." Plate E,9 shows a typical turbojet combustor liner; this particular liner is from the J33 engine. The air flow patterns inside this combustor liner have been determined using a transparent plastic model of the liner and balsa dust flow tracers [*28*]. Typical flow patterns determined from photographs are shown in Fig. E,9. A strong reversed flow pattern and several vortices are evident in the upstream end of this combustor liner. The combustor flow patterns were observed to vary considerably with time as indicated by the two views given in the figure; however, there always existed local low velocity regions and considerable recirculation near the upstream end of the liner. From this, it is evident that the flow patterns in the upstream end of a perforated liner are quite similar to the flow patterns existing in the wake of bluff-body flame holders.

The arrangement of the perforations in the liner controls the rate of air entry into the burning zone. It might therefore be expected that wider flame stability limits could be obtained with this device than are possible with the bluff-body flame holders which give no control over this air

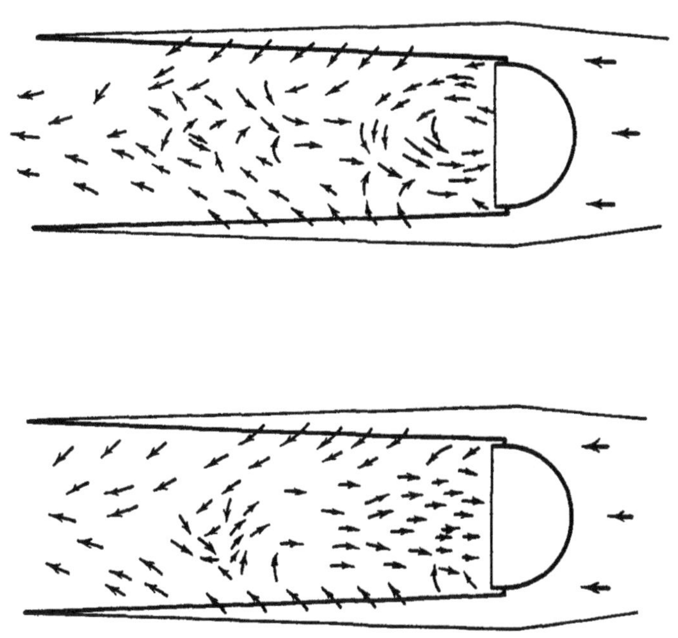

Fig. E,9. Isothermal flow patterns in J33 combustor [*28*]. Different patterns are for different times in same combustor at same operating conditions.

entry rate. Perforated liners, or "can-type" combustors, are not used in afterburners, however, because the pressure drop across them would be too great. In the primary combustor, where the pressures are higher and the flow velocities are lower, the higher pressure loss coefficient of these can-type combustors can be tolerated.

A large amount of development work has been done by the various jet engine companies to obtain high performance can-type combustors for both turbojet and ramjet application. Various turbojet primary combustors are described in more detail in Sec. H. For the purposes of this section, it is sufficient to point out that all of these combustors employ the same basic principle for flame stabilization; in all cases the flame is stabilized in a low velocity region where flow recirculation occurs.

**E,10. Effect of Flow and Design Variables on Flame Stabilization.** White and Lewis [29] have measured flame stability limits for various can-type combustors having a single row of holes for entry of the fuel-air mixture, as shown in Fig. E,10a. A prevaporized kerosene-air mixture was supplied to the combustors. The combustor of Fig. E,10a represents the upstream end, or piloting zone, of a combustor such as the one shown in

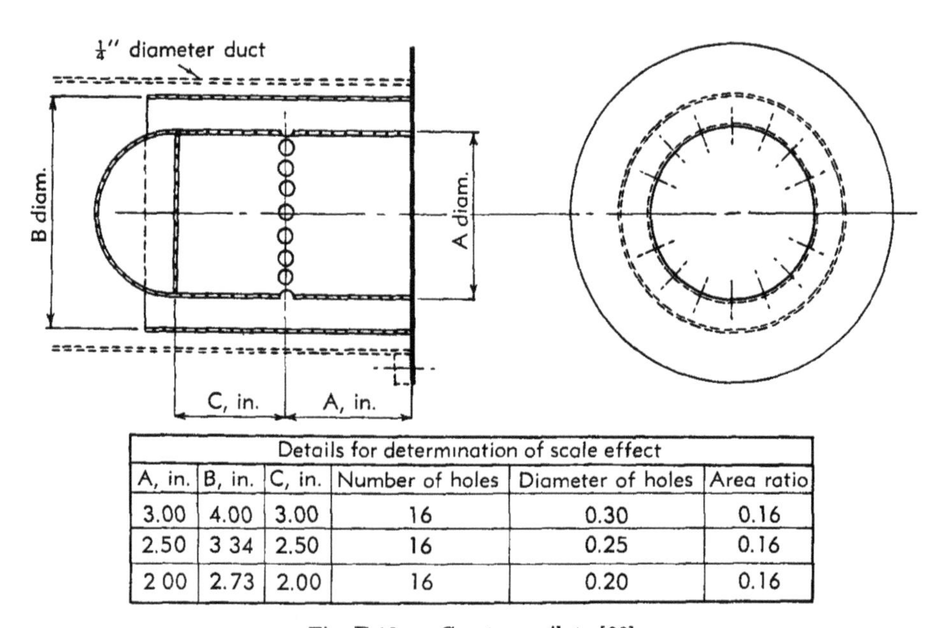


| Details for determination of scale effect | | | | | |
|---|---|---|---|---|---|
| A, in. | B, in. | C, in. | Number of holes | Diameter of holes | Area ratio |
| 3.00 | 4.00 | 3.00 | 16 | 0.30 | 0.16 |
| 2.50 | 3.34 | 2.50 | 16 | 0.25 | 0.16 |
| 2.00 | 2.73 | 2.00 | 16 | 0.20 | 0.16 |

Fig. E,10a. Can-type pilots [29].

Fig. E,9a. The principal results from White and Lewis' work are as follows:

1. For a fixed total area of the holes in the liner, there was no measurable effect on stability of changing from a large number of small holes to one half as many large holes.
2. An increase in the total area of the holes in the liner resulted in a slight improvement in the lean stability limit.
3. An increase in the recirculation zone depth (the distance from the upstream end of liner to the location of the holes) improved the lean stability limit until a recirculation zone depth of one can diameter was reached. A further increase in recirculation zone depth gave only a very minor improvement in stability.
4. A reduction in pressure caused a narrowing of both the rich and lean limits. Fig. E,10b shows this effect for a 2.5-in.-diameter can. The maximum velocity limits near stoichiometric fuel-air ratio (0.067 for kerosene) were not obtained because of inadequate exhauster capacity.

For both the rich and lean limits the increase in blowoff velocity with pressure was such that

$$w_{a,max} \sim p^{1.5}$$

or

$$u_{r,max} \sim p^{0.5}$$

where $u_r$ is the average velocity in the combustor obtained by dividing the volume flow rate by the maximum combustor cross-sectional area.

5. An increase in can size produced wider stability limits at most conditions investigated. For the lean limit, $w_{a,max} \sim D^{2.4}$.
6. A correlation of the blowoff data was obtained by plotting $w_{a,max}/p^{1.5}D^{2.4}$ against the fuel-air ratio; this is shown in Fig. E,10c. The correlation is good for the lean limits, but considerable data-scatter exists along the rich-limit curve at low velocity conditions. The parameter $w_{a,max}/p^{1.5}D^{2.4}$ is equivalent to $u_{r,max}/p^{0.5}D^{0.4}$.

Longwell [*30*, pp. 48–56] has also measured the stability limits of can-type combustors of the type shown in Fig. E,10a. For these data

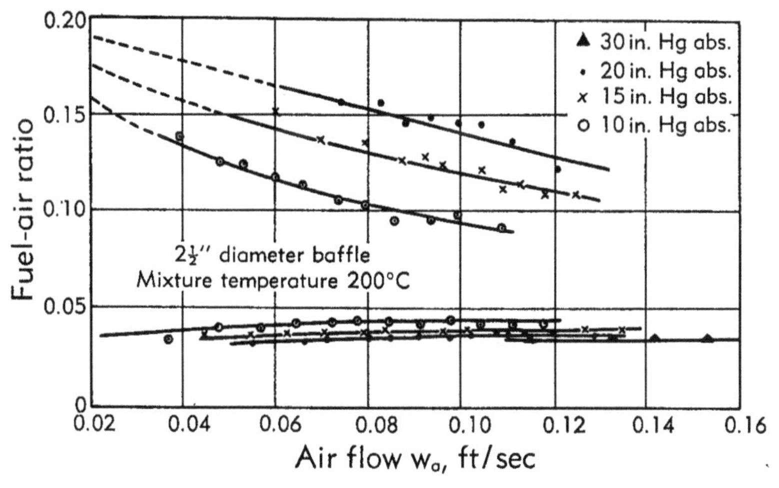


Fig. E,10b. Effect of pressure on stability limits of can-type pilots [*29*]. Fuel, kerosene.

the stability limits correlate as a function of $W_{a,max}/p^2$, as shown in Fig. E,10d. $w_{a,max}/p^2$ is equivalent to $u_{r,max}/p$. The blowout data of Childs, et al. [*31*] for a turbojet combustor with an annular perforated linear correlate better with $w_{a,max}/p^2$ than with the parameter $w_{a,max}/p^{1.5}$ of White and Lewis. Some doubt therefore exists as to the proper relation between $w_{a,max}$ and $p$ for can-type combustors, and further work is required to resolve the discrepancies in the existing literature.

Because of these uncertainties in the existing data and because the qualitative effects of the various flow and system variables are the same

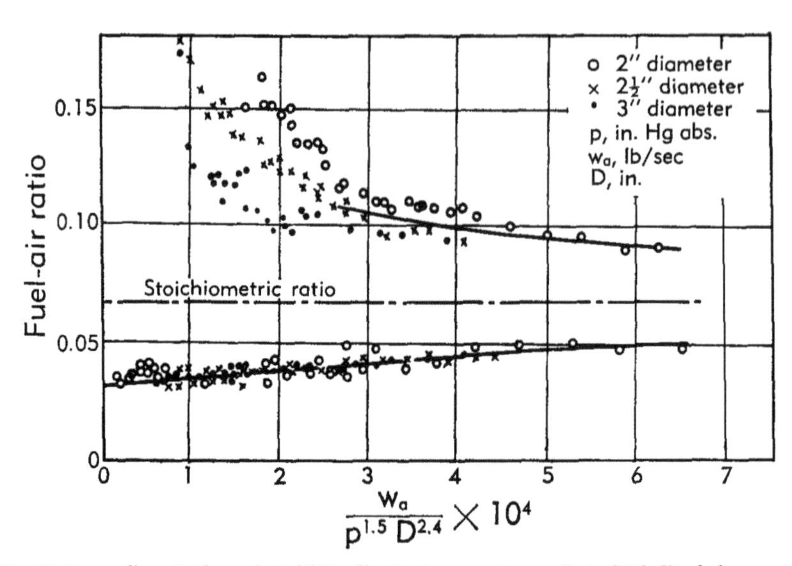


Fig. E,10c. Correlation of stability limits for can-type pilots [*29*]. Fuel, kerosene.

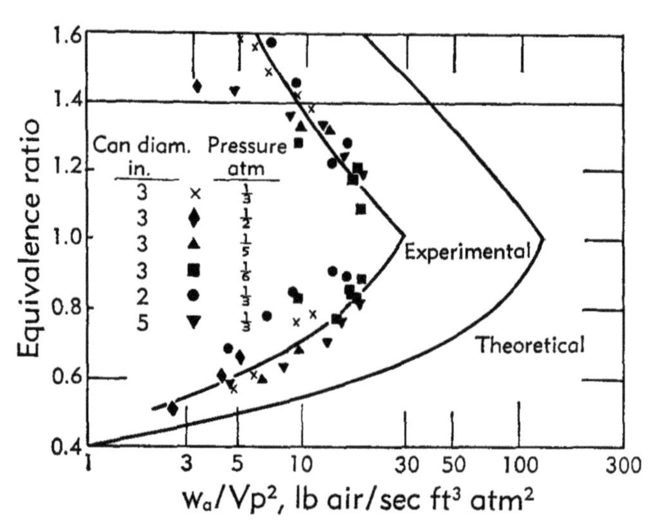


Fig. E,10d. Comparison of experimental and theoretical stability limits for can-type pilots [*30*, pp. 48–56]. Inlet temperature, 200°F; fuel, number 1 solvent naphtha.

for can-type combustors as for bluff-body flame holders, it appears reasonable to assume (until such time as additional data are available) that the same stability relations apply to both. The relation

$$\frac{u_{\max}}{D^{0.5}p^{0.95}T^{1.2}} = \phi(f) \tag{10-1}$$

which was derived for bluff-body flame holders agrees reasonably well with most of the data for can-type combustors. For can-type combustors $u_{r,max}$ is substituted for $u_{max}$ in Eq. 10-1.

**E,11. Theory of Flame Stabilization.** Longwell [*30*, pp. 48–56] suggests that the chemical reaction rate in the piloting zone determines the stability limits. This theory is the same as the Longwell theory presented in Art. 4 for bluff-body flame stabilization. For the can-type combustor, however, the arguments in favor of the Longwell theory are considerably stronger than for the case of bluff-body flame holders. The jets of fresh mixture entering through the liner holes produce rapid mixing of the incoming charge and the burning gases.

Fig. E,10d shows a theoretical curve computed by Longwell for a homogeneous (infinite mixing) reactor. The theoretical curve is similar to the curve through the experimental data but is displaced to values of $w_{a,max}/Vp^2$ that are higher by a factor of about 4. This difference implies that present-day, can-type combustors do not make the most effective use of the available volume within the liner. If the mixing were instantaneous and all of the volume were filled with flame, then the limiting factor would be the chemical reaction rate and the experimental data should lie on the theoretical curve. A high rate of mixing within the combustor requires a high pressure drop. Because of the requirement of reasonably low pressure losses in jet engine combustors, perhaps the optimum combustor design will never approach the theoretical curve of Fig. E,10d.

**E,12. Possible Relation between Combustion Efficiency and Flame Stability.** For a given combustor, flame stability is a function of the parameter $u_r/p^{0.95}T^{1.2}$, which is quite similar to the parameter $u_r/pT$ that has been used to correlate combustion efficiency data for some turbojet combustors [*32,33*]. The $u_r/pT$ parameter was derived from theory [*32,33*] by assuming that chemical reaction kinetics control the over-all combustion rate in turbojet combustors. However, the marked similarity between the flame stability parameter and the combustion efficiency parameter invites the speculation that turbojet combustion efficiency may be a direct function of flame stability in some cases. Most turbojet combustors have unstable eddies created in the combustion zone (Art. 9). If it is assumed that these eddies create many individual "pockets" of combustible mixture that burn as they flow through the combustor, then it becomes possible to apply the same theoretical concepts to each of these pockets as were applied in the analysis of combustion in a flame holder wake. With this picture of the process, the combustor is then filled with a large number of burning pockets of varying size. The smallest of these individual burning pockets will be quenched first, and progressively larger pockets will be quenched as operating conditions become more severe. The decay in combustion efficiency that is observed experimentally [31,32] could then be the result of this extinction of some of the burning pockets.

If the assumed model of the combustion process is correct, then turbojet combustion efficiency would be expected to correlate as a function of the stability parameter $u_r/p^{0.95}T^{1.2}$, provided that the size distribution of the eddies within the combustor does not change with operating conditions. Some justification exists for assuming no effect of flow conditions on eddy size for Reynolds numbers above some minimum value [34,35].

The difference between the stability parameter $u_r/p^{0.95}T^{1.2}$ and the parameter $u_r/pT$ is not sufficient to change the efficiency correlations of [32] appreciably.

# *CHAPTER 4. FLAME STABILIZATION IN BOUNDARY LAYERS*

**E,13. Characteristics of Bunsen Flames.** No discussion of flame stabilization would be complete without consideration of flame stabilization in boundary layers. Even though the mean stream velocity in a duct may be high, there always exists a low velocity boundary layer at the wall of the duct. Under certain conditions, it is possible for the flow velocity at some point in the boundary layer to equal the flame speed. If this occurs, then a flame can be stabilized at this location. This is the mechanism by which a flame is anchored on the port of a Bunsen burner.

To be able to understand this flame stabilization process it is desirable to first consider the basic characteristics of Bunsen burner flames. Many investigations (e.g. [7; 8, pp. 3–21; 36]) have measured the velocity and concentration regions for which Bunsen flames are stable. Fig. E,13 is a typical map of Bunsen flame behavior [8, pp. 3–21]. When the approach velocity to a seated open flame is decreased until the flame velocity exceeds the approach velocity over some portion of the burner port, the flame flashes back into the burner. The flashback always occurs in the unshaded area under the flashback curve in Fig. E,13. If, on the other hand, the approach velocity is increased until it exceeds the flame velocity at every point outside the thin layer near the tube wall where no flame can exist, then the flame will either be extinguished completely (when conditions fall in the unshaded region to the left of the blowoff curve) or, for fuel-rich mixtures, it will be lifted above the burner until a new stable position in the gas stream above the port is reached as a result of turbulent mixing with secondary air. The lift curve is a continuation of the blowoff curve beyond a critical percentage of the fuel gas at point *A*. The blowout curve corresponds to the gas velocity required to extinguish a lifted flame. Once the flame has been lifted above the port, the approach velocity must be decreased to well below the lift velocity before the flame will drop back and be seated on the burner rim. Between fuel concentrations *A* and *B*, the blowout of the lifted flame occurs at a lower velocity than the flame blowoff from the port. Such a lifted flame at constant composition can be produced only by ignition from above the port.

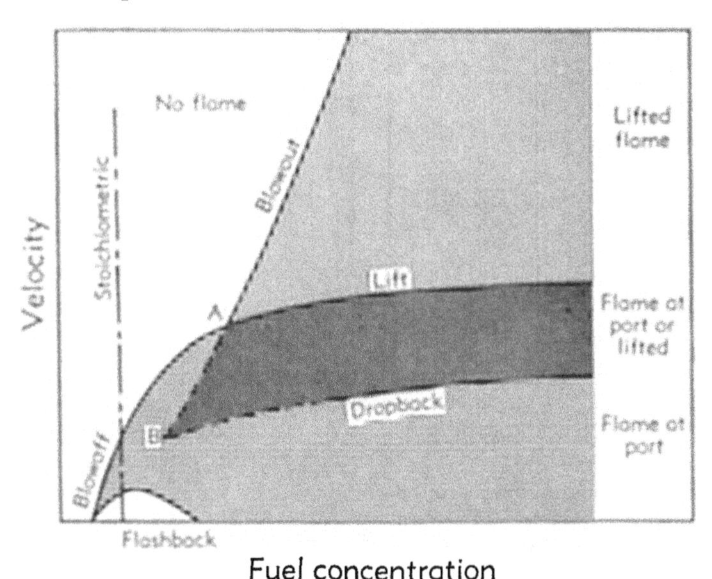


Fig. E,13. Characteristic stability diagram for open Bunsen flames [*8*, pp. 3–21].

**E,14. Mechanism of Flame Stabilization.** The conditions for stability may be described in terms of laminar flow regardless of whether the flow in the tube is laminar or turbulent because in either case there is a laminar sublayer at the stream boundary [*8*, pp. 3–21]. Any point of equality between flow velocity and flame velocity must lie within the laminar sublayer because the gas velocity at the boundary between the sublayer and the turbulent core is much greater than the flame velocity. The velocity gradient near the tube wall may be assumed constant if the width of the region is small compared with the tube diameter. Curves for flashback of natural gas flames [*7*] indicate that this assumption is satisfactory for correlating flashback data obtained with tube diameters larger than the flame-quenching diameter. The critical boundary velocity gradients for flashback and blowoff have been used quite successfully to correlate the flashback and blowoff data obtained with burners of various sizes and shapes [*5*, pp. 695–701; *7*] and to correlate turbulent with laminar blowoff data [*36*,*37*].

The interaction of the flame velocity and the critical boundary velocity gradient for a Bunsen flame can be understood by reference to Fig. E,14a and E,14b. Fig. E,14a illustrates how the flame position shifts with increasing gas flow, the flow increasing from flame position *A* to *C*. As the flame moves away from the burner port, the fringe of the flame moves closer to the stream boundary. The reason for this can be seen by

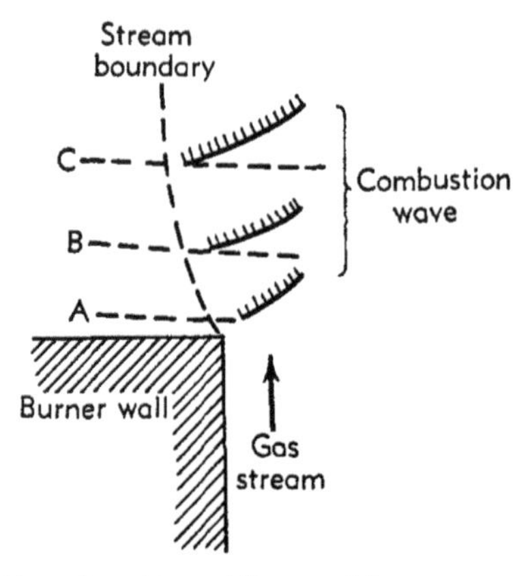


**Fig. E,14a. Location of flame with respect to burner at various distances above port [7].**

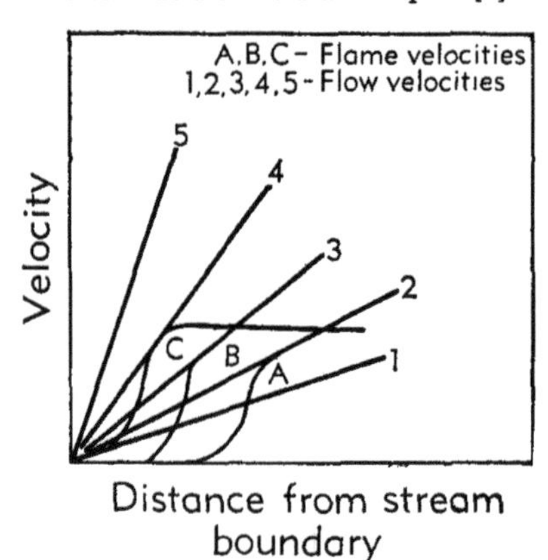


Fig. E,14b. Relation between flame and stream velocity as function of distance from stream boundary [7].

also considering curves *A*, *B*, and *C* of Fig. E,14b, which represent the variation of flame velocity with distance from the stream boundary for flame positions *A*, *B*, and *C*. In all these curves, the flame velocity is constant far from the stream boundary, but at smaller distances the solid burner rim exerts a quenching effect on the flame by extracting heat and destroying reactive chain carriers, and the flame velocity decreases. At some small distance from the stream boundary, the quenching becomes complete and the flame velocity falls to zero [7]. The height of the flame fringe above the rim (Fig. E,14a) has been called the dead space above the rim. As the flame moves up from position $A$ to position $B$ or $C$, the loss of heat and active particles diminishes, the region of constant flame velocity extends closer to the stream boundary, and the fringe of the flame approaches the stream boundary. For lean mixtures, there is a limit to this movement toward the stream boundary, because there is superimposed on the quenching effect a flame velocity reduction due to the diffusion of external air into the unburned gas. This diluting effect increases as the flame moves away from the burner. Thus, flame velocity curve $C$ may represent the limit of approach of the flame fringe toward the stream boundary for a lean mixture. For a rich mixture, secondary-air dilution may increase the flame velocity; for a very rich mixture, a stable lifted flame may be obtained several tube diameters above the port. Conversely, curve $A$ of Fig. E,14b may represent the limit due to quenching to which a stable flame can approach the burner rim.

For any flame velocity curve between the limiting curves $A$ and $C$ (Fig. E,14b), there is a gas flow for which the straight line representing the boundary velocity gradient is tangent to the flame velocity curve (lines 2 to 4). Any line such as 1, which represents a smaller gradient than line 2, intersects curve $A$, and there will be a region where the flame velocity exceeds the gas velocity. As a result, the flame flashes back into the tube. Any gradient larger than line 4 causes the gas velocity to be everywhere greater than the flame velocity, and the flame blows off. Hence, gradients 2 and 4 are called the critical boundary velocity gradients for flashback and blowoff, respectively.

The critical boundary velocity gradients for Bunsen flames are calculated from equations for the frictional drag imposed by a wall. For fully developed laminar flow through a long cylindrical tube (Reynolds number $< 2100$), the critical boundary velocity gradient may be found by differentiating the Poiseuille equation [*7*, p. 279], which gives

$$g = \frac{8u_{av}}{d_t} \tag{14-1}$$

The more general equation, which holds for either laminar or turbulent flow, is

$$g = \frac{F\rho u_{av}^2}{2\mu} = \frac{Fu_{av}Re}{2d_t} \tag{14-2}$$

where $F$ = friction factor from empirical Fanning equation [*37*]. For the laminar flow condition through long cylinders considered in Eq. 14-1, $F = 16/Re$. The following expressions (Table E,14) are given in [*5*, pp. 695–701] for the friction factor in laminar flow through various types of burners. In all these cases, the diameter used in computing $Re$ is the hydraulic diameter, that is, twice the cross-sectional area of the channel divided by the perimeter.

*Table E,14*

| Type of burner | Friction factor F |
|---|---|
| Short circular port (orifice) | $8.5/Re^{0.85}$ |
| Long square channel | $18.9/Re^{1.11}$ |
| Long rectangular channel | $40.2/Re^{1.27}$ |
| Long triangular channel | $29.8/Re^{1.29}$ |

For turbulent flow in long, smooth, cylindrical tubes, several equations have appeared in the literature. The empirical equation of Blasius [38] is used in [5, pp. 695–701; 36; 37]. For $3000 < Re < 100,000$, this equation gives

$$F = \frac{0.080}{Re^{0.25}} \tag{14-3}$$

Others (e.g. [8, pp. 3–21]) have used the empirical equation of Koo [27], which for the range $5000 < Re < 200,000$ gives

$$F = \frac{0.046}{Re^{0.2}} \tag{14-4}$$

For flow in the transition region between laminar and turbulent flow ($2100 < Re < 3000$ to $5000$), there is no available equation for $F$. Some estimates of $F$ are obtained from arbitrary curves connecting the curves for laminar and turbulent flow [27].

A number of investigators have found that flashback and blowoff limits of open flames burning from tubes of nozzles at room conditions are independent of tube diameter when plotted as fuel concentration (or fuel-air ratio) against the critical boundary velocity gradient. Lewis and von Elbe [7, pp. 282–300] plotted flashback limit data for natural gas-air, hydrogen-air, hydrogen-oxygen, acetylene-oxygen, methane-oxygen-nitrogen, and propane-oxygen-nitrogen flames in this manner. They found the critical boundary velocity gradients for flashback $g_F$ to be independent of tube diameter as long as the diameter was somewhat larger than the quenching diameter but not so large that flashback was preceded by a severe tilting of the flame, in which case the limit was not clearly distinguished.

The critical boundary velocity gradients for blowoff $g_B$ of natural gas-air mixtures in the laminar flow region were also independent of diameter, both for ordinary Bunsen flames and for some inverted flames stabilized at the ends of wires mounted in the axes of the tubes. Successful correlations were obtained with $g_B$ for several other fuel-oxygen-nitrogen mixtures in laminar flow. [*36*] presents a correlation extending from the laminar flow range well into the turbulent flow range on a continuous curve for propane-air flames for a range of tube diameters as shown in Fig. E,14c. This figure shows that the curves for the small tubes deviate toward higher velocity gradients only above a propane mole fraction of 0.15. [*8*, pp. 3–21] also presents continuous curves from laminar into turbulent flow for butane-air flames. Furthermore, it was shown that the hydrogen-air data of [*7*, pp. 292–293], when properly calculated, gave a similar smooth curve extending into the turbulent region.

A large amount of research has been conducted on the stability of Bunsen flames, including the effect of various flow and system variables

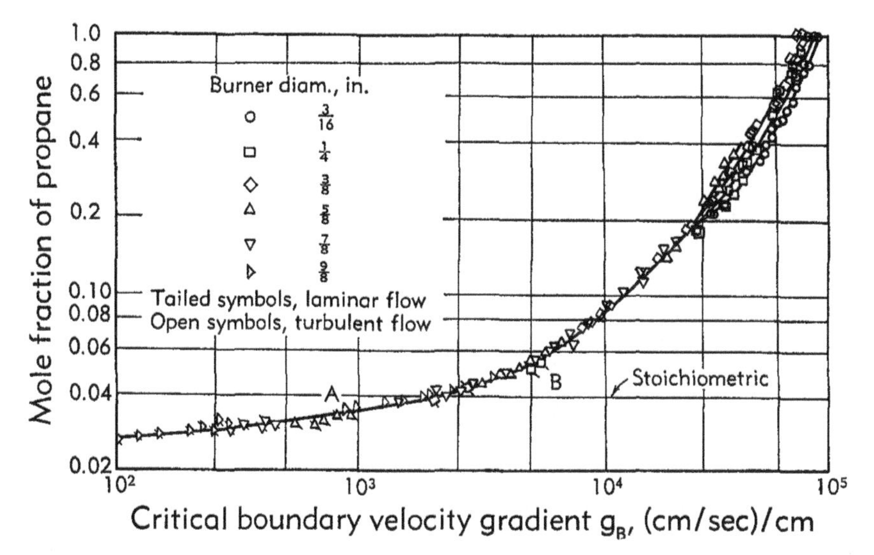


Fig. E,14c. Correlation of Bunsen burner flaskback data, showing that velocity gradient at wall is unique function of mixture composition [*36*].

[*5*, pp. 68–69, 527–535, 545–552, 695–701, 708–714; *7*; *8*, pp. 3–21, 245–253; *36* to *50*]. A detailed discussion of this work is considered beyond the scope of this section, however, since the primary emphasis here is on flame stabilization of practical use in present-day jet engine combustors. At the present time, no production-model, jet engine combustors utilize the principle of flame stabilization in low velocity boundary layers.

A definite possibility exists, however, that flame stabilization in boundary layers may be feasible in some future jet engines. This is particularly true if special fuels of high reactivity are employed. The high flame speed of such fuels could conceivably make it possible to stabilize flame at the combustor walls or in the boundary layer of streamlined fuel injectors.

## E,15. Cited References.


1. Haddock, G. W. *Calif. Inst. Technol. Jet Propul. Lab. Rept. 3-24,* May 1951.
2. Zukoski, E. E. Flame stabilization on bluff bodies at low and intermediate Reynolds numbers. *Calif. Inst. Technol. Jet Propul. Lab. Rept. 20-75,* June 1954.
3. Goldstein, S., ed. *Modern Developments in Fluid Dynamics.* Clarendon Press, Oxford, 1938.
4. Younger, G. C., Gabriel, D. S., and Mickelsen, W. R. Experimental study of isothermal wakeflow characteristics of various flame-holder shapes. *NACA Research Mem. E51K07,* 1952.
5. *Fourth Symposium on Combustion.* Williams & Wilkins, 1953.
6. Scurlock, A. C. Flame stabilization and propagation in high-velocity gas streams. *Mass. Inst. Technol. Meteor Rept. 19, NOrd Contract 9661,* May 1948.
7. Lewis, B., and von Elbe, G. *Combustion, Flames and Explosions of Gases.* Academic Press, 1951.
8. *Third Symposium on Combustion, Flame and Explosion Phenomena.* Williams & Wilkins, 1949.
9. Westenberg, A. A., Berl, W. G., and Rice, J. L. Studies of flow and mixing in the recirculation zone of baffle-type flameholders. *Proc. Gas Dynamics Symposium on Aerothermochemistry, Northwestern Univ.,* Aug. 1955. Multicopy Corp., Evanston, Ill., 1956.
10. McManus, H. N., Jr., and Cambel, A. B. The variation of flameholding characteristics of open vee-gutters with apex angle. *Proceedings of the Iowa Thermodynamics Symposium, Apr. 27–28, State Univ. Iowa,* 1953.
11. Caldwell, F. R., Ruegg, F. W., and Olsen, L. O. 71st report of progress on the combustion chamber research program for the quarter ending Sept. 30, 1951. *Natl. Bur. Standards,* 1951.
12. DeZubay, E. A. Characteristics of disk-controlled flame. *Aero. Digest 61,* 54-56, 102–104 (1950).
13. Mestre, A. *Recherche Aéronaut. Paris 41,* 23–26 (1954).
14. Zukoski, E. E., and Marble, F. E. Experiments concerning the mechanism of flame blow-off from bluff bodies. *Paper presented at Gas Dynamics Symposium on Aerothermochemistry, Northwestern Univ.,* Aug. 1955.
15. Zukoski, E. E., and Marble, F. E. The role of wake transition in the process of flame stabilization on bluff bodies. *Combustion Researches and Reviews, 1955.* Butterworths Scientific Publs., London, 1955.
16. Weir, A., Rogers, D. E., and Cullen, R. E. *Univ. Mich. Willow Run Research Center Rept. UMM-74,* Sept. 1950.
17. Longwell, J. P., Frost, E. E., and Weiss, M. A. *Ind. Eng. Chem. 45,* 1629 (1953).
18. Dugger, G. L. Effect of initial mixture temperature on flame speed of methane-air, propane-air, and ethylene-air mixtures. *NACA Tech. Note 2374,* 1951.
19. 66th report of progress on the combustion chamber research program. *Natl. Bur. Standards.* Apr.–June 1950.
20. 67th report of progress on the combustion chamber research program. *Natl. Bur. Standards,* July–Sept. 1950.
21. 65th report of progress on the combustion chamber research program. *Natl. Bur. Standards,* 1950.
22. Blackshear, P. L. Unpublished data, *NACA Lewis Lab.*
23. Shonerd, D. E. *Calif. Inst. Technol. Jet Propul. Lab. Rept. 3-18,* June 1952.
24. Williams, G. C. Basic studies on flame stabilization. *J. Aeronaut. Sci. 16,* 714–722 (1949).
25. Mullins, B. P. Studies on the spontaneous ignition of fuels injected into a hot air stream. *Natl. Gas Turbine Establishment, England, Rept. R90,* Sept. 1951.
26. Marble, F. E., and Adamson, T. C., Jr. Ignition and combustion in a laminar mixing zone. *Calif. Inst. Technol. Jet Propul. Lab. Progress Rept. 20-204,* 1954.
27. McAdams, W. H. *Heat Transmission,* 2nd ed. McGraw-Hill, 1942.
28. Olson, W. T., Childs, J. H., and Jonash, E. R. The combustion-efficiency problem of the turbojet at high altitude. *Trans. Am. Soc. Mech. Engrs. 77,* 605–615 (1955).

29. White, S. W., and Lewis, W. G. E. The performance of a multi-stage combustion system. Part I: Flame stability of recirculation zones. *Natl. Gas Turbine Establishment, England, Rept. R165*, May 1955.
30. *Fifth Symposium on Combustion.* Reinhold, New York, 1955.
31. Childs, J., McCafferty, R. J., and Surine, O. W. Effect of combustor-inlet conditions on performance of an annular turbojet combustor. *NACA Rept. 881*, 1947.
32. Childs, J. H. Preliminary correlation of efficiency of aircraft gas-turbine combustors for different operating conditions. *NACA Research Mem. E50F15*, 1950.
33. Childs, J. H., and Graves, C. C. Relation of turbine-engine combustion efficiency to second-order reaction kinetics and fundamental flame speed. *NACA Research Mem. E54G23*, 1954.
34. Rosenhead, L., and Schwabe, M. An experimental investigation of the flow behind circular cylinders in channels of different breadths. *Proc. Roy. Soc. London A129*, 115–135 (1930).
35. Fage, A., and Johnsen, F. C. The structure of vortex sheets. *Phil. Mag. and J. Sci. 5*, 417–441 (1928).
36. Bollinger, L. M., and Williams, D. T. Experiments on stability of Bunsen-burner flames for turbulent flow. *NACA Rept. 913*, 1948. (Supersedes *NACA Tech. Note 1234.*)
37. Grumer, J., and Harris, M. E. Flame-stability limits of methane, hydrogen, and carbon monoxide mixtures. *Ind. Eng. Chem. 44*, 1547–1559 (1952).
38. Garside, J. E., Hall, A. R., and Townsend, D. T. A. The stability of aerated burner flames. *Trans. Inst. Gas. Eng. 91*, 81–110 (1941–1942).
39. Kurz, P. F. The stability of isobutane-ethylene flames on a Bunsen burner shielded to exclude ambient air. *Battelle Memorial Inst. Tech. Rept. 15036-18, Contract AF 33(038)-12653, E. O. 460-35, S.R.-8*, Feb. 1953.
40. Polanyi, M. L., and Markstein, G. H. Phenomena in electrically and acoustically disturbed Bunsen burner flames. *Cornell Aeronaut. Lab. Project Squid Tech. Rept. 5*, Sept. 1947.
41. Calcote, H. F., and Pease, R. N. Electrical properties of flames—Burner flames in longitudinal electric fields. *Ind. Eng. Chem. 43*, 2726–2731 (1951).
42. Wohl, K., and Kapp, N. M. Flame stability at variable pressures. *United Aircraft Corp. Research Dept. Meteor Rept. UAC-42*, Oct. 1949. (*Bur. Ord. Contract NOrd 9845 with Mass. Inst. Technol.*)
43. Culshaw, G. W., and Garside, J. E. Recent studies of aerated burner flames. *Inst. Gas Engrs. (London), Inst. Gas Research Fellowship Rept.*, 1946–1947. (Reviews papers from period 1943–1946, including those of Delbourgh, Heiligenstaedt, and Vasilesco.)
44. Sachsse, H. Uber die Temperaturabhangigkeit der Flammengeschwindigkeit und das Temperaturgefalle in der Flammenfront. *Z. Physik. Chem. A180*, 305–313 (1935).
45. Kurz, P. F. Flame-stability studies with mixed fuels. *Ind. Eng. Chem. 45*, 2072–2078 (1953).
46. Walker, P. L., Jr., and Wright, C. C. Stability of burner flames for binary and tertiary mixtures of methane, carbon monoxide and water vapour. *Fuel 31*, 37–44 (1952).
47. Walker, P. L., Jr., and Wright, C. C. Stability and burning velocity of Bunsen flames with propane-carbon monoxide mixtures. *Fuel 31*, 45–49 (1952).
48. Kapp, N. M., Snow, B., and Wohl, K. The effect of water vapor on the normal burning velocity and on the stability of butane-air flames burning above tubes in free air. *United Aircraft Corp. Meteor Rept. UAC-30*, Nov. 1948. (*Bur. Ord. Contract NOrd 9845 with Mass. Inst. Technol.*)
49. Markstein, G. H. Experimental and theoretical studies of flame-front stability. *J. Aeronaut. Sci. 18*, 199–209 (1951).
50. Simon, D. M., and Wong, E. L. Burning velocity measurement. *J. Chem.Phys. 21*, 936–937 (1953).

# SECTION F

# *MIXING PROCESSES*

CHARLES C. GRAVES
WILFRED E. SCULL

**F,1. Introduction.** The importance of mixing in the over-all combustion process was recognized in some of the earliest research on turbojet engines [*1*]. With the development of high output combustors over the succeeding years the problem of obtaining adequate mixing has become increasingly difficult as combustor reference velocities and heat release rates have increased. The mixing problem is complicated further by design requirements that the combustor have a low total pressure loss, be of light weight and simple and durable construction, and be capable of operating efficiently over wide ranges of inlet conditions.

For purposes of discussion, the turbojet engine combustor may be divided into two regions: (1) the primary zone where the combustible fuel-air mixture preparation and flame-spreading steps occur, and (2) the secondary zone where the combustion products are diluted. This division is arbitrary since the complexity of the over-all process in the actual combustor usually precludes any clear-cut distinction between the individual steps or between the primary and secondary zones.

In the primary zone, the rapid mixing required for a high heat release rate is usually obtained by means of high velocity air jets issuing into the combustion space from swirlers and openings in the dome and liner. The resultant turbulence levels are appreciably higher than those found for fully developed turbulent pipe flow. However, the production of a high turbulence level is but one and not necessarily the most important criterion for adequate mixing. Above a certain level of turbulence (measured in terms of combustor total pressure loss), the effects of turbulence may be secondary to effects produced by changes in location of liner wall openings [*2*].

Other factors impose limitations on the manner in which the rapid mixing in the primary zone is achieved. Since the reference velocities are appreciably greater than normal flame velocity (1 to 2 ft/sec), a sheltered, low velocity or reverse flow region must be provided in the primary zone for flame stabilization. The fuel spray also contributes to the mixing process since the various size drops penetrate various distances into the flame. Accordingly, the design of the primary zone also involves the matching of fuel and air flow patterns for efficient mixing and a rapid spreading of the flame into the unburned fuel-air mixture.

In the secondary zone, the hot combustion products are diluted to reduce the temperature to levels tolerated by the turbine. In order to accomplish this mixing process within a reasonable distance, the dilution air is usually directed into the combustion chamber through large openings in the liner wall. Here the designer is interested in obtaining adequate penetration of the dilution air jets into the hot gas stream. Intimate mixing may not be required provided that (1) the time-averaged values of the local combustor outlet temperatures can be adjusted to the desired temperature profile and (2) the outlet temperature profile is stable.

It is evident that the over-all mixing process in turbojet engine combustors involves the mixing and penetration of air jets and fuel sprays in high temperature, turbulent gas streams. Mixing in the actual combustor is extremely complex. However, a number of investigations have been reported for simpler systems that are more amenable to analysis. While much of this basic work may not be applicable in detail to the study of mixing in combustion chambers, many of the principles and trends involved may apply. In this section, some of the pertinent theoretical and experimental investigations of liquid-air mixing and mixing in simple jet systems will be treated. Mixing in combustion chambers will be discussed primarily in terms of gross air flow patterns, fuel-air ratio distribution, and mixing devices.

**F,2. Mixing Theory for Simple Jet Systems.** Simple jet mixing systems may be divided into various types, depending on:

1. The mixing region entered by the jet, that is,
   (a) Laminar or turbulent
   (b) At rest (the case of a free jet), or in motion
   (c) Same, or differing from the jet in velocity, temperature, density, or chemical composition.
2. Direction of entry of jet into a moving medium—in the same direction, normal, or oblique.
3. Jet type, divided into two-dimensional plane jets and axially symmetric jets. (Hottel [*3*, pp. 97–113] lists some special cases of the two broad groups for purposes of fuel-air mixing.)
4. Property diffusing—momentum, mass, and heat.

Excellent bibliographies and discussions of analytical and experimental research on jet mixing are included in [*4*,*5*,*6*].

As shown in Fig. F,2a, a jet consists essentially of a uniform stream of fluid emanating from an orifice. Within a certain region (the potential cone) outside the orifice, the jet parameters of temperature, velocity, and jet fluid concentration are maintained approximately constant. Outside and downstream of the potential cone, a mixing region exists, in which the jet expands and interaction may occur between the jet and the surrounding fluid. Jet parameters of interest are usually the velocity, temperature, and jet fluid concentration at any position within the jet, and any interaction of the jet with adjacent jets.

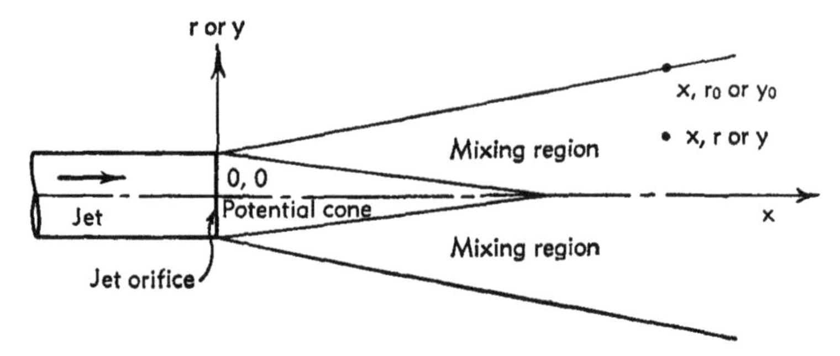


Fig. F,2a. Simplified sketch of typical jet.

Prominent explanations of an abundance of jet-mixing data available in the literature are based on the following theories or equations:

1. The momentum transfer theory
2. The vorticity transfer theory
3. The theory of constant exchange coefficients of the property diffusing
4. The von Kármán similarity theory
5. The Reichardt theory
6. The von Kármán momentum equations (integral equations of motion)
7. Point source diffusion of momentum, mass, or heat.

These theories will be discussed briefly as they apply to various types of jet mixing.

*Free jets.* The momentum transfer, vorticity transfer, constant exchange coefficient, and von Kármán similarity theories of jet mixing are similar in that each uses a boundary layer form of the fundamental equation of motion. Essential differences occur in the expression of an apparent shear stress due to turbulent momentum exchange. The fundamental equation for two-dimensional steady flow is

$$u\frac{\partial u}{\partial x} + v\frac{\partial u}{\partial y} = \frac{1}{\rho}\frac{\partial \tau}{\partial y} - \frac{1}{\rho}\frac{\partial p}{\partial x} \tag{2-1}$$

and for axially symmetric steady flow is

$$u\frac{\partial u}{\partial x} + v\frac{\partial u}{\partial r} = \frac{1}{\rho}\left(\frac{\tau}{r} + \frac{\partial \tau}{\partial r}\right) - \frac{1}{\rho}\frac{\partial p}{\partial x} \tag{2-2}$$

where

$p$ = instantaneous absolute static pressure
$r$ = radial coordinate
$u$ = longitudinal component of mean velocity
$v$ = lateral component of mean velocity
$x$ = longitudinal coordinate
$y$ = lateral coordinate
$\rho$ = fluid density
$\tau$ = apparent turbulent shear stress

Apparent shear stresses are expressed in the various theories as shown in Table F,2a. The quantities $k$, $l_P$, $l_T$, and $\epsilon$ are defined in the following paragraphs.

*Table F,2a*

| Theory | Apparent turbulent shear stress | |
|---|---|---|
| | Two-dimensional | Axially symmetric |
| Momentum transfer | $\rho l_P^2 \left\|\frac{\partial u}{\partial y}\right\| \frac{\partial u}{\partial y}$ | $\rho l_P^2 \left\|\frac{\partial u}{\partial r}\right\| \frac{\partial u}{\partial r}$ |
| Vorticity transfer | $\frac{\rho}{2} l_T^2 \left(\frac{\partial u}{\partial y}\right)^2$ | |
| Constant exchange coefficient | $\rho\epsilon \left(\frac{\partial u}{\partial y}\right)$ | $\rho\epsilon \left(\frac{\partial u}{\partial r}\right)$ |
| von Kármán similarity | $\rho k^2 \left(\frac{\partial u}{\partial y}\right)^4 \left(\frac{\partial^2 u}{\partial y^2}\right)^{-2}$ | $\rho k^2 \left(\frac{\partial u}{\partial r}\right)^4 \left(\frac{\partial^2 u}{\partial r^2}\right)^{-2}$ |

The momentum transfer theory of jet mixing, based essentially on the mixing length of Prandtl [7], assumes that the momentum of a fluid particle is transferred unchanged by turbulent mixing through a mixing length $l_P$ to a new position at which point turbulent mixing occurs. For a first approximation, a negligible pressure gradient in the $x$ direction is generally assumed. Solutions of Eq. 2-1 and 2-2, when the appropriate turbulent shear stresses are inserted, have involved the following assumptions:

1. The flow phenomena are geometrically and mechanically similar, and the mixing length $l_P$ is constant at any transverse jet cross section. This results in

$$l_P \sim (y_0 \text{ or } r_0) \sim x \tag{2-3}$$

where $y_0$ or $r_0$ are outer boundaries of jets.

2. The momentum flux of the jet in the $x$ direction at any cross section is constant. This results in

$$u_{max} \sim x^{-\frac{1}{2}} \text{ (two-dimensional jets)} \tag{2-4}$$

$$u_{max} \sim x^{-1} \text{ (axially symmetric jets)} \tag{2-5}$$

where $u_{max}$ is the velocity on the jet axis.

The momentum transfer theory was utilized by Tollmien [8], who investigated the mixing of two-dimensional and axially symmetric free jets and mixing along a boundary.

For axially symmetric free jets, the differential equation of steady state heat transfer [9] is

$$u\frac{\partial T}{\partial x} + v\frac{\partial T}{\partial r} = \frac{1}{r}\frac{\partial}{\partial r}\left[l_P^2\left|\frac{\partial u}{\partial r}\right| r\frac{\partial T}{\partial r}\right] \tag{2-6}$$

where $T$ is the absolute static temperature. The same boundary conditions are applied for velocity and temperature distributions based on the momentum transfer theory. Therefore, since Eq. 2-6 differs from Eq. 2-2 (using the momentum transfer theory) only in the substitution of $T$ for $u$, the solution of Eq. 2-6, based on the momentum transfer theory, yields radial temperature profiles identical with those of velocity at any transverse cross section in axially symmetric jets.

The vorticity transfer theory of jet mixing is based on the concept of Taylor [10] that the vorticity of a fluid element is conserved in turbulent transport of the element from one point to another, at which time vorticity is dissipated in mixing. Abramovich [11] points out that the assumption of no large pressure gradients in the $x$ direction (momentum transfer theory) does not need to be applied in this theory. In addition, various references show that a comparison of the turbulent shear stresses for two-dimensional flow with the momentum and vorticity transfer theories yields

$$l_T = l_P\sqrt{2} \tag{2-7}$$

where $l_P$ is the Prandtl mixing length and $l_T$ the Taylor mixing length. However, this simple relationship is not apparent for the case of axially symmetric jets.

The same assumptions are used in the solution of Eq. 2-1 and 2-2 with the vorticity transfer theory as with the momentum transfer theory. From Eq. 2-7, it is then apparent that the vorticity and momentum transfer theories for two-dimensional jets will give similar velocity profiles, which is not true for axially symmetric jets. Temperature profiles based on vorticity transfer in axially symmetric heated jets are based on the same differential equation of heat transfer utilized in the momentum transfer theory (Eq. 2-6). However, with the vorticity transfer theory, the boundary conditions used to determine temperature distributions are unequal to those used for velocity distributions. Thus, in axially symmetric jets, the velocity and temperature distributions are not identical as in the case of the momentum transfer theory.

The constant exchange coefficient theory is based on a constancy of the turbulent exchange coefficients across the jet wake at any transverse cross section. Prandtl suggested expressing $\epsilon$, the turbulent exchange coefficient or coefficient of eddy kinematic viscosity, as

$$\epsilon = kb(u_{max} - u_{min}) \tag{2-8}$$

where $b$ is the width of the mixing zone, $k$ a dimensionless proportionality factor, and $u_{min}$ the minimum velocity at the jet transverse cross section.

Pai [12] indicates that

$$\epsilon = \epsilon_0 \left(\frac{x}{l}\right)^n \tag{2-9}$$

where $l$ is a characteristic length, $n$ an exponent having a value between 0 and 1, and $\epsilon_0$ an empirical constant, determinable by experiment and having the dimensions of $\epsilon$. The value of $n$ depends upon the mixing condition. For the mixing of two uniform streams, $n$ equals 1. For the mixing of the flow from a two-dimensional jet with a surrounding stream of slightly different temperature and mean velocity, $n = 0$.

*Table F,2b*

| Property transferred | Laminar | Turbulent |
|---|---|---|
| Momentum | $\frac{\tau}{\rho} = \nu \frac{\partial u}{\partial y}$ | $\frac{\tau}{\rho} = \nu_t \frac{\partial u}{\partial y}$ |
| Heat | $\frac{Q}{\rho c_p} = \frac{k}{\rho c_p} \frac{\partial T}{\partial y}$ | $\frac{Q}{\rho c_p} = \frac{k_t}{\rho c_p} \frac{\partial T}{\partial y}$ |
| Mass | $N = D \frac{\partial C}{\partial y}$ | $N = D_t \frac{\partial C}{\partial y}$ |

Table F,2b from [4] compares the exchange coefficients of momentum, heat, and mass transfer in turbulent and laminar flow.

The quantities in the table are defined as follows:

$c_p$ = specific heat at constant pressure (per unit mass)
$C$ = concentration per unit volume
$D$ = coefficient of mass diffusion
$k$ = thermal conductivity
$N$ = diffusion mass flow, concentration per unit area per unit time
$Q$ = heat flow, energy per unit area per unit time
$\nu$ = kinematic viscosity of fluid

The subscript $_t$ indicates turbulent conditions, and all exchange coefficients $\nu$, $k/\rho c_p$, and $D$ have the units of $(L^2T^{-1})$. Now,

$$\frac{\text{Momentum exchange coefficient}}{\text{Mass exchange coefficient}} = \frac{\nu}{D} = \frac{\mu}{\rho D} \tag{2-10}$$

and is known as Schmidt number $Sc$. Also,

$$\frac{\text{Momentum exchange coefficient}}{\text{Temperature exchange coefficient}} = \frac{\nu}{k/\rho c_p} = \frac{\mu c_p}{k} \tag{2-11}$$

and is known as the Prandtl number $Pr$ where $\mu$ is the absolute viscosity of the fluid. The corresponding ratios of turbulent exchange coefficients are known as the turbulent Schmidt and Prandtl numbers. The Schmidt and Prandtl numbers have been defined as properties of the fluids in laminar flow. However, the turbulent process and flow dynamics determine the turbulent Schmidt and Prandtl numbers.

Von Kármán's similarity theory is regarded [*13*] essentially as an extension of the momentum transfer theory. The fundamental assumption is that the mechanism of turbulence at all locations in the fluid is similar, differing only in the units of length and time used, that is, the eddying motions referred to axes moving with local mean velocities are similar at all points. From this assumption, the mixing length $l$ is expressed as

$$l = k\left|\frac{\partial u}{\partial y}\left(\frac{\partial^2 u}{\partial y^2}\right)^{-1}\right| \tag{2-12}$$

where $k$ is an empirical constant of universal character.

Reichardt [*14*] suggested from experimental results that the momentum profiles in axially symmetric jets could be correlated by an error function or probability curve of the form of

$$\overline{u^2} = \frac{K}{b^2}\,e^{-(r/b)^2} \tag{2-13}$$

where $b$ is a function proportional to jet width (dependent only on $x$), $K$ a constant depending on the jet diameter and initial velocity, and the bar indicates a time-averaged value. In investigating the assumptions necessary that Eq. 2-13 be a solution for the equation of motion, Reichardt considered a point source, axially symmetric, free turbulent jet, and indicated that

$$\overline{\rho v u} = -\frac{b}{2}\frac{db}{dx}\frac{\partial(\overline{\rho u^2})}{\partial r} \tag{2-14}$$

where $u$, $v$, and $\rho$ are local, instantaneous factors; that is, the momentum flux density in a radial direction is assumed proportional to the radial gradient of the axial momentum flux density. The theory was generalized

[5] to include the transport of heat and mass in free jets. For axially symmetric jets, the flux transport could be specified as

$$\frac{\partial(\overline{\psi u})}{\partial x} + \frac{1}{r}\frac{\partial}{\partial r}(r\overline{\psi v}) = 0 \tag{2-15}$$

where $\psi$ represents a local, instantaneous flux such that

| Flux transported | $\psi$ |
|---|---|
| Momentum | $\rho u$ |
| Heat | $\rho c_p(T - T_0)$ |
| Mass | $\rho w_i$ |

where $T$ is the local, instantaneous absolute temperature; $T_0$ the ambient absolute temperature; and $w_i$ the local, instantaneous concentration of jet fluid. A solution of Eq. 2-15 is

$$\frac{(\overline{\psi u})_{x,r}}{(\overline{\psi u})_{0,0}} = \frac{d_j^2}{4C_\psi^2 x^2}\, e^{-(r/c_\psi x)^2} \tag{2-16}$$

where $C_\psi$ is a spreading coefficient for the flux and $d_j$ the jet nozzle diameter. This solution is based on the assumptions of the validity of Eq. 2-15, a point source jet, and a similarity of flux distribution profiles at any transverse cross section in the jet.

Various investigators have used the von Kármán momentum equation or the von Kármán integral form of the boundary layer equation in analyses of jet mixing. The integral momentum equation may be written for a two-dimensional plane jet [15] as

$$\int_{a(x)}^{b(x)} \frac{\partial(\rho u^2)}{\partial x}\, dy = -[b(x) - a(x)]\frac{\partial p}{\partial x} + [\tau]_{a(x)}^{b(x)} - [\rho uv]_{a(x)}^{b(x)} \tag{2-17}$$

where $a(x)$ and $b(x)$ are variable limits. For axially symmetric jets, Eq. 2-17 becomes

$$\int_0^{r(x)} \frac{\partial(\rho u^2)}{\partial x}\, rdr = -r(x)\frac{\partial p}{\partial x} + [r\tau]_0^{r(x)} - [r\rho uv]_0^{r(x)} \tag{2-18}$$

The von Kármán or integral momentum equations have been especially useful in application to coaxial jets.

*Point source diffusion of momentum, mass, or heat.* The differential equations of diffusion may be derived from the empirical Fick and Biot-Fourier laws by equating the amount of the diffusing quantity entering an infinitesimal volume to the time-rate of change of the total amount of the diffusing quantity in that infinitesimal volume. For the case of combined turbulent and molecular diffusion of mass in a homogeneous, isotropic, turbulent field without local sources or sinks, the differential equation of diffusion is

$$\frac{\partial C}{\partial t} = (D_m + d_m) \sum_i \frac{\partial^2 C}{\partial x_i} \qquad (i = 1, 2, 3) \tag{2-19}$$

where $d_m$ = coefficient of molecular diffusion of mass, a constant; $D_m$ = coefficient of turbulent diffusion of mass; $t$ = time; and $x_i$ = rectangular coordinates. In a similar manner, the differential equation of the combined molecular and turbulent transport of heat in a turbulent field that is homogeneous, isotropic, and source- and sink-free may be written as

$$\frac{\partial h}{\partial t} = \left(D_h + \frac{k}{\rho c_p}\right) \sum_i \frac{\partial^2 h}{\partial x_i^2} \qquad (i = 1, 2, 3) \tag{2-20}$$

where $D_h$ is the coefficient of turbulent heat transport and $h$ the specific enthalpy. The derivation of Eq. 2-19 and 2-20 depends on the experimentally verified fact [*16*] that, in a homogeneous turbulence, the displacements of fluid particles have a Gaussian probability density.

Exact solutions of the preceding differential equations can be made for a variety of transport problems only if the functional forms of the diffusion coefficients are known. The effect of temperature and pressure on the molecular transport coefficient $d_m$ and $k/\rho c_p$ for most gaseous systems can be found in the literature. The turbulent diffusion coefficients $D_m$ and $D_h$ can be determined experimentally, or may be predicted theoretically from the knowledge of two fundamental parameters of the turbulent field.

The solution for an instantaneous point source of diffusion of mass injected into a stream having a velocity $u_0$ is [*17*]

$$C = \frac{K^*}{(4\pi\omega)^{\frac{3}{2}}} e^{\frac{-[(x-u_0t)^2+r^2]}{4\omega}} \tag{2-21}$$

where

$C$ = concentration of diffusing quantity at point $x$, $r$ and time $t$, lb/ft³

$K^*$ = strength of point source, lb

$r$ = radial distance from streamline passing through point source, ft

$x$ = distance downstream of point source, ft

$\omega$ = standard square deviation of fluid particles, or spreading coefficient, ft²

$\omega$, which is a function of time, may be determined from [*17*]

$$\omega = \int_0^t (D_m + d_m)dt \tag{2-22}$$

The solution for the diffusion of heat from a point source would be similar.

The instantaneous point source solution may be used to solve for the case of a continuous point source [18]. The concentration $C$ at a point $x, r$ may be obtained by a summation of contributions to that point by instantaneous point sources located at the point $(x - u_0 t), 0, 0$:

$$C = \frac{m^*}{(4\pi)^{\frac{3}{2}}} \int_0^\infty \frac{1}{\omega^{\frac{3}{2}}} e^{-\frac{[(x-u_0 t)^2 + r^2]}{4\omega}} dt \tag{2-23}$$

where $m^*$ is the weight-flow rate (lb/sec) of material issuing from the point source. With the exception of the limiting cases of very small and very large distances from the point source, the functional form of $\omega$ does not permit direct integration of Eq. 2-23. However, graphical integration may be possible. An approximate solution of Eq. 2-23 is [19]

$$\frac{C}{m^*} = \frac{1}{4\pi u_0 \omega} e^{-r^2/4\omega} \tag{2-24}$$

where $\omega$ is evaluated at $t = x/u_0$.

*Coaxial jets.* Many data, both theoretical and experimental, have been reported for jets entering moving fluids. Such systems may be somewhat more applicable to the mixing problem in turbojet engine combustors in which the primary and secondary air enter through openings in the combustor liners.

In a coaxial jet system, the jet fluid emanates into a fluid moving parallel to the jet axis. The presence of the external stream complicates the specification of the boundary conditions at the jet boundary. For coaxial jets, the Kármán momentum equations have proved practical in obtaining solutions [4,20]. One of the major differences in using the Kármán momentum equations to determine the mixing in coaxial jets and using the momentum or vorticity transfer theory with free jets is the specification of a velocity profile at any transverse cross section of the jet. For example, Squire and Trouncer [20] specified the lateral velocity profile in the mixing region of two coaxial isothermal jets within the axial length of the potential cone as

$$u = u_0 + \left(\frac{u_j - u_0}{2}\right)\left[1 - \cos \pi \left(\frac{r_0 - r}{r_0 - r_p}\right)\right] \tag{2-25}$$

In the fully developed jet, the lateral velocity profile was specified as

$$u = u_0 + \left(\frac{u_{max} - u_0}{2}\right)\left(1 + \cos \pi \frac{r}{r_0}\right) \tag{2-26}$$

where all velocities are in the $x$ direction, and

$r$ = radial distance from jet axis to point where $u$ is measured
$r_0$ = radius of fully developed jet
$r_p$ = radius of potential cone
$u$ = velocity at any position in jet
$u_j$ = velocity of jet
$u_0$ = velocity of coaxial stream

In addition, the turbulent shear stress was written as with the momentum transfer theory in free jets and the mixing length $l$ was assumed proportional to the width of the mixing region. Thus, within the potential cone,

$$l = c(r_0 - r_p) \tag{2-27}$$

and in the fully developed flow region,

$$l = cr_0 \tag{2-28}$$

where $c$ is a proportionality constant, characteristic of the turbulent motion under consideration. The equations of motion for the region of fully developed flow and the region containing the potential cone were solved by use of proper boundary conditions with the requirement that continuous values of jet radii and velocities exist at the intersection of the two regions. Additional assumptions made in the solution were:

(1) Incompressible flow
(2) Pressure gradients near jet small enough to be neglected
(3) Constant velocity throughout potential cone
(4) Constant momentum in $x$ direction at any jet section
(5) Mixing length parameter $c$ equal to $(0.0067)^{1/2}$. This differs slightly, approximately 14 per cent, from a value found in another reference [*21*].

*Jets penetrating a moving stream.* A major portion of the air entering the combustion zone of a turbojet engine combustor enters a medium differing in velocity, temperature, density, chemical composition, and general direction of motion. The designer of turbojet combustors, therefore, is primarily interested in jet penetration and spreading in terms of velocity, temperature, and concentration profiles.

Jets entering a moving stream may enter at any angle from 0 to 180°. As such, the complex mixing problem does not lend itself to a ready solution. However, Ehrich [*22*] has analyzed the problem by means of a simplified two-dimensional potential flow analysis. Essentially, Ehrich assumed that a jet entering a moving stream at an angle is separated from the main stream by a streamline and has a constant velocity. Thus no discontinuities exist across the streamline. The jet wake downstream of the jet was assumed to be an area of dead fluid separated from the jet by a constant pressure vortex sheet.

These assumptions oversimplify the problem of jet penetration into moving streams. However, the analysis leads to a convenient method of calculating the relation between geometrical and velocity parameters.

Hawthorne, Rogers, and Zaczek [23] present a theoretical analysis for penetration of normal jets into flowing gas streams. Velocity and temperature distributions for free jets and a negligible pressure gradient at the mixing section were assumed. Results of the analysis indicated that penetration of normal jets could be expressed as

$$\text{Relative penetration } \frac{l}{h} = f\left[\left(\frac{\text{Momentum of jet}}{\text{Momentum of fluid stream}}\right)\left(\frac{\text{Volume flow of fluid stream}}{\text{Total volume flow of fluid stream and jet}}\right)^2\right] \quad (2\text{-}29)$$

where $h$ is the duct depth and $l$ the penetration of the jet.

For penetration of circular normal jets into fluid streams in rectangular ducts of width $d'$ and depth $h$, penetration at a distance $x/h$ from the orifice is suggested as

$$\text{Relative penetration } \frac{l}{h} = f\left(\frac{\frac{d_j}{h}\frac{u_j}{u_0}\sqrt{\frac{T_0}{T_j}}}{1 + \frac{\pi}{4}\frac{d_j^2}{hd'}\frac{u_j}{u_0}}\right) \quad (2\text{-}30)$$

where $d_j$ is the jet orifice diameter, $T_j$ the absolute temperature of jet fluid, $T_0$ the absolute temperature of fluid stream, $u_j$ the jet velocity, and $u_0$ the fluid stream velocity.

Two expressions for the penetration of multiple jets into fluid streams are also presented in [23] from theoretical considerations. For the penetration of $n$ jets entering a fluid stream at right angles through $n$ circular orifices spaced laterally across a rectangular duct of width $d'$ and depth $h$, penetration at a distance $x/h$ from the orifice is suggested as

$$\text{Relative penetration } \frac{l}{h} = f\left(\frac{\frac{d_j}{h}\frac{u_j}{u_0}\sqrt{\frac{T_0}{T_j}}}{1 + n\frac{\pi}{4}\frac{d_j^2}{hd'}\frac{u_j}{u_0}}\right) \quad (2\text{-}31)$$

For the penetration of $n$ jets entering a fluid stream at right angles through $n$ circular orifices equally spaced circumferentially around a duct of diameter $d_{dt}$, the expression suggested for a distance $x/d_{dt}$ from the orifice is

$$\text{Relative penetration } \frac{l}{d_{dt}} = f\left[\frac{\frac{d_j}{d_{dt}}\frac{u_j}{u_0}\sqrt{\frac{T_0}{T_j}}}{1 + n\left(\frac{d_j}{d_{dt}}\right)^2\frac{u_j}{u_0}}\right] \quad (2\text{-}32)$$

*Pressure changes due to mixing.* In all the mixing methods discussed, the mixing does not occur without any energy change. This change in energy is reflected in a change in total or stagnation pressure in the mixing streams. In [24], Hawthorne and Cohen present a theoretical analysis for the pressure losses due to heat release and mixing in frictionless, compressible flow. Complete mixing of the fluid streams after their junction was assumed. $\gamma$, the specific heat ratio for gases, and $\mathcal{R}$, the gas constant, were assumed to be constant throughout the mixing process. In addition to mixing due to heat release, three types of jet mixing—normal, oblique, and parallel jets—were considered. These mixing types and their nomenclature are shown in Fig. F,2b.

| Mixing, gas streams | Arrangement | B | C |
|---|---|---|---|
| Normal | 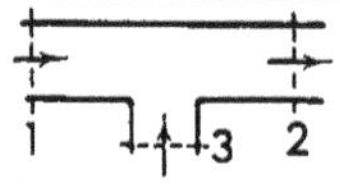 1 3 2 | 1 | $\sqrt{\left(1+\frac{m_3}{m_1}\right)\left(1+\frac{m_3}{m_1}\frac{T_3^0}{T_1^0}\right)}$ |
| Oblique | 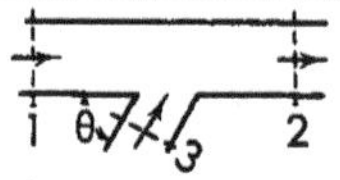 1 θ 3 2 | $1+\frac{m_3}{m_1}\frac{u_3}{u_1}\cos\theta$ | $\sqrt{\left(1+\frac{m_3}{m_1}\right)\left(1+\frac{m_3}{m_1}\frac{T_3^0}{T_1^0}\right)}$ |
| Parallel | 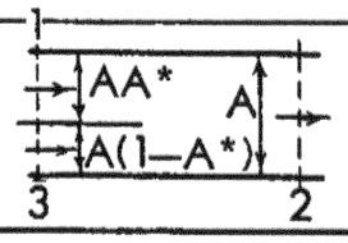 1 AA* A A(1−A*) 3 2 | $A^*\left(1+\frac{m_3}{m_1}\frac{u_3}{u_1}\right)$ | $A^*\sqrt{\left(1+\frac{m_3}{m_1}\right)\left(1+\frac{m_3}{m_1}\frac{T_3^0}{T_1^0}\right)}$ |

Fig. F,2b. Arrangements for mixing of gas streams [24].

For mixing of gas streams, it was determined that expressions similar to those defining pressure changes due to total or stagnation temperature changes in constant area ducts might be utilized. These expressions are [25]

$$\frac{T_2^0}{T_1^0} = \left[\frac{M_2(1+\gamma M_1^2)}{M_1(1+\gamma M_2^2)}\right]^2 \left(\frac{1+\frac{\gamma-1}{2}M_2^2}{1+\frac{\gamma-1}{2}M_1^2}\right) \tag{2-33}$$

$$\frac{p_2^0}{p_1^0} = \left(\frac{1+\gamma M_1^2}{1+\gamma M_2^2}\right)\left(\frac{1+\frac{\gamma-1}{2}M_2^2}{1+\frac{\gamma-1}{2}M_1^2}\right)^{\frac{\gamma}{\gamma-1}} \tag{2-34}$$

The subscripts $_1$ and $_2$ represent before and after stagnation temperature change, respectively, and the superscript $^0$, stagnation conditions. These equations are expressed in Fig. F,2c for an increase in stagnation temperature.

For mixing of normal gas streams, Eq. 2-33 and 2-34, or Fig. F,2c (if applicable), can be used to determine stagnation pressure changes, if $T_2^0/T_1^0$ is replaced by

$$\frac{T_2^0}{T_1^0} = \left(1 + \frac{m_3}{m_1}\right)\left(1 + \frac{m_3 T_3^0}{m_1 T_1^0}\right) \tag{2-35}$$

where $m_1$ is the weight flow rate in one gas stream, $m_3$ the weight flow rate in the other gas stream, $T_1^0$ the stagnation temperature in the stream having a flow rate $m_1$, and $T_3^0$ the stagnation temperature in the stream

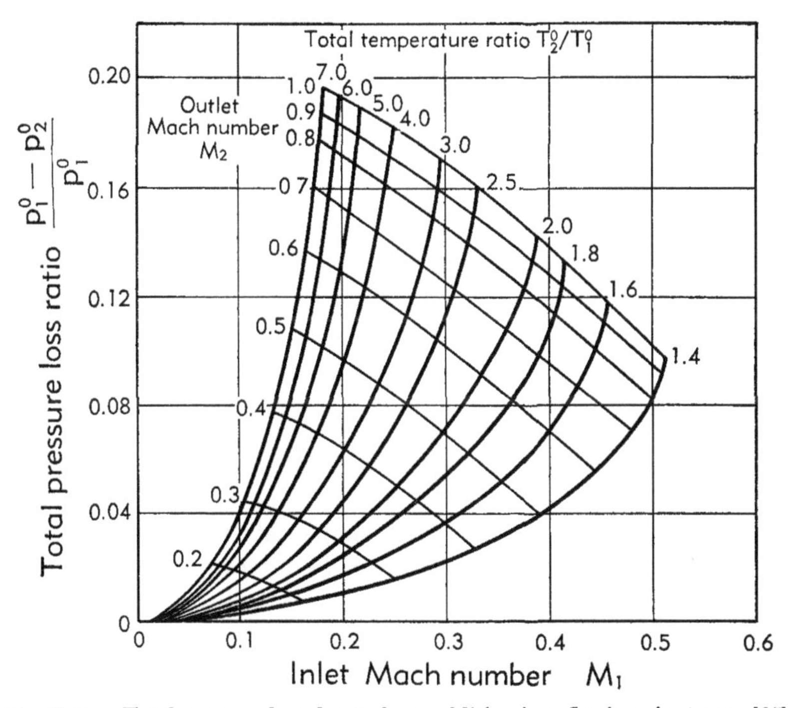


Fig. F,2c. Total pressure loss due to heat addition in a flowing air stream [25].

having a flow rate $m_3$; the subscript $_2$ indicates after full mixing of streams 1 and 3.

The mixing of oblique and parallel gas streams is somewhat more complicated. Additional parameters derived in [24] include $B$ and $C$, which are defined for oblique and parallel mixing jets as

| *Jets* | $B$ | $C$ |
|---|---|---|
| Oblique | $1 + \frac{m_3 u_3 \cos\theta}{m_1 u_1}$ | $\sqrt{\left(1 + \frac{m_3}{m_1}\right)\left(1 + \frac{m_3 T_3^0}{m_1 T_1^0}\right)}$ |
| Parallel | $A^*\left(1 + \frac{m_3 u_3}{m_1 u_1}\right)$ | $A^*\sqrt{\left(1 + \frac{m_3}{m_1}\right)\left(1 + \frac{m_3 T_3^0}{m_1 T_1^0}\right)}$ |

where

$A^*$ = ratio of jet flow area $A_1$ to mixed-jet flow area $A_2$
$u_1$ = velocity in stream having a flow rate $m_1$
$u_3$ = velocity in stream having a flow rate $m_3$
$\theta$ = included angle between two gas streams

$M_2$, the Mach number after complete mixing, can be determined from Eq. 2-33, or Fig. F,2c (if applicable), if $T_2^0/T_1^0$ is replaced by

$$\frac{T_2^0}{T_1^0} = \left[C\left(\frac{1 + \gamma M_1^2}{1 + B\gamma M_1^2}\right)\right]^2 \tag{2-36}$$

The stagnation pressure change due to mixing of oblique or parallel gas streams can then be found by substitution of the values of $B$ and $M_2$ in

$$\frac{p_2^0}{p_1^0} = \left(\frac{1 + B\gamma M_1^2}{1 + \gamma M_2^2}\right)\left(\frac{1 + \frac{\gamma - 1}{2} M_2^2}{1 + \frac{\gamma - 1}{2} M_1^2}\right)^{\frac{\gamma}{\gamma - 1}} \tag{2-37}$$

where $M_1$ is the Mach number in a stream having a flow rate $m_1$. Similar expressions accounting for variations in $\gamma$ and $\mathcal{R}$ during mixing of gas streams have been derived. However, the expressions are more complicated than the preceding ones.

**F,3. Experiments on Mixing and Penetration in Simple Jet Systems.** In this article, some of the experimental results obtained in studies of mixing and penetration in simple jet systems are discussed. The jet mixing studies are treated primarily in terms of the jet velocity, temperature, and concentration profiles.

*Free jet velocity profiles.* Much of the experimental data relative to gas jets concern free jets mixing with air. In general, data are presented in terms of velocity distributions in both the potential cone in axially symmetric jets and the region of fully developed jet flow. Equations written to fit the experimental data for isothermal free jets mixing with air are presented in [*26*]. Data are expressed in terms of momentum flux velocity ratios, in which

$$\text{Momentum flux velocity} = \sqrt{\frac{\overline{\rho u^2}}{\bar{\rho}}} \tag{3-1}$$

where $u$ is the instantaneous velocity in jet in $x$ direction at any radius $r$ and longitudinal distance $x$, $\rho$ the instantaneous density of fluid, and the bar indicates time-averaged values. The value of $\overline{\rho u^2}$ is determined from impact tube measurements, and $\bar{\rho}$ is determined from static temperature and pressure. For incompressible flow, the momentum flux velocity is

$\sqrt{\overline{u^2}}$. Momentum flux velocity ratios of an isothermal free jet are presented in Fig. F,3a (1) for the potential cone, a transition region at the end of the potential cone, and the region of fully developed flow. Many of the data are presented, as in many references, in terms of a half radius, that is, the jet radius at which the momentum flux velocity equals one half the momentum flux velocity on the jet axis at the same longitudinal distance from the jet orifice. The experimental data indicate that: (1) The potential cone within which the flux velocity ratios are constant radially is approximately 3.5 jet orifice diameters in length. (2) The radius of the potential cone could be expressed as

$$\frac{r_p}{d_j} = \frac{3.5 - (x/d_j)}{7.0} \tag{3-2}$$

where $d_j$ is the diameter of jet, $r_p$ the potential cone radius at any longitudinal distance $x$, and $x$ the longitudinal distance from the jet orifice. Velocity ratios within the potential cone (less than 3.5 jet diameters downstream) closely approached a single generalized curve. Within the transition region, however, a different curve could be drawn than for distances farther downstream. Radial flux velocity ratios, covering a range of jet velocities from 166 to 801 ft/sec and distances from 10 to 30 jet diameters downstream of the jet orifice, were correlated by the following expression:

$$\sqrt{\frac{\overline{u^2}}{\overline{u^2_{max}}}} = e^{-\ln 2(r/r_{\frac{1}{2}})^2} \tag{3-3}$$

where $r_{\frac{1}{2}}$ is the half radius at longitudinal distance $x$ and $u_{max}$ the velocity in $x$ direction on the axis at longitudinal distance $x$.

Some data have been reported [*4*] from an investigation of momentum and mass transfer in coaxial jets. Isothermal experiments were conducted with homogeneous mixtures of air and helium flowing from $\frac{1}{4}$-in.- and 1-in.-diameter nozzles into a coaxial stream of air in a 4-in.-diameter duct. Extrapolation of the experimental results to the case of a free jet, that is, zero coaxial stream velocity, shows that: (1) The potential cone is 4.0 jet orifice diameters in length. (2) The velocity in the fully developed flow region (downstream of the potential cone) can be represented by

$$\frac{u}{u_{max}} = \frac{1}{2}\left(1 + \cos\frac{\pi r}{2r_{\frac{1}{2}}}\right) \tag{3-4}$$

The potential cone length is slightly larger than the experimental value (3.5 diameters) of [*26*] and a theoretical value (3.1 diameters) of [*20*]. However, the lateral velocity distribution agrees exactly with the distribution assumed by Squire and Trouncer (Eq. 2-26) and fairly well with the experimental curve of [*26*], as shown in Fig. F,3a(3).

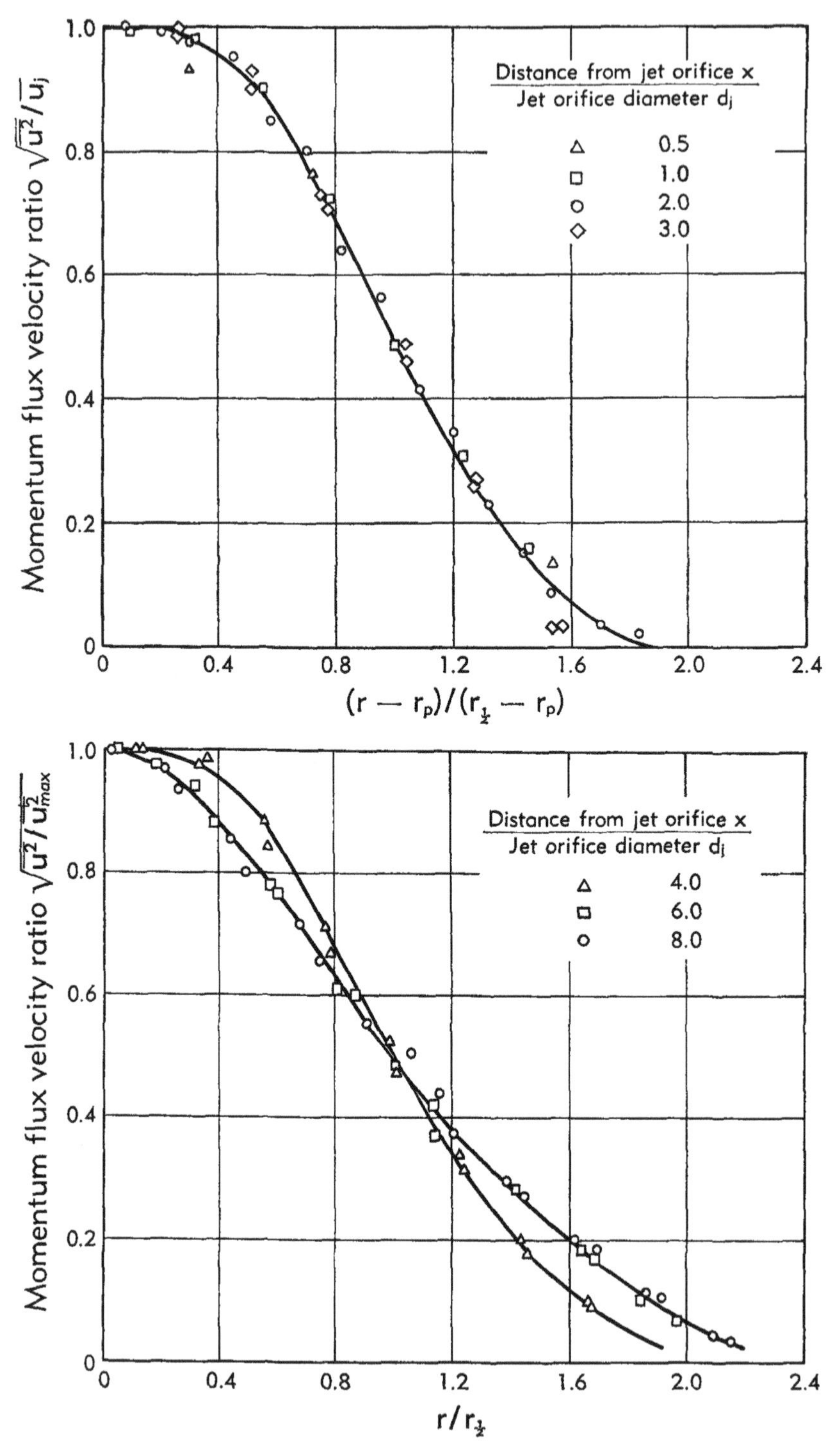


**Fig. F,3a.** Radial distribution of momentum flux ratios in isothermal axially symmetric free jets of air. 0.9-in.-diameter, long throat nozzle, $\bar{u}_j = 390$ ft/sec. (1) potential cone region [26]; (2) transition region [26]; (3) fully developed flow region [26].

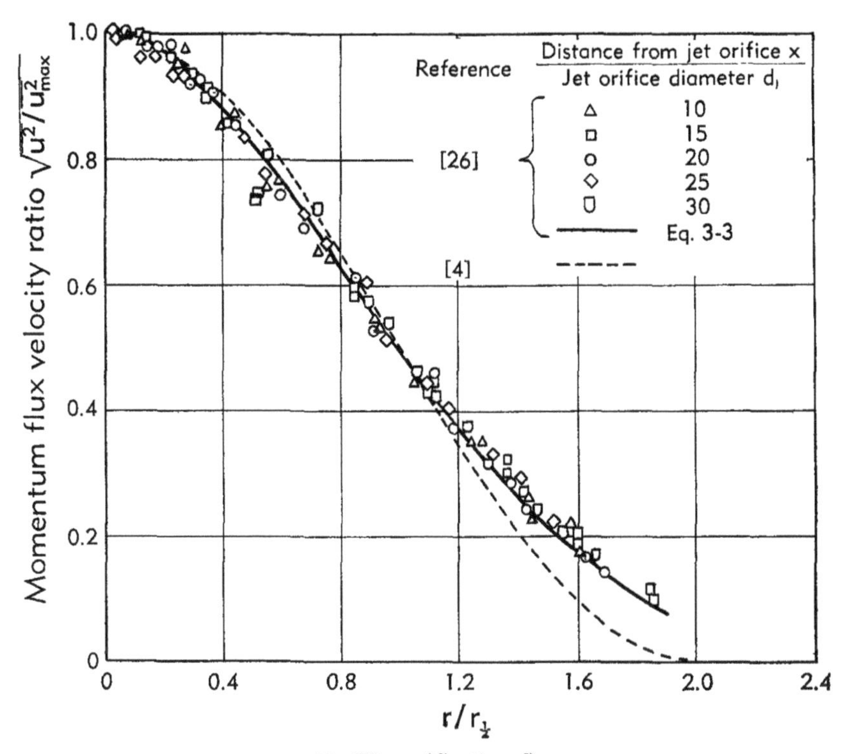


Fig F3a. (*Continued*)

Axial velocity distribution within the fully developed flow region of the circular isothermal free jets [*26*] could be closely approximated by

$$\frac{\sqrt{\overline{u^2_{max}}}}{\overline{u_j}} = \frac{6.63}{(x/d_j)} \tag{3-5}$$

where $u_j$ is the velocity of the jet. The curve of Fig. F,3b and the experimental data agree closely with the previously discussed theoretical analysis of a symmetrical jet in which the assumption of constancy of momentum at any jet section resulted in

$$u_{max} \sim \frac{1}{x}$$

Eq. 3-5 for axial velocity distribution agrees closely with an expression of Squire [*27*], who recommended the use of the relation

$$\frac{u_{max}}{u_j} = \frac{6.5}{(x/d_j)} \tag{3-6}$$

Extrapolation of the data of [*4*] for coaxial jets to a free jet gives the axial velocity distribution as

$$\frac{u_{max}}{u_j} = \frac{4}{(x/d_j)} \tag{3-7}$$

The large discrepancy between Eq. 3-7 and Eq. 3-5 and 3-6 may be due in part to the experimental configuration of [4].

Many data relative to axially symmetric free jets in air have been summarized in [27]. Half-velocity ratio radii in the fully developed flow region of a free jet in air are presented in Fig. F,3c. Experimental data from [26,28] and theoretical data from [20,14] are also included. Reichardt [14] expressed the half radius at any longitudinal position as

$$r_{\frac{1}{2}} = 0.0725x\sqrt{\ln 4} \tag{3-8}$$

This expression and the experimental data of [26] agree closely with the data of Fig. F,3c. For distances greater than 10 jet diameters from the

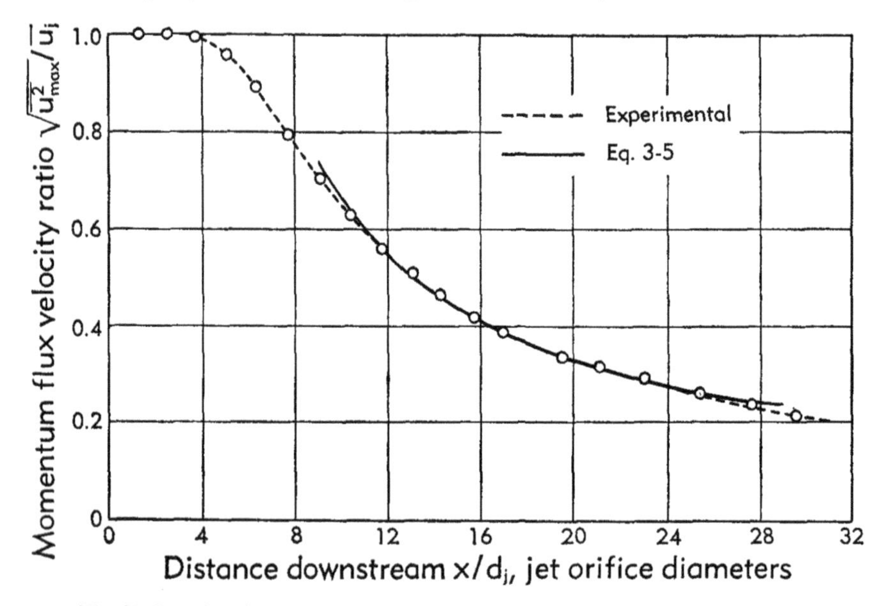


Fig. F,3b. Axial velocity distribution in an axially symmetric isothermal free jet in air. 0.9-in.-diameter, long throat nozzle, $\bar{u}_j$ = 390 ft/sec [26].

orifice, the data indicate that the lines of half-velocity ratio lie on a cone of half angle of 5°, with the apex of the cone at the center of the jet orifice.

A small amount of experimental data included in [27] for 0.1-velocity ratio lines is also presented in Fig. F,3c. For distances greater than 10 jet diameters from the orifice, the 0.1-velocity ratio lines are straight lines emanating from the center of the jet orifice at an angle of 9.2°.

A small amount of experimental data pertaining to the velocity distribution in two-dimensional jets is available. For two-dimensional and axially symmetric free jets, theoretical analyses have indicated that: (1) for two-dimensional jets, $u_{max} \sim x^{-\frac{1}{2}}$; (2) for axially symmetric jets, $u_{max} \sim x^{-1}$. Using these relations, Squire [27] infers that expressions applicable to axially symmetric jets may be applied to two-dimensional jets by replacing $x/d_j$ with $(x/y)^{\frac{1}{2}}$ wherever $x/d_j$ appears. In this case, $y$ is the initial width of the two-dimensional jet. Howarth [9] presents some experimental data from a previous investigator on the velocity profiles in a two-dimensional free jet and compares them with the theoretical distribution derived from the momentum-transfer and modified

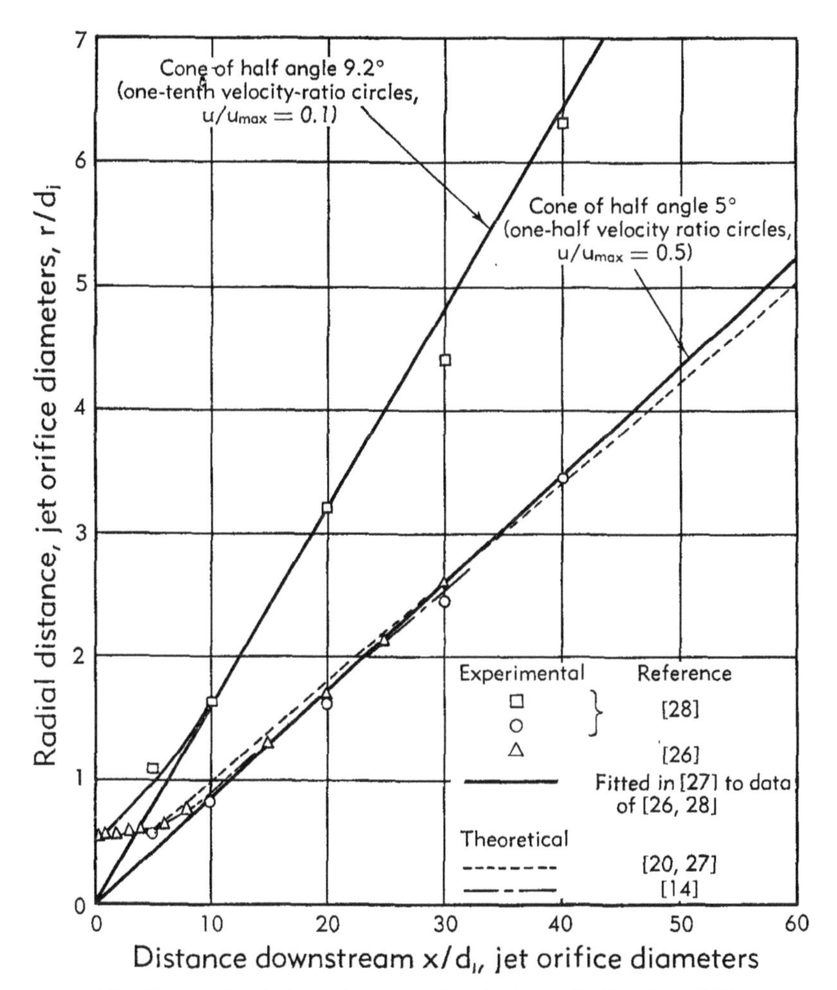


Fig. F,3c. One-half- and one-tenth-velocity ratio lines in axially symmetric free jets in air. Original figure from [27].

vorticity-transfer theories, both of which give the same theoretical velocity profile. Excellent agreement of experimental and theoretical data is indicated.

The large amount of data on free jets includes many expressions for velocity distributions. For free jets, recommended expressions for the radius of the potential cone, and lateral and axial velocity distributions,

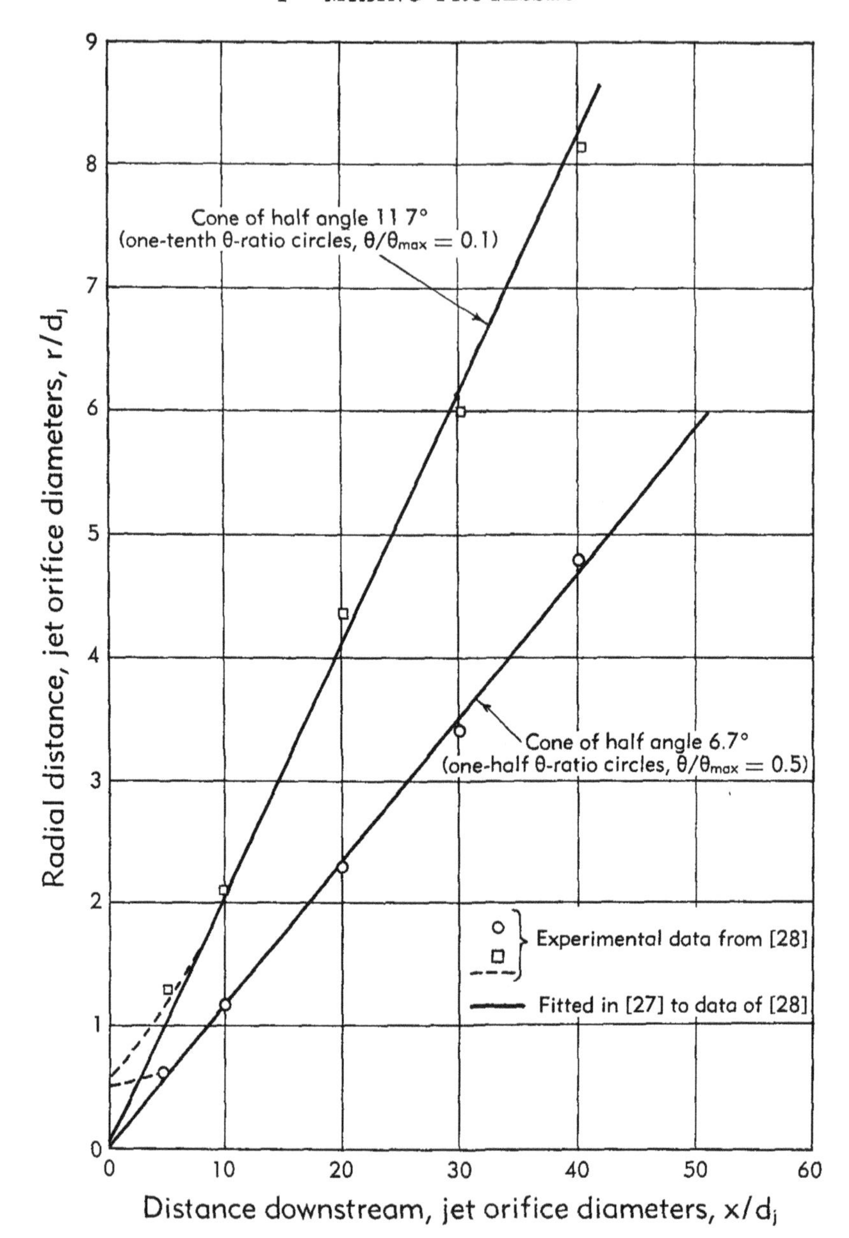


Fig. F,3d. One-half- and one-tenth-temperature coefficient ratio lines in axially symmetric free jets in air. Figure from [27].

are Eq. 3-2, 3-3, and 3-5, respectively. One-half- and 0.1-velocity ratio radii can be determined from Fig. F,3c.

*Free jet temperature profiles.* Experimental distributions of temperature in axially symmetric free jets are given in Fig. F,3d. Results are expressed in terms of half-temperature and 0.1-temperature coefficient ratio lines. In the same manner as the velocity ratios, the temperature coefficient ratios are defined as the ratio of the temperature coefficient $\theta$ at any point to the temperature coefficient $\theta_{max}$ at a point on the axis in the same cross-sectional plane. The temperature coefficient $\theta$ is defined [*11,27*] as

$$\theta = \frac{T - T_0}{T_j - T_0} \tag{3-9}$$

where $T$ is the temperature at any point in the jet-mixing region, $T_j$ the temperature at any point in the region of undisturbed flow in the jet, or in the potential cone, and $T_0$ the temperature of the fluid in the space

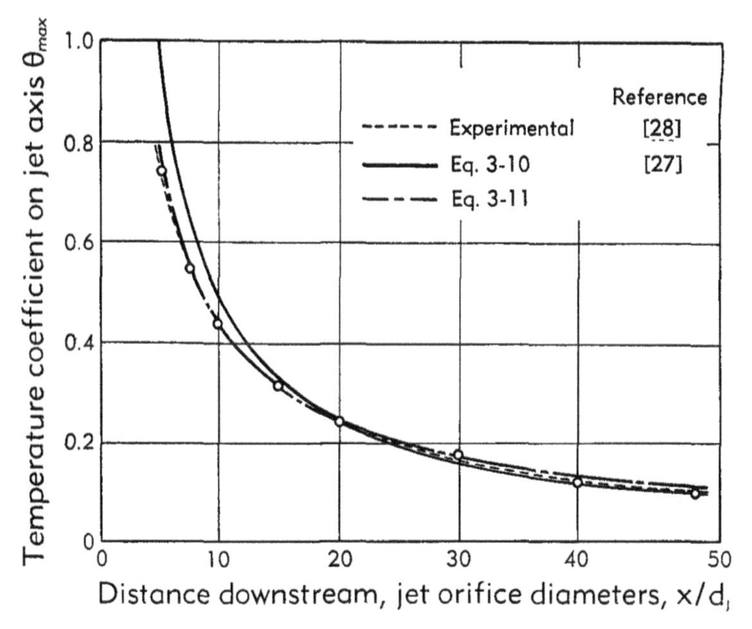


Fig. F,3e. Axial temperature distribution in axially symmetric free jets in air.

surrounding the jet. For distances greater than 5 jet orifice diameters downstream, straight lines of half-temperature coefficient ratios lie on a cone of half angle of 6.7° with the apex of the cone at the center of the jet orifice. For 0.1-temperature coefficient ratios, the distance downstream must be greater than 10 jet orifice diameters, and the half angle is 11.7°.

The longitudinal distribution of temperature in axially symmetric free jets is shown in Fig. F,3e. The temperature of the jet remains approximately constant for a distance of about 5 diameters, or the length of the potential cone. For distances from 10 to 50 jet orifice diameters from the orifice,

$$\theta_{max} = \frac{4.8}{(x/d_j)} \tag{3-10}$$

This equation is recommended by Squire [*27*] as representative of the temperature distribution on the jet axis. A better expression for axial temperature distribution, for distances from 5 to 20 jet orifice diameters from the orifice, is

$$\theta_{max} = 3\left(\frac{x}{d_j}\right)^{-0.845} \tag{3-11}$$

No experimental data showing the temperature distribution in two-dimensional free jets were found. However, Howarth [*9*] assumes that the agreement of experimental and theoretical temperature distributions in two-dimensional free jets is as good as with axially symmetric jets.

Experimental lateral temperature distribution in axially symmetric free jets [*28*] is shown in Fig. F,3f. Data are for a 1-in.- and a 3-in.-diameter heated jet of air entering ambient atmosphere. In the central portion of the jet, the curve of $T/T_{max}$ is lower than $u/u_{max}$ for $(x/d_j)$'s less than 7, and greater than $u/u_{max}$ for $(x/d_j)$'s greater than 7. Corrsin suggests that in the region upstream of the potential cone, the flow may be approximated by the two-dimensional case of a single mixing region between two semi-infinite bodies—one, a moving body of heated air; the other, a stationary body of cooler air. For this case, the assumption is made that within the region of the potential cone the effective heat transfer coefficient is greater than the effective shear coefficient. Thus, plots of $u/u_{max}$ and $T/T_{max}$ show temperatures lower than velocity in the region of the higher values of the velocity and temperature ratios, and temperatures higher than velocity at the low ratios. However, for the fully developed jet, the assumption of the effective heat transfer coefficient greater than the shear coefficient gives temperature profiles greater than velocity profiles at all radial positions.

Fig. F,3g [*29*] compares the axial distribution of velocity and temperature in heated axially symmetric free jets discharging into a medium having approximately twice the initial jet density. From Fig. F,3g, it is apparent that the lower density (hotter) jet has shorter potential cones of velocity and temperature and a greater axial decrease of velocity and temperature than the higher density jet. From these figures, it is concluded that the lower density jet spreads more rapidly. Fig. F,3g and F,3h agree with this conclusion.

*Free jet concentration profiles.* Hawthorne, Weddell, and Hottel [*30*, pp. 266–288] present the following conclusions for axially symmetric free turbulent jets, at axial distances greater than 8 jet orifice diameters from the jet orifice:

(1) The concentration $C$ of jet fluid at any value of axial and radial distance, $x$ and $r$, in the turbulent jet is independent of jet orifice diameter and jet velocity, provided generalized dimensions $r/d_j$ and

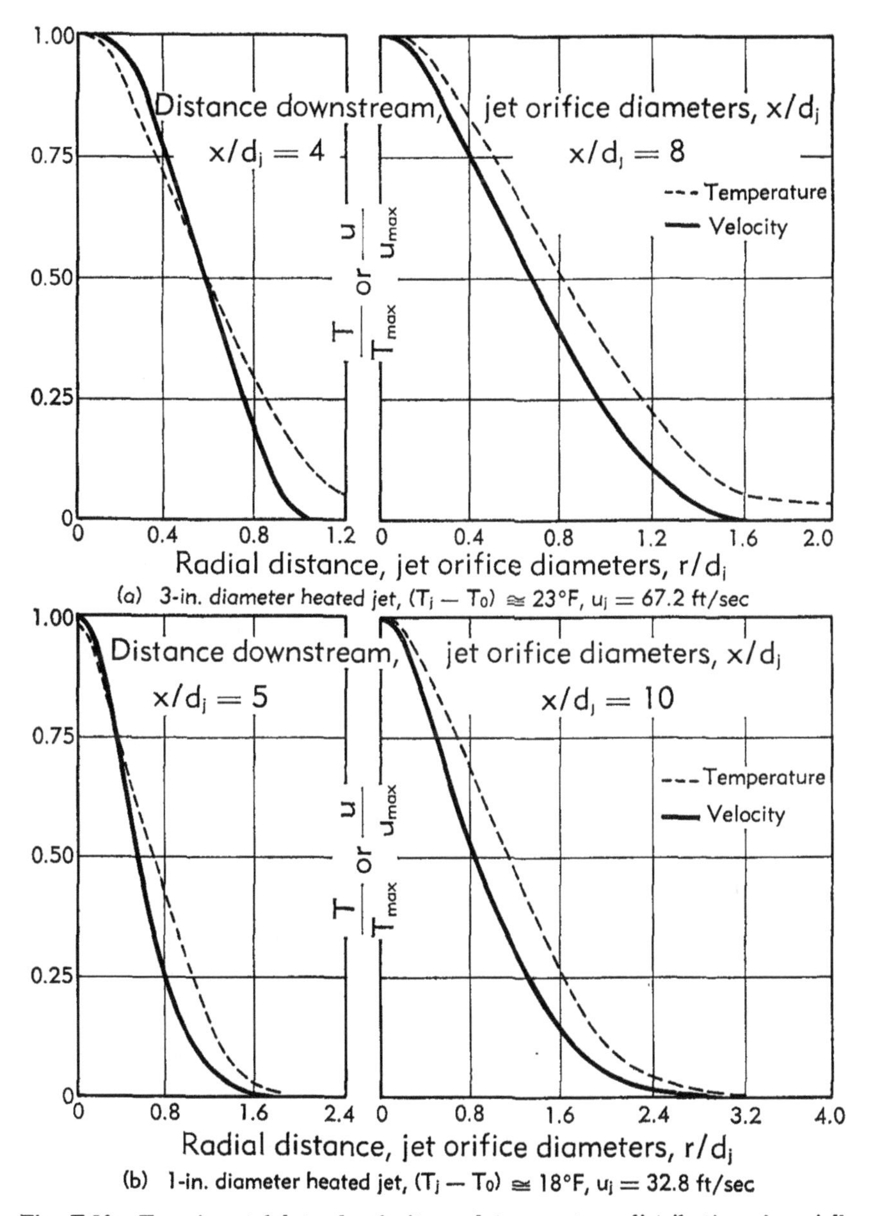


Fig. F,3f. Experimental lateral velocity and temperature distributions in axially symmetric heated jets in air [28]. (Top) 3-in.-diameter heated jet, $(T_j - T_0) \cong 23°F$, $u_j = 67.2$ ft/sec; (bottom) 1-in.-diameter heated jet, $(T_j - T_0) \cong 18°F$, $u_j = 32.8$ ft/sec.

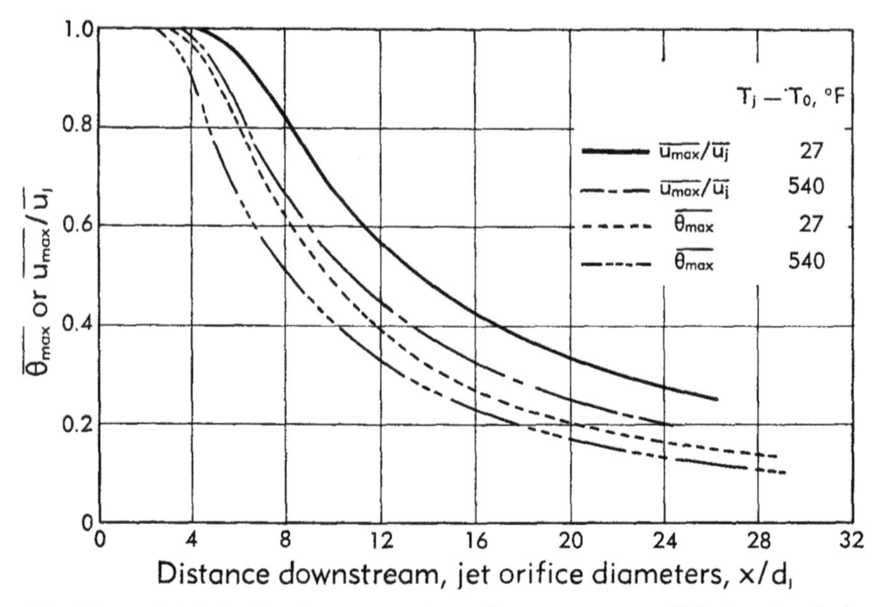


Fig. F,3g. Axial distribution of experimental temperature coefficient and velocity in axially symmetric heated jets in air [29]. Jet diameter = 1 in. Bars indicate mean values.

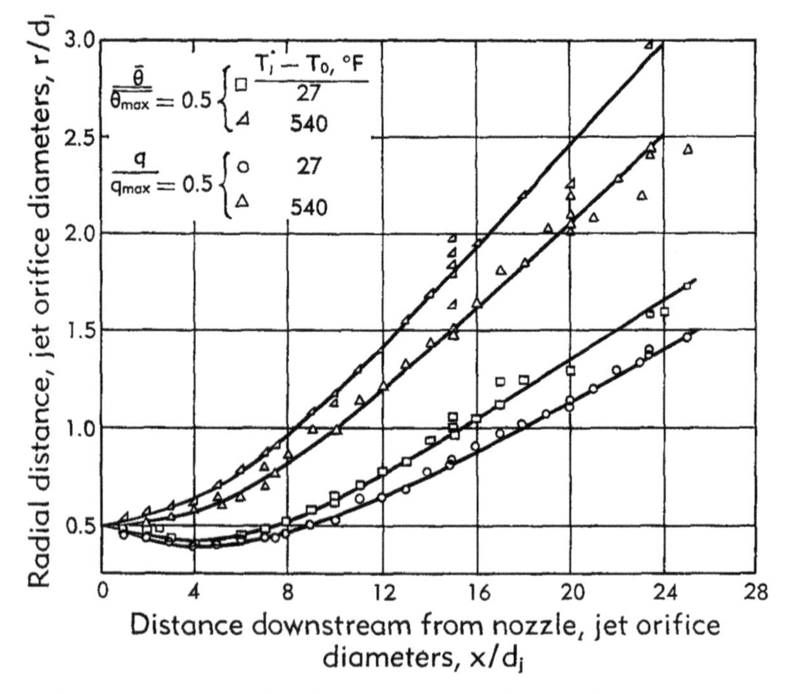


Fig. F,3h. Experimental half radii for spreading of heat and momentum in axially symmetric jets in air [29]. Diameter = 1 in. $q = \frac{1}{2}\bar{\rho}\bar{u}^2/2$. Bars indicate mean values.

$x/d_j$ are used. Essentially, the distance $x$, to achieve the same degree of mixing in jets, is a function only of jet diameter.

(2) Radial profiles of $C/C_{max}$ as a function of $r/r_0$, where $C_{max}$ is the concentration of jet fluid on the jet axis and $r_0$ is the jet outer radius, are similar at different transverse sections of the jet, for all values of jet diameter and velocity.

(3) The jet boundary, where $C/C_{max}$ remains a small constant value, spreads conically approximately from the center of the jet orifice.

These conclusions, determined from a survey of several theoretical and experimental studies, indicate that concentration of jet fluid in turbulent jets may be generalized in much the same manner as velocity and temperature. In addition, for a simplified system, [*30*, pp. 266–288] indicates theoretically that

$$\frac{1}{C} = K\left(\frac{x}{d_j}\right) + b \tag{3-12}$$

where $b$ = const and $K$ = const. Although this expression is for a simplified system, various investigators have shown experimentally that

$$\frac{1}{C_{max}} = K\left(\frac{x}{d_j}\right) + b \tag{3-13}$$

*Coaxial jets.* In general, only a free jet spreads conically. Also, a boundary layer from the tube or nacelle containing the jet may have an important influence on the jet when the stream velocity is an appreciable fraction of the jet velocity. As a result, correlation of experimental data has led to general empirical rules concerning the characteristics of coaxial jets.

Many of the experimental data with coaxial jets have been correlated and compared with the theoretical analysis of Squire and Trouncer [*20*] that was discussed briefly in Art. 2. A diagram of the coaxial jets and the results of their analysis are presented in Fig. F,3i and F,3j. Results of the analysis indicated that: (1) the length of the potential cone decreases with decreasing ratios of stream-to-jet velocity to approximately 3.1 jet diameters for a free jet, and (2) similar to free jets, constant velocity lines are approximately straight, emanating from the center of the jet orifice.

Squire [*27*] states that

$$\left\{\begin{matrix}\text{Increase in radii of half-velocity or}\\ \text{0.1-velocity circles above radius of}\\ \text{jet orifice}\end{matrix}\right\} \sim \frac{u_j - u_0}{u_j} \tag{3-14}$$

This expression should not be applied for longitudinal distances greater than 30 jet orifice diameters downstream. For the same longitudinal

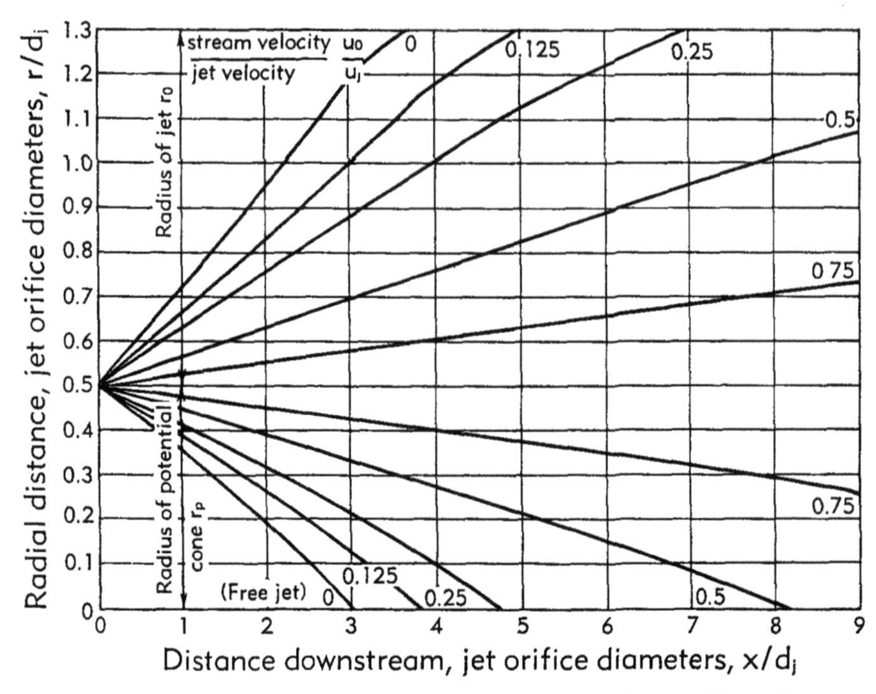


Fig. F,3i. Boundaries of jet and potential cone in circular coaxial jets [20].

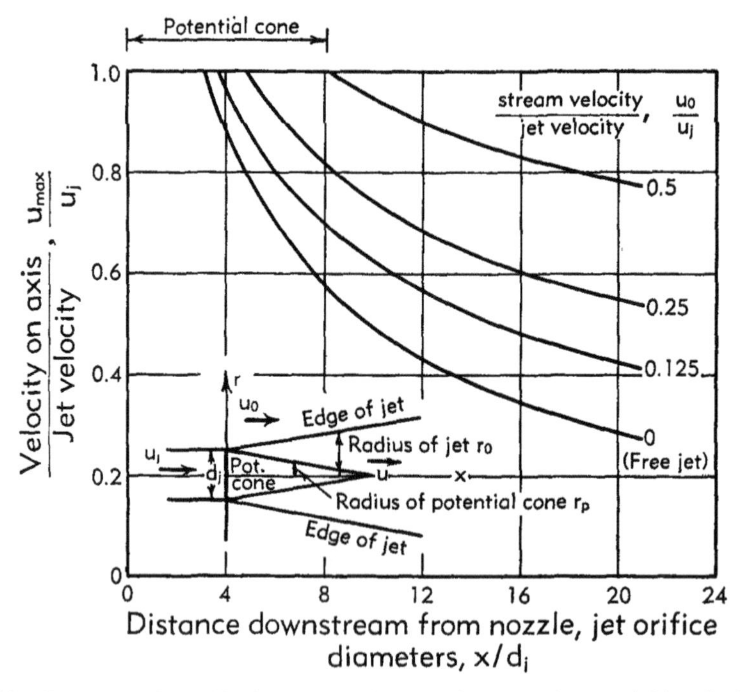


Fig. F,3j. Axial distribution of velocity on axis in circular coaxial jets [20].

distances and for values of $u_0/u_j$ less than 0.6, the velocity on the axis can be approximated by

$$\left(\frac{u_{max} - u_0}{u_j - u_0}\right)\frac{x}{d_j} = \frac{6.5}{1 - 0.6(u_0/u_j)} \tag{3-15}$$

An approximation for the length of the potential cone is

$$\text{Potential cone length} \sim \left(\frac{u_j}{u_j - u_0}\right) \tag{3-16}$$

Forstall and Shapiro [4] present the following empirical equations for coaxial jets:

$$\frac{L_p}{d_j} = 4 + 12\left(\frac{u_0}{u_j}\right) \tag{3-17}$$

$$\left(\frac{u_{max} - u_0}{u_j - u_0}\right)\frac{x}{d_j} = 4 + 12\left(\frac{u_0}{u_j}\right) \tag{3-18}$$

$$\frac{r_{\frac{1}{2}}}{d_j} = 0.5\left[\frac{4 + 12\left(\frac{u_0}{u_j}\right)}{\left(\frac{x}{d_j}\right)}\right]^{\left(\frac{u_0 - u_j}{u_j}\right)} \tag{3-19}$$

$$\frac{u - u_0}{u_{max} - u_0} = 0.5\left[1 + \cos\frac{\pi}{2}\left(\frac{r}{r_{\frac{1}{2}}}\right)\right] \tag{3-20}$$

$$r_{\frac{1}{2}} \sim x^{\left(\frac{u_j - u_0}{u_j}\right)} \tag{3-21}$$

$$r_0 \sim x^{\left(\frac{u_j - u_0}{u_j}\right)} \tag{3-22}$$

where $L_p$ = potential cone length, $r$ = radial distance from jet axis, $r_0$ = outer radius of jet, and $r_{\frac{1}{2}}$ = radius at which $u = 0.5(u_{max} + u_0)$. Discrepancies exist between these equations and the empirical correlations of [27]. However, Eq. 3-20 approximates a similar theoretical equation of [20].

From their experimental data, Forstall and Shapiro verified that the assumption of the cosine velocity profile by Squire and Trouncer was justified, and showed that the same theoretical analysis can suitably predict the half-velocity and concentration boundaries of mixing with mixing length parameter (see Eq. 2-28) $c^2 \cong 0.007$ for velocity and $c^2 \cong 0.0100$ for concentration (Fig. F,3k). However, the Squire and Trouncer theoretical analysis did not predict the center line values of velocity and concentration as well as the Forstall and Shapiro empirical formula (Eq. 3-18). Additional conclusions of [4] were:

(1) The fully normalized shapes of the velocity and concentration profiles are substantially alike, and are independent of velocity ratio and axial distance from the potential cone.
(2) Mass transport is more rapid than momentum transport. (This is similar to findings of other investigations that temperature spreads faster than momentum.)

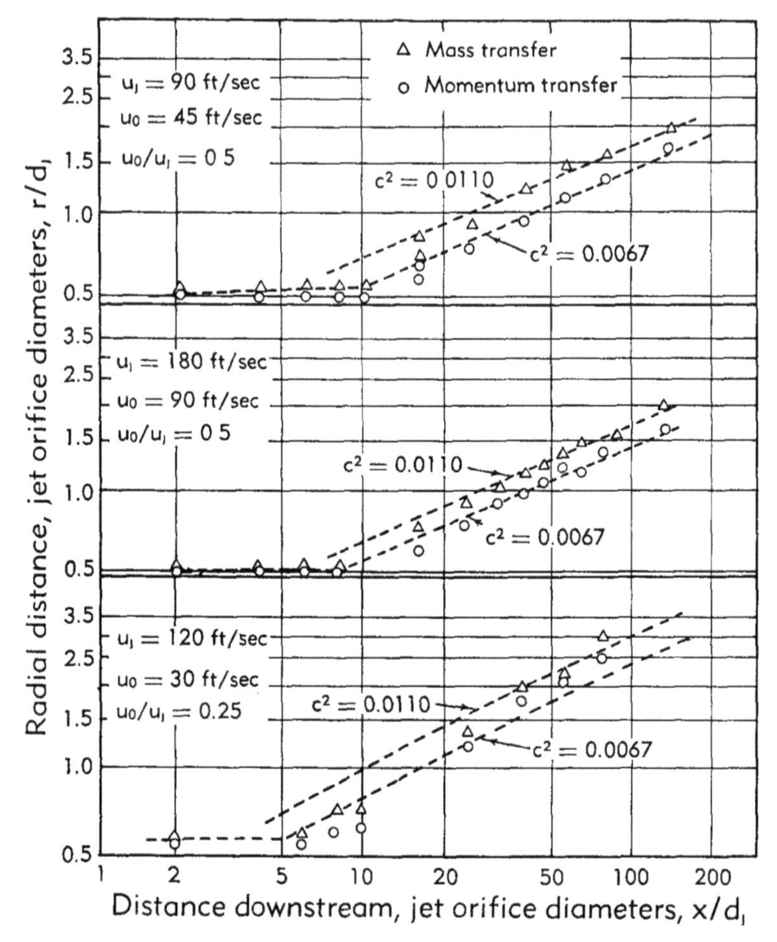


Fig. F,3k. Half radii for mass and momentum transfer in coaxial jets. (Half radii, where velocity or concentration = $0.5(u_{max} + u_0)$ or $0.5(C_{max} + C_0)$ respectively.) From [4].

Forstall and Shapiro have summarized some experimental data on the diffusion of momentum, heat, and mass in Table F,3. $Pr_t$ and $Sc_t$ were defined in the section on free jets (constant exchange theory). $Sc_t$ remained approximately constant at 0.70 although $Sc$ was varied over two-hundredfold. The data indicate that the transport of momentum occurs approximately 70 per cent as rapidly as the transport of mass or temperature. Similar results were found experimentally by Corrsin and Uberoi [*29*] who determined an effective turbulent Prandtl number $Pr_t$ from mean temperature and velocity distributions in a 1-in.-diameter hot jet (temperature, approximately 306°F greater than ambient). $Pr_t$ was roughly constant at 0.7 across the main part of the jet.

*Table F,3*

| Fluid flowing | Diffusion material | Laminar flow | | Turbulent flow | |
|---|---|---|---|---|---|
| | | $Pr$ | $Sc$ | $Pr_t$ | $Sc_t$ |
| Air | Warm air | 0.74 | | 0.76 | |
| Air | Helium | | 0.34 | | 0.70 |
| Water | Salt water | | 785 | | 0.64 |

A different analytical approach was used by Schlinger and Sage [*31*] in their study of the mixing of coaxial jets of natural gas and air under nonburning conditions. In their investigation the natural gas issued from a 1-in.-diameter nozzle into a 4-in.-diameter duct. Air issued from the annular space between the nozzle and duct. The area-weighted average approach velocity and the temperature of the natural gas and air streams were kept equal during all tests. Measurements of radial concentration profiles at various axial distances were taken at velocities of 25, 50, and 100 ft/sec. Since the duct was horizontal, buoyancy forces resulted in serious distortion of the radial profiles at the lower velocities. Accordingly, in the analysis of the data, the axis of symmetry was changed to agree with the maximum point of the concentration curve.

For this case of equal coaxial stream velocity, the analysis involves the solution of the differential equation for material balance. With the assumptions of (1) steady state, (2) axial symmetry, (3) constant axial velocity, (4) negligible radial velocity, (5) negligible turbulent mass transfer in the axial direction, and (6) constant total diffusivity, the equation for material balance of the natural gas is

$$\frac{\partial^2 C_f}{\partial r^2} + \frac{1}{r}\frac{\partial C_f}{\partial r} = \frac{u}{D_t}\frac{\partial C_f}{\partial x} \tag{3-23}$$

where

$C_f$ = mole fraction of gas in mixture
$D_t$ = total diffusivity in the radial direction, the sum of turbulent and molecular diffusion coefficients
$r$ = radial distance
$u$ = axial velocity
$x$ = axial distance

For the boundary condition,

$$\frac{\partial C_f}{\partial r} = 0 \text{ at duct wall where } r = a$$

$$C_f = 1 \text{ at } x = 0 \text{ for } 0 \leqq r \leqq \frac{d_j}{2}$$

where $d_j$ = nozzle diameter

$$C_f = 0 \text{ at } x = 0 \text{ for } \frac{d_j}{2} \leqq r \leqq a$$

where $a$ = outer duct radius

$$C_f = \frac{1}{a^2} \sum_{i=0}^{\infty} \frac{d_j J_0(\beta_i r) J_1\left(\beta_i \frac{d_j}{2}\right) e^{-(\beta_i^2 D_t x/u)}}{\beta_i [J_0(\beta_i a)]^2} \tag{3-24}$$

where $J_0$ = Bessel function of first kind, zero order; $J_1$ = Bessel function of first kind, first order; and $\beta_i$ = roots of $J_1(\beta_i a) = 0$.

An iterative procedure was used to obtain values of $D_t/u$ required for satisfactory agreement of experimental compositions with values predicted by Eq. 3-24. A comparison of experimental and predicted composition is presented in Fig. F,31. For an average velocity of 100 ft/sec, $D_t/u = 1.55 \times 10^{-4}$ ft. At average velocities of 50 and 25 ft/sec, values of $D_t/u$ of $1.9 \times 10^{-4}$ ft and $2.3 \times 10^{-4}$ ft, respectively, were required for satisfactory agreement between predicted and experimental values.

Ruegg and Klug [32] describe an experimental investigation of mixing in the dilution zone of a simplified gas turbine combustor. The products of combustion of an approximately stoichiometric mixture of propane and air exhausted from a $3\frac{9}{16}$-in.-diameter liner into a 6-in.-diameter duct. A coaxial stream of dilution air issued from the annular space between the liner and the duct. The ratio of dilution air velocity to combustion product velocity, $u_0/u_j$, was varied from approximately 1.5 to 0.1.

This coaxial jet system differs principally from the systems of Squire and Trouncer [20] and Forstall and Shapiro [4] in that (1) there is a large difference between the temperature and density of the two streams and (2) the outer wall is close to the mixing zone. Since there is a finite quantity of dilution air, the final mixture approaches an average temperature $T_{av}$ rather than the temperature $T_0$ of the outer stream [20].

The experimental variations in temperature $T_{max}$ along the duct center line were first treated in terms of the following empirical equation involving the probability integral

$$\frac{T_{max} - T_{av}}{T_j - T_{av}} = \frac{2}{\sqrt{\pi}} \int_0^{z=f(x)} e^{-z^2} dz \tag{3-25}$$

where $x$ is the axial distance from the end of the liner and $T_j$ is the temperature of the combustion products at the end of the liner. Eq. 3-25 is similar in form to an equation by Pai [12] for mixing in a two-dimensional jet. Application of Eq. 3-25 to the data indicated that $z$ followed the relation

$$z = Ke^{-2c(x/d_j)} \tag{3-26}$$

where $d_j$ is the diameter of the liner. For $u_0/u_j < 1$, $K$ was a constant. For $u_0/u_j > 1$ no definite relation for $K$ was found. The term $c$ in Eq. 3-26

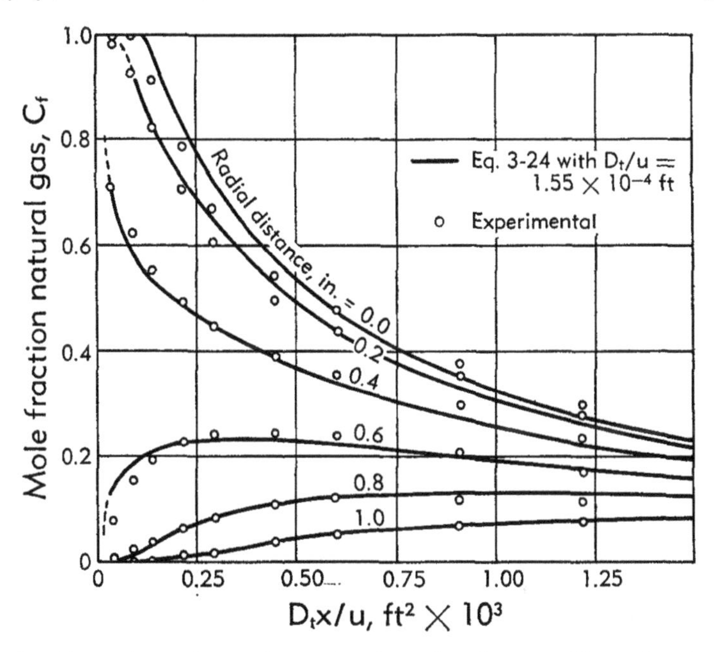


Fig. F,31. Comparison of experimental and predicted compositions of natural gas issuing from a 1-in.-diameter nozzle into a coaxial 4-in.-diameter air duct [31]. Area-weighted average approach velocity of natural gas and air streams = 100 ft/sec. Temperature of both streams equal.

is an inverse measure of the degree of mixing, that is, the distance required for mixing decreases with increase in $c$. An approximate correlation of $c$ with velocity ratio was found as shown in Fig. F,3m after Ruegg and Klug. As would be expected, the least effective mixing occurs when $u_0/u_j$ equals unity.

Comparison of the experimental data with the results of [4,20] would require the use of axial temperature in the dimensionless form $\theta_{max} = (T_{max} - T_0)/(T_j - T_0)$ rather than the form given in Eq. 3-25. $\theta_{max}$ approaches a finite value in contrast to the limiting value of zero for the analysis of [12]. Accordingly, the comparison can be made only over a limited range of $\theta_{max}$.

At positions downstream of the potential cone, $\theta_{max}$ was found to vary inversely with axial distance as in [20]. In contrast to the results of [20], however, the length of the potential cone remained approximately constant at $5.5d_j$ for $u_0/u_j < 1$. For $u_0/u_j > 1$, limited data indicated that the length of the potential cone decreased with increase in $u_0/u_j$.

Analysis of axial velocity data indicated that velocity was transferred at a lower rate than temperature. For this case of a low density inner stream, the degree of mixing was less than that found in [4] for the mixing of equal density streams. This might be expected from the analysis of Hawthorne, et al. for the free jet [30, pp. 266–288].

Radial temperature profile data agreed well with the cosine curve of Squire and Trouncer [20] at an axial distance of $4.5d_j$. However, at

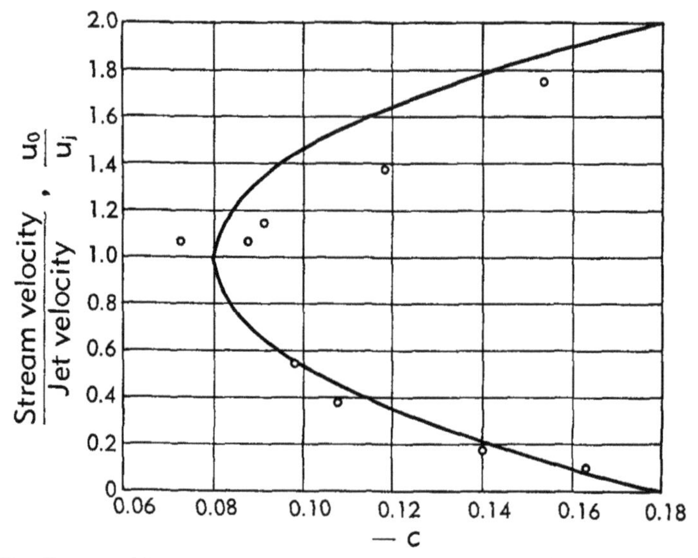


Fig. F,3m. Effect of velocity ratio on mixing in coaxial jet system [32].

greater axial distances the data tended to approach the profile predicted by Tollmien [8] for a free jet.

*Parallel and impinging jets.* The linearization of the differential equation of momentum flux by application of the Reichardt hypothesis [14] suggested that the solution for flow fields other than for point source free jets might be obtained by superposition of solutions for point sources. Baron and Alexander [33] demonstrated that this method is satisfactory for treating the flow characteristics in regions near a free jet of finite diameter. Grimmett [34] also applied this method to regions well downstream of two parallel jets issuing into stagnant air. From the principal of superposition of solutions, the mean momentum flux $(\overline{\rho u^2})_m$ operating together is given by

$$(\overline{\rho u^2})_m = (\overline{\rho u^2})_A + (\overline{\rho u^2})_B \tag{3-27}$$

where $(\overline{\rho u^2})_A$ and $(\overline{\rho u^2})_B$ are the mean values of momentum flux at the same point for the jets *A* and *B* operating separately. Fig. F,3n gives a comparison of experimental data in the plane of the center line of two parallel jets with values predicted from Eq. 3-27. The general agreement is satisfactory. However, the comparisons for all the data indicated that the method of superposition tends to give values lower than the data in the region between the jets and values higher than the data in the outer regions [*34,35*]. For this case of parallel jets, the disagreement was not excessive. However, Baron and Bollinger [*35*] have found that the method

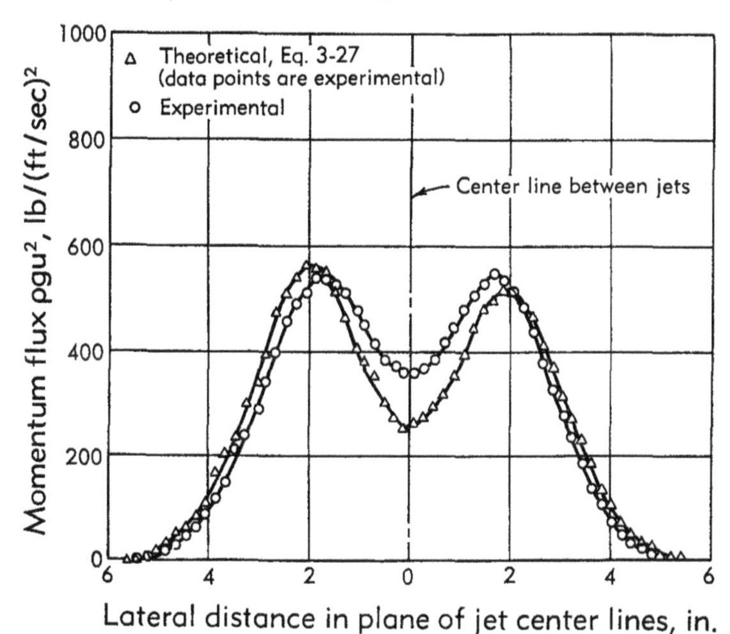


Fig. F,3n. Momentum flux distribution in plane of center lines of two parallel jets, 23.6 in. downstream of jet orifices [*35*]. Orifice diameters = 0.899 in. Distance between center lines = 2.99 in. Jet velocities approximately equal.

is unsatisfactory for the case of two impinging jets. The discrepancy between experimental and predicted values increased rapidly with increase in the included angle between the jet axes. Fig. F,3o illustrates the disagreement between predicted and experimental values for an included angle of 14.1°. Mutual entrainment and variation of static pressure in the region between the jets were attributed as probable causes for the large discrepancies between predicted and experimental values. Rummel [*36*] has shown that the distance required to achieve a given degree of mixing is decreased markedly with increase in the included angle between the jet axes.

*Jet penetration into normal stream.* The penetration of gas jets into streams moving normal to the jet are of special interest in the design of combustion chambers of turbojet engines in which the openings for air admission are normal to the flow direction of the hot gases. In general the definition of jet penetration and the degree of mixing have been arbitrarily specified by different investigators. As a result, different correlations of experimental data have been developed.

Experimental data indicating the relative penetration and mixing of jets of cold air (temperature, approximately 115°F) with a normal flowing stream of hot air (temperature, approximately 890°F) are presented in [*23*]. Penetration was defined as the normal distance from the

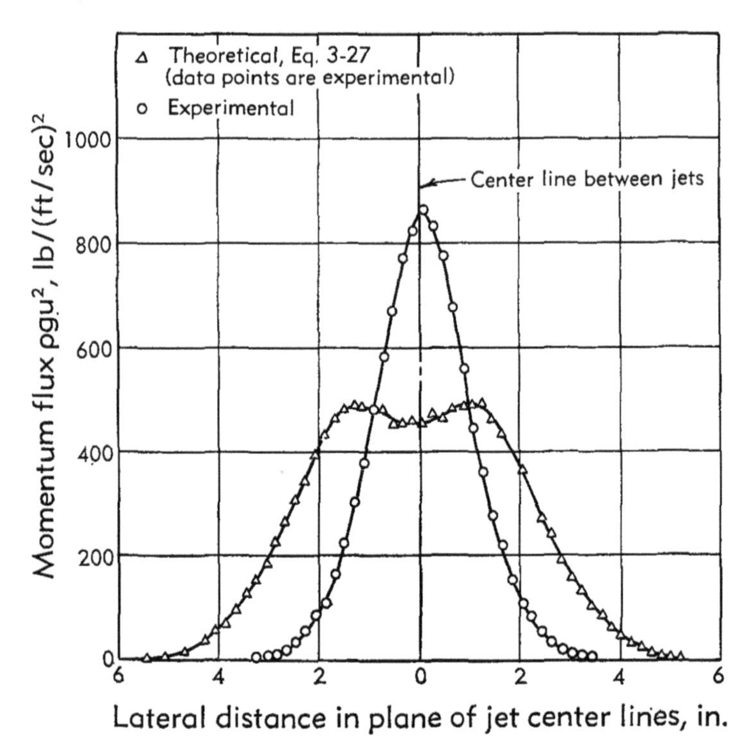


Fig. F,3o. Momentum flux distribution in plane of center lines of two nonparallel intersecting jets, 23.6 in. downstream of jet orifices [*35*]. Orifice diameters = 0.899 in. Distance between center lines at plane of orifices = 2.99 in. Included angle between jets = 14.1°. Jet velocities approximately equal.

plane of the cold-air orifice to the point of minimum temperature 1.5 duct widths downstream of the orifice center line. It is concluded in [*23*] that:

1. For circular cold-air jets of different diameters and a range of velocities and temperatures of the two streams,

   Relative penetration, or

$$\frac{\text{Penetration}}{\text{Duct depth}} = k_1 + k_2 \left[ \frac{\left(\dfrac{\text{Momentum, cold stream}}{\text{Momentum, hot stream}}\right)^{\frac{1}{2}}}{\left(\dfrac{\sum \text{Volumetric flow rate}}{\text{Volume flow rate, hot stream}}\right)} \right] \tag{3-28}$$

where $k_1$ and $k_2$ are constants. This relation holds until the relative penetration becomes approximately 0.6, at which time the edge of the cold-air jet approaches the opposite wall of the duct.

2. For the same open area, circular and square holes give approximately the same penetration, while longitudinal and transverse rectangular holes give better and worse penetration, respectively (Fig. F,3p).
3. In general, jet penetration is maximized with the least random turbulence in the fluid streams.
4. Pressure losses are less for a given penetration by use of a small number of large holes than by use of a large number of small holes.
5. Jet penetration is maximized with lowest pressure losses by the use of openings with rounded entrances.

Results from a similar investigation in a $1\frac{7}{8}$-in.-square duct in which hot-air jets (temperature, approximately 580°F) entered normal flowing streams of cold air (temperature, approximately 140°F) through different shaped openings have been presented in [*37*]. Penetration was defined in the same manner as in [*23*], except that no limitations were imposed on the distance downstream of the orifice center line. Relative penetration data were satisfactorily correlated in terms of either mass or momentum ratios of the hot stream to the cold stream. Relative penetration increased with increases in either mass or momentum ratios. The effect of hole shape on relative penetration was generalized by dividing the relative penetration of jets from equal-area holes of circular, square, and longitudinally rectangular shape [*23*] by the longitudinal length of the hole (Fig. F,3p).

Callaghan and Ruggeri [*38*] investigated the normal penetration of heated circular jets of air (temperature, approximately 400°F) into a stream of cold air (temperature, ambient) flowing through a 2 by 20-in. duct. Penetration was defined as the distance to the point at which the temperature was 1°F greater than the free stream total temperature. Experimental data expressing the penetration were correlated by the expression

$$\left(\frac{l}{d_j}\right)^{1.65} = 2.91\left(\frac{\rho_j u_j}{\rho_0 u_0}\right)\sqrt{\frac{x}{d_j}} \tag{3-29}$$

where

$d_j$ = jet orifice diameter
$l$ = penetration of jet into normal stream
$u_j$ = velocity of jet at vena contracta
$u_0$ = velocity of normal stream
$x$ = longitudinal distance from jet orifice center line
$\rho_j$ = density of jet fluid at vena contracta
$\rho_0$ = density of normal stream

Penetration was unaffected by variations of Reynolds number from $0.6 \times 10^5$ to $5.0 \times 10^5$, viscosity ratios of the two streams from 1.5 to 1.9, and ratios of duct width to jet diameter from 3.2 to 8.0.

A similar relation was found [*39*] for the penetration of liquid jets of water injected normally from small, simple, orifice-type nozzles into a

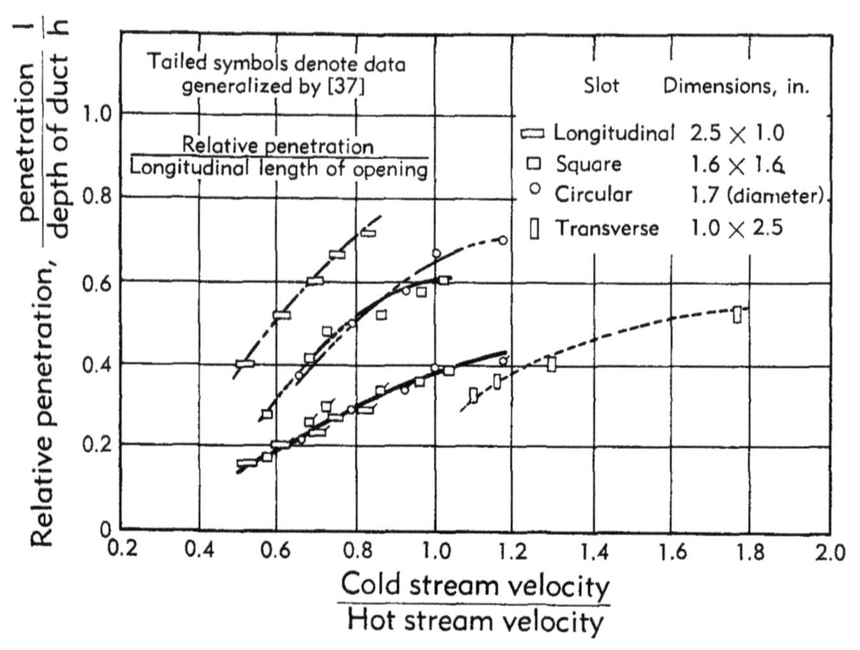


Fig. F,3p. Effect of opening shape on penetration of cold air jets (temperature = 115°F) into normal hot air stream (temperature = 890°F) [*23*]. Dimensions of hot air duct, 3 × 3 in. Velocity of hot air stream ≅ 280 ft/sec.

high velocity stream of air (velocity, approximately 740 ft/sec). Penetration, defined as the maximum penetration of liquid at any longitudinal position, was determined photographically. The data were correlated by the following empirical expression:

$$\left(\frac{l}{d_j}\right) = 0.450 \left(\frac{u_j}{u_0}\right)^{0.95} \left(\frac{\rho_j}{\rho_0}\right)^{0.74} \left(\frac{x}{d_j}\right)^{0.22} \tag{3-30}$$

*Jet penetration into oblique stream.* Data expressing the penetration of oblique jets are meager. However, Squire [*27*] presents general experimental penetration data in terms of temperature distribution in a hot jet entering an external cold stream from a 1-in.-diameter circular orifice in the wall of a 3-in.-diameter duct. The angle between streams was 15°. Representative data for this configuration are presented in Fig. F,3q. With increasing main stream velocities, the spread of the jet is reduced and the center line of the temperature profiles is bent downstream. In

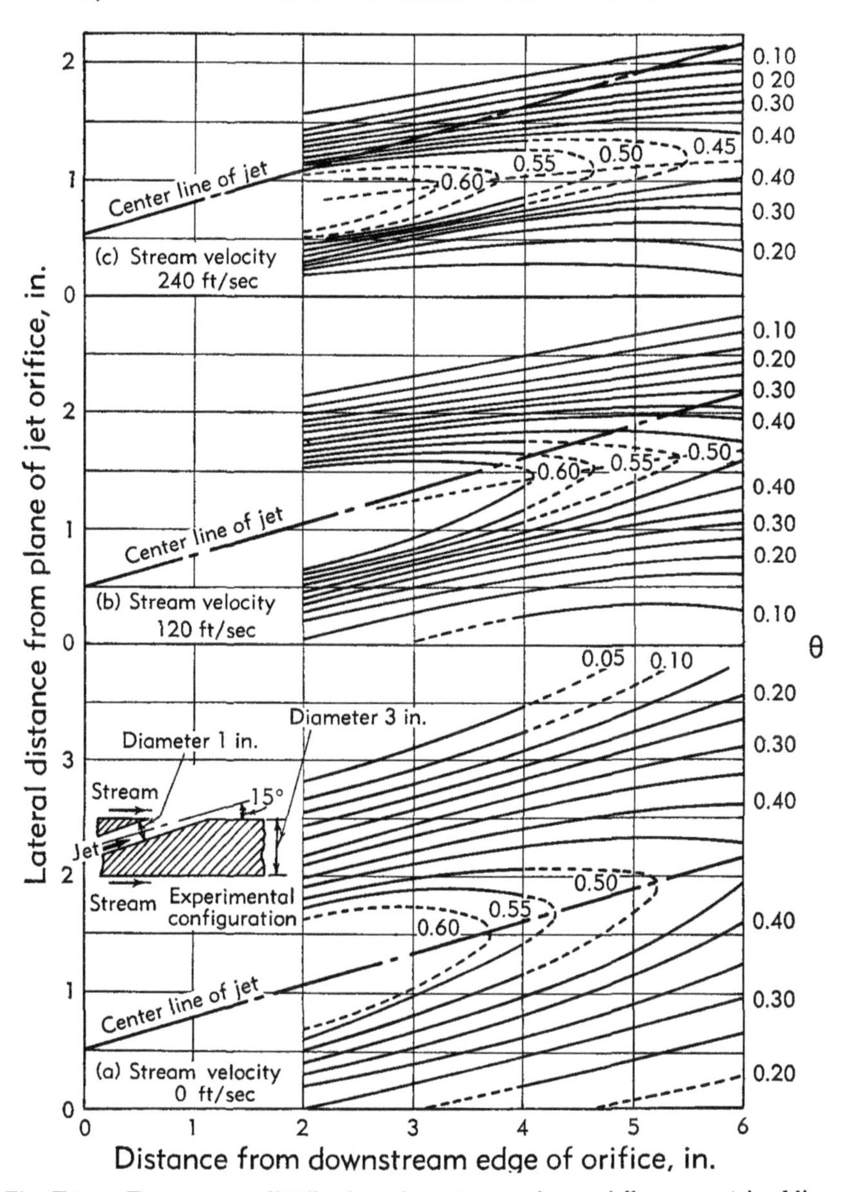


Fig. F,3q. Temperature distributions downstream of an axially symmetric oblique jet. Jet velocity = 615 ft/sec; jet temperature = 374°F. From [27].

addition, at a stream-to-jet velocity ratio of approximately 0.4, the center line of the jet becomes parallel to the main stream flow at approximately 8 inches downstream of the center of the jet orifice.

The effect of the angle between fluid jets on jet penetration has been discussed in [37]. The investigators of [37] varied the entrance angle of hot circular jets to include angles of $22\frac{1}{2}$°, 45°, 90°, and 135° with the direction of flow of the main stream. In general, maximum and minimum penetrations were obtained with jets at 90° and $22\frac{1}{2}$° to the direction of flow of the main stream for all momentum ratios and distances downstream of the jet orifice center line, respectively. For angles of 135° or 45°, the better angle for penetration was dependent upon the position downstream and the momentum ratio of the two streams. Representative data are presented in Fig. F,3r in terms of relative penetration of the jets.

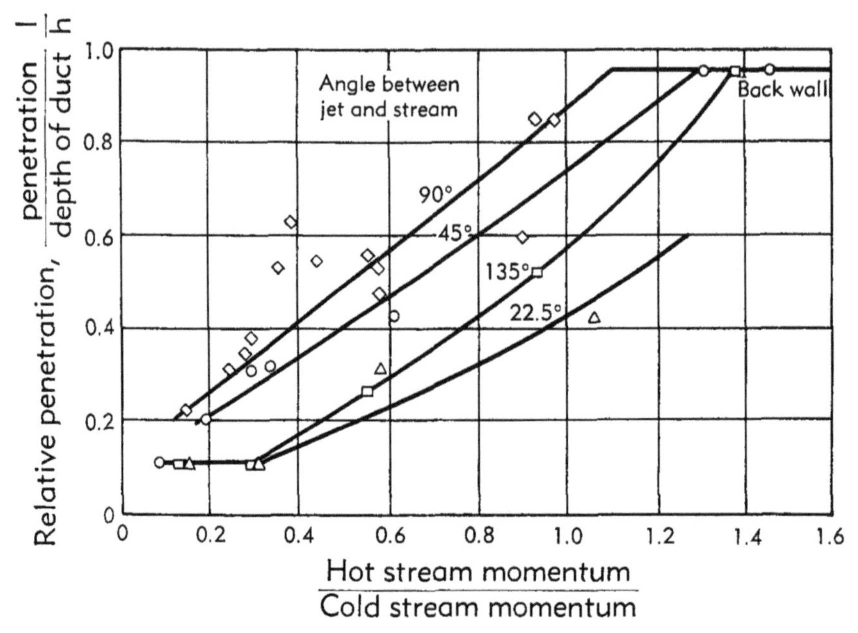


Fig. F,3r. Effect of angle between hot-air jets (temperature = 580°F) and cold-air streams (temperature = 140°F) on penetration of 0.824-in.-diameter jets [37].

## F,4. Mixing of Gases with Burning.

*Free jet turbulent diffusion flames.* The main body of work reported in the literature on gaseous diffusion flames has been concerned with the flames from free fuel jets issuing vertically from circular nozzles. Fig. F,4a illustrates the characteristics of such diffusion flames. In the laminar region there is a progressive increase in flame height with increase in port velocity. Both experiment and theory indicate that the height of laminar flames is a function primarily of fuel properties and volume flow rate of the fuel [3, pp. 97–113]. In the transition region, turbulent eddies form at the flame tip. With increase in velocity, the break point (point where the eddies start) approaches the nozzle and the flame length decreases. In the fully developed turbulent region, the break point has receded to a position close to the nozzle and the flame height is essentially independent of velocity. In this region, experiment indicates that flame height is a function primarily of fuel properties and tube diameter [3, pp. 97–113]. At still higher port velocities, there is some further increase in flame length. The reasons for this trend are not clear. Comparison of theory and experiment in the fully developed turbulent flame region indicates that the chemical reaction may be assumed to occur instantaneously and that the flame progress is controlled primarily by macroscopic mixing. At these higher velocities, however, the local heat release rates may be sufficiently high so that the chemical reaction rate begins to impede flame progress. Another possible explanation would be the limitations imposed by molecular mixing for this type of system.

There have been essentially two analytical approaches to the complex mixing processes involved in the fully developed turbulent region of free jet diffusion flames. Yagi and Saji [*3*, pp. 771–781] and Wohl, et al. [*30*, pp. 288–300] use equations developed for the laminar diffusion

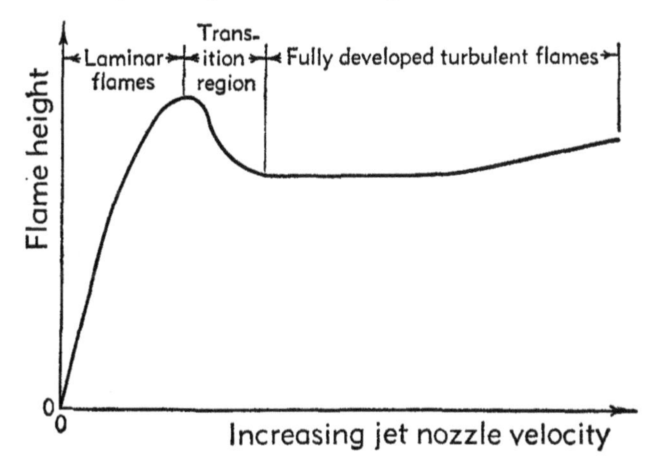


Fig. F,4a. Flame heights of free jet diffusion flames [*30*, pp. 254–266].

flame, but with the molecular diffusion coefficient replaced by a turbulent eddy diffusion coefficient. Hawthorne, et al. [*30*, pp. 266–288] and Baron [*40*] treat the turbulent diffusion flame in terms of simplified jet-mixing equations. Of the two approaches, the jet-mixing analyses are more pertinent to the present paper and are considered herein.

Hawthorne, et al. considered a vertically oriented, conical jet with flat velocity, temperature, and concentration profiles at all transverse cross sections. With these assumptions and the use of the perfect gas law and continuity equation, the following form of the momentum equation may be derived for a small horizontal section of the jet

$$d\left\{\frac{\frac{T}{\alpha T_j C^2}\left[C + (1 - C)\frac{\mathfrak{M}_0}{\mathfrak{M}_j}\right]}{Y^2}\right\} = \left\{\frac{\rho_0}{\rho_j} - \frac{\alpha T_j}{T}\left[C + (1 - C)\frac{\mathfrak{M}_0}{\mathfrak{M}_j}\right]\right\} Y^2 dY \quad (4\text{-}1)$$

where

$C$ = time-mean molal concentration of nozzle fluid in the sample converted to its unreacted constituents

$\mathfrak{M}$ = molecular weight

$T$ = absolute temperature

$\alpha$ = moles of gas in sample converted to its unreacted constituents divided by actual moles in sample

$\rho$ = density of fluid

$Y = \left(\frac{d_j g}{2u_j^2 \tan\theta}\right)^{\frac{1}{3}} \left(\frac{d_0}{d_j}\right)$

$d_j$ = jet nozzle diameter

$d_0$ = jet outer diameter

$g$ = gravitational constant

$u_j$ = jet velocity

$\theta$ = half angle of spread of jet

The subscripts $_0$ and $_j$ refer to the surrounding fluid and nozzle fluid, respectively.

In Eq. 4-1, the change in momentum across a differential transverse section of the jet is equated to the net buoyancy force acting on the section. Solutions of Eq. 4-1 for a hydrogen jet mixing isothermally with air and a hydrogen jet burning in air are shown in Fig. F,4b after

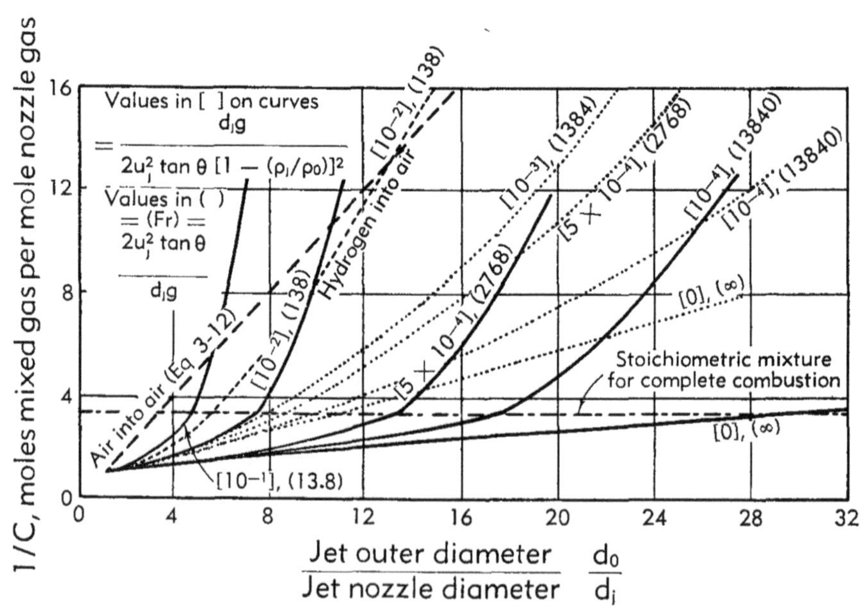


Fig. F,4b. Simplified analysis of flames and jets of hydrogen in air. Variation, with flame width, of concentration along direction of flow. Dotted curves, cold; solid curves, burning [*30*, pp. 266–288].

Hawthorne, et al. The solution for an air jet mixing isothermally with air is also presented. The ordinate, $1/C$, is a measure of the degree of entrainment of the surrounding fluid; that is, higher values of $1/C$ are associated with increased entrainment.

Comparison of the curves for the cold hydrogen and the air jets shows that a decrease in jet density results in a decrease in entrainment. The effects of buoyancy on mixing are indicated by the relative positions of the cold hydrogen jet curves for various values of the modified Froude number, $2u_j^2 \tan \theta/d_j g$. A decrease in buoyancy (increase in modified Froude number) is seen to produce a decrease in entrainment. The deviation of the cold hydrogen jet curves from the linear relation between $1/C$ and $d_0/d_j$ found for the cold air jet is evident, even for the case of negligible buoyancy $[(2u_j^2 \tan \theta/d_j g) \rightarrow \infty]$. The upward curvature may be attributed to the increase in density along the jet with entrainment of the air of higher molecular weight.

Comparison of the curves for the burning and cold hydrogen jet indicates the effect of heat release on the mixing process. Near the nozzle, the decrease in density with combustion results in a decrease in entrainment. At the point where combustion ends, there is an increase in entrainment accompanying the increase in density with dilution of combustion products. At some distance from the nozzle where the jet velocity has decreased sufficiently, the increase in entrainment due to buoyancy may offset the decrease due to low density. Accordingly, the entrainment of the burning jet may eventually exceed that of the cold jet. Data obtained by Hawthorne, et al. for an unenclosed hydrogen jet burning in air exhibited trends similar to those shown in Fig. F,4b.

If the buoyancy forces are neglected, the right-hand side of Eq. 4-1 equals zero and the bracketed left-hand side term is constant at its value at the nozzle. At the nozzle, $C = \alpha = T/T_j = 1$. At the point where combustion ends, it is assumed that $T$ equals the stoichiometric flame temperature $T_f$, and $C$ and $\alpha$ have the values $C_f$ and $\alpha_f$ corresponding to a stoichiometric mixture. Accordingly, one can solve for the jet diameter $d_f$ at the point where combustion ends to obtain

$$\frac{d_f}{d_j} = \frac{1}{C_f}\sqrt{\frac{T_f}{\alpha_f T_j}\left[C_f + (1 - C_f)\frac{\mathfrak{M}_0}{\mathfrak{M}_j}\right]} \tag{4-2}$$

The qualitative agreement found between the experimental axial concentration curves of an unenclosed hydrogen flame and the curves of Fig. F,4b suggested that Eq. 4-2 might be used to correlate flame length data. Assuming that the spreading angle of the flame is constant for all fuels and velocities, the turbulent flame length $L_f$ should be directly proportional to $d_f$. This relation was found to correlate the turbulent flame length data for a number of fuels as shown in Fig. F,4c. The equation of the correlation line is

$$\frac{L_f}{d_j} = \frac{5.3}{C_f}\sqrt{\frac{T_f}{\alpha_f T_j}}\left[C_f + (1 - C_f)\frac{\mathfrak{M}_0}{\mathfrak{M}_j}\right] \tag{4-3}$$

Eq. 4-3 predicted turbulent flame lengths with an average and maximum error of 10 and 20 per cent, respectively.

Baron [*40*] extended a previous analysis for free jets to the turbulent diffusion flame. Buoyancy forces were neglected and combustion was assumed to occur instantaneously. In terms of the molal concentration $C$ of [*30*, pp. 266–288], Baron obtained the relation

$$C = \frac{c_x d_j}{2c_m^2 x}\sqrt{\frac{1}{\alpha^2}\left(\frac{T}{T_j}\right)\left(\frac{\mathfrak{M}}{\mathfrak{M}_j}\right)}\, e^{-\left(\frac{r}{x}\right)^2\left(\frac{1}{c_m^2} - \frac{1}{2c_x^2}\right)} \tag{4-4}$$

where

$c_m$ = spreading coefficient for mass transfer
$c_x$ = spreading coefficient for momentum transfer
$\mathfrak{M} = \alpha[C\mathfrak{M}_j + (1 - C)\mathfrak{M}_0]$
$r$ = radial coordinate

If, as in [*30*, pp. 266–288], it is assumed that both the stoichiometric "nozzle fluid" concentration and the stoichiometric flame temperature occur at all points in the flame front, Eq. 4-4 may be used to compute the flame height and flame shape. The flame height is obtained from Eq. 4-4 for $r = 0$, and is given by

$$\frac{L_f}{d_j} = \left(\frac{c_x}{2c_m^2}\right)\left\{\frac{1}{C_f}\sqrt{\frac{T_f}{\alpha_f T_j}}\left[C_f + (1 - C_f)\frac{\mathfrak{M}_0}{\mathfrak{M}_j}\right]\right\} \tag{4-5}$$

where the subscript $_f$ refers to the stoichiometric conditions at the flame front. The term in the brackets is identical in form with Eq. 4-3. For values of the constants $c_x$ (= 0.075) and $c_m$ (= 0.0855) obtained from an analysis of cold-flow jet experiments, the term $c_x/2c_m^2$ of Eq. 4-5 is equal to 5.2. This agrees within experimental error with the empirical constant of Eq. 4-3. From Eq. 4-4 and 4-5 and the above values of $c_m$ and $c_x$, the following relation for flame diameter $d_f$ may be obtained:

$$d_f = 0.29x\sqrt{\ln\left(\frac{L_f}{x}\right)} \tag{4-6}$$

where $x$ is the axial distance from the nozzle.

*Coaxial jet turbulent diffusion flames.* Little work has been done on coaxial jet turbulent diffusion flames. Berry, et al. [*41*] investigated the mixing of coaxial jets of natural gas and air under burning conditions. Tests were made at average approach velocities of 10, 25, and 50 ft/sec.

The ratio of gas and air flow rates was held constant at approximately the stoichiometric ratio for all tests.

Fig. F,4d, after Berry, et al., gives the temperatures indicated by a shrouded thermocouple at various points in the vertical plane through the center line of the duct. Buoyancy forces caused marked asymmetry of the flow. It is seen that the peak temperatures indicated in the figure are well below stoichiometric flame temperature. Aside from deviations associated with thermocouple errors, this trend might be attributed to

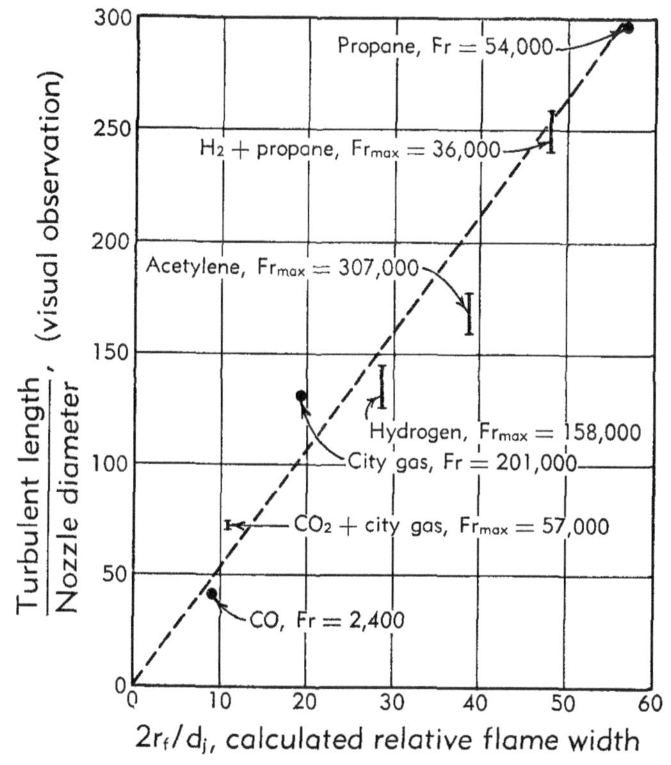


Fig. F,4c. Correlation of turbulent free flame lengths with calculated flame spread [*30*, pp. 266–288]. Abscissa is calculated width at point of completion of combustion, ignoring buoyancy.

"unmixedness" (which is defined in a subsequent article) and high heat transfer rates to the water-jacketed outer wall of the apparatus. This trend points up the deficiencies of some of the assumptions made in the analyses of turbulent diffusion flames from free fuel jets.

There was no significant axial shift in the position of the 1600°F profile with increase in approach velocity from 10 to 25 ft/sec. This might be expected from work on the length of turbulent diffusion flames from free jets. However, for an approach velocity of 50 ft/sec, the axial distance from the nozzle to the 1600°F line was nearly twice that for the lower velocities. In this region, the "combustion velocity" was independent of approach velocity. The reason for this shift in trend at the high approach velocity is not apparent.

No experimental data are available for the burning of coaxial fuel and air jets when the jet velocities are unequal. However, from the results of the cold-flow coaxial jet tests, it might be expected that the flame length would decrease with increase in the difference between the jet velocities.

*Heat release rates in turbulent flames.* It would be of interest to compare the heat release rates in the turbulent free jet diffusion flames with the typical heat release rates for turbojet engine combustors (3 to

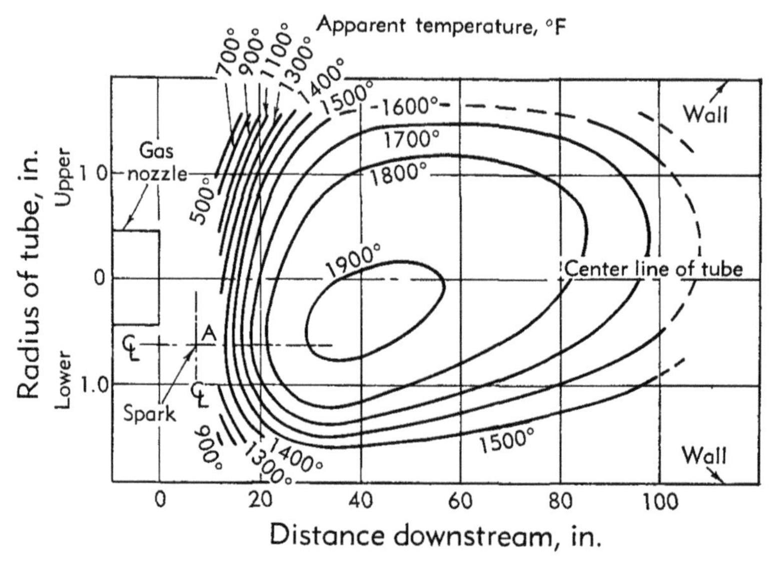


Fig. F,4d. Apparent temperature distribution in combustion tube at average velocity of 10 ft/sec [41].

$6 \times 10^6$ BTU/ft³-hr-atm). A rough estimate of the heat release rate in the turbulent diffusion flame may be obtained from Baron's equation for flame envelope shape and the correlation for flame height. From Eq. 4-6, it may be shown that the volume, $V$, within the flame envelope equals $0.00734L_f^3$. The heat release rate $q$ is then given by

$$q = 3.85 \times 10^5 \frac{hu_j}{d_j}\left(\frac{L_f}{d_j}\right)^{-3} \text{ BTU/ft}^3\text{-hr} \tag{4-7}$$

where

$d_j$ = jet diameter, ft
$h$ = lower heat of combustion at standard conditions, BTU/ft³
$L_f$ = length of flame, ft
$u_j$ = jet velocity, ft/sec

Since $L_f/d_j$ is approximately constant for a fully developed turbulent flame, $q$ is inversely proportional to jet diameter.

In Fig. F,4e, $q$ is plotted against jet velocity for a jet diameter of 0.375 in. The curves are for a fully developed turbulent propane diffusion flame ($L_f/d_j \cong 296$ [*30*, pp. 266–288]). Even at the higher jet velocities, the heat release rates are lower than those found for typical turbojet engine combustors. Values of the heat release rate for turbulent Bunsen-type flames are also presented. The values were obtained from the flame height data of Bollinger and Williams [*42*] for a stoichiometric propane

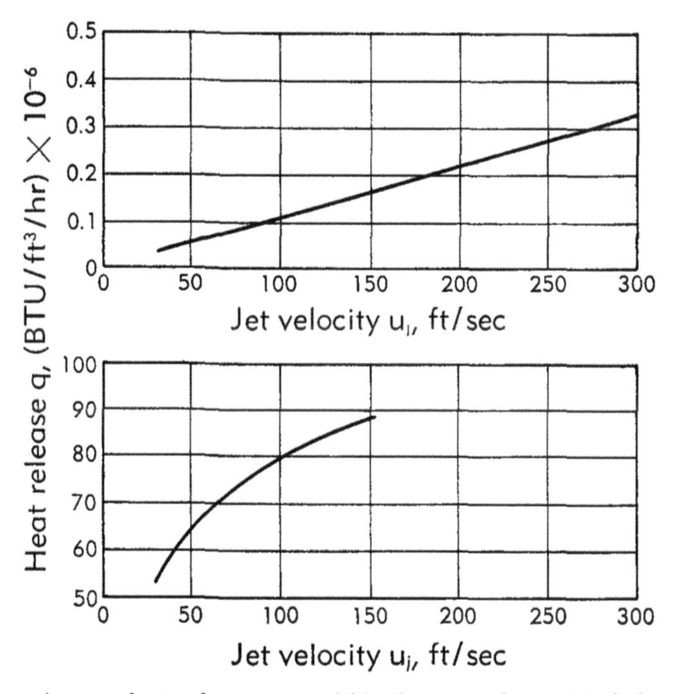


Fig. F,4e. Average heat release rates within flame envelope of turbulent flames. Fuel, propane; nozzle diameter, 0.375 in. (Top) free jet turbulent diffusion flame. $L_f/d_j = 296$ [*30*, pp. 266–288]; (bottom) turbulent Bunsen flame. Stoichiometric fuel-air ratio. Calculated from data of [*42*].

flame, by assuming the flame envelope to have the form of a right circular cylinder. It is seen that the heat release rates for the premixed turbulent flame are appreciably higher than those for the turbulent diffusion flame. However, the values are still well below the estimated heat release rates of $700 \times 10^6$ BTU/ft³-hr-atm cited in [*43*] for the reaction zone of a laminar propane diffusion flame.

*Unmixedness.* Comparison of the turbulent diffusion flame length data with predicted values has indicated that the low heat release rates are primarily the result of limitations imposed by the turbulent mixing process (macroscopic scale mixing). However, combustion involves the combination of fuel and oxygen molecules. Accordingly, it would be worthwhile to consider the role of molecular mixing in the turbulent diffusion flames.

The turbulent diffusion process in fuel jets can be considered to be the intermingling of discrete volumes of fuel and air, with molecular mixing occurring across the interface between the volumes. At a given point in the flame, then, a sampling probe will ingest a succession of fuel-rich and air-rich volumes. The instantaneous values of the fuel-air ratio at this point can be represented as the sum of a time-averaged value and a time-fluctuating value. This picture is analogous to treatments of the fluctuating velocities in turbulent flow. Although the time-averaged value of the fuel-air ratio at a point may be stoichiometric (assumed to indicate complete combustion in some of the turbulent flame theories), combustion may be far from complete. Sampling tests in some turbulent diffusion flames [*30*, pp. 266–288] have indicated that roughly twice the stoichiometric quantity of air was present at the point where combustion was 99 per cent complete. If the chemical reaction rate is assumed to be infinite, these tests are indicative of the delay time required for the completion of the molecular mixing process.

The transition from macroscale to molecular scale mixing in turbulent diffusion flames has been treated analytically by a number of investigators. Hawthorne, et al. [*30*, pp. 266–288] assumed that the frequency of occurrence of various nozzle fluid concentrations for the volumes approaching the sampling probe could be represented by a Gaussian distribution function. Reaction was assumed to be complete in each volume. Further reaction of the mixed sample in the probe was assumed to be suppressed by the use of a chilled probe. With these assumptions, the "unmixedness factor," $\sqrt{\overline{C'^2}}$, can be calculated in terms of the time-averaged values of the nozzle fluid concentration and the reacted fractions of fuel or oxygen. In the unmixedness factor, $C'$ is the fluctuating component of the instantaneous concentration of nozzle fluid at a given point in the flame. The unmixedness factor, then, is the root-mean-square value of the deviation from the local mean composition. Higher values of $\sqrt{\overline{C'^2}}$ are associated with less intimate mixing and lower combustion efficiency. In an extension of this analysis, Richardson, et al. [*3*, pp. 814–817] used a different distribution function which was considered to give a more realistic representation of the fluctuating concentration. For the case of poor mixing, there was a significant difference in the results of the two analyses. For the case of intimate mixing, similar results were obtained. A different analytical approach has been used by Wohlenberg. In [*44*], Wohlenberg developed the concept of reaction interface extension; that is, the area between fuel-rich and air-rich zones per unit volume of combustion space was used as the common factor in the development of equations for diffusion and reaction rates. The approach of Wohlenberg appears to be a promising method of analyzing the turbulent diffusion flame and has resulted in some intriguing calculations on limiting heat release rates [*3*, pp. 796–806].

Measurements in a number of free jet turbulent diffusion flames [*30*, pp. 266–288] have showed a rapid decay in the unmixedness factor with increase in distance along the flame axis. In general, the effects of jet velocity, nozzle diameter, and nozzle type on unmixedness appear to be minor. The similarity of the unmixedness factor to the root-mean-square velocity fluctuation in turbulent flow led Thring and Newby [*3*, pp. 789–796] to use turbulence intensity measurements as a means of predicting unmixedness. On the flame axis, the intensity of turbulence $\sqrt{\overline{u'^2}}/u_{max}$ was equated to $\sqrt{\overline{C'^2}}/C_{max}$. Here, $\sqrt{\overline{u'^2}}$ is the local root-mean-square velocity fluctuation and $u_{max}$ and $C_{max}$ are the local average velocity and nozzle fluid concentration on the flame axis. From turbulence intensity measurements of Corrsin [*28*] for a free jet and the analysis of Hawthorne, the fraction of unburned fuel at various points along the flame axis could be predicted. Close agreement was found between predicted and experimental values of unburned fuel at the point where the time-averaged fuel-air ratio was stoichiometric.

*Flame-generated turbulence in enclosed channels.* Turbulence in enclosed channels may be generated from the velocity gradients caused by the combustion process. Williams, et al. [*30*, pp. 21–40] analyzed the velocity profiles in high velocity gases burning in ducts. Velocity profiles were plotted as a function of the fraction of gas burned, the ratio of unburned gas density to burned gas density, and the distance from the duct center line. The investigators concluded that the effective flame velocity was increased by:

(1) Approach-stream turbulence, which generally had less effect than internally caused disturbances, at turbulent intensities as great as 6.5 per cent.

(2) Turbulence generated downstream of a flame stabilizer by velocity gradients across the flame front. This turbulence is more likely with large values of

$$\left(\frac{u_0 n}{\nu}\right)^{0.1 \text{ to } 0.3} \left[\frac{u_0}{S_{lam}\left(\frac{\rho_1}{\rho_2} - 1\right)}\right]$$

(3) Turbulence caused by velocity gradients across the flame front produced by the pressure decrease resulting from combustion. This turbulence is more likely with large values of

$$\frac{u_0 w}{\nu} (F) \left(\frac{\rho_1}{\rho_2} - 1\right) \frac{u_0}{S_{lam}}$$

where

$F$ = fraction of gases burned
$n$ = characteristic dimension of flame stabilizer
$S_{lam}$ = laminar flame speed
$u_0$ = approach velocity of unburned gases
$w$ = rectangular duct width
$\nu$ = kinematic viscosity of unburned gases
$\rho_1$ = density of unburned gases
$\rho_2$ = density of burned gases

**F,5. Liquid-Air Mixing.** In most jet engine applications the fuel is injected into the combustion chamber as a liquid spray. Among other factors, the evaporation and burning of such sprays is a function of the mixing of the various drops with the combustion air. The complexity of the gross flow patterns in the vicinity of the fuel nozzle has made analytical treatments of drop-air mixing prohibitively difficult for the turbojet engine combustor. However, good results have been obtained in studies of the mixing of liquid sprays with air streams flowing uniformly through ducts. Such systems are similar to those found in some ramjet and afterburner installations.

Some factors of interest in the study of drop-air mixing are the drop drag coefficients, relative drop-air velocities, and the diffusion of drops in a turbulent field. Segregation of the various size drops in vortex flow and the deposition of the spray on bluff bodies and duct walls may have a significant effect on the evaporation and burning characteristics of the spray.

*Drag coefficients.* With the exception of limiting conditions such as occur during the initial period of spray formation or during drop breakup, the drops are approximately spherical in form. Accordingly, some information on the drag forces for liquid drops may be inferred from experiments on solid spheres. The total drag on a solid sphere may be divided into two components, the friction drag associated with viscous shear stress tangential to the surface and the form drag associated with the variation in pressure over the surface. At low drop Reynolds numbers, $Re_d$, which is based on the relative velocity between the spheres and the main body of the fluid, friction drag predominates. At high $Re_d$, form drag predominates as the result of the local decrease in static pressure in the separation region.

For sphere diameters large compared to the mean free path of the gas stream and for $Re_d$ up to approximately 2, the drag for solid spheres follows Stokes' law (where the drag coefficient $C_D$ equals $24/Re_d$). At Reynolds numbers above approximately 2, $C_D$ is greater than that given by Stokes' law as the result of the increased importance of the form drag. Provided the relative velocity is subsonic, however, the drag coefficient remains a function only of $Re_d$ for $Re_d$ up to approximately 10,000 [*45*].

If the conditions are outside of the above restrictions on Reynolds number, particle size, and relative velocity, other factors affect the drag coefficient. For particles in gases, the Cunningham correction for Stokes' law should be applied if the particle diameter is less than approximately 3 microns [*46*]. On the other hand, stream turbulence and sphere rotation affect the drag coefficient for Reynolds numbers greater than approximately 10,000 [*45*]. At relative velocities approaching the local speed of sound, there is a pronounced increase in drag coefficient [*45*]. However, the drop size and relative velocities for jet engine fuel sprays are usually such that these complicating effects can be neglected.

Under the usual test conditions, there is a rapid acceleration of the fuel spray to the mean air stream velocity. An indication of the effects of acceleration may be obtained from the analysis of the movement of a sphere through a perfect fluid [*47*]. In this case, the drag at constant relative velocity is zero. However, a force applied to the sphere is found to equal

$$\left(1 + \frac{1}{2}\frac{\rho_0}{\rho_d}\right) m_d \frac{du_d}{dt}$$

where $\rho_d$, $m_d$, and $du_d/dt$ are the density, mass, and acceleration of the spheres, respectively, and $\rho_0$ is the fluid density. The correction term $1 + \frac{1}{2}(\rho_0/\rho_d)$ is required, since the fluid stream, as well as the sphere, is accelerated by the applied force. If the density of the fluid stream approaches that of the sphere, it is apparent that the acceleration of the fluid stream should be considered in the equation of motion or the drag coefficient should be modified to account for the acceleration of the fluid stream. Hughes and Gilliland [*48*] have presented graphs indicating the effects of acceleration on the drag coefficient at very low Reynolds numbers. For $\rho_0/\rho_d > 1$, the effects are appreciable. However, for sprays in air streams where $\rho_0/\rho_d \ll 1$, the effect of acceleration on the drag coefficient would appear to be minor.

In the evaluation of drag coefficients for liquid drops, the effects of (1) drop circulation, (2) drop distortion, and (3) drop evaporation should be considered in addition to those factors considered for the solid sphere.

In the low Reynolds number range where viscous drag predominates, the circulation currents within a fluid sphere might be expected to produce a decrease in the drag coefficient. If Stokes' method is applied to the case of fluid spheres of zero surface tension [*47*], the drag coefficient is given by

$$C_D = \frac{24}{Re_d}\left[\frac{1 + \frac{2}{3}\left(\frac{\mu_0}{\mu_d}\right)}{1 + \frac{\mu_0}{\mu_d}}\right] \tag{5-1}$$

where $\mu_d$ and $\mu_0$ are the absolute viscosities of the sphere and surrounding fluid, respectively. For gas bubbles moving in a viscous liquid $\mu_0/\mu_d \to \infty$, and the drag coefficient is $\frac{2}{3}$ that of a solid sphere. However, for a typical jet fuel in an air stream, $\mu_0/\mu_d$ is in the order of 0.01, and the drag coefficient for the drop is within 1 per cent that of the solid sphere. At the higher Reynolds numbers where form drag constitutes the major part of the total drag, the effect of circulation would be expected to be even less pronounced.

Data in the literature for freely falling drops of water, nitrobenzene, *n*-propanol, and methylsalicylate (summarized in [*45*]) indicate that the

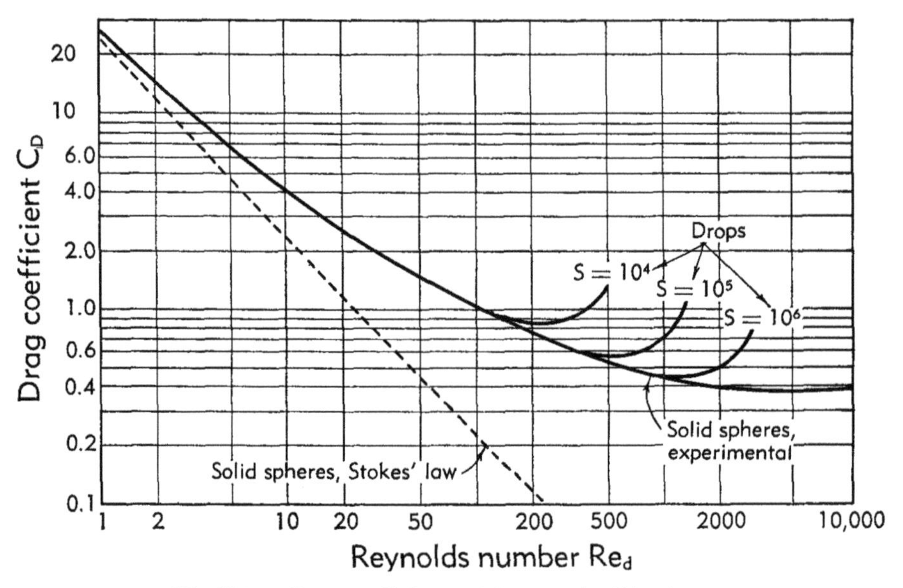


**Fig. F,5a. Drag coefficients of drops and solid spheres. Effect of distortion on drops [*48*]. $S = \sigma_d d\rho_0/g_c\mu_0^2$.**

drag coefficients exceeded those for solid spheres at Reynolds numbers greater than approximately 100. This increase was attributed to drop distortion. Since the drop oscillates in flight, there is a time variation in the drag coefficient. In an analysis of the data of laws for freely falling water drops, Williams [*45*] found maximum time variations in the order of ±20 per cent of the average drag coefficient. Hughes and Gilliland [*48*] treated the drag coefficient of liquid drops in terms of an average value of distortion. From an analysis of data for drops falling at terminal velocity, they obtained the calculated curves for drag coefficient shown in Fig. F,5a. The figure indicates an increase in drag coefficient with decrease in the factor $S = \sigma_d d\rho_0/g_c\mu_0^2$, where $d$ is drop diameter, $g_c$ is the conversion factor or dimensional constant in Newton's law, and $\sigma_d$ is surface tension of the drop. For a typical jet engine fuel ($\sigma_d = 14.3 \times 10^{-5}$ lb/in.), air at standard conditions, and a representative average drop diameter of 100 microns, $S = 9600$. Then from Fig. F,5a, drop distortion would affect the drag coefficient at Reynolds numbers greater than approximately 250.

The previously discussed data on freely falling drops were obtained under conditions where the evaporation rate was relatively low. At the lower Reynolds numbers, the drag coefficients for these drops approach values for solid spheres. However, data obtained with burning drops or drops with relatively high evaporation rates have indicated a substantial reduction in the drag coefficient.

In a study of an *n*-hexane spray evaporating in a high velocity air stream Fleddermann and Hanson [*49*] obtained drag coefficients as low as 0.05 those of the Stokes-law drag coefficient ($= 24/Re_d$) for solid spheres. Ingebo [*50*] used a high speed photographic system to determine drag coefficients for an iso-octane spray accelerating in a high velocity air stream. The average drop size was approximately 45 microns. For this case of a volatile fuel and high relative velocity, the evaporation rate was appreciable during the acceleration period. The drop acceleration data for all drop sizes could be expressed by the equation

$$\frac{d(u_d)}{dt} = 333(u_0 - u_d) \tag{5-2}$$

where $u_d$, $u_0$, and $(u_0 - u_d)$ are the drop velocity, air velocity, and relative air-drop velocity, respectively, in feet per second. Eq. 5-2 indicates that the acceleration of the evaporating drops was independent of drop diameter. On the basis of drag coefficients for solid spheres, the drop acceleration would have been expected to decrease markedly with increase in drop diameter.

The instantaneous drag on an evaporating drop may be equated to the force required to accelerate both the drop and diffusing vapor to obtain

$$C_D \frac{\pi}{4} d^2 \frac{\rho_0(u_0 - u_d)^2}{2} = \rho_d \frac{\pi}{6} d^3 \frac{d(u_d)}{dt} + \frac{(u_0 - u_d)}{g} \frac{dw_d}{dt} \tag{5-3}$$

where $\rho_d$ and $dw_d/dt$ are the density and vaporization rate of the liquid drop, respectively. In the derivation of Eq. 5-3, the vaporized liquid is assumed to be accelerated to stream velocity. The vaporization rate can be expressed in terms of the heat transfer equation

$$\frac{dw_d}{dt} = \frac{\pi d k_0 (Nu)(\Delta T)}{h_v} \tag{5-4}$$

where

$h_v$ = latent heat of vaporization
$k_0$ = thermal conductivity of gas stream
$Nu$ = Nusselt number for heat transfer
$\Delta T$ = difference between temperature of the gas stream and wet-bulb temperature of drop

Substituting Eq. 5-2 and 5-4 in Eq. 5-3 and solving for the drag coefficient

$$C_D = 444 \frac{\rho_d}{\rho_0} \frac{d}{(u_0 - u_d)} + \frac{8k_0(\Delta T)(Nu)}{gh_v d\rho_0(u_0 - u_d)} \tag{5-5}$$

where the constant 444 has the units reciprocal seconds. Curves of the drag coefficient obtained from this type of analysis are presented in

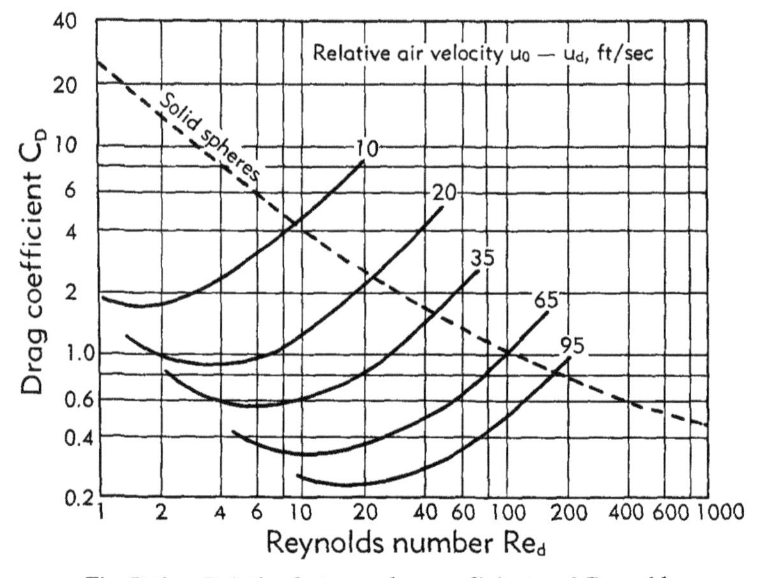


Fig. F,5b. Relation between drag coefficient and Reynolds number for iso-octane droplets and solid spheres [*50*].

Fig. F,5b after Ingebo. Values of $\Delta T$ and $Nu$ were calculated from a previous correlation [*51*]. It is seen that the drag coefficient for the evaporating drops may be as low as 0.1 that for the solid sphere. At a constant relative velocity, the drag coefficient for the evaporating drop exceeds that for the solid sphere at the higher Reynolds numbers (larger drop diameters). This effect might be attributed to drop distortion (Fig. F,5a). It is noted that Eq. 5-5 is not a general equation and that the curves of Fig. F,5b hold only for the particular conditions investigated.

A decrease in the drag coefficient would also be expected when the drop burns. An indication of this effect may be obtained from Spalding's experiments on cylindrical liquid fuel surfaces, burning with an open wake flame [52]. The air stream was at right angles to the axis of the cylinder. It was shown that the open wake flame produced an increase in the static pressure at the rear of the cylinder and a reduction in the drag coefficient. The ratio of the drag with flame to drag without flame could be correlated with the Reynolds number. Spalding cites the similar experimental results by Khudyakov for spheres of gasoline and kerosene burning with open wake flames.

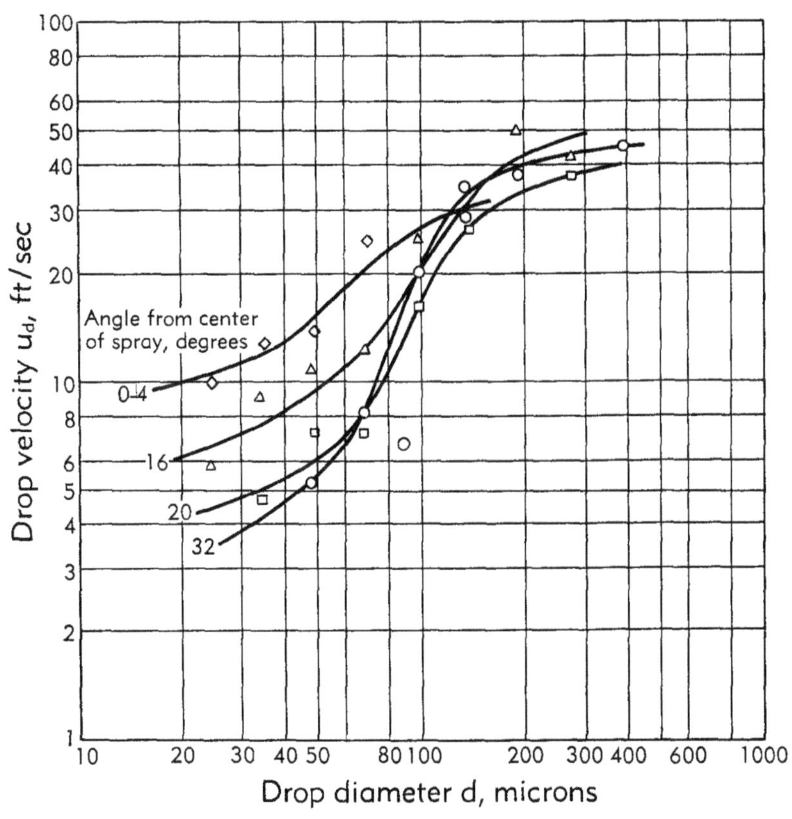


Fig. F,5c. Average velocities of drops of water sprayed from hollow-cone type, pressure-atomizing nozzle [53].

*Relative drop-air velocities.* Some measurements of drop velocity for a fixed configuration, hollow-cone-type nozzle spraying water into quiescent room air have been reported by York, et al. [53]. Fig. F,5c illustrates the variation in drop velocity with angle from the spray axis and drop diameter. It is seen that the drop velocity data tend to pass through a maximum with variation in spray angle and that there is a marked increase in velocity with increase in drop diameter. Calculation of the relative drop-air velocity for the various drop sizes would require values of the unknown induced air velocity resulting from the transfer of spray momentum to the air. If, as a first approximation, the induced air velocity is neglected, the drop Reynolds number for the data of Fig. F,5c would vary from approximately 0.8 for the 20-micron drops to 6.7 for the 90-micron drops. This corresponds to about a 3.5-fold variation in the Nusselt number for heat or mass transfer.

The data of Fig. F,5c are of interest in the study of atomization for swirl-type nozzles. However, the data are not necessarily indicative of conditions to be found for a nozzle spraying into a combustion chamber. During combustion, for example, the effect of the blast of highly turbulent, high temperature gases from the reverse flow region of a tubular combustor (Fig. F,6a) would probably overshadow trends such as shown in Fig. F,5c. This blast would affect not only the relative velocity between the drops and gas stream but also the initial drop size distribution.

For the case of fuels injected into a high velocity air stream, as in some ramjet combustors and turbojet afterburners, the relative drop-air velocities may be treated in three stages: (1) the maximum velocities at the fuel injector, (2) the rapidly changing velocities in the regions immediately downstream of the fuel injector, and (3) the velocities resulting from turbulent velocity fluctuations in regions well downstream of the fuel injector.

A rough estimate of the probable maximum relative drop-air velocity and drop Reynolds number at the fuel injector may be obtained from relations for drop breakup. Lane [*54*] confirmed the prediction of Hinze [*55*] that drop breakup is associated with a critical value of the Weber number,

$$We = \frac{\rho_0 d(u_0 - u_d)^2}{2\sigma_d}$$

where the symbols have been previously defined. Then the drop Reynolds number at breakup would be given by

$$\frac{2(We)\sigma_d}{\mu_0(u_0 - u_d)}$$

Taking the critical value of the Weber number as 10 [*55*] and assuming $(u_0 - u_d)$ equals the air stream velocity, the Reynolds number for the larger drops of a typical jet fuel spray in an air stream at standard conditions would vary from approximately 900 at an air velocity of 100 ft/sec to 300 at 300 ft/sec. It is noted that the initial Reynolds numbers for the average drop size would be appreciably lower than the above values. A closer estimate of the maximum drop size and Reynolds number has been obtained from the use of experimental values of the drag coefficient and an equation relating the drag forces on the drop to the surface tension forces [*50*].

A fuel sprayed into a high velocity air stream quickly accelerates to the air velocity. Accordingly, there is a rapid decrease in the drop Reynolds number. Fig. F,5d, after Ingebo [*50*], illustrates the variation in drop axial velocity with distance downstream of a plain orifice fuel injector. At a distance of 1 inch downstream of the fuel injector, the drops had accelerated to approximately one half of the air velocity. Within a distance of 18 inches, the drops had accelerated to approximately the air stream velocity. Although the acceleration distance was small, the average drop Reynolds number during acceleration was such that approximately 50 per cent of the spray evaporated.

Even after the relative axial velocity during acceleration has decreased to a negligible value, the drop would be subject to a relative velocity as a result of the turbulent flow. Since drop inertia prevents the drop from

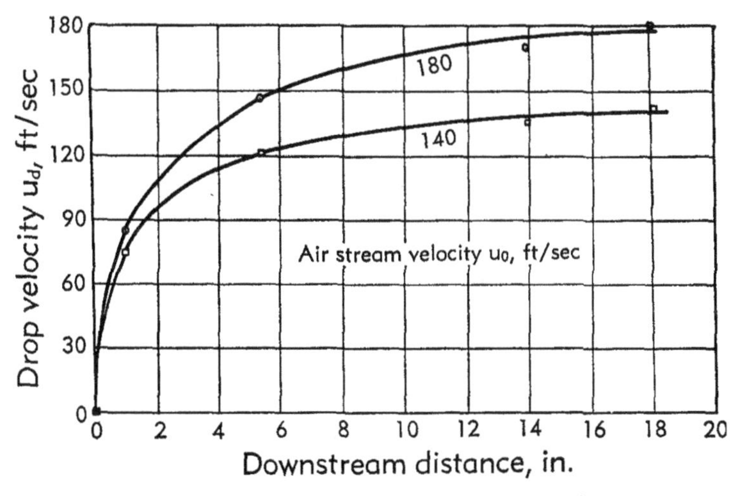


Fig. F,5d. Velocity of iso-octane droplets at given distances downstream of fuel-injector orifice [*50*].

following the turbulent velocity fluctuation exactly, a fluctuating relative velocity occurs. This picture was used by Longwell and Weiss [*56*] to explain the low values of eddy diffusivity found for fuel sprays spreading in high velocity air streams. A rough estimate of the root-mean-square value of drop Reynolds number may be obtained from the approach of Longwell and Weiss. It is assumed that the drag coefficient is given by Stokes' law and that the turbulent velocity fluctuations have the sinusoidal form $u' = u'_0 \cos \omega t$ where $u'$ and $u'_0$ are respectively the instantaneous and maximum turbulent velocity fluctuations of air relative to its time-average velocity, and $\omega$ the frequency of turbulent fluctuations. Then it can be shown that the steady state root-mean-square drop Reynolds number $Re'_d$ is given by

$$Re'_d = \frac{\rho_0 d \sqrt{\overline{u'^2}}}{\mu_0} \left( \frac{K}{\sqrt{K^2 + \omega^2}} \right) \tag{5-6}$$

where $K$ is $18\mu_0/\rho_d d^2$ and $\rho_d$ is the drop density. For fully developed pipe turbulence and an air velocity of 200 ft/sec in a 1-ft-diameter duct, the estimated values of $\sqrt{\overline{u'^2}}$ and $\omega$ are 6 ft/sec and 200 rad/sec, respectively. For a 50-micron fuel drop and air at standard condition, $K/\sqrt{K^2 + \omega^2}$ equals 0.69 and the value of $Re'_d$ would be approximately 4.4. This corresponds to about a 60 per cent increase in the evaporation rate over that for still air.

*Diffusion of a cloud of drops.* There are two principal factors determining the spreading of a liquid fuel spray in a high velocity air stream. The momentum imparted to the fuel by the injector may result in an initial radial dispersion of the spray. After the initial spreading occurs, a further transport of the spray results from eddy diffusion.

For contrastream injection from a simple orifice, the initial spreading occurs as the spray is forced back on itself by the air stream. For cross stream injection or injection from swirl-type nozzles, the initial spreading is affected by the radial component of the initial fuel momentum. The initial spreading is also affected by the turbulent air flow patterns produced by the injector. In systems where the allowable duct length is relatively small, the initial spreading may be the major factor determining the spray distribution.

The turbulent spreading of the spray downstream of the injector has generally been treated in terms of approximate solutions for spreading from a point source in a general stream (see Art. F,2). No distinction has been made between liquid drops and fuel vapor in the analysis of fuel spreading.

Longwell and Weiss [*56*] reported an extensive analytical and experimental study of fuel spray spreading in circular ducts. For the case of low velocity fuel injection from a simple orifice, the initial spreading was small. Accordingly, an equation for diffusion from a single point source was found to be applicable to the radial fuel concentration profiles at various stations downstream of the injector. Data for high velocity contrastream injection or injection from swirl-type nozzles were treated in terms of solutions for spreading from a disk source. Solutions for more complex systems could be obtained by summing up the contributions from the various elements of the system. The use of these methods for predicting fuel spray spreading has been discussed in detail by Longwell in II,J.

Longwell treated the extreme conditions of fuel spreading: (1) the spreading of a highly volatile fuel where the major portion of fuel could be considered to have vaporized and (2) the spreading of a very low volatility fuel where the amount of vaporization could be considered negligible. At high air stream velocities, the effective eddy diffusion coefficient for the high volatility fuel was more than twice that for the low volatility fuel. Longwell explained this deviation in terms of drop inertia effects. Since the drops are unable to follow closely the high frequency, turbulent velocity fluctuations, the effective eddy diffusion coefficient for the drops is reduced. On the basis of a simplified analysis for the drop motion, Longwell showed that the effective eddy diffusion coefficient for the drops should decrease with increase in drop size and frequency of the turbulent velocity fluctuations (proportional to air stream velocity). The data and analysis of Longwell show that spray evaporation is a major factor in the spreading of fuel sprays in high velocity air streams.

Bahr [57] investigated the spreading of a liquid iso-octane spray in high velocity air streams. Contrastream injection from a single orifice

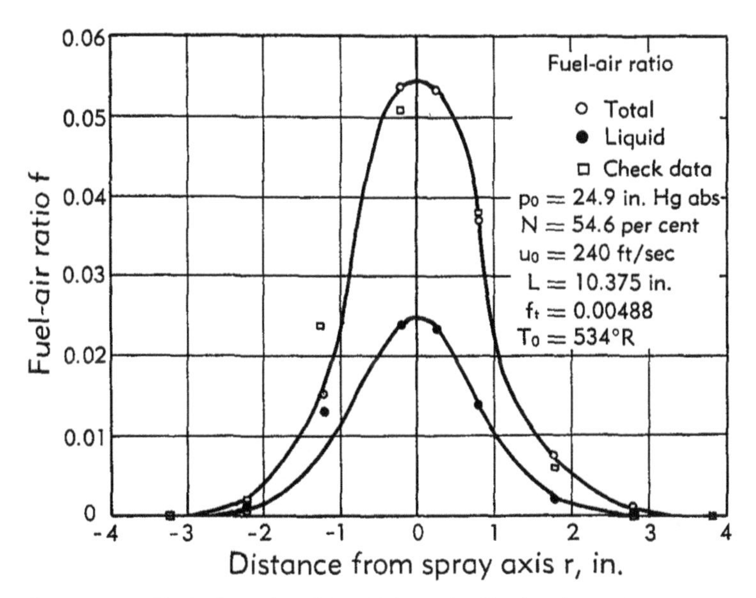


Fig. F,5e. Typical total and liquid fuel distribution for contrastream injection of iso-octane from simple-orifice fuel injector [57].

was used for all tests. A typical example of the data obtained by Bahr is shown in Fig. F,5e. Here the fuel-air ratio is plotted against the radial distance from the duct axis. The upper curve is based on the total of liquid and vapor fuel; the lower curve is the fuel-air ratio based on liquid fuel alone. The test conditions for these data are indicated on the figure. Even for this case of low air temperature and high air velocity, approximately 55 per cent ($N = 54.6$ per cent) of the spray was evaporated at a distance of approximately 10 inches downstream of the fuel injector.

The fuel-air ratio profiles based on the total liquid and vapor fuel were treated by use of the following simplified equation for spreading from a point source

$$f_t = \frac{f_0 R^2}{M} e^{-r^2/M} \tag{5-7}$$

where

$f_0$ = over-all, fuel-air ratio (weight fuel per sec/weight air per sec)
$f_t$ = total fuel-air ratio at a given point in the duct
$M$ = spreading index
$r$ = radial distance from spray axis
$R$ = duct radius

$M$ is a measure of the spreading; that is, higher values of $M$ are associated with increased spreading. Bahr obtained a correlation of $M$ in terms of the experimental variables as shown in Fig. F,5f. The correlation is given by the following relations

$$M = 0.0598\Phi + 0.00042 \tag{5-8}$$

$$\Phi = \left(\frac{T_0}{1000}\right)^{0.67} L^{0.76} p_f^{0.49} D^{0.79} \left(\frac{u_0}{100}\right)^{-0.85} p_0^{-0.57} \tag{5-9}$$

where

$D$ = fuel orifice diameter (0.024 to 0.041 in.)
$L$ = axial distance downstream of the fuel injector (5 to 18 in.)
$p_f$ = fuel injection pressure differential (25 to 85 lb/in.$^2$)
$p_0$ = air pressure (20 to 30 in. Hg abs)
$T_0$ = air temperature (545 to 850°R)
$u_0$ = air velocity (100 to 350 ft/sec)

The absolute values of $M$ are subject to such factors as fuel type, fuel injector design, duct size, and turbulence level of the air stream. However, the relative effects of the various experimental variables indicated by Eq. 5-9 might be expected to hold for similar systems and for gasoline-type fuels.

The spreading index $M$ is equivalent to $4EL/u_0$, where $E$ is the effective eddy diffusion coefficient for the total liquid and vapor fuel. If the small second term of Eq. 5-8 is neglected and $M$ is replaced by $4EL/u_0$, it is seen that the term $E/u_0$ decreases with increase in air velocity. This trend is consistent with the data of [*56*]. The increase in $E/u_0$ with rise in air temperature might be expected as a result of the higher vaporization rates and corresponding increase in the fraction of the more rapidly diffusing fuel vapor. However, most of the effects of the air temperature and pressure and the fuel injector orifice diameter and pressure drop on $E/u_0$ could probably be attributed to changes in upstream penetration and initial radial dispersion of the spray. Part of the trends might also be attributed to changes in the atomization of the spray. However, it is noted that the drag coefficient data of Ingebo [*50*] suggest that, for rapidly vaporizing sprays, the spreading rate is the same for all drop sizes.

*Fuel evaporation and flame progress.* In analytical treatments of the mixing and combustion of liquid sprays, the problem arises as to the effect of liquid drop evaporation on the flame progress. If the flame for the liquid fuel spray were entirely equivalent to the turbulent gaseous diffusion flame, the location of the constant fuel-air ratio surfaces resulting from fuel spreading might be used to estimate flame progress. One would assume, as with the gaseous diffusion flame, that the chemical reaction occurs instantaneously and that the flame front exists at the stoichiometric fuel-air ratio surface. If complete vaporization occurs at

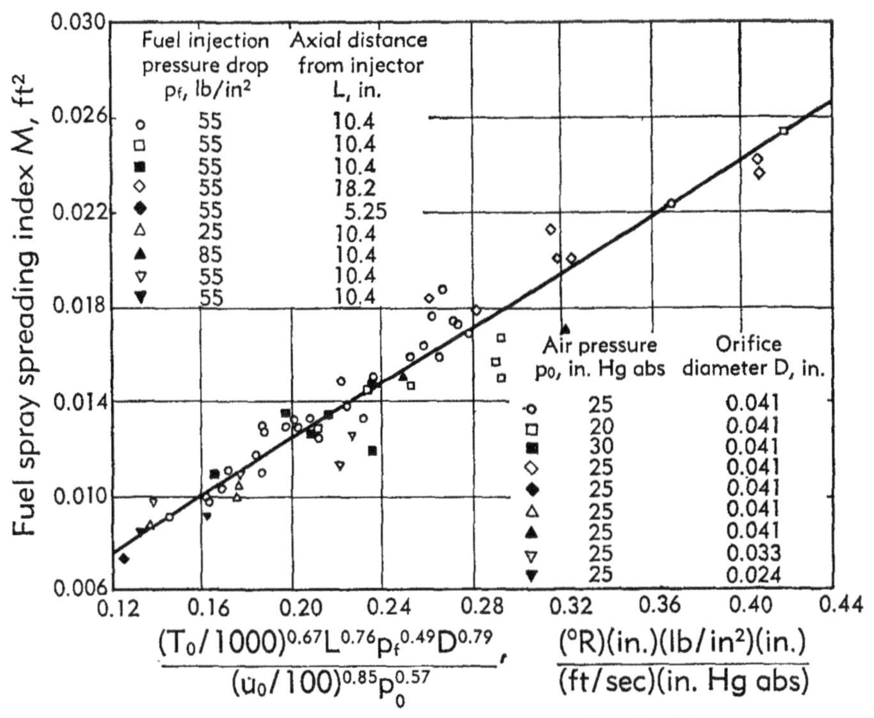


Fig. F,5f. Correlation of index of fuel-spray spreading. Fuel, iso-octane; air velocity = 100 to 350 ft/sec; air temperature = 545 to 850°R [57].

the flame front, the fuel-air ratio surface considered from the cold flow tests would be that based on the total of liquid and vapor fuel. If complete vaporization does not occur, the delay time required for vaporization should result in an effective decrease in the rate of flame progress.

An indication of the drop sizes for which vaporization time may become significant may be obtained from some investigations of the combustion properties of uniform clouds of fuel drops. Jones, et al. [58] investigated the flammability limits of a condensed mist of JP-1 fuel. The flammability limits at 32°F ranged from 0.043 to 0.23 fuel-air ratio. These limits are similar to those for a vaporized JP-1 fuel at 300°F, if suitable correction is made for the effect of initial temperature on the flammability limits. Similarly, the data of Haber and Wolff [*59*] for spatial flame speeds of condensed fuel mists agree with flame speeds for vapor-air mixtures provided a temperature correction is made. The drop sizes for the fuel mists of [*58*] were approximately 10 microns.

Wolfhard and Parker [*60*] investigated the evaporation of kerosene drops approaching the flame front to a butane-air, Bunsen-type flame. Drops from approximately 9 to 30 microns in diameter evaporated before they reached the luminous inner cone of the flame. Evidently, then, the drops in the fuel mist experiments can evaporate and diffuse in the preheating zone with sufficient enough rapidity that the reaction zone "sees" only a vapor-fuel-air mixture. However, the time for complete vaporization of a drop varies as the square of the initial diameter. Accordingly, it might be expected that the effect of evaporation on combustion properties of mists and the flame progress of a fuel spray would become noticeable as drop size increases much above 20 or 30 microns. For the large drop diameters (100 microns or above), the spray would probably burn as a number of individual drops. In such a case, the location of the stoichiometric fuel-air ratio surface would probably have little significance.

Estimates of the effect of evaporation on flame progress and mixing for a particular system would require correlations for predicting the degree of vaporization at various points along the duct. Correlations are available for the forced convection evaporation of single drops over a wide range of operating conditions, e.g. [*51*]. Unfortunately, there have been but a limited number of extensive studies of spray evaporation.

Bahr [*57*] correlated the evaporation data for the iso-octane sprays by the following relation

$$\frac{N}{100 - N} = 9.35 \left(\frac{T_0}{1000}\right)^{4.4} \left(\frac{u_0}{100}\right)^{0.8} p_0^{-1.2} p_i^{0.42} L^{0.84} \tag{5-10}$$

where $N$ is the per cent of the fuel spray evaporated at a given station and the same symbols and range of variables apply as in Eq. 5-9. From the inspection of the exponents of Eq. 5-10, it is seen that the air stream temperature is the major factor determining the evaporation rate.

The driving force for evaporation is the temperature difference $\Delta T$ between the air stream temperature and the corresponding wet bulb temperature for the iso-octane drops. Examination of the data indicates that the term $N/(100 - N)$ of Eq. 5-10 would be roughly proportional to $\Delta T$ to a power somewhat less than unity. Over the range of air stream temperatures investigated by Bahr, there was roughly a tenfold increase in $\Delta T$ for a 40 per cent increase in the absolute air stream temperature. However, the per cent increase in $\Delta T$ for a given per cent increase in the absolute air stream temperature drops off rapidly at the higher air stream temperatures. Accordingly, the exponent of the temperature term in

Eq. 5-10 might be expected to decrease appreciably for air stream temperatures much greater than those investigated by Bahr.

It is seen that the evaporation at a given station increased with increase in air stream velocity. Evidently the finer atomization and greater heat transfer coefficients for the drops at the higher air velocities offset the decrease in residence time. At the higher air pressures there should have been an improvement in atomization [*61*] and a corresponding increase in evaporation. However, the decrease in $\Delta T$ at the higher pressures may have offset this effect and produced the observed decrease in evaporation with the rise in air pressure.

Correlations such as obtained by Bahr are of interest in the study of the mixing and combustion of liquid fuel sprays in high velocity air streams. From such correlations, both the constant fuel-air ratio surfaces and per cent fuel evaporated may be estimated at various stations downstream of the fuel injector. More general relations might be obtained, however, if the spray evaporation data could be related to single-drop evaporation tests. The data of Sacks [*62*] indicate some of the problems to be found in attempts to treat evaporation rates for the entire spray in terms of single-drop theory. The actual evaporation rate of a kerosene spray in a low velocity air stream was about one per cent of that predicted from the use of Langmuir's equation [*63*] in Probert's relations for spray evaporation [*64*]. This large difference between predicted and experimental values probably resulted from the existence of local high fuel vapor concentrations about a large portion of the drops in the spray.

Ingebo [*50*] has shown that the evaporation rate of an iso-octane spray accelerating in a high velocity air stream may be related closely to single-drop data. The spray was treated in terms of a mean drop size, $D_{20}$ defined as the drop having an area equal to the ratio of total spray area to total number of drops in the spray. Values of $D_{20}$ and relative velocities were obtained by the use of high speed photographic equipment. The spray evaporation data obtained from these values and a previously determined relation for single-drop evaporation [*51*] agreed closely with the evaporation rates predicted from Bahr's correlation. Although the data of Ingebo were obtained over a limited range of conditions and for but one fuel, the results offer hope that a general relation for the evaporation rate of sprays in high velocity streams may eventually be obtained. Such a relation requires generalized expressions for predicting both the drag coefficients of evaporating drops and the drop size distribution.

*Drop segregation.* In turbojet afterburners and ramjet combustors, the liquid fuel may be in the form of a fine spray introduced longitudinally into a high velocity gas stream. Much of the spray may be deposited upon the flame holders, the amount depending on the fineness of the spray and the gas stream velocity. It is suggested in [*65*] that the flame stability limits depend, among other things, upon the amount of liquid fuel that collects upon the flame holder. Deposition of the spray on duct walls of turbojet combustors should affect flame patterns, combustion efficiency, and carbon deposition. In addition to spray deposition, the segregation of drops remaining in the air stream should be considered. Force fields set up by vortex flow or flow around bluff bodies may produce large changes in local fuel-air ratio.

A number of analytical investigations of the impingement of drops on bodies in high velocity air streams have been reported. Some of the

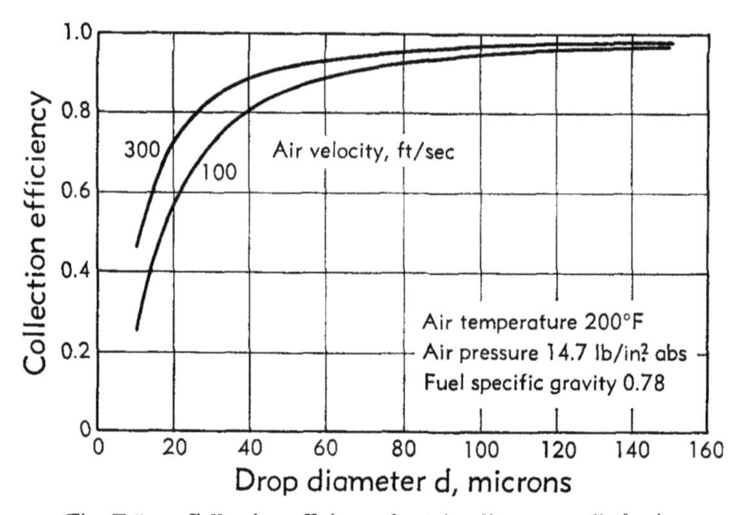


Fig. F,5g. Collection efficiency for 2-in.-diameter cylinder in high velocity air stream. Calculated from data of [*67*].

body shapes considered have been spheres and ribbons [*66*], right circular cylinders [*66,67*], airfoils [*68*], and ellipsoids [*69*]. The analyses have generally involved the following major assumptions:

1. Uniform diameter and spatial distribution of drops in spray
2. Zero relative drop-air velocity at large distances ahead of body
3. Air flow streamlines given by potential flow theory
4. Drop drag coefficients following solid sphere data
5. Negligible drop vaporization.

Calculations have been presented for such factors as the collection efficiency, the angle from the forward stagnation point beyond which no droplets strike the body, and impact velocities. The collection efficiency may be defined as the ratio of the drops striking the body to the drops contained in the volume of air swept out by the body. Fig. F,5g presents an example of the variation of collection efficiency with drop size for the impingement of a fuel spray on a right circular cylinder. The curves were calculated from the data of [*67*] for the following conditions: cylinder diameter, 2 inches; air stream temperature, 200°F; air stream pressure, one atmosphere; and fuel specific gravity, 0.78. Since the actual spray contains a range of drop sizes, such curves serve mainly for qualitative comparisons of the collection efficiencies obtained with various flame holder sizes and operating conditions. If the drop size distribution is known, a weighted average collection efficiency for the entire spray may be determined [*67*].

The presence of the body also results in a difference in the concentration of drops at various points in the air stream. Dorsch and Brun [*69*] have shown that the drop concentration may vary from zero to several times the free values within a short distance from the surface of an ellipsoid.

Investigations of spray deposition in straight ducts have been reported by Alexander and Coldren. In [*70*] water was atomized in a small air-atomizing nozzle and injected coaxially into an air stream flowing through a straight circular duct. The calculated mean drop diameter was 27 microns. In the region far downstream of the nozzle, the water concentration profile was essentially flat over most of the duct cross section, with the major resistance to mass transfer occurring near the duct walls. Mass transfer coefficients at the duct walls were from 10 to 20 times greater than those for gases under similar conditions. This could probably be attributed to drop inertia which permitted the drops to penetrate the boundary layer.

In [*71*] Alexander and Coldren report values for the fraction of a water spray deposited on nonwetted walls when the water is injected downstream from a tube at the duct axis. The duct diameter was 1.81 inches. An example of the deposition data for an air stream velocity of 180 ft/sec and a high turbulence level is presented in Fig. F,5h. The data indicate negligible deposition within the first 6 inches downstream of the nozzle. At greater axial distances there was a rapid increase in deposition. The deposition at a given axial distance decreased with increase in air stream velocity. Additional data indicated large effects of the water velocity on deposition.

The presence of the duct walls may have a noticeable effect on the fuel distribution in the air stream regardless of the amount of fuel deposited on the walls. This effect was treated by Alexander and Coldren [*71*] by suitable choice of boundary conditions in the solution of the differential equation for mass transfer. For the case of negligible spray deposition, Longwell and Weiss [*56*] used simple graphical procedures to correct the solution for spreading in an infinite duct for the effect of duct walls.

Segregation effects may also be imposed by centrifugal forces in the combustion chamber. Several studies have been made of vortex or cyclone combustion systems. The regular flow pattern in a vortex combustor seems to offer a better opportunity for obtaining a controlled mixing length and flame stabilization than in conventional turbojet combustors. Theoretically, if gravitational effects are neglected, a cylindrical combustor with air introduced tangentially at the outer radius at one end would have a helical flow pattern, such as shown by the solid line in Fig. F,5i. However, at any longitudinal transverse cross section, the vortex-type flow imposes a radial pressure gradient with lowest pressures at the combustor center. In addition, a decrease in velocity caused by viscous effects as the air flows through the combustor results in an increasing static pressure in the longitudinal direction. As a result, a recirculatory

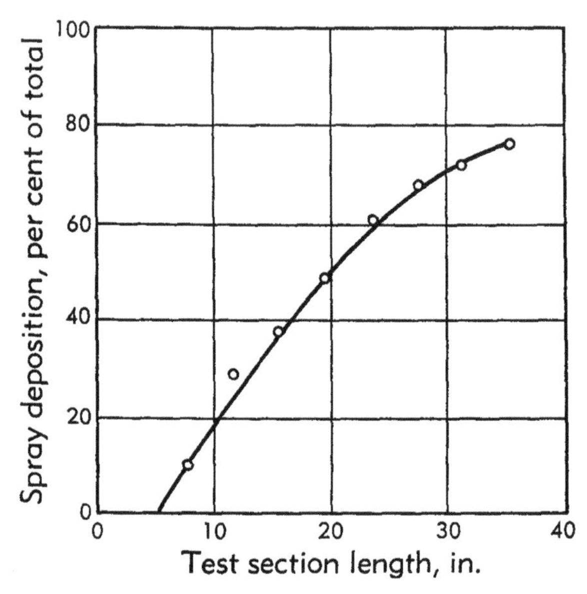


Fig. F,5h. Spray deposition on walls of straight duct [*71*]. Fluid, water. Duct diameter = 1.81 in. Air velocity = 180 ft/sec.

flow pattern (dotted lines) may be initiated with a reverse flow in the center of the combustor.

Hottel and Person [*3*, pp. 781–788] determined the performance of a vertical vortex combustor in which air was introduced tangentially at the base of a $5\frac{3}{4}$-in.-diameter chamber. Gaseous fuel entered at the center of the combustor through a 2-inch circular opening. Isothermal tests showed the existence of a combined vortex. The outer part exhibited velocities inversely proportional to the radius (theoretical vortex), while the center part had velocities approximately proportional to the radius (fixed vortex). During combustion, the gaseous fuel entering at the center of the base, was thrown outward by the centrifugal effect of the fixed vortex. Most of the reaction occurred in a turbulent cylindrical annulus between the fixed and free vortex regions. Very little recirculatory flow

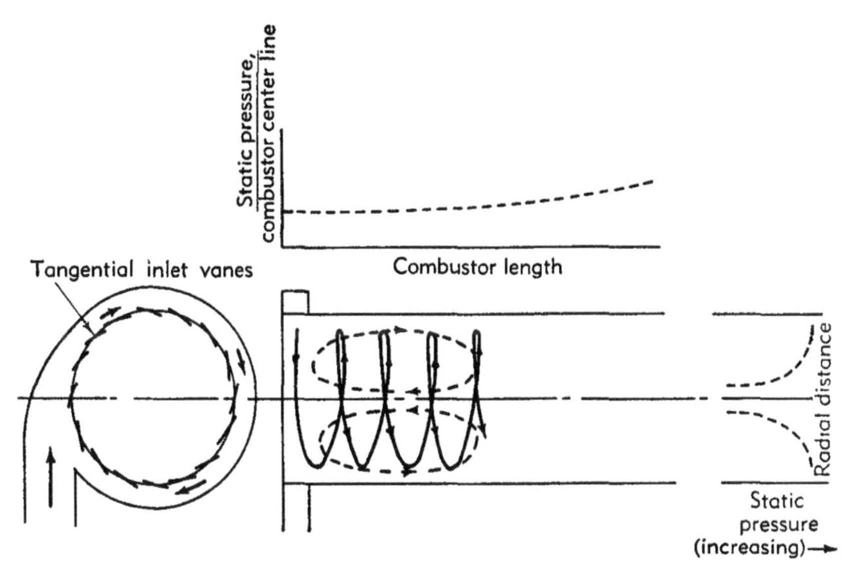


Fig. F,5i. Recirculatory flow in a tubular combustion chamber having tangential air admission [*3*, pp. 781–788].

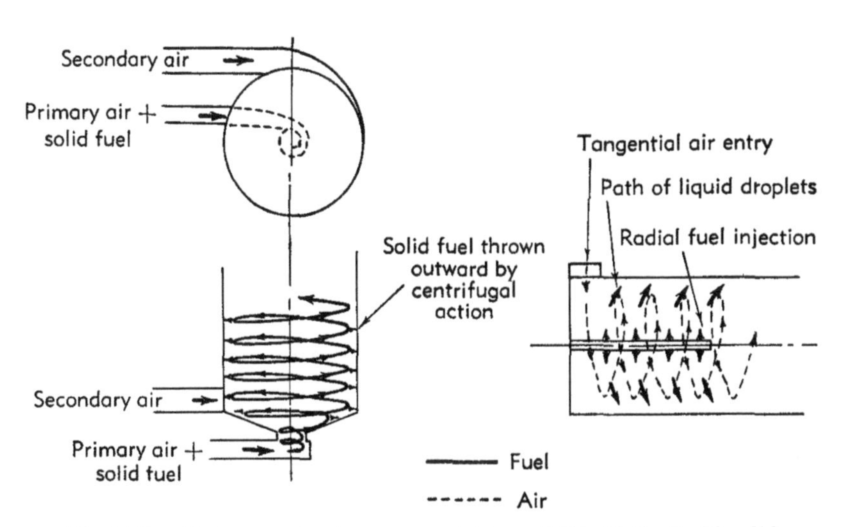


Fig. F,5j. Vortex combustion systems [*46*, pp. 1036, 1605]. (Left) solid fuel-burning furnace; (right) liquid fuel-burning combustor.

was observed in the center of the combustor. The combustion efficiency of the combustor was substantially independent of velocities of the systems, depending only on combustor size and fuel-air ratio.

Liquid or solid fuel entering a vortex combustor would undergo a similar centrifugal section. In a vortex, solid fuel-burning furnace (Fig. F,5j), primary air and solid fuel enter through a small swirler at the center of one end of a cylinder. Secondary air enters tangentially at the outer radius. The solid fuel is thrown by centrifugal action to the wall of the furnace where it is held by a molten slag layer. High scrubbing velocities of the air result in efficient combustion. In the vortex combustion chamber in which liquid fuel is injected radially outward along the length of the combustor from a central fuel tube (Fig. F,5j), the fuel particles would be thrown radially outward and moved downstream much as in a spray gas scrubber.

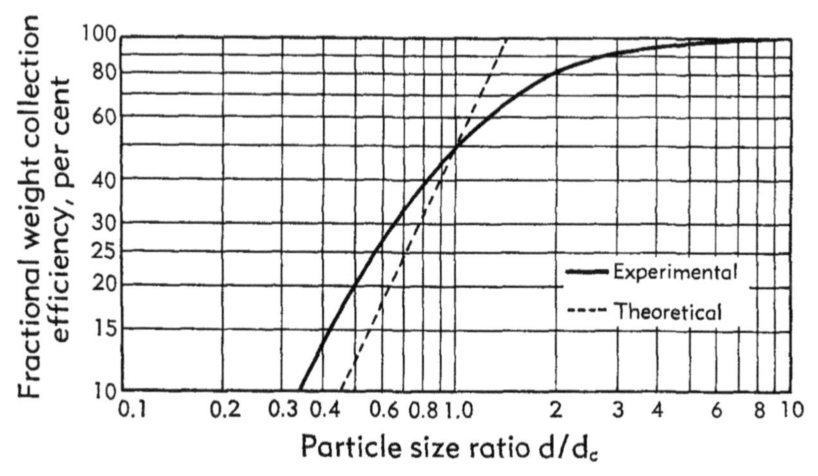


Fig. F,5k. Separation efficiency of typical cyclones [*46*, p. 1026].

In [*46*, pp. 1021–1028], the following equation is presented for the minimum diameter of a particle that should be completely segregated from the gas stream in a cyclone separator

$$d_{\min} = \sqrt{\frac{9\mu_0 B_c}{\pi N_t u_c(\rho_d - \rho_0)}} \tag{5-11}$$

where

$A_c$ = cyclone inlet area
$B_c$ = width of cyclone rectangular inlet duct
$N_t$ = number of turns made in cyclone
$u_c$ = average cyclone inlet velocity based on $A_c$
$\mu_0$ = fluid (air) viscosity
$\rho_d$ = liquid drop density
$\rho_0$ = fluid (air) density

Eq. 5-11 is based on Stokes' law and assumes no mixing or turbulence during a fixed number of turns at constant spiral velocity. The removal of smaller particles is proportional to the initial distance of the particles from the wall. The relation of Rosin, Rammler, et al. [*46*, pp. 1021–1028] is shown in Fig. F,5k in terms of a collection efficiency and a cut size diameter $d_c$.

$$d_c = \sqrt{\frac{9\mu_0 B_c}{2\pi N_e u_c(\rho_d - \rho_0)}} \tag{5-12}$$

$N_e$ represents the effective number of turns made in cyclone, and the same symbols apply as in Eq. 5-11. For the theoretical curve, $N_e = N_t$. For no reentrainment, the curve and $N_e$ should be unique for a cyclone of given geometric proportions. An experimental curve is included in Fig. F,5k for an $N_e \cong 5.0$. However, this value may be conservative, since $N_e$'s of 10.0 were given by some data. Use of an inlet guide vane caused appreciable reentrainment, and $N_e$ became approximately 2.0 for inlet velocities of 50 ft/sec at atmospheric pressure.

**F,6. Mixing in Combustion Chambers.** As discussed previously the need for high heat release rates and reference velocities has complicated the problem of achieving adequate mixing in typical turbojet-engine combustion chambers. A satisfactory mixing system in such

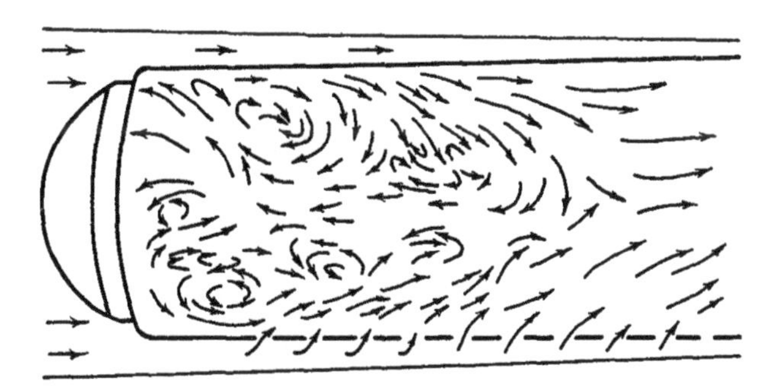

Fig. F,6a. Air flow pattern in a tubular turbojet combustor under nonburning conditions [*72*].

combustors requires, among other factors, (1) high turbulence levels, (2) flame-stabilizing flow patterns, (3) matching of fuel and air flow patterns, and (4) effective dilution of combustion products within a short distance. As the result of such requirements, the gross air flow patterns and fuel-air ratio distribution patterns in typical turbojet engine combustors are quite complex.

*Gross air flow patterns.* An example of the turbulent air flow patterns existing without combustion in a tubular turbojet engine combustor is shown in Fig. F,6a [*72*]. The pattern shown by the arrows was sketched from a high speed flash photograph of balsa dust particles injected into the flow in the inlet ducting of a transparent model of the combustor. The arrows in the sketch are not proportioned in length to indicate relative velocities, but serve principally to indicate the general trends of the flow patterns at the various stations. Although these patterns were obtained without combustion, they might be considered as giving an approximate indication of the general trend in flow pattern during combustion.

It is seen that there is a region of reverse flow and large scale vortices in the upstream end of the combustor. Such flow patterns may result from vortex flow, the entrainment of gases in high velocity air jets issuing in the axial direction from the dome, and the impingement of opposing air jets from the liner. The patterns have been found effective both for flame stabilization and mixing. A highly turbulent region exists along the center line of the combustor as a result of the impingement of the

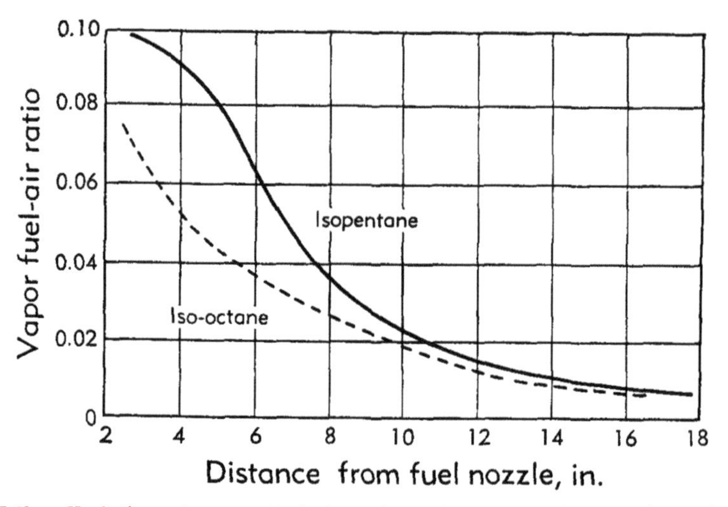


Fig. F,6b. Variation of vapor fuel-air ratio with distance from fuel nozzle [73]. Inlet conditions: simulated high altitude; over-all fuel-air ratio = 0.007.

various opposing air jets. Although not evident in the balsa dust photographs, mixing regions would be expected to occur at the boundaries of the air jets as in the case of free and coaxial jets.

In the secondary zone of the combustor, a deep penetration of the dilution air jets into the hot stream of combustion gases leaving the primary zone would be required in order to achieve adequate mixing within a short distance. The penetration of these jets is illustrated in Fig. F,6a. As a result of the relatively high axial velocities in the annulus, the various air jets along the combustor may enter the combustion space at an angle to the liner wall.

*Fuel-air ratio distribution in turbojet engine combustor.* Examples of the vapor fuel-air ratio distribution to be found during combustion in a tubular-type turbojet engine combustor are shown in Fig. F,6b and F,6c [73]. The data were obtained by means of a multiple point, radial sampling probe.

Fig. F,6b illustrates the change in average vapor fuel-air ratio along the combustor for a low volatility and high volatility fuel. As the result of the liner, open-hole area distribution along the combustor, vapor fuel-air ratios richer than stoichiometric can exist at the upstream end of the combustor, even for a lean over-all, fuel-air ratio. Since the over-all, fuel-air ratio is the same for both fuels, the relative positions of the curves indicate the relative degrees of vaporization at various distances along the combustor. At the upstream end of the combustor there is a pronounced change in vapor fuel-air ratio as the result of the difference in vaporization rate of the two fuels. Near the end of the combustor the two

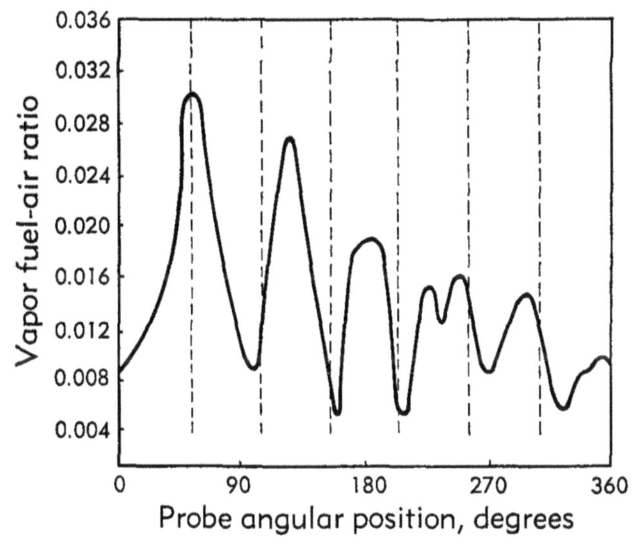


Fig. F,6c. Variation of fuel-air ratio with angular position of sampling probe [*73*]. Fuel, iso-octane; over-all fuel-air ratio = 0.014.

curves approach each other as both fuels approach the point of complete vaporization.

An example of the vapor fuel-air ratio variations to be found at a given station along the combustor is shown in Fig. F,6c. The data were obtained by sweeping the multiple point probe around the axis of the combustor. Here the average vapor fuel-air ratio along a radius is plotted against the angular position of the probe. The dashed lines indicate the angular position of the center lines of the liner wall openings. For this case, there was as much as sixfold variation in the average radial vapor fuel-air ratio. This variation could probably be attributed primarily to the effect of the air entering from the various liner wall openings. However, the asymmetrical form of the distribution curve indicates that other factors, such as an uneven fuel spray, liner wall distortion, and unsymmetrical flow patterns in the primary zone, may have had an important effect on fuel-air ratio distribution at various stations along the combustor. It is noted, of course, that a much wider variation in vapor fuel-air ratio might be expected if the data were for point-to-point sampling rather than for radial average sampling. The local vapor fuel-air ratios at a given cross-sectional plane are affected by the penetration and quantity of air entering from the various liner wall openings. Accordingly, the problem is not amenable to analysis such as that used in the treatment of flow from the simpler jet configurations.

The flow pattern and vapor fuel-air ratio distribution trends of Fig. F,6a, F,6b, and F,6c illustrate the difficulty of attempting to apply approaches such as those used for the simple free jet diffusion flames to the prediction of flame lengths for the turbojet engine combustor. Aside from difficulties associated with the complex air flow patterns, it is evident that the fuel vaporization process has a pronounced influence on the availability of vapor fuel and, hence, on the distance required for the fuel and air molecules to meet in stoichiometric properties.

*Mechanical mixing devices.* In the turbojet combustor, the Reynolds numbers are sufficiently high for the existence of normal pipe turbulence. However, mixing of parallel gas streams by such random turbulence requires relatively long distances [*74*; *52*, p. 485]. Accordingly, mechanical mixing devices are used to increase the transfer of momentum, heat, and mass by (1) producing large scale mixing patterns and (2) increasing the turbulence level in the fluid stream. Essentially, mechanical mixers try to increase the surface area of the interface between the components of the mixture and to distribute the interface throughout the mixture. As such, mechanical mixers may be classified as:

1. Mixers dependent on increasing interstream surface directly such as orifices, swirlers, stub tubes, air scoops, and "flower pot" mixers.
2. Mixers dependent on eddy generation such as "mixing rolls."

The simplest type of mixer listed under 1 is an orifice in the combustor liner. Turbulent mixing in both the reaction zone and dilution zone depends on complex flow patterns, which are a function of the penetration and spreading of individual air jets normal to the mean flow. As shown previously, interaction of the jets results in a high degree of turbulence.

Zettle and Mark [*75*] found that the shape of orifice used for air entry can influence mixing and over-all combustor performance. For example, long axial openings were used to admit primary air into the reaction zone of two different annular combustor liners. With these combustors, marked improvements in combustion efficiency, stability, altitude operating limits, and combustor outlet temperature profile were noted. The improvements in performance were felt to result from the introduction of alternate longitudinal air-rich regions which could mix with longitudinal fuel-rich regions in the primary zone.

Recirculatory motion in the primary combustion zone can be induced not only by peripheral air entry through orifices, but also by the use of swirlers. Mechanical swirlers, consisting of a series of vanes placed symmetrically around the axis of a tubular combustor and at an angle to the entering air, give the air a rotary motion. As discussed previously under segregation effects, the rotary motion combines with the forward air velocity component to form a vortex, causing the air near the center of the combustor liner to travel upstream in a reverse direction to the general flow (Fig. F,5i).

Stub tubes have been used in combustors to obtain required penetration of air jets, when adequate penetration could not be obtained with orifices alone. One of the first combustors to use the stub tube device was the Rolls Royce "Welland" engine. A major objection to the use of stub tubes is their susceptibility to damage by combustion.

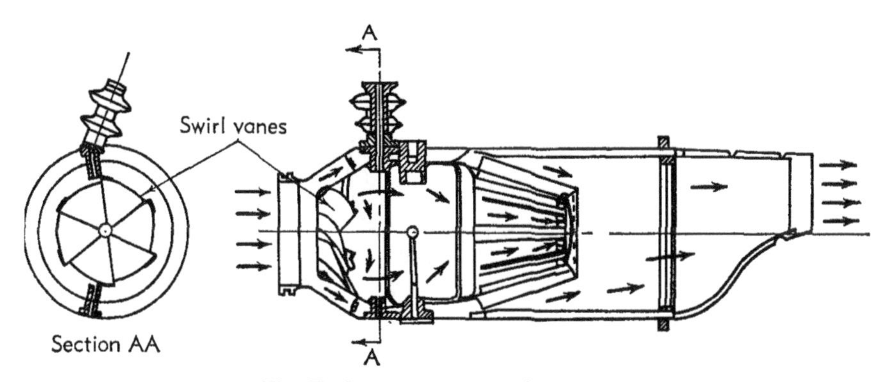


Fig. F,6d. Jumo 004 combustor [*76*].

The stub tube system of mixing was used in a somewhat modified form on the Jumo 004 combustor (Fig. F,6d). The mixing assembly consisted of ten chutes welded to a ring at the forward end and to a dished 4-in.-diameter baffle at the downstream end. Air was diverted from the secondary stream and forced down the chutes for mixing. Mixing of the air and the combustion gases was aided by the dished baffle. The hot gases were forced out between the chutes into the cold secondary air stream, where additional mixing might occur.

The abrupt change in direction of the air stream when orifices or stub tubes were used to promote mixing was thought to result in inefficient use of the velocity pressure of the secondary air [*76*]. Improvements in mixing, together with reduced pressure losses, seemed obvious if the air could be introduced gradually into the mixing region without abrupt directional changes.

Air scoops which protrude into the high velocity secondary air streams have been used to deflect the air gradually into the mixing region. One of the early designs of this type was the British "daisy" mixer (Fig. F,6e). The mixer consisted of a series of flutes staggered around the combustor liner and used to direct the dilution air into the secondary mixing zone. This device suffered from the same fault as stub tubes; the petals or

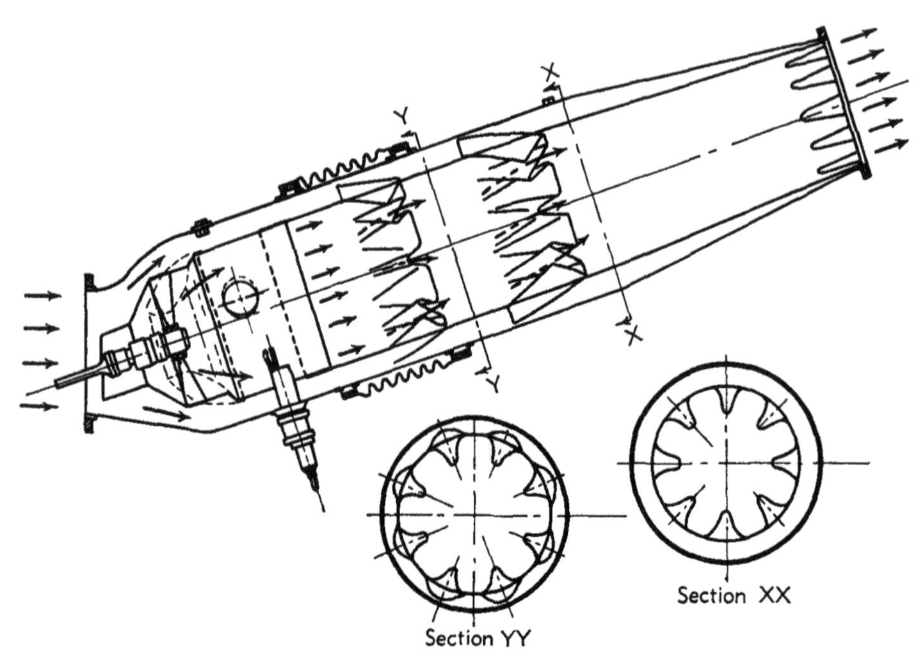


Fig. F,6e. Combustor with "daisy mixer" for introducing secondary air [76].

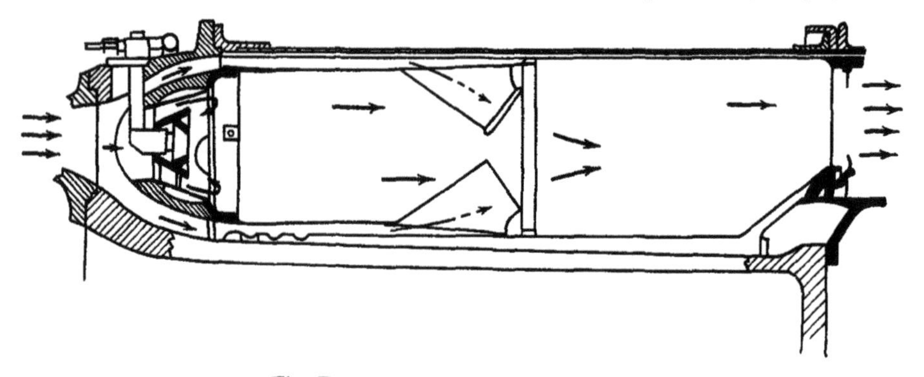

Fig. F,6f. BMW 003 combustor [76].

flutes of the mixer exposed to the combustion gases were damaged by combustion.

Air scoops essentially attempt to create mixing by interleaving streams of gases differing in temperatures. The "sandwich scoop," used in a modified form on the B.M.W. 003 combustor (Fig. F,6f), interleaved streams of cold secondary air between parallel streams of hot combustion products. This type resulted in lower pressure losses than other mixers used in contemporary combustors. However, air scoops, like stub tubes, are subject to deterioration by the combustion gases.

A device known as the "flower pot mixer" has been used in several combustor designs. Essentially the "flower pot" combustor (Fig. F,6g) features a generally divergent combustor liner. Truncated cones make up the downstream portions of the liner. Secondary air enters through rectangular slots in the conical surfaces. Thus arranged, the slots tend to reduce the over-all total pressure losses of the combustor by converting part of the secondary air velocity energy to pressure while still obtaining a desired combustor outlet temperature distribution.

Minchin [*52*, p. 485] noted that the temperature profile of two parallel hot and cold streams mixing in a common duct was similar to half the profile obtained in the wake of a heated wire. In general a temperature

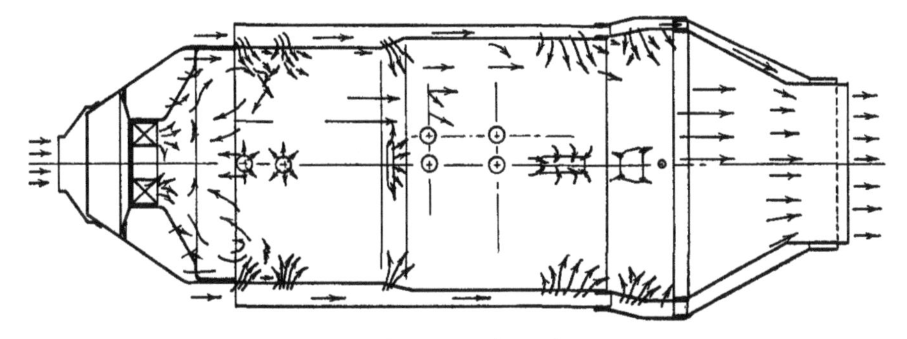

Fig. F,6g. "Flower pot" combustor [*76*].

profile similar to curve *A* of Fig. F,6h was obtained. Increasing the macroturbulence of the streams by upstream obstacles leveled the temperature profile to one similar to *B*. However, use of a cylindrical obstacle at the junction of the two streams caused formation of a Kármán vortex trail with interleaving of alternate hot and cold eddies immediately downstream. With the "mixing roll," an almost uniform temperature profile *C* could be obtained within 2 to 3 roll diameters. However, further downstream, a reversed temperature profile *D* was obtained. According to Minchin, complete mixing had not actually occurred, the profile *C* being only a constant statistical average of evenly distributed eddies. The confining effect of the duct caused the temperature reversal by forcing the vortices to cross instead of remaining in parallel Kármán trails. According to Lloyd [*74*], the mixing roll is effective only with gas streams flowing at similar velocities, without excessive macroturbulence.

*Model tests.* The simplicity of the free jet and coaxial jet systems permitted approximate analytical treatments and the presentation of experimental data in terms of relatively simple equations. However, most combustion systems are so complex that analytical treatment is prohibitively difficult. In such cases, the use of models as a means of studying trends in flow patterns and mixing may be warranted.

The principal use of model tests has been in the study of fluid flow without combustion. For the case where inertial and viscous forces alone are considered, mechanical similarity of the flow in two geometrically

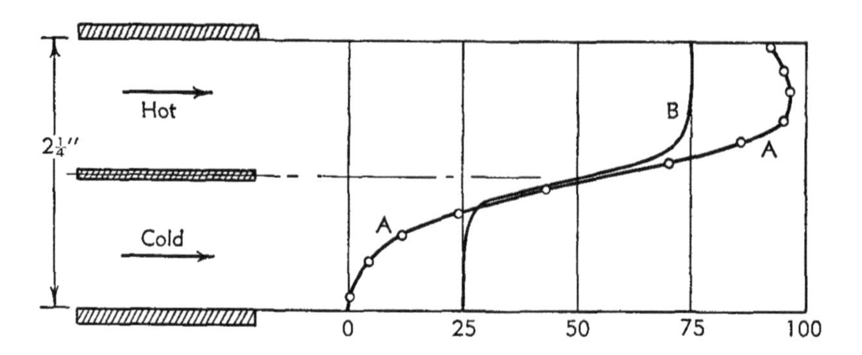


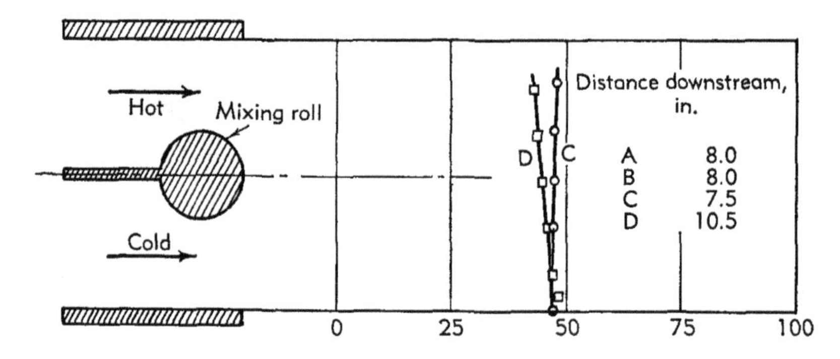


Fig. F,6h. Effect of mixing roll on mixing of hot and cold parallel gas streams [*52*, p. 485]. (Top) no mixing roll; (bottom) with mixing roll.

similar systems is obtained if the Reynolds number is the same for the two systems. For the case where inertial and buoyancy forces alone are considered, the criterion for mechanical similarity of flow is equality of the Froude number $u/\sqrt{Lg(\Delta\rho/\rho)}$. In some studies of jet mixing, however, it may not be necessary to achieve equality of the Reynolds and Froude numbers. If the Reynolds number is sufficiently high, the mixing pattern may become independent of the Reynolds number. This was indicated by the data for the flame length of free jet diffusion flames. As seen in Fig. F,4a, the flame height over much of the fully developed turbulent flame region was independent of changes in port velocity. For high output combustion systems, the average flow velocity may be sufficiently high so that buoyancy forces and hence Froude number equality may be neglected.

Both air and water have been used as the fluid medium in the model tests. With air, the flow patterns may be determined by the use of high speed photography and fine particles such as aluminum powder [*30*, pp. 44–68] or balsa dust. If slit lighting is used, the general flow pattern in a particular plane of the model may be obtained. The patterns are indicated by the streak traces of the particles on the photograph (e.g. Fig. F,6a). Smoke may be used to a limited extent to determine flow patterns. For the highly turbulent flow in combustion chambers, the smoke should be injected in short bursts since it diffuses rapidly throughout the chamber. Fine tufts attached to wires have also been used to indicate direction of local air currents in a turbojet combustor.

Since the kinematic viscosity of water is appreciably smaller than that of air, the average flow velocity may be reduced appreciably in the model tests, while maintaining equality of the Reynolds number. For some cases the resulting velocity is sufficiently low for the flow patterns to be determined from visual observation. Aluminum powder or small air bubbles [*77*] have been used successfully as flow indicators.

The determination of turbulent mixing patterns in models operating with air has generally involved the use of a tracer gas and samplings. The gaseous fuel is simulated by air containing a small quantity of the tracer gas. From analysis of samples obtained at various positions in the model, the spatial location of constant fuel-air ratio surfaces may be obtained. Since the eddy diffusion coefficient may be in the order of 100 times the molecular diffusion coefficient, a tracer gas with a high molecular diffusion coefficient may be used without significant error. The turbulent eddy diffusion coefficient at various stations may also be determined from the concentration profiles at stations downstream of a point source injection of the tracer gas. The success of the method depends on the accuracy of the detection means used to determine the concentration of the tracer gas. Some of the tracer gases and detection means reported in the literature have been (1) hydrogen and a thermal conductivity cell [*36*], (2) carbon dioxide and an infrared analyzer [*78*], and (3) radon and an electroscope [*79*].

In [*3*, pp. 97–113] and [*3*, pp. 806–814], liquid models were used to simulate the mixing pattern in turbulent diffusion flames. A dilute caustic, colored by a suitable indicator, simulates the gaseous fuel, while a dilute acid simulates the air. The flame front is given by the point where the acid is neutralized and the color disappears. The normality ratio of base to acid may be adjusted to simulate the stoichiometric volume fuel-air ratio for a given fuel. The effects of unmixedness on the flame length found in actual flames have also been found in the liquid model tests [*3*, pp. 97–113].

It is to be noted, of course, that the model tests cannot adequately simulate the actual system during combustion. In the actual system the velocity and density change markedly over the combustion space. Thus, even for gaseous fuels, the model tests are not completely adequate. For the case of liquid fuels sprayed from pressure-atomizing nozzles into combustion chambers, the penetration and evaporation of the various drops is not duplicated. However, model tests would still appear to be of value in determining the general effects of variation in combustion chamber geometry on the flow patterns and mixing process.

## F, 7. Cited References.

1. Whittle, F. *The Aeroplane 69*, 503–507 (1945).
2. Olson, W. T., Childs, J. H., and Jonash, E. R. The combustion efficiency problem of the turbojet at high altitudes. *Am. Soc. Mech. Engrs. Preprint 54-SA-24*, 1954.
3. *Fourth Symposium on Combustion.* Williams & Wilkins, 1953.
4. Forstall, W., Jr., and Shapiro, A. H. Momentum and mass transfer in coaxial gas jets. *Mass. Inst. Technol. Project Meteor Rept. 39*, 1949.
5. Alexander, L. G., Baron, T., and Comings, E. W. Transport of momentum, mass, and heat in turbulent jets. *Univ. Illinois Bull. 50*, 66 (1953).
6. *Proceedings of Third Midwestern Conference on Fluid Mechanics.* Univ. Minnesota, 1953.
7. Prandtl, L. Turbulent flow. *NACA Tech. Mem. 435*, 1927.
8. Tollmien, W. Calculation of turbulent expansion processes. *NACA Tech. Mem. 1085*, 1945.
9. Howarth, L. *Proc. Cambridge Phil. Soc. 34*, 185–203 (1938).
10. Taylor, G. I. *Proc. Roy. Soc. London A135*, 685–702 (1932).
11. Abramovich, G. N. The theory of a free jet of a compressible gas. *NACA Tech. Mem. 1058*, 1944.
12. Pai, S. I. *J. Aeronaut. Sci. 16*, 463–469 (1949).
13. von Kármán, Th. *J. Aeronaut. Sci. 1*, 1–20 (1934).
14. Reichardt, H. *Z. angew. Math. u. Mech. 21*, 257–264 (1941).
15. Liepmann, H. W., and Laufer, J. Investigations of free turbulent mixing. *NACA Tech. Note 1257*, 1947.
16. Batchelor, G. K. *Australian J. Sci. Research A2*, 437–450 (1949).
17. Mickelsen, W. R. An experimental comparison of the Lagrangian and Eulerian correlation coefficients in homogeneous isotropic turbulence. *NACA Tech. Note 3570*, 1955.
18. Beeton, A. B. P. A theoretical expression for point-source diffusion in turbulent flow. *Natl. Gas Turbine Establishment, England, Rept. 152*, 1954.
19. Towle, W. L., and Sherwood, T. K. *Ind. Eng. Chem. 31*, 457–462 (1939).
20. Squire, H. B., and Trouncer, J. Round jets in a general stream. *Brit. Aeronaut. Research Council Repts. and Mem. 1974*, 1944.
21. Kuethe, A. *J. Appl. Mech. 2*, A87–A95 (1935).
22. Ehrich, F. F. *J. Aeronaut. Sci. 20*, 99–104 (1953).
23. Hawthorne, W. R., Rogers, G. F. C., and Zaczek, B. Y. Mixing of gas streams—the penetration of a jet of cold air into a hot stream. *Roy. Aircraft Establishment Tech. Note Eng. 271*, 1944.
24. Hawthorne, W. R., and Cohen, H. Pressure losses and velocity changes due to heat release and mixing in frictionless, compressible flow. *Roy. Aircraft Establishment Rept. E3997*, 1944.

25. Bohanon, H. R., and Wilcox, E. C. Theoretical investigation of thrust augmentation of turbojet engines by tail-pipe burning. *NACA Research Mem. E6L02*, 1947.
26. Taylor, J. F., Grimmett, H. L., and Comings, E. W. *Chem. Eng. Progr. 47*, 175–180 (1951).
27. Squire, H. B. *Aircraft Eng. 22*, 62–67 (1950).
28. Corrsin, S. Investigation of flow in an axially symmetrical heated jet of air. *NACA Wartime Rept. 94*, 1943.
29. Corrsin, S., and Uberoi, M. S. Further experiments on the flow and heat transfer in a heated turbulent air jet. *NACA Rept. 998*, 1950.
30. *Third Symposium on Combustion, Flame and Explosion Phenomena.* Williams & Wilkins, 1949.
31. Schlinger, W. G., and Sage, B. H. *Ind. Eng. Chem. 45*, 657–661 (1953).
32. Ruegg, F. W., and Klug, H. J. Analytical and experimental studies with idealized gas turbine combusters. *Natl. Bur. Standards Research Paper 2365*, 1952.
33. Baron, T., and Alexander, L. G. *Chem. Eng. Progr. 47*, 181–185 (1951).
34. Grimmett, H. L. *Entrainment in Air Jets. Ph.D. Thesis*, Univ. Illinois, 1950.
35. Baron, T., and Bollinger, E. H. Mixing of high-velocity air jets. *Univ. Illinois Eng. Expt. Sta. Tech. Rept. CML-3*, 1952.
36. Rummel, K. Der einfluss des mischvorganges auf die verbrennung von gas und luft in feuerungen. *Arch. Eisenhuttenw. 11*, 67–80 (1937).
37. Pfeiffer, A., Murati, G. T., and Engel, A. B. *Mixing of Gas Streams. M.S. Thesis*, Mass. Inst. Technol., 1945.
38. Callaghan, E. E., and Ruggeri, R. S. Investigation of the penetration of an air jet directed perpendicularly to an air stream. *NACA Tech. Note 1615*, 1948.
39. Chelko, L. J. Penetration of liquid jets into a high-velocity air stream. *NACA Research Mem. E50F21*, 1950.
40. Baron, T. *Chem. Eng. Progr. 50*, 73–76 (1954).
41. Berry, V. J., Mason, D. M., and Sage, B. H. *Ind. Eng. Chem. 45*, 1596–1602 (1953).
42. Bollinger, L. M., and Williams, D. T. Effect of Reynolds number in turbulent-flow range on flame speeds of Bunsen burner flames. *NACA Rept. 932*, 1949.
43. Hibbard, R. R., Drell, I. L., Metzler, A. J., and Spakowski, A. E. Combustion efficiencies in hydrocarbon-air systems at reduced pressures. *NACA Research Mem. E50G14*, 1950.
44. Wohlenberg, W. J. *Trans. Am. Soc. Mech. Engrs. 70*, 143–160 (1948).
45. Williams, G. W. *Chem. Eng. Sci. D. Thesis*, Mass. Inst. Technol., 1941.
46. Perry, J. H. *Chemical Engineers' Handbook*, 3rd ed. McGraw-Hill, 1950.
47. Lamb, H. *Hydrodynamics*, 6th ed. Dover, 1932.
48. Hughes, R. R., and Gilliland, E. R. *Chem. Eng. Progr. 48*, 497–504 (1952).
49. Fleddermann, R. G., and Hanson, A. R. The effects of turbulence and wind speed on the rate of evaporation of a fuel spray. *Univ. Mich. Eng. Research Inst. Rept. CM667*, 1951.
50. Ingebo, R. D. Vaporization rates and drag coefficients for isooctane sprays in turbulent air streams. *NACA Tech. Note 3265*, 1954.
51. Ingebo, R. D. Vaporization rates and heat-transfer coefficients for pure liquid drops. *NACA Tech. Note 2368*, 1951.
52. *Proceedings of the General Discussion on Heat Transfer.* Inst. Mech. Eng. London, Sept. 1951.
53. York, J. L., and Stubbs, H. E. *Trans. Am. Soc. Mech. Engrs. 74*, 1157–1162 (1952).
54. Lane, W. R. *Ind. Eng. Chem. 43*, 1312–1317 (1951).
55. Hinze, J. O. *Appl. Sci. Research A1*, 273–288 (1949).
56. Longwell, J. P., and Weiss, M. A. *Ind. Eng. Chem. 45*, 667–677 (1953).
57. Bahr, D. W. Evaporation and spreading of isooctane sprays in high-velocity air streams. *NACA Research Mem. E53I14*, 1953.
58. Jones, G. W., et al. Research on the flammability characteristics of aircraft fuels. *Wright Air Develop. Center Tech. Rept. 52-35*, 1952.

59. Haber, F., and Wolff, H. *Z. angew. Chem. 36*, 373–377 (1932).
60. Wolfhard, H. G., and Parker, W. G. Combustion processes in flames. Part II: Evaporation process in a burning kerosene spray. *Roy. Aircraft Establishment Rept. Chem. 437*, 1947.
61. Nukiyama, S., and Tanasawa, Y. (Transl. by E. Hope.) Experiments on the atomization of liquids in an air stream. *Can. Defence Research Board, Dept. Natl. Defence Rept. 2*, Ottawa, 1950.
62. Sacks, W. The rate of evaporation of a kerosene spray. *Can. Natl. Aero. Establishment Note 7*, 1951.
63. Langmuir, I. *Phys. Rev. 12*, 368–370 (1918).
64. Probert, R. P. *Phil. Mag. 37*, 95–167 (1946).
65. Scurlock, A. C. Flame stabilization and propagation in high-velocity gas streams. *Mass. Inst. Technol. Project Meteor Rept. 19*, 1948.
66. Langmuir, I., and Blodgett, K. B. A mathematical investigation of water droplet trajectories. *Army Air Force Tech. Rept. 5418*, 1946.
67. Brun, R. J., and Mergler, H. W. Impingement of water droplets on a cylinder in an incompressible flow field and evaluation of rotating multicylinder method for measurement of droplet-size distribution, volume-median droplet size, and liquid water content in clouds. *NACA Tech. Note 2904*, 1953.
68. Brun, R. J., Gallagher, H. M., and Vogt, D. E. Impingement of water droplets on NACA $65_1$-208 and $65_1$-212 airfoils at 4° angle of attack. *NACA Tech. Note 2952*, 1953.
69. Dorsch, R. G., and Brun, R. J. Variation of local liquid-water concentration about an ellipsoid of fineness ratio 5 moving in a droplet field. *NACA Tech. Note 3153*, 1954.
70. Alexander, L. G., and Coldren, C. L. *Ind. Eng. Chem. 43*, 1325–1331 (1951).
71. Alexander, L. G., and Coldren, C. L. Deposition of a spray on the walls of a straight duct. *Conference on Fuel Sprays*, Univ. Michigan, 1949.
72. Straight, D. M., and Gernon, J. D. Photographic studies of preignition environment and flame initiation in turbojet-engine combustors. *NACA Research Mem. E52I11*, 1953.
73. Graves, C. C. and Gerstein, M. Some aspects of combustion of liquid fuel. *AGARD Mem. AG16/M10*, 1954.
74. Lloyd, P. Combustion in the gas turbine—a survey of war-time research and development. *Brit. Aeronaut. Research Council Repts. and Mem. 2579*, 1946.
75. Zettle, E. V., and Mark, H. Simulated altitude performance of two annular combustors with continuous axial openings for admission of primary air. *NACA Research Mem. E50E18a*, 1950.
76. Watson, E. A., and Clarke, J. S. *J. Inst. Fuel. 21*, 1–34 (1947).
77. Chesters, J. A., Howes, R. S., Halliday, I. M., and Philip, A. R. *J. Iron and Steel Inst. 162*, 385 (1949).
78. Collins, R. D., and Tyler, J. D. *J. Iron and Steel Inst. 162*, 457 (1949).
79. Mayorcas, R., and Thring, M. W. *Nature 95*, 66 (1943).

# SECTION G

# *FUELS FOR AIRCRAFT GAS TURBINE ENGINES*

LOUIS C. GIBBONS

**G,1. Introduction.** Fuels for aircraft powered with gas turbine engines must meet three primary requirements. They must give optimum performance in aircraft engines, they must be suitable for operation in aircraft fuel systems at low ambient air pressures and temperatures, and they must be available in relatively large quantities.

Investigations of fuel performance in gas turbine engines, and in combustors for such engines, have shown that the optimum fuel for an engine may not be optimum for an aircraft fuel system. In addition, a compromise fuel that is satisfactory for both the engine and the fuel system, may not be practical for service because of limited availability or high cost. It is the purpose of this chapter to discuss the information that is available on these three topics and to point out the compromises that have been made in arriving at fuel specifications for gas turbine engines.

**G,2. Fuel Performance in Engines.**

Combustion Efficiency. The efficiency with which fuels can be burned in engines is obviously an important factor. Combustion efficiency is a function of engine design, engine operating conditions, and fuel variables. The effect of operating conditions on efficiency can usually be described in terms of a correlating parameter developed in [*1*]. The parameter is $pT/V$, where $p$ and $T$ are combustor inlet-air pressure and absolute temperature, respectively, and $V$ is the reference air velocity through the combustor. Operating conditions become more severe as $pT/V$ decreases, that is, combustion efficiency decreases with decreasing pressure and temperature and with increasing air velocity.

Examples of the effect of fuel variables on the efficiencies obtained at varying operating conditions are shown in Fig. G,2a and G,2b. Fig. G,2a shows results obtained with 14 fuels in a full scale engine at mild operating conditions. The investigation was conducted at sea level and the pressure in the combustion chamber varied with the engine speed from 27 to 49 pounds per square inch. The results are plotted as engine thrust in

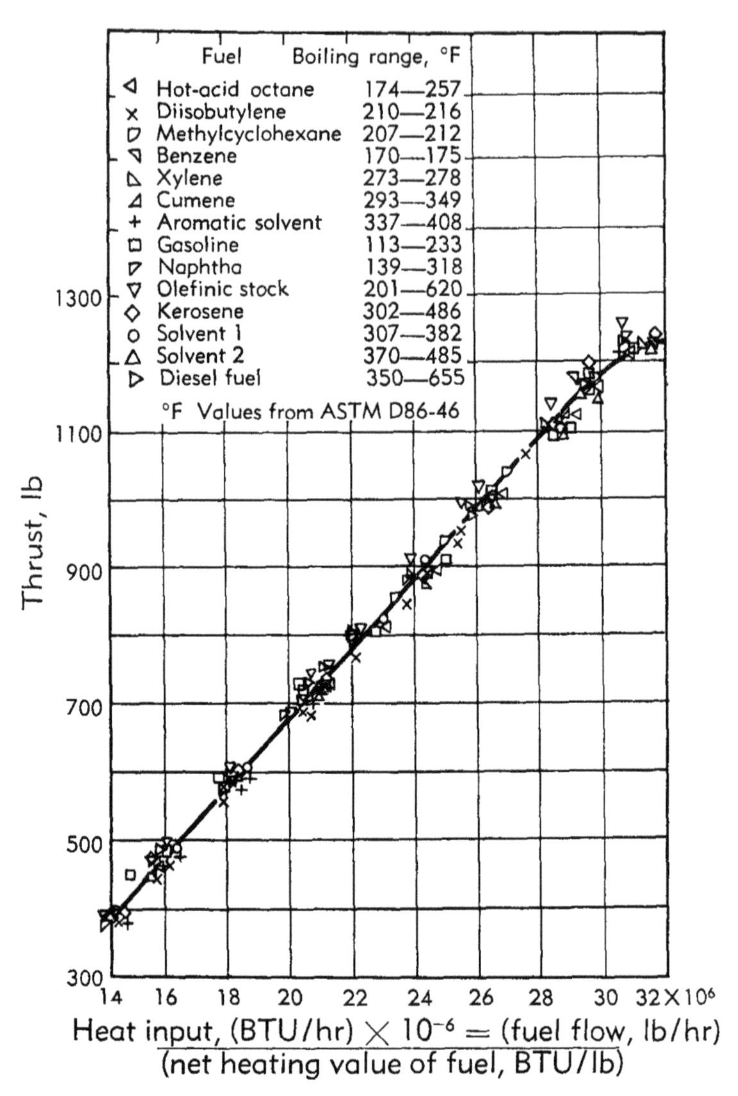


**Fig. G,2a.** Variation of thrust with heat input for several fuels in a turbojet engine at static sea level conditions. Conditions of inlet air to combustors increased with increasing heat input as follows: pressure = 27 to 49 lb/in.$^2$; temperature = 180 to 300°F; velocity = 124 to 128 ft/sec.

pounds against heat input in BTU/hr. This term gives the theoretical quantity of heat that would be released per hour from each fuel. Each fuel burned equally efficiently. The physical properties of the fuels varied from gasoline through diesel oil, and the molecular structure of the fuels included paraffinic, olefinic, cycloparaffinic, and aromatic types. Thus neither the fuel volatility nor the molecular structure influenced the combustion efficiency under the conditions investigated. However, if a gas turbine combustor is operated at low air pressures, such as might be encountered at high altitudes, then the fuel volatility and composition affect combustion efficiency. Such an effect is shown in Fig. G,2b where combustion efficiency is plotted against altitude for three fuels: gasoline, kerosene, and diesel oil. At sea level operating conditions the combustion efficiency was 100 per cent for the three fuels. However, as the altitude was increased the combustion efficiency of the kerosene was lower than the gasoline, and the efficiency of the diesel oil was lower than for the kerosene. Because of this trend most investigations of fuel performance

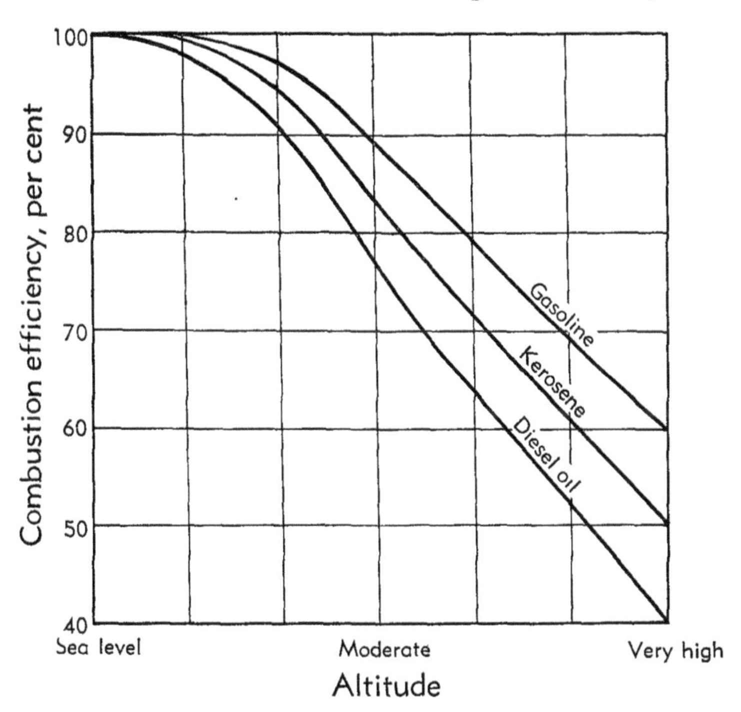


Fig. G,2b. Effect of altitude on combustion efficiency of three fuels.

have been conducted at conditions simulating high altitude operation. Also, because of the obvious economies involved, most fuel investigations have been conducted in single combustors.

*Effect of volatility and fuel injection.* The effect of volatility on the combustion efficiency of an atomizing-type combustor is shown in Fig. G,2c where the 50 per cent ASTM distillation temperature is used as a measure of volatility [*2*]. Efficiencies at two engine conditions are shown. The higher engine-rotating speed delivers air to the combustor at higher pressures and temperatures, and therefore higher efficiencies are obtained. At the lower engine speeds the reverse is true. It is seen that efficiency decreases with increasing 50 per cent temperature of the fuel, but that the effect is small over a wide range of temperature. A sharp drop in efficiency is not obtained until the 50 per cent temperature exceeds about 480°F. Current jet fuels of the JP-1, JP-3, JP-4, and JP-5 types have 50 per cent temperatures ranging between about 300 and 425°F [3]; for these the effect of varying volatility would be small in this engine.

Presumably the low boiling fuels vaporize quickly and under these conditions of operation give combustible fuel-air mixtures in the low velocity region of the combustor. Thus such fuels have time to burn more completely than high boiling fuels that vaporize more slowly.

The effect of fuel spray is just as important as the effect of fuel volatility on combustion efficiency. Radcliffe discusses in Sec. D the atomizing characteristics of fixed geometry, variable area, and swirl-type

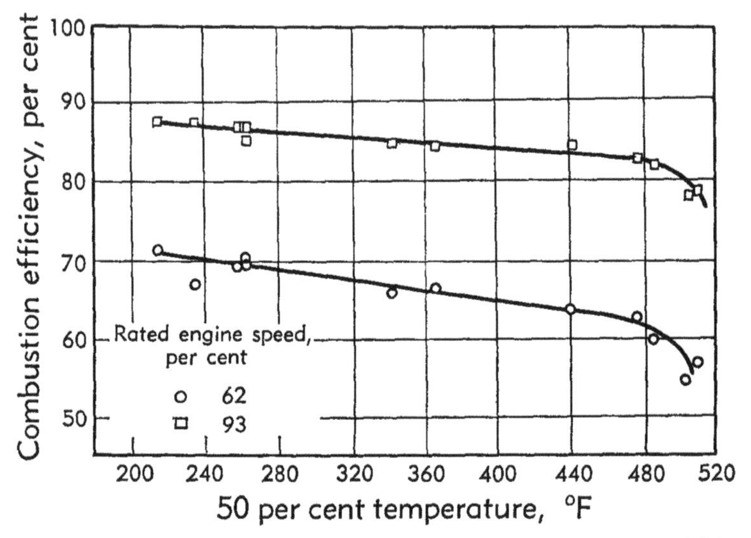


Fig. G,2c. Effect of fuel volatility on combustion efficiency. (From [2].)

nozzles. He points out that nozzles must be capable of handling about a 100 to 1 range of fuel flows between sea level take-off and high altitude cruise conditions. Simple nozzles sized to handle the high fuel flows required for take-off give a very poor spray at high altitude [4]. Fig. G,2d and G,2e show efficiencies obtained at various fuel flows with such nozzles.

In Fig. G,2d is shown the combustion efficiency obtained over a range of fuel flows when two different fuel nozzles were used. The lower curve shows data obtained with a nozzle that gave practically no atomization at low fuel flows. The upper curve shows greatly improved combustion efficiency at low fuel flow due to improved atomization of the spray [4].

The effect of the fuel nozzle on the relative combustion efficiencies of two fuels is shown in Fig. G,2e. The combustion efficiencies of gasoline and diesel oil were compared in a combustor provided with fuel nozzles

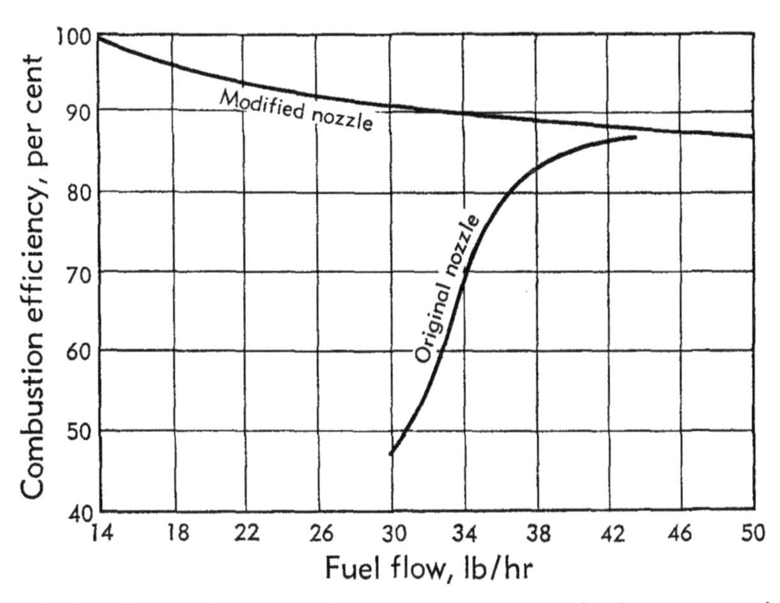


Fig. G,2d. Effect of fuel nozzle design on performance with kerosene-type fuel.

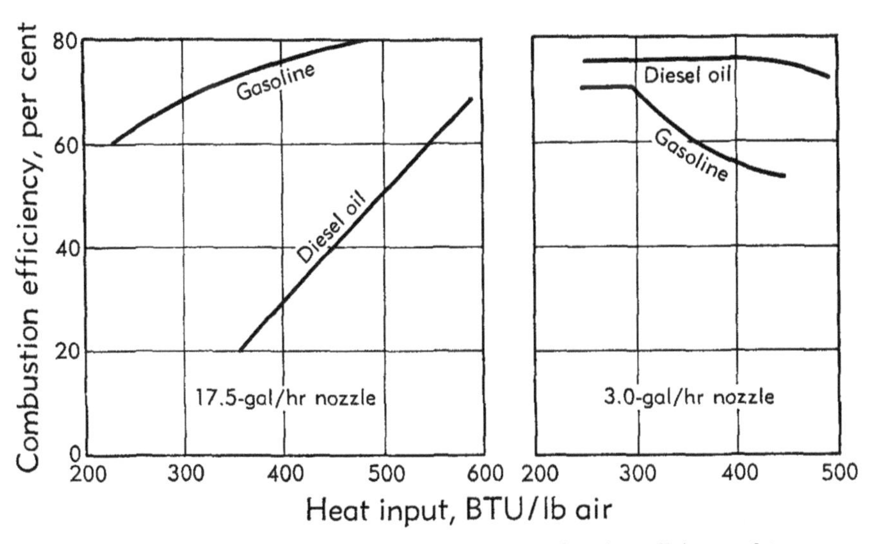


Fig. G,2e. Effect of fuel nozzle capacity on combustion efficiency of two fuels. Inlet conditions to combustor: simulated high altitude.

rated to deliver 17.5 gallons per hour at 100 pounds per square inch pressure. At the conditions of operation the fuel flow was low and the fuel pressure was much less than 100 pounds per square inch. Consequently the fuel droplets were very large and a hollow cone spray was not well developed. It is shown in Fig. G,2e that under these conditions the combustion efficiency with gasoline was much higher than that obtained with diesel oil. When the large nozzles were replaced with nozzles rated at only 3 gallons per hour, the combustion efficiency with diesel oil was higher than with gasoline. These results are attributed to the fact that the small fuel nozzle delivered a finely atomized spray which allowed the diesel oil to evaporate sufficiently to burn well in the combustor. However, the finely atomized drops of gasoline evaporated so rapidly that the primary zone of the combustor was too rich for efficient combustion. When the fuel moved downstream where there was sufficient air for combustion, there was insufficient time for complete burning before the combustion reaction was partially quenched by cooling air, and the gases were swept out of the combustor. This point of view is substantiated by the fact that the combustion efficiency with gasoline decreased as the fuel flow was increased.

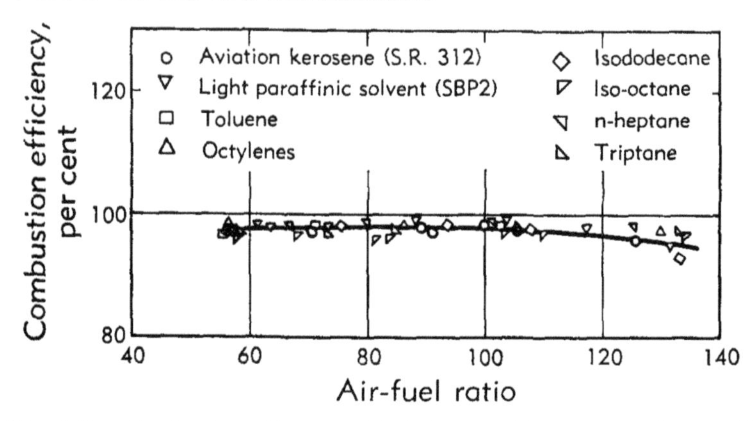


Fig. G,2f. Combustion efficiency of fuels in vaporizing-type combustor [*5*].

These data indicate that fuel volatility, fuel nozzle design, and combustor design are all interrelated. Optimum combustion efficiency will be obtained only when the fuel injection system and the combustor are developed on the fuel to be used in service. However, recently designed gas turbine engines use quite complex nozzle systems which give adequate atomization at both high and low fuel flows (see Sec. D). For these engines the effects of varying fuel flows and fuel types are much less than those shown in Fig. G,2d and G,2e.

It is of interest to know if fuel volatility has an effect on combustion efficiency in a vaporizing-type combustor, or whether volatility effects are encountered only with combustors in which the fuel is introduced as a liquid spray. Various designs of both vaporizing and atomizing combustors are described in Sec. H.

Unfortunately, only limited information has been published on the subject of fuel performance in vaporizing combustors. Sharp [*5*] shows a plot of combustion efficiency against air-fuel ratio for eight fuels evaluated in a vaporizing-type combustor. The results are shown in Fig. G,2f. All of the fuels studied gave approximately the same combustion efficiencies over the air-fuel range. However, the highest boiling fuel investigated was isododecane which boils only slightly above the average boiling temperature for a kerosene-type fuel, for which the combustor was probably designed. It seems possible that a vaporizing tube designed to properly vaporize aviation kerosene might completely vaporize more volatile fuels, but would not completely vaporize higher boiling fuels.

Consequently, it seems likely that a vaporizing burner would have to be designed specifically for the task of efficiently utilizing a high boiling fuel.

*Effect of molecular structure.* In addition to knowing the effects of fuel volatility on combustion efficiency, it is important to know if the molecular structure of hydrocarbons may influence the combustion efficiency of turbojet combustors. That is, will combustion efficiency differ with a paraffinic, an olefinic, cycloparaffinic, or aromatic-type fuel?

A comparison of the combustion efficiencies of iso-octane (2,2,4-trimethylpentane) and toluene is reported in [*6*]. At an atomizer flow number of 1.35 the iso-octane gave an appreciably higher combustion efficiency than the toluene. At an atomizer flow number of 0.55 the differences in combustion efficiency were small. The flow number is defined as gallons per hour divided by the square root of the fuel pressure; thus the low flow number indicates better atomization. The higher combustion efficiency of the iso-octane at low fuel pressure may be attributed to differences in molecular structure since the difference in boiling temperature is relatively small. The iso-octane boils at 210°F and the toluene at 231°F. Other investigators have also attributed lowered combustion efficiencies to aromatic hydrocarbons [*7*].

In results reported by Scott, Stansfield, and Tait [*7*] a comparison of the combustion efficiencies at a simulated high altitude condition of naphthenic, paraffinic, and aromatic fuels boiling in the kerosene range is given. Their data are shown in Fig. G,2g as a plot of combustion efficiency versus heat input in BTU per pound of air. Over most of the range investigated, the paraffinic fuel gave a higher efficiency than the naphthenic fuel and the latter gave a higher efficiency than the aromatic fuel. Some properties of the fuels are shown in Table G,2a. Complete inspection data may be found in [*7*].

The NACA Lewis Laboratory has confirmed [*7*] in that, at severe inlet conditions to the combustor, high boiling aromatics tend to burn less efficiently than paraffinic fuels. However, the NACA data indicate that low boiling aromatics burn as efficiently as paraffins. Data to show this trend are presented in Table G,2b. Combustion efficiencies of four fuels were determined in a turbojet combustor at one operating condition. The lower boiling fuels, benzene and mixed heptanes, which represent aromatic and paraffinic types, respectively, gave essentially the same

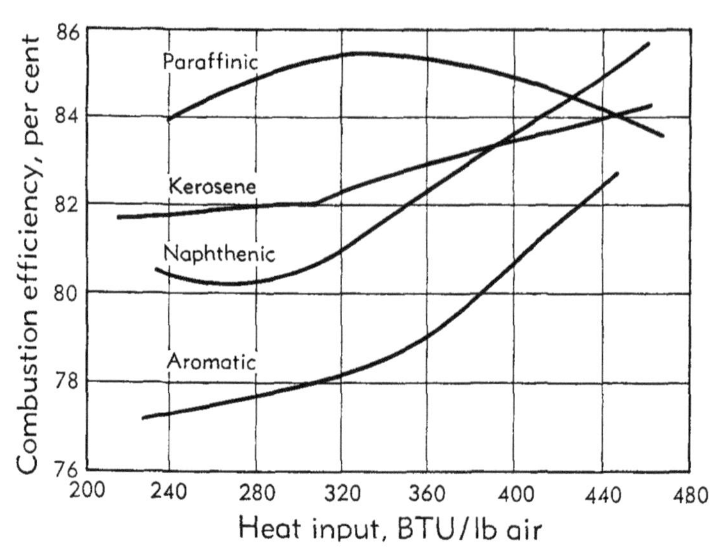


Fig. G,2g. Effect of fuel type on combustion efficiency [7].

*Table G,2a.* Properties of mixed fuels. (Data from [7, Table VI]).

| | Reference kerosene | Naphthenic material | Paraffinic material | Aromatic material |
|---|---|---|---|---|
| Specific gravity at 60/60°F | 0.7940 | 0.7910 | 0.7475 | 0.8645 |
| I.B.P., °F | 312 | 320 | 318 | 331 |
| Per cent recovered | | | | |
| 10 | 352 | 363 | 344 | 354 |
| 30 | 386 | 388 | 370 | 372 |
| 50 | 413 | 410 | 394 | 390 |
| 70 | 441 | 433 | 419 | 414 |
| 90 | 480 | 474 | 467 | 449 |
| F.B.P. | 520 | 519 | 509 | 487 |
| Hydrocarbon-type analysis | | | | |
| Aromatics, weight per cent | 19.4 | 1.4 | 2.6 | 75.2 |
| Naphthenes, weight per cent | 39.2 | 51.9 | 5.2 | 11.4 |
| Paraffins, weight per cent | 41.4 | 46.7 | 92.2 | 13.4 |
| Viscosity at 100°F, centistokes | 1.54 | 1.53 | 1.30 | 1.16 |
| at 0°F | 4.29 | 4.58 | 3.97 | 3.34 |
| Net heating value, BTU/lb | 18,584 | 18,651 | 18,816 | 18,000 |

combustion efficiencies, whereas the aromatic solvent gave considerably lower combustion efficiency than the kerosene which was a mixture of predominantly paraffinic and cycloparaffinic materials.

Thus the available data indicate that aromatic hydrocarbons boiling in the kerosene range, or higher, tend to burn less efficiently than paraf-

*Table G,2b.* Combustion efficiencies of paraffinic and aromatic fuels in turbojet combustor.

| Fuel | Hydrocarbon type | ASTM 50 per cent evaporated temperature, °F | Combustion efficiency, per cent |
|---|---|---|---|
| Mixed heptanes | Paraffinic | 180 | 90 |
| Benzene | Aromatic | 172 | 89 |
| Kerosene | Paraffinic and cycloparaffinic +10% aromatics | 370 | 78 |
| Aromatic solvent | Aromatic | 328 | 57 |

finic or naphthenic types at high altitude operating conditions, but low boiling aromatics burn as efficiently as paraffins.

The effect of branching of a paraffinic hydrocarbon has been investigated to a limited extent [*8*]. The combustion efficiencies of normal heptane and 2,2,4-trimethylpentane, which boil at the same temperature, have been compared by the NACA in a turbojet combustor at a severe operating condition, with the result shown in Fig. G,2h. At a relatively low velocity the combustion efficiencies of the two fuels were about the same. As the inlet air velocity was increased the combustion efficiency of the 2,2,4-trimethylpentane fell below that of the heptane. It has been suggested that heptane burns at a higher velocity than the 2,2,4-trimethylpentane and, in a combustor where the time available for burning is very short, the heptane burns more completely than the 2,2,4-trimethylpentane.

A second comparison in Fig. G,2h shows combustion efficiencies of hexane and 2,3-dimethylbutane. These two compounds have the same number of atoms in the molecule but differ in boiling temperature by 18°F. Here again the straight chain paraffin, hexane, gives higher combustion efficiencies than the branched compound, although the 2,3-dimethylbutane has a lower boiling temperature.

In order to find significant differences in combustion efficiency between normal and branched paraffins it is necessary to operate at more severe operating conditions than are encountered in present-day engines. Therefore, with the present types of gas turbine engines, there are no advantages to be realized in the use of straight chain paraffins in preference to branched chain compounds, if the engine is operated well below the altitude operational limit.

Altitude Operational Limits. Aircraft, when flown to extremely high altitudes, may encounter a situation where the engine loses speed in spite of a wide-open throttle. This condition is called an altitude operational limit. Such limits were studied in the early days of gas turbine engines [9] and the cause was found to be in the combustion chamber. It was learned that at high altitude conditions the temperature rise across the combustor varied with fuel flow as shown in Fig. G,2i. At some fuel flow rate a limiting temperature rise was reached and further increases in fuel flow caused a decrease in the exhaust gas temperature. A further increase in fuel flow caused the flame to blow out of the combustor. This was called "rich blowout" and was attributed to the

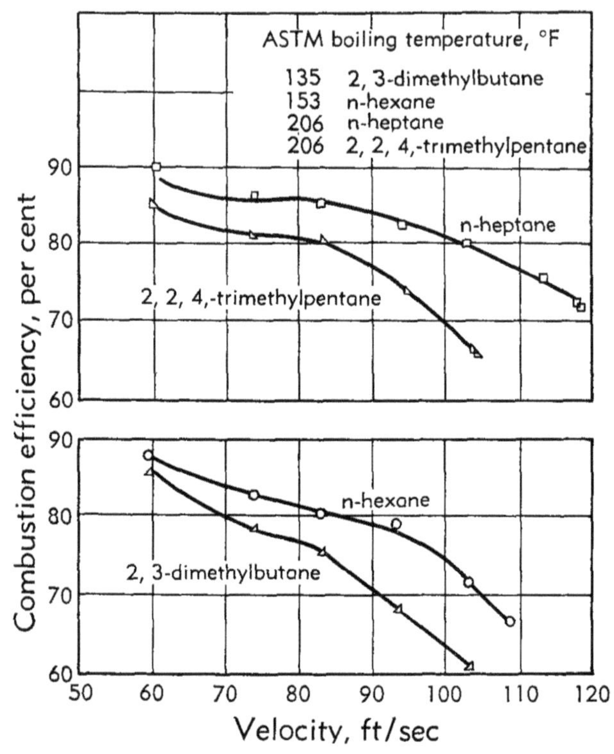


Fig. G,2h. Variation of combustion efficiency of pure hydrocarbon fuels with reference velocity in a turbojet combustor. Combustor inlet conditions: low pressure and temperature.

presence of so much fuel vapor in the primary zone of the combustor that the resultant fuel-air mixture would not burn.

If the concept expressed above is correct, then it seems likely that fuels of different volatility have an influence on altitude operational limits. A volatile fuel would be more likely to form a rich mixture near the fuel nozzle than a nonvolatile fuel. Experiments with different fuels have shown that volatility does influence the altitude operational limits of a combustor. A plot of the altitude operational limits is shown in Fig. G,2j as a function of simulated engine speed [10]. It is shown that at 100 per cent rated engine speed the JP-1 or kerosene-type fuel gave

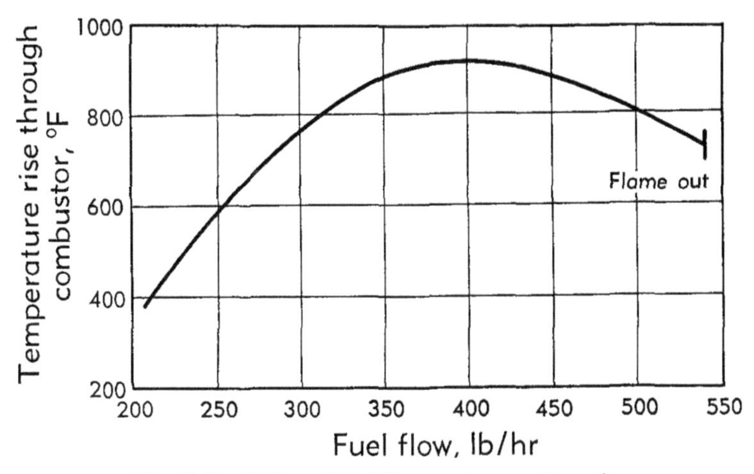


Fig. G,2i. Effect of fuel flow on temperature rise.
Inlet air conditions: simulated high altitude.

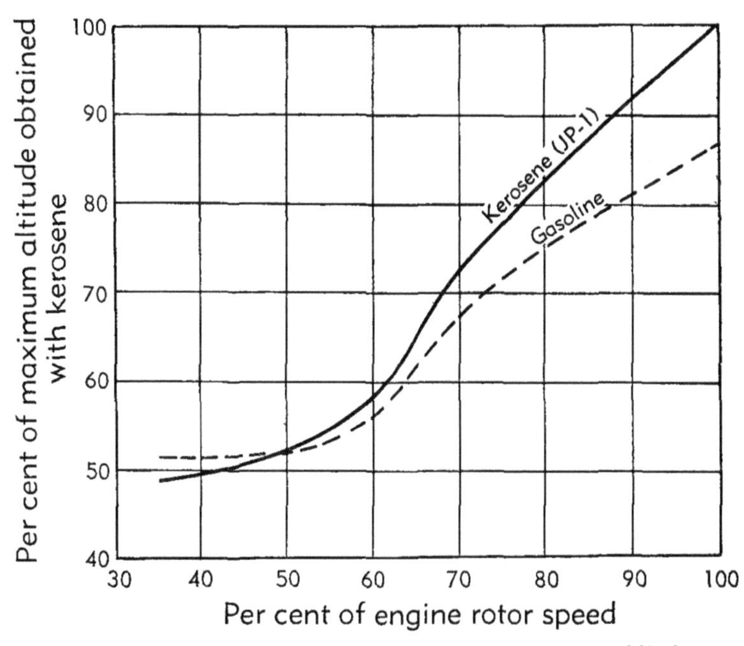


Fig. G,2j. Effect of fuel volatility on altitude operational limits.

altitude limits considerably above those obtained with gasoline. Presumably this phenomenon is due to the fact that the gasoline vaporizes more readily than the JP-1 fuel and at high fuel flow rates gives an overly rich mixture in the primary zone of the combustor. The JP-1 fuel vaporizes more slowly and is less likely to form an overly rich mixture. To further evaluate the idea that a rich mixture may establish the altitude operational limit, gasoline was used in a combustor and the limits established with two different fuel nozzles. The results are shown in Fig. G,2k. The smaller fuel nozzle when operated in the combustor gave a finely atomized spray which gave a vapor-air mixture that was rich in fuel. The larger nozzle gave a less finely atomized spray and was less conducive to the formation of an overly rich mixture. Thus it was possible to operate with the larger nozzle to considerably higher altitudes before reaching the altitude operational limits of the combustor.

The supposition that volatile fuels tend to vaporize near the fuel nozzle and produce rich fuel-air mixtures in that region has been evaluated qualitatively by gas sampling inside a turbojet combustor at the NACA Lewis Laboratory. A water-cooled gas sampling probe was

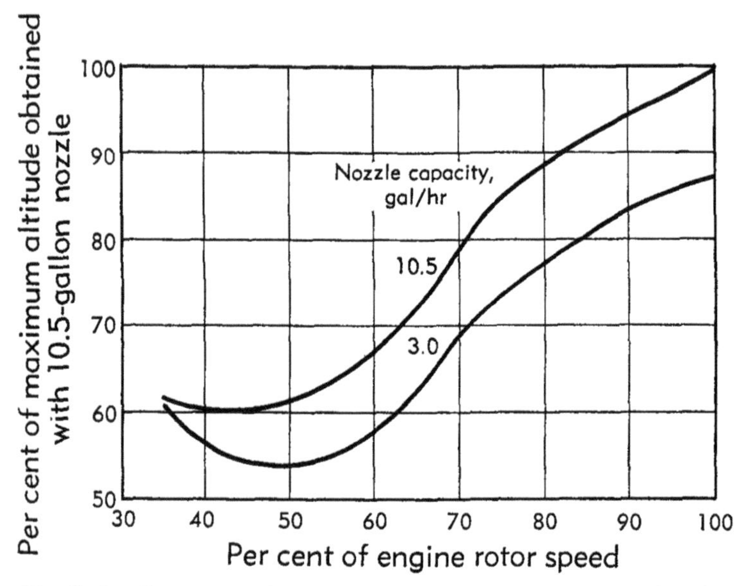


Fig. G,2k. Effect of fuel nozzle capacity on altitude operational limits.

inserted into a tubular turbojet combustor during burning and average fuel-air mixtures were determined across the diameter of the combustor. Typical results of a preliminary study are plotted in Fig. G,2l as fuel-air ratio against distance from the fuel nozzle. It is shown that isopentane gave a richer fuel-air ratio near the fuel nozzle than did the 2,2,4-trimethylpentane. These results tend to corroborate the suggestion that rich mixtures from volatile fuels may impose altitude operational limits that are lower than operational limits with less volatile fuels.

Some turbojet combustors allow a large part of the air required for combustion to enter the combustor near the fuel nozzle. Under very adverse conditions it is possible to blow out the flame due to the presence of excess air and this phenomenon has been called "lean blowout." Some investigations have been conducted to determine the effect of fuels on lean blowout [*5,7*]. Scott [*7*] reported, for fuels boiling in the kerosene range, that a paraffinic fuel would allow operation at leaner fuel-air mixtures than would an aromatic fuel. Sharp [*5*] has reported that with a vaporizing combustor, light gasoline gave much leaner operation than did aviation kerosene or gas oil. The differences in this case were probably due to the fact that the kerosene and gas oil were not completely vaporized before introduction into the combustor.

Carbon Deposits. The combustors that have seen service in turbojet engines can burn thousands of gallons of fuel with practically no deposit left in the combustion chamber. However, with marginal types of fuels a few grams of carbon are deposited on the liner or fuel nozzle with practically all combustion chamber designs. Carbon deposits are

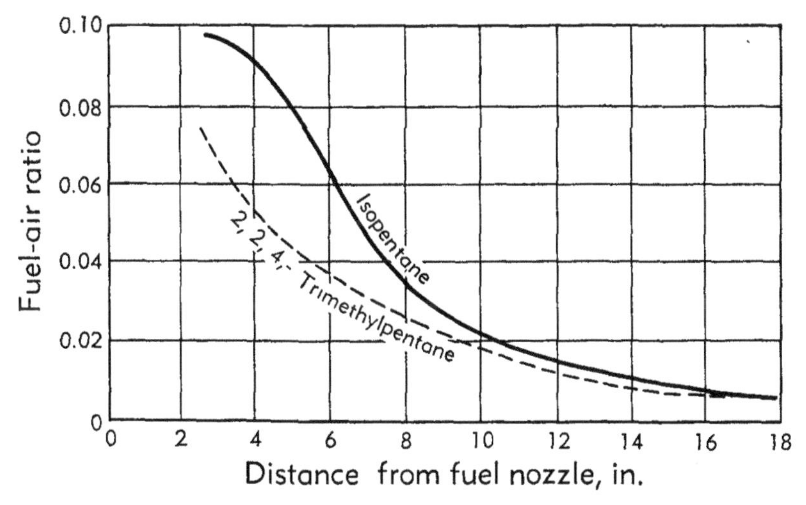


Fig. G,21. Effect of fuel volatility on fuel-air ratio inside the combustor during burning. Inlet air conditions: simulated high altitude. Over-all fuel-air ratio 0.007.

considered undesirable because they may interfere with fuel spray or air flow patterns and thereby cause hot spots on the combustion chamber liners. In some cases, carbon deposits have fouled spark plugs, and fuel sprays have been distorted to the extent that ignition did not occur.

The physical appearance of the two distinct types of carbon deposits usually formed in American-type, fuel-atomizing combustors has led to the classification of soft carbon and hard carbon. The soft carbon is usually formed near the fuel nozzle and in the dome of the combustor, whereas the hard carbon is usually formed on the walls of the combustion chamber liner a few inches from the upstream end. These two types of deposits have been examined by means of electron microscopy, X-ray diffraction, and chemical analysis [*11*]. The results show that the soft carbon consists of soot, soaked with residual high boiling fuel components and partially polymerized products. Presumably the soot is formed due to incomplete combustion and is deposited on the burner walls where fuel then soaks into the soot.

The hard carbon deposits were shown to be a material very similar to petroleum coke. The chemical analyses reported in [*11*] are compared in Table G,2c with the analysis reported in [*12*] for a petroleum refining residue coked at 800°C for 120 minutes. Coking processes are described in [*13*].

*Table G,2c.* Chemical analyses of carbon deposits and smoke.

| | Per cent carbon | Per cent hydrogen | Hydrogen-carbon ratio |
|---|---|---|---|
| Soft carbon from combustor dome | 80.0 | 2.0 | 0.025 |
| Hard carbon from combustor liner | 92.4 | 1.6 | 0.018 |
| Coked petroleum [*12*] | 93.3 | 1.9 | 0.020 |
| Smoke from wick lamp with JP-4 fuel | 96.2 | 0.80 | 0.008 |
| Smoke from combustor with JP-4 fuel | 88.9 | 0.70 | 0.008 |

*Table G,2d.* Effect of fuel volatility and composition on carbon deposits.

| Fuel | Volumetric average boiling temperature, °F | Hydrogen-carbon ratio | Specific gravity | Carbon deposit, grams |
|---|---|---|---|---|
| Isoheptanes | 182 | 0.177 | 0.725 | 1 |
| Paraffinic solvent | 348 | 0.179 | 0.775 | 2 |
| Benzene | 172 | 0.084 | 0.882 | 34 |
| Methylnaphthalenes | 459 | 0.079 | 1.016 | 134 |

Presumably the analyses do not add to 100 per cent because of the presence of some oxygen in the deposits examined. It is interesting to note the very close similarity in chemical composition between the petroleum coke and the hard carbon from a turbojet combustor. The analysis of the soft carbon gave a considerably different result, although about the same hydrogen-carbon ratio.

Investigators in many laboratories have shown that both fuel composition and volatility influence the quantities of carbon that are deposited in the combustion chambers of turbojet engines. The relative quantities of carbon deposited by various types of fuels are shown in Table G,2d and G,2e. The first four fuels in Table G,2d show typical results obtained at the NACA Lewis Laboratory with relatively pure hydrocarbons.

The fuels were burned in a turbojet combustor for a constant time and at constant operating conditions and the deposits obtained in the combustor are listed in Table G,2d. The data show the effects of both

*Table G,2e.* Effect of fuel volatility and composition on carbon deposits. Mixed fuels from [5]

| | Volumetric average boiling temperature, °F | Hydrogen-carbon ratio | Specific gravity | Relative carbon deposition |
|---|---|---|---|---|
| Diesel fuel | 439 | 0.159 | 0.828 | 6.8 |
| Hydrogenated creosote | 448 | 0.146 | 0.871 | 18.0 |
| Hydrogenated gas oil | 457 | 0.166 | 0.819 | 5.7 |
| Diesel fuel | 446 | 0.148 | 0.862 | 12.3 |
| Pool gas oil | 572 | 0.148 | 0.859 | 13.9 |
| Aviation kerosene | 417 | 0.166 | 0.805 | 3.7 |
| Hydrogenated gas oil | 457 | 0.166 | 0.819 | 5.7 |
| Diesel fuel | 446 | 0.148 | 0.862 | 12.3 |
| Typical cracked wax | 163 | 0.148 | 0.695 | 0.6 |

volatility and fuel composition as expressed by the hydrogen-carbon ratio.

A commercial mixture of isoheptanes gave 1 gram of carbon under the conditions of operation. Paraffinic solvent, a fuel of similar hydrogen-carbon ratio but with a higher boiling temperature, gave 2 grams of carbon. Benzene, which boils in the same range as the isoheptanes, gave 34 grams of carbon. The increased quantity of carbon may be attributed to the aromatic character of the benzene. One method of expressing aromatic character is by the hydrogen-carbon ratio, which for benzene is 0.084. The other aromatic material listed in Table G,2d, a mixture of methylnaphthalenes, has a hydrogen-carbon ratio similar to that of benzene, but a higher boiling temperature, and gave 134 grams of carbon. Thus as the boiling point of the aromatic hydrocarbon increases the carbon deposits increase.

Table G,2e contains data taken from Sharp [5] which illustrates the effects of volatility and the hydrogen-carbon ratio, as illustrated by complex hydrocarbon mixtures, and indicates materials that might be used as fuels.

Comparison of the first three fuels shows that, as the hydrogen-carbon ratio is decreased from 0.166 to 0.146, the relative carbon deposits increased from 5.7 to 18.0 grams, while the volumetric average boiling temperature of the fuels changed less than 20°F. The other fuels are grouped into three pairs of two fuels each, to compare carbon deposits where the hydrogen-carbon ratio was maintained constant for each pair of fuels but the volumetric average boiling temperature was varied. In each case it is shown that as the boiling temperature was increased the quantity of carbon deposit was increased. Thus it is shown that both the hydrogen-carbon ratio and the boiling temperature influence the tendency of fuels to form carbon deposits.

The relative influence of the H/C ratio and the boiling temperature of the fuel has been expressed [*14*] in equation form as

$$K = (t + 600)(0.7)\,\frac{\text{H/C} - 0.207}{\text{H/C} - 0.259}$$

where $K$ is the carbon factor, $t$ the volumetric average boiling temperature in °F, and H/C the hydrogen-carbon ratio.

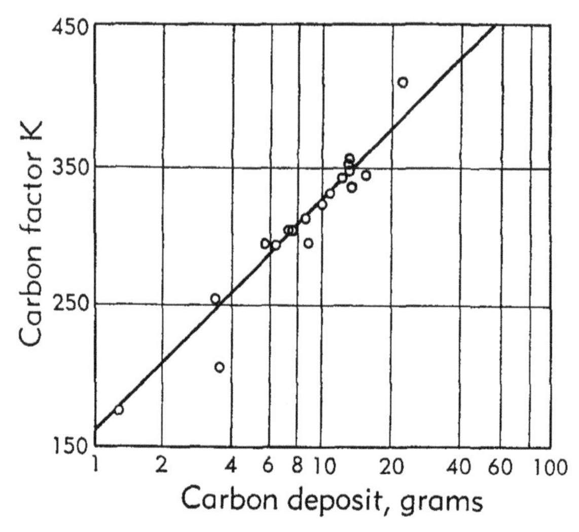


Fig. G,2m. Correlation of carbon factor K, with carbon deposits obtained in a turbojet combustor.

A plot of $K$ against carbon deposits is shown in Fig. G,2m for a series of fuels.

An equation to predict carbon deposits has also been published by Cattaneo and coworkers in [*15*]. The equation is expressed as

$$W = \frac{\ln(0.83R - 1.5)}{0.54} + \frac{T}{225} - 3.0$$

where $W$ = carbon deposit, grams; $R$ = carbon-to-hydrogen weight ratio; $T$ = boiling point for a pure compound or 10 per cent ASTM distillation temperature for a normal mixture of hydrocarbons, °F.

Many other correlations have been attempted to predict the relative carbon-forming tendencies of fuels. Probably the simplest parameter to use is specific gravity. Most authors who have published on the subject show a satisfactory correlation between carbon-forming tendencies and specific gravity of the fuel [*5,6,14,15*]. Such a plot is shown in Fig. G,2n for some data obtained by the NACA for a series of 19 mixed fuels. It is shown that as the specific gravity increases the carbon-forming tendencies of the fuels increase. Some investigators [*5,6*] have shown better correlations with specific gravity than that shown in Fig. G,2n.

The discussion above pertains to the effect of fuel properties on carbon deposits. It has also been shown that combustor design, fuel spray, conditions of operation, and time of operation all have an important effect on the quantity of carbon that is formed in a combustor.

The effect of altitude operation on carbon deposits has been studied at the NACA Lewis Laboratory. In this case air pressure, temperature, velocity, and fuel flow all changed as the altitude was increased. Thus the

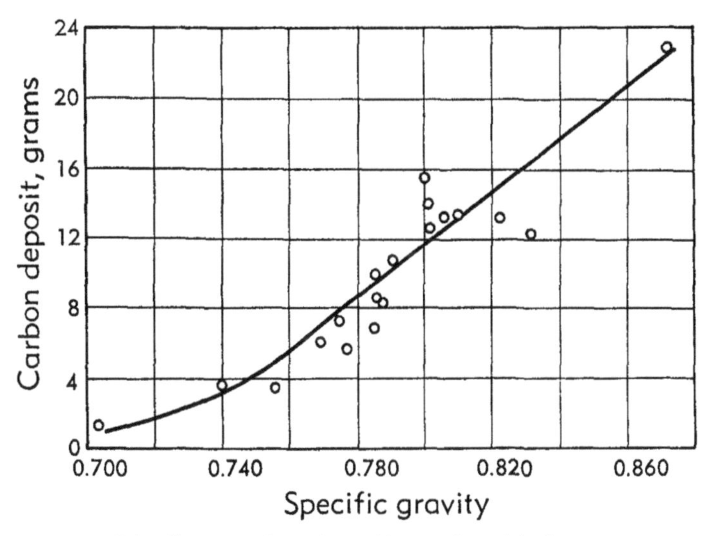


Fig. G,2n. Effect of specific gravity of fuels on carbon deposits in a turbojet combustor.

effects noted are the gross effects of changing altitude. The data obtained are shown in Fig. G,2o. A combustor was operated for a constant time at several simulated altitudes with three different fuels. There was a marked decrease in carbon deposits with all of the fuels as altitude was increased. It is of interest to determine whether the decreased deposits with increasing altitude were due to reduced fuel flow or to an effect of pressure on the combustion process. The pounds of fuel required to produce 1 gram of carbon were calculated for the data shown in Fig. G,2o and the results are given in Table G,2f as data for combustor *A*. The three fuels showed similar trends but the results with kerosene were the most striking. At low altitude, 21 pounds of fuel were burned to produce 1 gram of carbon in the combustor. At high altitude, 97 pounds of fuel were burned to give 1 gram of carbon. In this case there was a pronounced reduction in carbon-depositing tendency of each fuel with increased altitude.

The effect of changing altitude on carbon deposits was also investigated on a combustor of a different design with four different fuels. Deposits were measured at low and moderate simulated altitudes. The results are shown in Table G,2f as data for combustor *B*. In this case deposits were reduced as the altitude was increased, but within a variation of 5 per cent the reduction was in direct proportion to the reduction

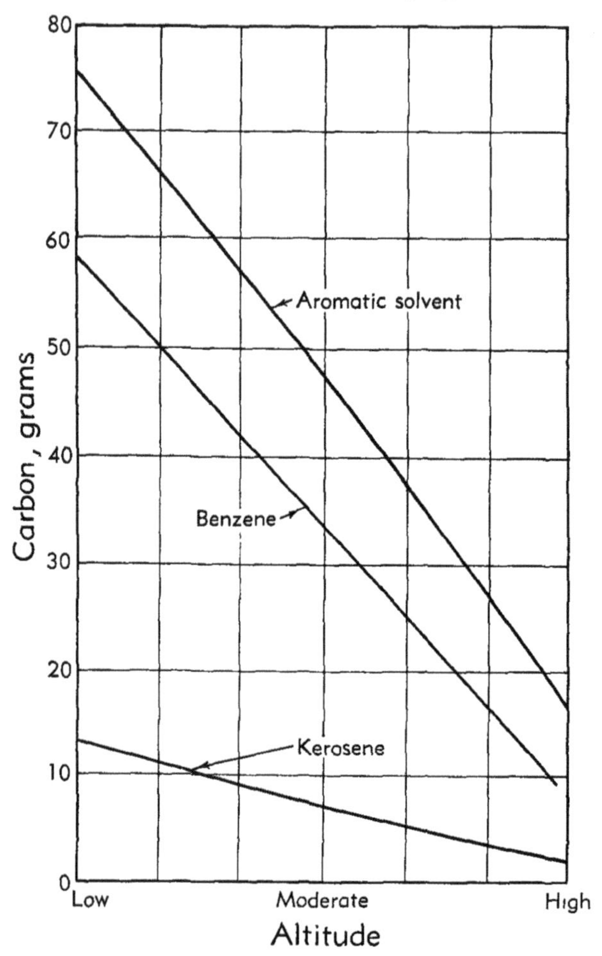


Fig. G,2o. Effect of altitude on carbon deposits.

in fuel flow. Therefore, there was no apparent effect of altitude on carbon deposits. The contradictory results indicate that fuel spray characteristics and combustor design, rather than a pressure effect, may have determined the carbon-depositing rate.

It has been shown in [*15*] that in a 2-in.-diameter combustor, the carbon deposits are influenced by inlet air pressure, temperature, and air flow.

*Table G,2f.* Pounds of fuel required to produce 1 gram of carbon at various altitudes

| Altitude | Pounds of fuel to give 1 gram of carbon in combustor *A* | | |
|---|---|---|---|
| | Aromatic solvent | Benzene | Kerosene |
| Low | 4 | 5 | 21 |
| Moderate | 5 | 8 | 43 |
| High | 13 | 16 | 97 |

| Altitude | Pounds of fuel to give 1 gram of carbon in combustor *B* | | | |
|---|---|---|---|---|
| | JP-3 | High boiling JP-3 | High aromatic JP-3 | JP-1 |
| Low | 84 | 66 | 39 | 45 |
| Moderate | 80 | 68 | 40 | 43 |

The effect of combustion chamber pressure on carbon deposits is also discussed in [*5*]. The effect of pressure was investigated over a range of about 15 to 49 pounds per square inch absolute. The quantity of carbon deposit increased rapidly with pressure to a maximum value of about 110 grams at 27 $lb/in.^2$. At higher pressures the carbon deposit decreased so that at 49 pounds per square inch the carbon deposit was only 5 grams which is about the same value obtained at a pressure of 15 pounds per square inch. It is not possible to determine from the published information whether mass flow or velocity in the combustor might have varied while the inlet air pressure was varied. Therefore it may be that two variables were changing simultaneously.

The important effect of fuel spray on carbon deposition is indicated by the fact that Buckland and Berkey [*16*] have reported that the use of an air-atomizing nozzle eliminated objectionable carbon deposits in a combustor when "Bunker C" fuel was used. The combustor was designed for a gas turbine locomotive with a combustion chamber pressure of several atmospheres. The fuel produced objectionable deposits without the air-atomizing nozzle.

In summarizing the published information on carbon deposits in aircraft gas turbine engines it is apparent that the quantities of carbon formed in the combustors depend upon the fuel properties, the combustor design, the method of fuel introduction, and the operating conditions. In a given engine at constant operating conditions a series of fuels give increased carbon deposits as the hydrogen-carbon ratio is decreased and the boiling temperature of the fuel is increased. In general, it may be expected that carbon deposits will increase as the combustion chamber pressure is increased.

SMOKE. Some turbojet engines under certain operating conditions give a smoky exhaust. Examination of the soot from the exhaust of a turbojet combustor [*11*] has shown this material to be partially burned carbon particles that are formed due to incomplete combustion. Chemical analysis and examination by electron microscopy and X-ray diffraction have shown the smoke from a turbojet combustor and from a wick lamp to be the same type of material [*11*]. The carbon-hydrogen analyses are shown in Table G,2c. Apparently in both cases there is insufficient air in the reaction zone of the flame to burn the fuel completely.

Some research has been done to determine the effect of fuel type on the tendency of a flame to smoke. This has been done by burning various fuels in laboratory wick lamps [*17,18*].

In [*17*] a total of 38 pure hydrocarbons were examined, and the fuel types included normal paraffins, isoparaffins, cycloparaffins, olefins, acetylenes, and aromatics. The results were obtained by determining the maximum rate of fuel flow that would burn without smoke formation. Some of the results of this investigation are shown in Fig. G,2p as a plot of the relative smoking tendency of the hydrocarbons against the number of carbon atoms. The aromatic hydrocarbons smoke much more than the other hydrocarbons with the acetylenes in an intermediate position. It is of interest to note that the isoparaffins smoke somewhat more than normal paraffins.

The results shown in Fig. G,2p may not be directly applicable to turbojet combustors, but the process that leads to smoke formation is presumably the same in a turbojet combustor as in a wick lamp.

STARTING. The starting of a turbojet engine provided with several tubular combustors can be said to occur in three phases, namely: (1) ignition of the fuel-air mixtures in those combustors provided with spark plugs, (2) propagation of flame through interconnector tubes to all combustors, and (3) acceleration of the engine to full speed. In engines provided with annular combustors, of course, the second step involves flame-spreading through the complete annulus.

The variables that influence the possibility of ignition include the fuel-air ratio, the pressure, and the gas velocity in the vicinity of the spark discharge. Investigation of the energy requirements to obtain spark ignition of a homogeneous, stoichiometric mixture of propane and air [*19*] has shown that the three variables are quite important. As pressure was decreased from one atmosphere to approximately 0.1 atmosphere, the energy required to ignite a quiescent mixture of fuel and air increased more than one hundred times.

It has also been shown in [*19*] that, as mixture velocity past the spark electrodes is increased from 10 feet to 55 feet per second, a twofold

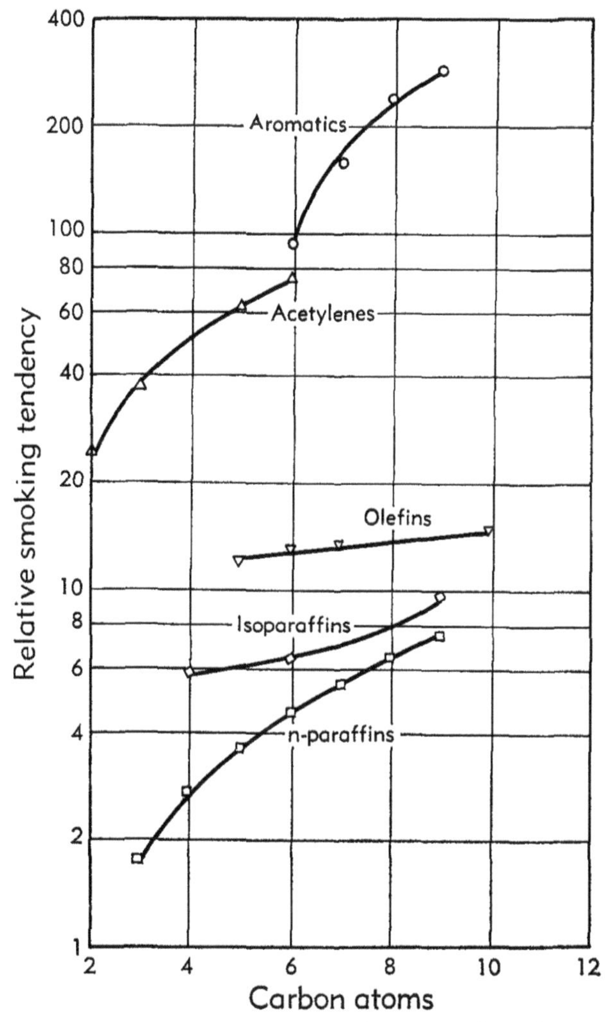


Fig. G,2p. Relative smoking tendency of hydrocarbons.

increase in energy is required for ignition. Fuel-air ratio also influences the energy required for ignition. Several investigators have shown that for most hydrocarbons, the minimum energy is required for ignition when the fuel-air ratios are slightly richer than stoichiometric [*20,21*]. Fuel-air mixtures that are appreciably richer or leaner than stoichiometric required considerably higher energies for ignition than the energies required at the optimum mixture ratio. It is also important to note that ignition, as usually determined, can only be obtained within a fairly limited range of fuel-air ratios. For liquid fuels such as hexane this range is from about 50 per cent to about 400 per cent of stoichiometric [22,23].

In attempting to apply information on the ignition of propane-air mixtures to ignition of liquid fuels in a turbojet combustor, it is apparent that the volatility and viscosity of the fuel influence the fuel-air ratio

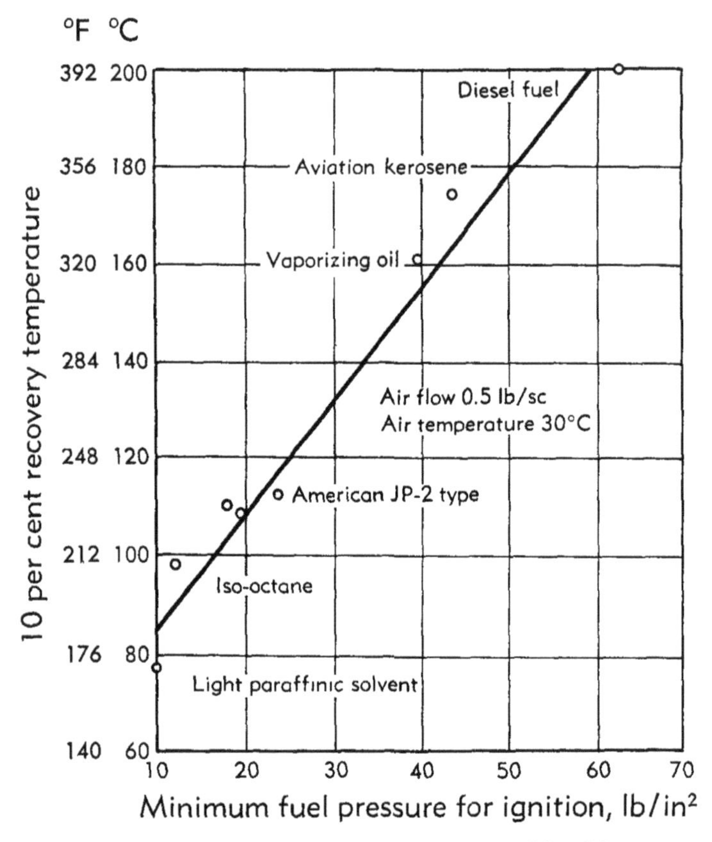


Fig. G,2q. Effect of fuel volatility on ignition [5].

that can be obtained in the vicinity of the spark electrodes. The viscosity influences the drop sizes that are obtained from a liquid injection nozzle and the volatility influences the rate at which vapor is formed from the liquid droplet. Fuels with a high boiling temperature must be injected into the combustion chamber at a higher pressure than volatile fuels in order to provide ignition. Quantitative data to illustrate this fact are shown in Fig. G,2q taken from [5]. The figure shows that, as the volatility of the fuel, indicated by the 10 per cent recovery temperature of the fuel, is increased, the minimum fuel pressure for ignition is decreased.Thus, when the fuel is rather volatile, a very low fuel pressure is sufficient for atomization and vaporization to produce a combustible mixture. As the boiling temperature of the fuel is increased, the fuel pressure must be increased in order to produce a finely atomized spray that will provide sufficient fuel vapor for the creation of a combustible mixture.

It has also been reported [*5,6,24*] that, by the use of a surface discharge spark plug, it is possible to obtain ignition in a turbojet engine at very high altitudes. Such results may be due to the creation of a combustible mixture near the spark plug by the high energy discharge itself. Thus fuel volatility is less critical if sufficient spark energy is available for ignition.

In summarizing the published information on the effect of fuel volatility on ignition in a turbojet combustor, the results indicate that volatile fuels are more readily ignited than nonvolatile fuels. However, high energy ignition systems with surface discharge plugs ignite kerosene-type fuels and, presumably, any fuels that are more volatile, at high altitudes.

## G,3. Fuel Behavior in Aircraft Fuel Systems.

*Volatility and fuel losses.* The vapor pressure of a fuel and the ease with which it evaporates are properties that are controlled to rather close limits depending upon the application for which the fuel is intended. For example, a gasoline for automotive use in the northern part of the United States in winter time is manufactured to have a Reid vapor pressure of 12 pounds per square inch whereas it is held to a Reid vapor pressure of 9 pounds per square inch in the summer months. An aviation gasoline is specified to have a maximum Reid vapor pressure of 7 pounds per square inch.

The fact that aircraft operate at high altitudes with attendant low ambient pressures is an important consideration in the selection of a fuel of the proper volatility. Fuel carried to altitude in an airplane tank boils when the fuel vapor pressure exceeds the ambient pressure. The fuel vapor losses that may be encountered at altitude can be calculated from [*3*] or [*25*]. The calculated vapor losses are shown in Fig. G,3a. It is shown that at a fuel temperature of 100°F, fuels with a Reid vapor pressure of 7 pounds per square inch give appreciable vapor losses at moderate altitudes. At lower fuel temperatures the vapor losses are lower, as shown in Fig. G,3b. It seems apparent that, because of the possibility of fuel vapor loss, it is desirable for an aircraft fuel to have as low a vapor pressure as is compatible with other requirements.

In addition to vapor losses encountered with volatile fuels, additional losses may occur due to "slugging" or foaming [*26*]. This phenomenon may occur when fuel boils so rapidly that liquid is entrained with vapor and is lost through the tank vent. Such losses of liquid fuel may be very large when an airplane climbs to high altitudes at a rapid rate.

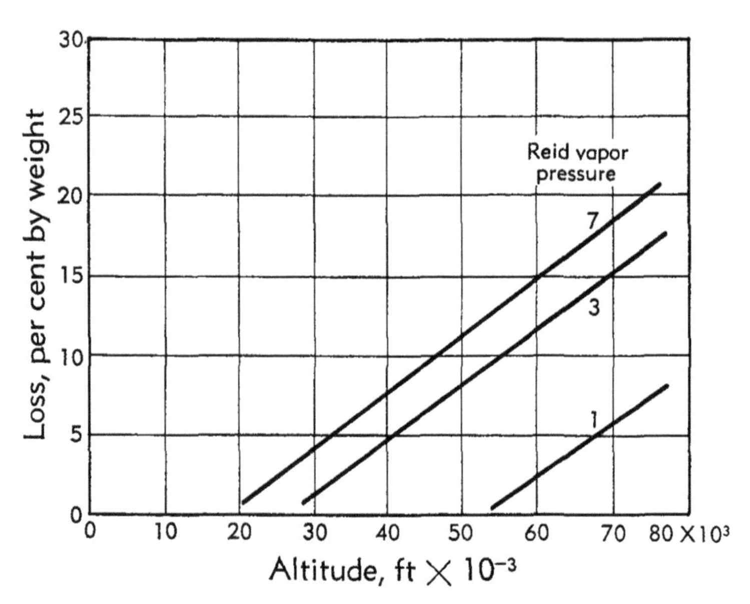


Fig. G,3a. Fuel vapor loss at altitude. Fuel temperature, 100°F; rate of climb, 3000 ft/min.

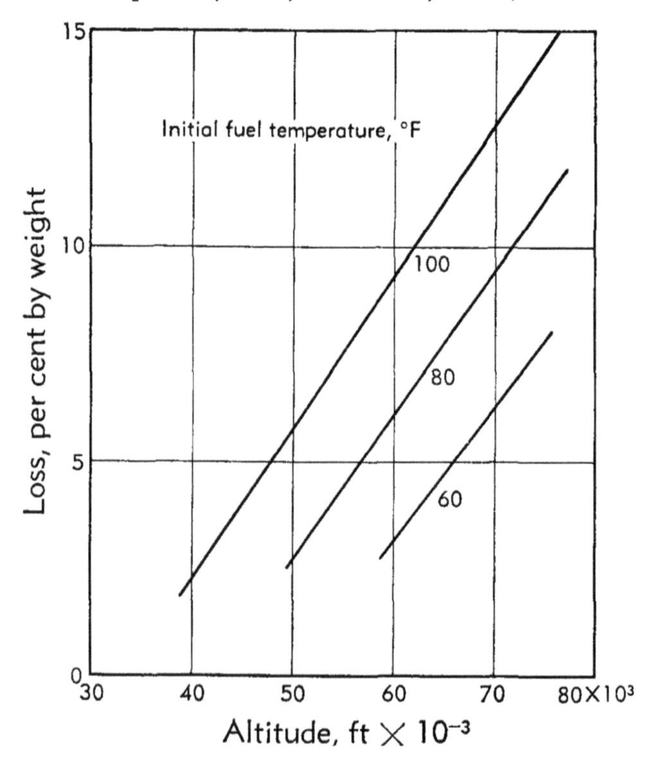


Fig. G,3b. Effect of fuel temperature on vapor loss with a 3-lb/in.[2] Reid vapor pressure fuel.

Here again the most obvious answer to the problem is to use a fuel with a low vapor pressure, so that boiling does not occur. If it is not feasible to supply a low vapor pressure fuel, then it may be necessary to provide a sufficient pressure in the fuel tank to eliminate boiling. In some cases, tank pressurization may require additional structural weight over the unpressurized fuel tank.

Fuel boiling may also be caused by the release of dissolved air from the fuel when an airplane climbs rapidly to high altitude. It is shown in [*26*] that a JP-1 type fuel dissolves air to the extent of 14 per cent by volume, and a JP-4 type fuel dissolves from 14 to 24 per cent by volume. Unfortunately there seems to be very little that can be done to eliminate air release except tank pressurization.

*Aerodynamic heating.* It is well known that flight at supersonic speeds is complicated due to aerodynamic heating. Skin temperatures may be expected to reach a value of 400°F or above at a Mach number of 2.5. Such temperatures require that the aircraft designer must use materials that maintain suitable strength characteristics at high temperatures. In addition, it has been necessary to provide special features for the comfort of the pilot.

It may also be necessary to consider the effect of aerodynamic heating on the fuel. The fuel temperatures to be encountered at high flight speeds depend upon many factors including the location of the fuel tanks with relation to the aircraft skin, the materials of construction, the fuel tank configuration, and the duration of the flight.

Sustained flight at supersonic speeds can result in severe fuel losses. As an example Keusch [*27*] has calculated tank losses of JP-4 as a function of flight time for a hypothetical bomber flying at Mach 2 and 40,000–50,000 feet. These losses are shown in Fig. G,3c, both with and without tank pressurization, and with and without tank insulation. With neither pressurization nor insulation, fuel losses can approach 20 per cent. However, a small amount of pressurization and the addition of only $\frac{1}{16}$ inch of cork reduces the loss to negligible proportions. If less volatile fuels than JP-4 are used, JP-1 or JP-5, for example, then the use of insulation alone would keep vaporization losses to a low level at this flight condition.

In a subsequent article on freezing point it is stated that, during flights at subsonic speeds at high altitude, very low fuel temperatures have been measured and the fuel must have a freezing point low enough to avoid trouble. Thus the temperatures to which the fuel may be subjected are very different during subsonic flight than during supersonic flight.

The petroleum fuels as normally manufactured, that have a suitable freezing point to meet subsonic flight conditions, are volatile enough so that they might boil or require pressurization at supersonic flight. There are several possible solutions to this problem. One possibility is to specify a different fuel for the use of aircraft that will always fly subsonically than the fuel for the use of supersonic aircraft. In such a case, airfields in cold climates might have to provide special storage and handling facilities so that the fuel for supersonic aircraft could be heated.

A second possibility is to provide the means for heating the fuel on board subsonic aircraft and to use the same fuel that is provided for supersonic flight.

If the total fuel requirements for supersonic flight were relatively small, then a high boiling fuel could be provided with a low freezing point. JP-1 and JP-5 are examples of such materials. Murray in [*28*] indicates

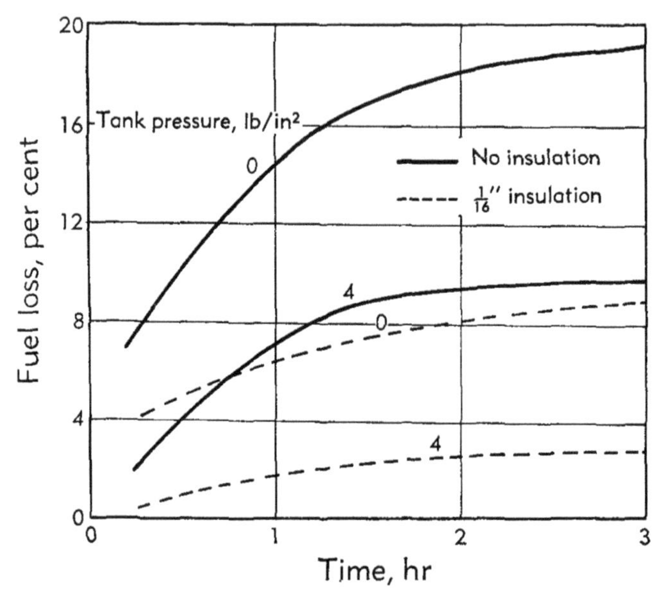


Fig. G,3c. Calculated loss of JP-4 during flight at Mach 2 and 50,000 ft. (From [*27*].)

that a fuel oil with a −40°F freezing point could be provided in the United States in a quantity of 225,000,000 barrels per year.

Another possibility is to use one fuel for all flight regimes and to pressurize and insulate the fuel tanks of supersonic aircraft. It has been suggested by some designers that aircraft built for supersonic flight must be so strong that pressurizing fuel tanks would not necessitate additional aircraft weight.

*Volatility on pump performance.* The tendency for volatile fuels to form large volumes of vapor at high altitudes influences the performance of fuel pumps. The flows of an aviation gasoline and a JP-1 or kerosene-type fuel delivered by the same fuel pump are shown in Fig. G,3d [*29*]. It is shown that the quantity of gasoline delivered by the pump is considerably lower than the quantity of JP-1 delivered. In fact, at a fuel pressure at the pump inlet of 22 inches of mercury absolute, the flow of gasoline is less than one fourth the flow of the JP-1 fuel.

Thus, a nonvolatile fuel allows the use of smaller pumps than those required for volatile fuels and possibly the elimination of booster pumps.

*Viscosity on pumping and atomization.* The viscosity of potential aircraft fuels is of importance because of pumping problems, fuel atomization, and for some engines, pump lubrication. Plots of viscosity against temperature are shown in Fig. G,3e for various aviation fuels and five grades of fuel oil [*3*]. The Commercial Standard, CS 12-48, allows a wide spread in viscosity for various fuel oils [*30,31*] and the plots shown in

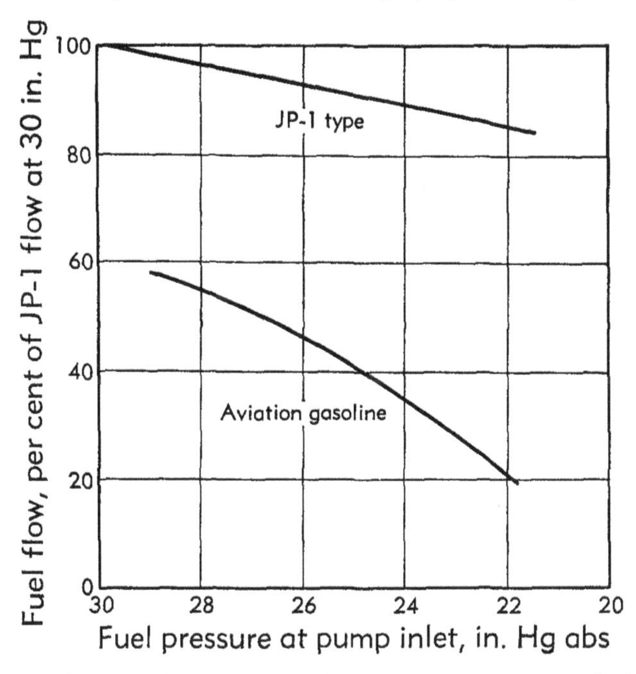


Fig. G,3d. Effect of fuel volatility on pump performance [*29*].

Fig. G,3e are arbitrarily chosen as oils that fall within the specifications for the respective fuel oils. The variations in viscosity to be anticipated with the aircraft fuels are relatively minor compared to the possible variations with the fuel oils.

It is stated in [*6*] that fuels with viscosities up to 2000 centistokes can be pumped in current types of aircraft fuel systems. Most fuels would have viscosities lower than 2000 centistokes at anticipated operating temperatures.

The only fuels shown in Fig. G,3e that have a viscosity high enough to cause pumping difficulties are the No. 5 and 6 fuel oils. At temperatures below 75°F, their viscosities would be above 2000 centistokes.

Fuels with viscosities of 15 centistokes or below can be atomized satisfactorily in aircraft fuel systems [5,6]. Presumably it is assumed that a high pressure fuel system is used. Possibly about 10 centistokes would be a more practical limit for low pressure fuel systems. Thus the requirement of proper atomization is much more restrictive than the pumping requirements.

The data in Fig. G,3e show that JP-1 must be kept at a temperature above −45°F and JP-5 above −20°F in order to obtain satisfactory atomization. Higher boiling fuels must be kept relatively warm in order

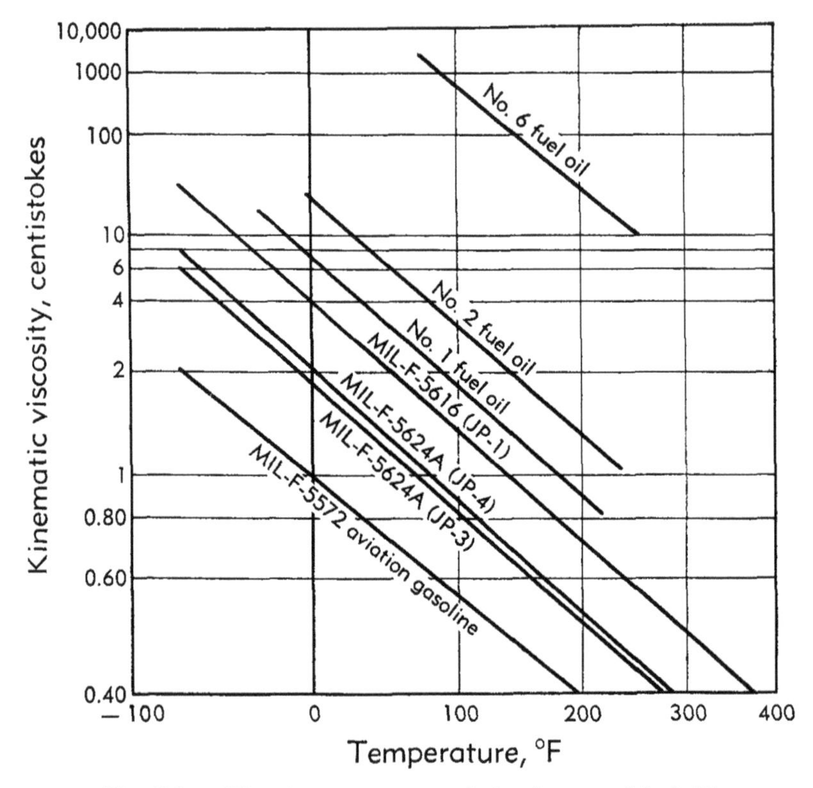


Fig. G,3e. Viscosity-temperature relation for several fuels [3].

to get satisfactory atomization. It would be necessary to keep a No. 1 fuel oil above a temperature of −15°F and the temperature of a No. 6 oil would have to be above 250°F to insure satisfactory atomization with a simple nozzle. However, an air-atomizing nozzle would probably give satisfactory atomization at lower temperatures.

*Filter plugging.* One of the items to be considered in the use of a fuel in aircraft is the possibility of filter plugging. Filters are placed in fuel lines to remove particles of dirt or rust that might have a harmful effect on fuel controls, pumps, or injection nozzles. Fuel composition, water in the fuel, and the filter design, all may have an influence on filter plugging.

Filter clogging could occur at low temperatures if the fuel had a composition such that some of the components became solid at the low temperatures encountered during high altitude flight. The filterability of fuels at low temperatures has been studied and a good correlation has been reported [*5,28,32*] between the cloud point of the fuel and temperature at which filter plugging occurs. Data reported by Sharp [*5*] are shown in Table G,3. A good correlation is shown between the cloud point and the temperature at which the flow through a filter is reduced. This factor may be eliminated by specifying fuels for aircraft use that have sufficiently low cloud points so that filter plugging does not occur.

*Table G,3.* Effect of cloud point on filterability. Data from [*5*].

| Fuel | Cloud point, °F | Temperature at which flow dropped by 20 per cent of value calculated from extrapolated viscosity, °F |
|---|---|---|
| Gas oil | 19 | 21 |
| Kerogasoil | −49 | −45 |
| Kerosine | −45 | −51 |
| Diesel fuel | −56 | −62 |

A second factor that may contribute to filter plugging is the fact that free water may be introduced into fuel tanks due to careless handling, or by condensation from air induced into fuel tanks during flight. In addition, fuels dissolve a small amount of water during normal manufacturing and handling procedures. Examples of the quantities of water dissolved are shown in Fig. G,3f, plotted as a function of temperature. The plot shows that a pure aromatic hydrocarbon, toluene, dissolves considerably more water than a pure paraffin such as *n*-octane. The aviation gasoline and the particular JP-3 fuel investigated dissolved about the same amount of water.

During normal processing and handling a fuel becomes saturated with water. If the fuel is cooled below 32°F, a considerable amount of the water that was originally dissolved comes out of solution. It may remain in the fuel as supercooled, finely divided water particles or it may freeze to ice.

It has been suggested [*28*] that, although the percentages of dissolved water in fuels are very low, the corresponding quantities of ice might very readily plug filters. For example, if water dissolved in a fuel to a value of 0.01 weight per cent, this would mean that over 0.6 pound of water would dissolve in 1000 gallons of fuel. If all of the water were converted into ice, then it seems possible that the ice might cause filter plugging difficulties on large aircraft that carry a huge quantity of fuel.

The filterability of four fuels that dissolved different quantities of water has been reported by Williams [*32*]. The data are presented in Fig. G,3g. When the fuels were passed through a filter, the fuels that dissolved 0.007 and 0.008 weight per cent water at 64°F gave a drop of 20 per cent in the flow at −10°F. The fuel that dissolved only 0.003 weight per cent water did not give a 20 per cent reduction in filtering rate until the fuel temperature was reduced to −38°F. A similar trend was observed for the 50 per cent reduction in flow. The data presented in

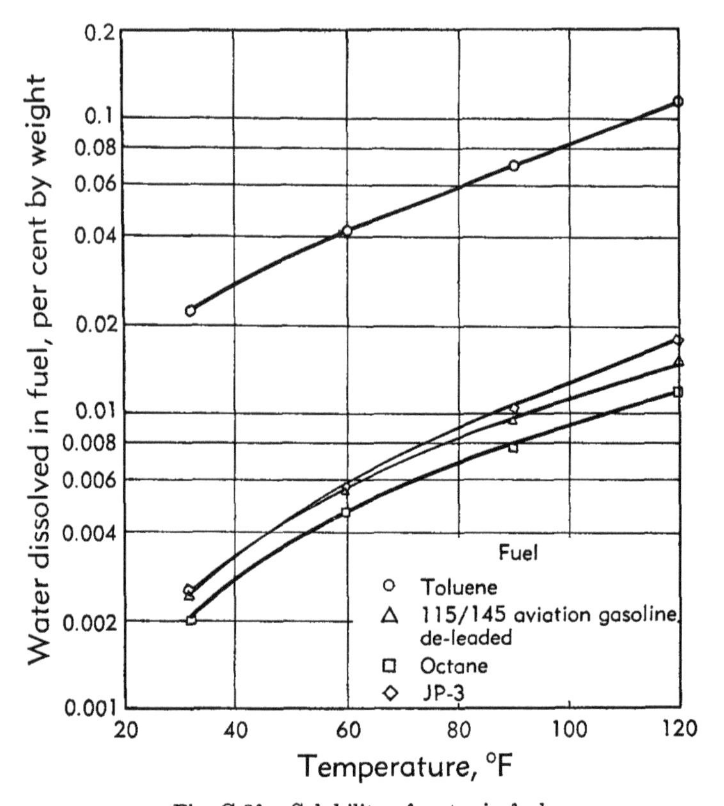


Fig. G,3f. Solubility of water in fuel.

Fig. G,3g indicate that filter plugging at low temperatures is most likely to occur with fuels that dissolve the most water. It has been suggested in [*29*] that ice formation is most likely to occur in a fuel that furnishes wax crystals as nuclei for ice formation.

Kerosene-like fuels suspend water particles fairly readily. In fuel transfers involving water displacement storage tanks, and in fueling aircraft, a serious operational problem results. Not only must time be allowed to settle suspended water, but filtering is required. Carelessness leads to icing in the aircraft system.

*Problems arising from use of fuel as a heat sink.* While the aerodynamic heating of fuel in tanks will become increasingly important, a more immediate problem arises from the use of the fuel as a heat sink for turbojet engine cooling. In the newer turbojets the lubricant both lubricates and cools the engine. The heat picked up by the lubricant is then rejected to the fuel as the latter flows to the combustor. An idealized diagram of this type of system is shown in Fig. G,3h. The fuel flow is shown as a solid line and the lubricant flow as a broken line. The fuel is pumped from

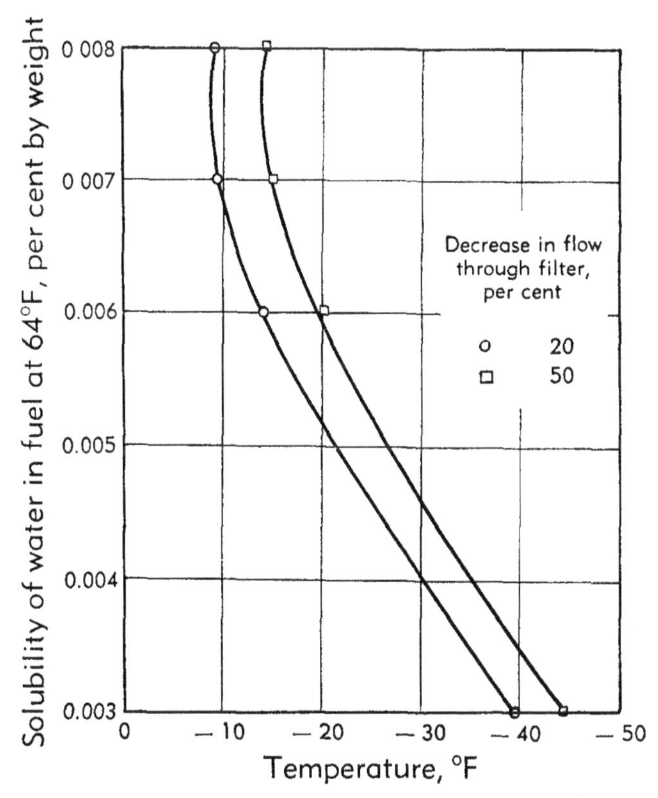


Fig. G,3g. Effect of water solubility of four fuels on filterability [*32*].

the tank through a lubricant-to-fuel heat exchanger where it may be rapidly heated to temperatures in excess of 250°F. At these temperatures the fuels may form insoluble products which can foul the heat exchanger or clog filter screens and atomizer orifices.

Jet fuels meeting current specifications vary widely in their stability under the temperature and residence time conditions encountered in these exchangers. The solids are probably formed by the condensation or polymerization of very small amounts of nonhydrocarbon impurities in the fuel. The removal of these impurities by chromatography [*33*] or in special cases by washing the fuel with dilute acid [*34*] greatly increased the stability of the fuel. Dissolved oxygen can also contribute to fuel instability [*34*].

The amount of insolubles formed in the fuel is very small and of the order of one milligram per gallon [*34*]. However, because of the very high fuel consumption rates in jet aircraft and the very close tolerances required for fuel metering and atomization, even this small concentration of solids can cause operational problems. In general the thermal instability of jet fuels appears to be the biggest single problem now being encountered with aircraft gas turbine fuels. The problem may become increasingly severe if additional heat loads are imposed on the fuel. Such loads might

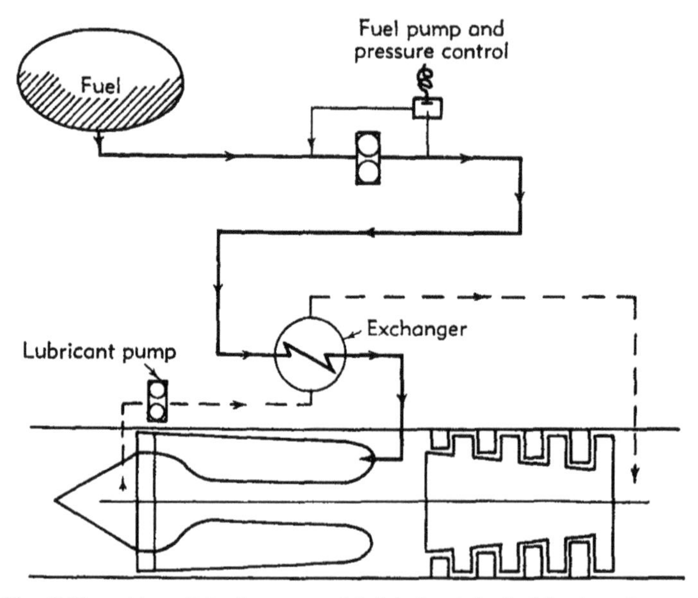


Fig. G,3h. Aircraft fuel system with lubricant to fuel heat exchanger.

be used to supply cooling for the pilot and for aircraft electronic and hydraulic systems.

*Fire hazard.* One of the items for consideration in the selection of a fuel for aircraft application is the relative fire and explosion hazard presented by fuels of different volatility.

A great deal has been written about the relative hazards of fuels of various volatilities and composition in an aircraft application [*20,26,28, 29,35,36*]. Most of the discussions have been concerned with one of three aspects of the problem: (1) the relative hazards of ground handling, (2) the possibilities of combustible mixtures in fuel tanks, and (3) crash fires. Probably the only clear-cut advantage of one fuel type over another is in relation to ground handling. It is fairly obvious that, in airport operations where spillage is a possibility, a kerosene-type fuel is less hazardous than a gasoline-type fuel.

With regard to combustible mixtures in fuel tanks, there are plots in [*3,25,26,29*] that show the conditions of altitude and fuel temperature that yield flammable mixtures for various types of fuels. No simple statement can be made concerning the relative flammable hazards with different fuels since so much depends on flight conditions. In prolonged subsonic flight the fuel is cooled and the less volatile fuels such as JP-1 and JP-5 may be preferred since their vapor pressure would be too low to yield flammable mixtures. At supersonic flight speeds the fuel may be heated to the extent that greatest safety would be obtained with the more volatile fuels since these would have sufficient vapor pressure to give fuel-air mixtures that are too rich to burn. In any case it may be unwise to assume, as is often done, that equilibrium vaporization conditions exist. The best approach towards minimizing tank explosion hazards appears to be through controlling the oxygen level to some low value by tank inerting.

The effect of fuel volatility on the relative severity of aircraft crash fires has been the subject of many discussions. It is easy to imagine a situation where a mild crash causes a spillage of fuel and where a gasoline would vaporize rapidly and present a greater fire hazard than kerosene. On the other hand, a more severe crash can release the fuel as a finely atomized spray. In that case the kerosene is just as hazardous as the gasoline, as shown by controlled crashes of full scale aircraft [*36*].

After consideration of the three aspects of the problem, it seems that the use of the less volatile fuels might tend to reduce the number and severity of aircraft fires but would certainly not eliminate them.

*Storage stability.* Aircraft fuels for both military and civilian use must be stable during relatively long-term storage. Some fuels, particularly those produced by thermal cracking, form gum during storage. It is assumed that the presence of gum in a fuel might cause malfunctioning of an aircraft gas turbine engine although no quantitative data have been published on the amount of gum that might be tolerated by an engine.

Aircraft fuel specifications carry three items to avoid the acceptance of a fuel that might produce gum during storage: (1) bromine number, (2) steam jet residue, and (3) accelerated gum. The bromine number indicates the concentration of olefinic double bonds in the fuel. Mono-olefins are satisfactory components for gas turbine fuels. However, the mono-olefins are often accompanied by small amounts of the very reactive diolefins which readily polymerize to form gum. For this reason the total olefin content of gas turbine fuels is limited to 5 per cent in current military specifications.

The accelerated gum test is a laboratory test designed to predict whether the fuel will form gum during long-term storage. The maximum value allowed by the military specification MIL-F-5624C in the United States is 14 milligrams per 100 milliliters of fuel. This specification tends to restrict the quantities of thermally and catalytically cracked components that can be included in the finished aircraft fuel.

The steam-jet residue test is to prevent acceptance of a fuel that may contain suspended material or gum at the time of its manufacture.

**G,4. Fuel Specifications and Availability.** In the previous articles, the performance of fuels of various volatilities and compositions has been discussed with regard to gas turbine engines and aircraft fuel systems. In this article, the specification requirements for the various grades of jet fuels, the development of these grades, and their availability are treated briefly. Mention is also made of other fuels which may some day be used in gas turbine-powered aircraft.

TYPICAL FUEL PROPERTIES. Generally, gas turbine fuel specifications have been set up by the military services. Most of the requirements for JP-1, JP-3, JP-4, and JP-5 are listed in Table G,4. Also listed in this table are data taken from [*3*] for average quality fuels; a typical 115/145 octane gasoline is included for comparison. A discussion of the significance of each control test to fuel performance can be found in [*3*].

In general, the different grades of jet fuel differ primarily in volatility. The final boiling point of average fuels of each grade is near 500°F but the initial boiling points range from 108°F for JP-3 to 359°F for JP-5 (Table G,4). Many of the other properties are much the same for all grades, exceptions being the Reid vapor pressure and flash point which are other measures of volatility, and the density, viscosity, and freezing point which are closely related to volatility for practical, petroleum-derived fuels.

GRADES AND AVAILABILITY OF JP FUELS. Military requirements for fuels can be so great in time of war that considerations of availability have always been a major factor in establishing fuel specifications. For this reason availability is discussed along with the development of the several grades of jet fuels.

*JP-1.* The first turbojets used kerosene, and *JP*-1 is primarily a kerosene-type fuel with a −76°F freezing point limit. As seen in Table G,4, the boiling range for this fuel is quite narrow and this factor, combined with the freezing point requirement, limits availability to about 5 per cent of the crude oil production. The quality of JP-1 is quite satisfactory from most standpoints, but the availability is insufficient for potential wartime requirements. Availability is probably great enough to meet peacetime requirements of both military and civilian aviation for many years, and a JP-1 type fuel was used in civilian turboprop operations in 1956. It has also been considered for future civilian turbojet uses [*37*].

*JP-3.* Since the amount of JP-1 that can be produced is limited, an attempt has been made to develop a fuel specification of much wider availability. This can be done by either raising the final boiling point or lowering the initial boiling point. The former path leads to fuels with undesirably high freezing points and greater tendencies to form combustor deposits. Therefore the final boiling point was raised but little, while the initial boiling was lowered considerably, thereby raising volatility. The result was JP-3, a fuel which can be produced to the extent of about 50 per cent of the crude.

However, the volatility of JP-3 proved too great and resulted in excessive tank vaporization losses as gas turbine-powered aircraft developed higher altitude capabilities. Furthermore the availability for JP-3 appears greater than is needed in any case, so that this fuel has been superseded by JP-4.

*JP-4.* Since the volatility of JP-3 was too high, the next obvious step was to lower this to some extent, which was done by reducing the Reid vapor pressure from 5–7 lb/in.$^2$ to 2–3 lb/in$^2$. This resulted in JP-4 which has an initial boiling point about 40°F higher than JP-3. The availability of JP-4 is somewhat less than JP-3 and is about 35 per cent of the crude which appears ample for all foreseeable needs.

JP-4 is now the fuel with the widest military usage. It is also being considered for civilian gas turbine operations [*37*].

*JP-5.* More recently there has been a need for an even less volatile fuel for use on aircraft carriers. It was required that the fuel have a flash point of at least 140°F. This material is JP-5. Two to three parts of JP-5 are blended on shipboard with one part of aviation gasoline to make a fuel quite similar to JP-4. The advantages to a carrier required to operate both turbine and piston engines are obvious.

Fuels having flash points as high as that required for JP-5 cannot be made without sacrificing some desirable low temperature characteristics. Therefore both freezing point and viscosity limits were raised.

The availability of JP-5 is quite limited and less than that of JP-1. Nevertheless it is the preferred fuel for carrier use and it, or a similar type of fuel, may be required for very high speed flight where aerodynamic heating makes it difficult to contain the more volatile fuels.

FREEZING POINT REQUIREMENTS AND EFFECT ON AVAILABILITY. The specifications for aircraft fuels for either military or commercial application have always required a low freezing point. Such a requirement has been considered necessary because of low temperatures that may be encountered during flight at high altitudes and because of low temperatures that may be encountered on the ground in arctic regions. The freezing point of an aircraft fuel is required to be below −76°F both in the United States and in Great Britain, although the British, until recently, specified only −40°F.

Because of the relatively mild climate in the British Isles, low temperatures are not encountered on the ground and, in the past, the −40°F

*Table G,4.* Specifications and typical properties of aircraft fuels.

| Property | Specification requirements | | | |
|---|---|---|---|---|
| Specification designation | MIL F-5616 | MIL F-5624C | | |
| Fuel grade | JP-1 | JP-3 | JP-4 | JP-5 |
| ASTM distillation<br>Per cent evaporated at T, °F<br>Initial point<br>10<br>20<br>50<br>90<br>End point | <br><br><br>410 (max)<br><br><br>490 (max)<br>572 (max) | <br><br><br><br>240 (max)<br>350 (max)<br>470 (max)<br> | <br><br><br><br>290 (max)<br>370 (max)<br>470 (max)<br> | <br><br><br>400 (max)<br><br><br><br>550 (max) |
| Freezing point, °F | −76 (max) | −76 (max) | −76 (max) | −40 (max) |
| Reid vapor pressure, lb/in² | | 5 to 7 | 2 to 3 | |
| Aromatics, per cent by volume | 20 (max) | 25 (max) | 25 (max) | 25 (max) |
| Bromine number | 3 (max) | 5 (max) | 5 (max) | 5 (max) |
| Total sulfur, per cent by weight | 0.2 (max) | 0.4 (max) | 0.4 (max ) | 0.4 (max) |
| Existent gum, mg/100 ml | 5 (max) | 7 (max) | 7 (max) | 7 (max) |
| Potential gum, mg/100 ml | 8 (max) | 14 (max) | 14 (max) | 14 (max) |
| Heat of combustion, Btu/lb | | 18,400 (min) | 18,400 (min) | 18,300 (min) |
| Specific gravity, 60/60°F | 0.85 (max) | 0.739 to 0.780 | 0.751 to 0.802 | 0.788 to 0.845 |
| Viscosity, centistokes<br>at −30°F<br>at −40°F | <br><br>10 (max) | | | <br>16.5 (max)<br> |
| Flash point, °F | 110 (min) | | | 140 (min) |

temperature has been considered well below fuel temperatures that would be experienced in flight. However, it is indicated in [6] that ambient temperatures at high altitudes may be as low as −139°F and that fuel temperatures as low as −49°F have been measured during flight. This indicates that, because aircraft powered by gas turbines fly at higher altitudes than usually encountered in the past, the requirement for a low freezing point may be more important. The actual fuel temperature to be

| Typical properties [3] | | | | |
|---|---|---|---|---|
| JP-1 | JP-3 | JP-4 | JP-5 | 115/145 |
| 326<br>347<br>353<br>370<br>407<br>448 | 108<br>164<br>201<br>303<br>437<br>497 | 144<br>216<br>250<br>319<br>425<br>487 | 359<br>390<br>404<br>428<br>475<br>511 | 116<br>141<br>154<br>198<br>248<br>327 |
| <−76 | <−76 | <−76 | −49 | <−76 |
| | 5.9 | 2.6 | | 6.2 |
| 14 | 11 | 11 | 16 | |
| 1 | 3 | 2 | 2 | |
| 0.04 | 0.09 | 0.08 | 0.15 | |
| 2 | 3 | 1 | 2 | |
| 3 | 7 | 2 | 4 | |
| 18,480 | 18,710 | 18,680 | 18,520 | 19,070 |
| 0.810 | 0.760 | 0.773 | 0.827 | 0.693 |
| 7.6 | 2.9 | 3.3 | 16.1 | |
| 117 | | | 147 | |

obtained during flight is a function of the flight speed and altitude and the duration of the flight.

Probably the most extensive information available on fuel temperatures experienced in flight has been reported by Walker [26]. Fuel temperatures were measured on the D.H. 106 Comet G-ALVG airplane during a $5\frac{1}{2}$-hour flight, of which $4\frac{1}{2}$ hours were spent at altitudes between 35,000 and 40,000 feet, cruising at a true air speed of between 450 and

500 miles per hour. The ambient air temperatures varied between −67°F and −87°F and the lowest fuel temperature recorded was −22°F. Walker concludes that in order to insure that the Comet operate satisfactorily at temperatures as low as −90°F the freezing point of the fuel should be at least −50°F.

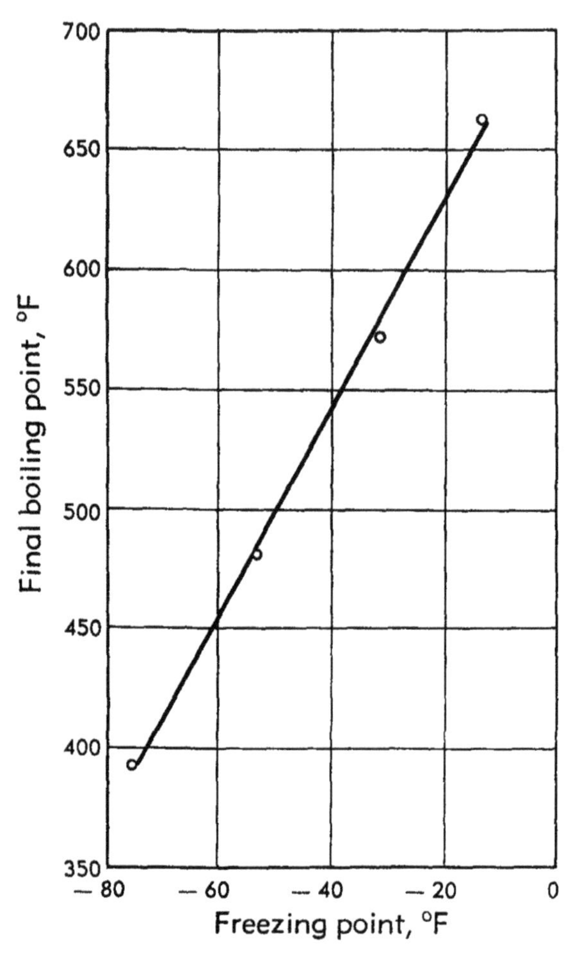


Fig. G,4a. Effect of final boiling point on freezing point of 100°F flash point fuel [*6*].

The freezing point requirement establishes to a large degree the extent to which high boiling components can be used in aviation fuels.

The effect of higher boiling components of fuels on the freezing point is discussed in [*6*] and the data are reproduced in Fig. G,4a. It is shown that, for derivatives from a particular crude oil, the freezing point is a direct function of the final boiling point of the fuel.

The JP-1 type fuel originally specified for use in the United States and Great Britain had a boiling temperature range of about 330°F to about 450°F and met the freezing point requirement of −76°F. However, in [38] it was stated that by raising the freezing point requirement to −50°F the fuel would be available as 17 per cent of a barrel of crude oil, without regard for the production of refined oil and diesel fuel, whereas the JP-1 was available as only 6 per cent of a barrel of oil, if the specification was held to a freezing point of −76°F. This case tends to illustrate the effect that the freezing point has on the availability of high boiling stocks as potential aircraft fuels.

If it is assumed that vapor pressure must be low for satisfactory aircraft operation, then the freezing point requirement of aircraft fuels probably restricts the quantity of fuel available more than any other item in the specification. Probably the next most restrictive item is the accelerated gum test. This restricts the quantities of cracked stocks that can be included in a finished fuel.

POTENTIAL LOW COST FUELS.

*Fuel costs.* It has been reported [28] that the fuel cost represents over 20 per cent of the direct operating cost of a commercial airline operating with reciprocating engines. It has been estimated that fuel cost may be as much as 33 per cent of the operating cost of airplanes powered with turbine propeller engines. Therefore a low fuel cost may represent the difference between profit and loss to an airline, and it is appropriate to consider low cost petroleum products as potential aircraft fuels.

Unfortunately, it is difficult to predict what costs might be for less highly refined fuels than those now used in aircraft. The price of petroleum products may vary, depending upon the demand and the advent of new refining methods. It has been estimated [37] that the jet fuel requirements will be about $2\frac{1}{2}$ per cent of the total crude refined for the year 1965. This is a relatively small figure and therefore the demands of commercial aviation should not change the price structure markedly.

Fuel prices at airports depend upon their location relative to refineries and bulk terminals, and consequently vary a great deal from one location to another. Some relative costs of petroleum products are presented in Fig. G,4b [39]. The costs are shown without the inclusion of taxes. The cost of aviation gasoline with a nominal rating of 100/130 is compared with other products. It is shown that the cost of kerosene is about 55 per cent and No. 6 fuel oil only 33 per cent of the cost of the gasoline, when prices are compared at refineries and bulk terminals. Tank wagon prices for most products show substantially smaller differences between aviation gasoline and the other products.

The price differentials between the various products have not been constant during the last 10 years. However, Fig. G,4b indicates the types of materials that are expected to have the lowest costs. It seems apparent that, if commercial aircraft were able to utilize distillate fuels such as No. 2 oil, savings in fuel costs could be anticipated. If residual fuel oils such as Nos. 5 or 6 could be used, then quite marked savings in fuel costs could be expected.

*Residual fuels for gas turbine application.* The cost advantage of using a residual fuel for commercial aircraft is so great that it is worthy of serious consideration. Some gas turbine engines for locomotives, marine application, and stationary power plants have been developed or are under development for the use of residual fuels. The experience reported

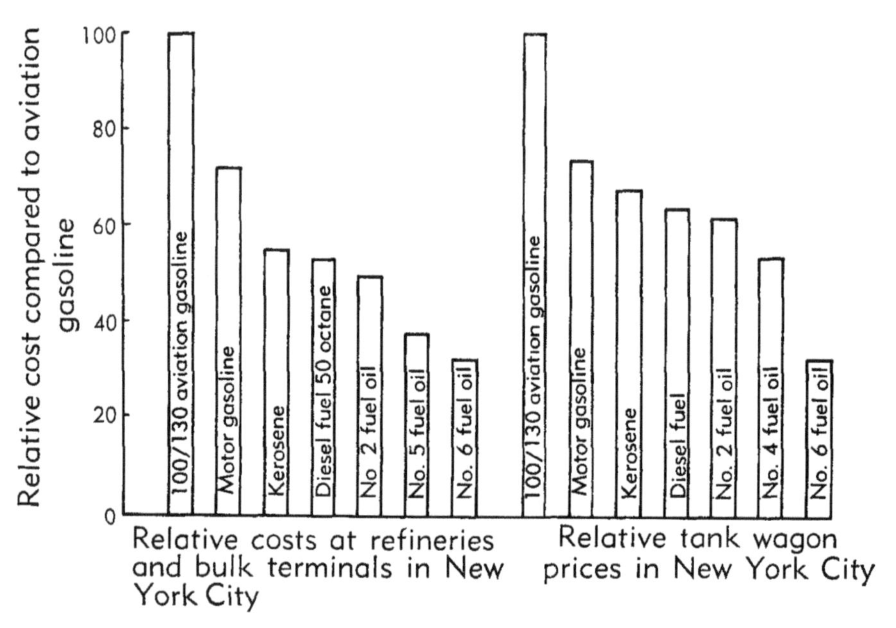


Fig. G,4b. Relative cost of fuels compared to aviation gasoline (March, 1952). (From [*39*].)

in [*7,16,40*] can perhaps be considered to indicate the potentialities of these fuels for aircraft use.

A residual fuel, of course, is a material that is left as a residue after distillation in refining operations and several qualities of such fuels are marketed. The chief differences in quality are expressed as viscosity although other properties, particularly ash and sulfur content, also vary widely. The high viscosity of residual oils means that the fuel must be heated if it is to be pressure-atomized and pumped.

Experience with residual fuels has indicated that they burn slowly [*7,16*] and that heat release rates common with aircraft combustors have not been achieved with residual fuels.

Heat release rates reported for aircraft combustors are usually of the order of $3.5 \times 10^6$ BTU per cubic foot per hour per atmosphere [*40*], whereas industrial gas turbines are usually reported to have heat release rates of the order of $5 \times 10^5$ to $1 \times 10^6$ [*7,16*] BTU per cubic foot per hour per atmosphere. This lower heat release rate would require a bigger engine for aircraft application than now used, a factor that of course would be of considerable importance to an aircraft designer.

Most satisfactory results in combustors utilizing residual fuels have been achieved by the use of air-atomizing spray nozzles. Lloyd and Probert [*40*] reported good results with an atomizer in which the air flow to the nozzle was one-tenth the fuel flow and the air pressure was twice the fuel pressure. The installation of an air pump to provide fuel atomization would complicate an aircraft installation, but certainly could be done if a major reduction in fuel cost were to be realized.

The most difficult problem in connection with the use of residual fuels is the ash that results from combustion. Although the solids that may be in a residual fuel may be removed by filtration or centrifuging, some ash-forming constituents are soluble in the fuel. Chief among these are compounds containing sodium and vanadium. Sodium sulfates, sodium vanadates, and vanadium oxides may be formed during combustion. The properties and composition of the deposits are discussed in [*7,16,40*]. If gas temperatures are high enough so that any of the ash is molten, then it may stick to nozzle vanes and turbine blades and thus lower turbine efficiencies in a relatively short time.

In addition to the physical problem of ash formation, the vanadium pentoxide tends to corrode the turbine blades. The corrosion becomes more severe as gas temperatures are raised. However, this corrosion can be minimized if calcium or magnesium compounds are added to the fuel in the ratio of about three parts of alkali to one part of vanadium [*41*].

Industrial gas turbine engines may be able to use residual fuels without serious corrosion problems by operating with exhaust gas temperatures below 1150°F. However, operation at such cycle temperature would be very detrimental to aircraft engine performance.

It may be concluded that some method of eliminating corrosion due to ash must be found before it can be feasible to use residual fuels in aircraft gas turbine engines. It is interesting to note in this connection that two investigators [*7,40*] report that a slight amount of carbon in the exhaust gases almost eliminates ash deposits. Lloyd and Probert [*40*] suggest the possibility of operating the combustor so as to create a slight amount of smoke, thereby minimizing ash deposits and corrosion.

In summarizing the possibilities of using residual fuels in aircraft gas turbines, it seems apparent that such fuels cannot be utilized until more developmental work has been conducted to correct or alleviate the following problems:

1. Fuels will have to be heated on board the aircraft.
2. It will probably be necessary to provide an air pump for air-atomizing fuel nozzles.
3. Present combustors for residual fuels have less than one third the heat release rate of aircraft engine combustors.
4. Methods must be developed to eliminate ash deposits.

In any case, it would not appear justified to use residual fuels in really high performance turbojet engines.

**G,5. Concluding Remarks.** In the preceding pages an attempt has been made to discuss the performance of various types of fuels in gas turbine combustors and in aircraft fuel systems, and to point out the relative availability and cost of various potential fuels. The results are summarized below.

The available data indicate that in the range of fuel volatility from 100 to 600°F, fuel composition and volatility have no important effect on combustion efficiency at the operating conditions encountered in aircraft gas turbine engines. Volatile fuels are more readily ignited at high altitude starting conditions than nonvolatile fuels. It is reported that such differences can be eliminated by proper combustor and ignitor design.

Fuel composition has a marked effect on carbon deposits in combustion chambers. High boiling fuels with low hydrogen-to-carbon ratios, such as some types of aromatic hydrocarbons, are prone to cause deposits.

An optimum fuel to meet the requirements of an aircraft fuel system would have a negligible vapor pressure under all conditions of aircraft operation. In addition, a low freezing point is necessary for present types of aircraft. These requirements tend to restrict the quantity of fuel that can be produced by present refinery methods. Thus in recent years the military specifications for aircraft gas turbine fuels have been compromises of engine, fuel system, and availability requirements.

**G,6. Cited References.**

1. Childs, J. H. Preliminary correlation of efficiency of aircraft gas-turbine combustors for different operation conditions. *NACA Research Mem. E50F15*, 1950.
2. Olson, W. T., Childs, J. H., and Jonash, E. R. The combustion-efficiency problem of the turbojet at high altitude. *Trans. Am. Soc. Mech. Engrs. 77*, 605 (1955).
3. Barnett, H. C., and Hibbard, R. R. Properties of aircraft fuels. *NACA Tech. Note 3276*, 1956.
4. Dittrich, R. T. Effects of fuel-nozzle carbon deposition on combustion efficiency of single tubular-type, reverse-flow, turbojet combustor at simulated altitude conditions. *NACA Tech. Note 1618*, 1948.
5. Sharp, J. G. Fuels for gas-turbine aero-engines. *Aircraft Eng. 23*, 1–7 (1951).
6. Bass, E. L., Lubbock, I., and Williams, C. G. The gas turbine and its fuels. *Proc. Third World Petroleum Congress, Sec. VII*, 99–118 (1951).

7. Scott, M. D., Stansfield, R., and Tait, T. Fuels for aviation and industrial gas turbines. *J. Inst. Petroleum 37*, 487–509 (1951).
8. Silverstein, A. Research on aircraft propulsion systems. *J. Aeronaut. Sci. 16*, 197–222 (1949).
9. Childs, J. H., McCafferty, R. J., and Surine, O. W. Effect of combustor-inlet conditions in turbojet engines. *SAE Quart. Trans. 1*, 266–278, 344 (1947).
10. Gibbons, L. C., and Jonash, E. R. Effect of fuel properties on the performance of the turbine engine combustor. *Am. Soc. Mech. Engrs. Paper 48-A-104*, 1948.
11. Clark, T. P. Examination of smoke and carbon from turbojet combustors. *NACA Research Mem. E52I26*, 1952.
12. Berry, A. V. G., and Edgeworth-Hohnstone, R. Petroleum coke formation and properties. *Ind. Eng. Chem. 36*, 1140–1144 (1944).
13. Thomas, C. L. Petroleum coke and coking. *Progress in Petroleum Technology, Am. Chem. Soc.*, Aug. 1951.
14. Jonash, E. R., Wear, J. D., and Hibbard, R. R. Relations between fuel properties and combustion carbon deposition. *NACA Research Mem. E52B14*, 1952.
15. Starkman, E. S., Cattaneo, A. G., and McAllister, S. H. Carbon formation in gas turbine combustion chambers. *Ind. Eng. Chem. 43*, 2822–2826 (1951).
16. Buckland, B. O., and Berkey, D. C. Combustion system for burning Bunker C oil in a gas turbine. *Am. Soc. Mech. Engrs. Preprint 48-A-109*, 1948.
17. Schalla, R. L., and McDonald, G. E. Variation in smoking tendency among hydrocarbons of low molecular weights. *Ind. Eng. Chem. 45*, 1497 (1953).
18. Hunt, R. A., Jr. The relation of smoke point to molecular structure. *Ind. Eng. Chem. 45*, 602 (1953).
19. Swett, C. C., Jr. Spark ignition of flowing gases. I: Energies to ignite propane-air mixtures in pressure range of 2 to 4 inches mercury absolute. *NACA Research Mem. E9E17*, 1949.
20. Scull, W. E. Relation between inflammables and ignition sources in aircraft environments. *NACA Rept. 1019*, 1951.
21. Swett, C. C., Jr. Effect of gas stream parameters on the energy and power dissipated in a spark and on ignition. *Third Symposium on Combustion, Flame and Explosion Phenomena.* Williams & Wilkins, 1949.
22. DiPiazza, J. T., and Gerstein, M. Flammability limits of hydrocarbon air mixtures. *Ind. Eng. Chem. 43*, 2721–2725 (1951).
23. Coward, H. F., and Jones, G. W. Limits of flammability of gases and vapors. *Bur. of Mines Bull. 503*, 1952.
24. Anon. What properties should jet fuels have? *Aviation Week 54*, 26–43 (1951).
25. *CRC Handbook.* Coordinating Research Council, Inc., 1946.
26. Walker, J. E. Fuel systems for turbine-engined aircraft. *J. Roy. Aeronaut. Soc. 56*, 657–680 (1952).
27. Keusch, R. B. Effect of supersonic flight on power-plant installation system. *Trans. Am. Soc. Mech. Engrs. 77*, 721 (1955).
28. Murray, H. A. Aircraft gas-turbine fuels. *Oil and Gas Journal*, Mar. 1951.
29. Droegemueller, E. A. Aircraft turbine engine fuel requirements. *Proc. Third World Petroleum Congress, Sec. VII*, 87–98 (1951).
30. Resek, J. V. Commercial burner oil specifications. *Am. Soc. Mech. Engrs. Paper 51-A-125*, 1951.
31. Faust, F. H., and Kaufman, G. T. *Handbook of Oil Burning.* Oil-Heat Inst. America, 1951.
32. Williams, C. G. Fuels and lubricants for aero gas turbines. *J. Inst. Petroleum 33*, 267–306 (1947).
33. Barringer, C. M. Stability of jet fuels at high temperatures. *SAE Preprint 526*, 1955.
34. McKeown, A. B., and Hibbard, R. R. Effect of dissolved oxygen on the filterability of jet fuels for temperature between 300° to 400°F. *NACA Research Mem. E55I28*, 1955.
35. Weiss, S., and Pesman, G. J. Bibliography of unclassified aircraft-fire literature. *NACA Research Mem. E9H03*, 1949.

36. Pinkel, I. I., Preston, G. M., and Pesman, G. J. Mechanism of start and development of aircraft crash fires. *NACA Research Mem. E52F06*, 1952.
37. McLaughlin, E. J., and Bert, J. A. Kerosene or JP-4 for commercial jets. *Petroleum Refiner 32*, (*6*), 112 (1955).
38. Sweeney, W. J., Blackwood, A. J., and Guyer, W. R. F. Report on aviation gas turbine fuels presented at *Natl. Aircraft Propulsion Meeting, Inst. Aeronaut. Sci.*, Cleveland, Ohio, March 1947.
39. Anon. *Natl. Petroleum News 44*, (*12*), 16 (1952).
40. Lloyd, P., and Probert, R. R. The problem of burning residual oils in gas turbines. *Proc. Inst. Mech. Eng. 163*, 206–220 (1950).
41. Buckland, B. O., and Sanders, O. G. Modified residual fuel for gas turbines. *Trans. Am. Soc. Mech. Engrs. 77*, 1199 (1955).

# SECTION H

## *COMBUSTION CHAMBER DEVELOPMENT*

WALTER T. OLSON

**H,1. Introduction.** Designs for high speed combustion systems for aircraft have had to join what nature dictates that fuel, air, and flame must do on the one hand with demands imposed by flight and by the engine on the other hand. Lloyd (Sec. B) has indicated the requirements to be met and the difficulties to be surmounted. Development of combustion systems to meet these requirements has been an outstanding achievement in engineering.

As in other areas of engineering, combustor development has advanced with the help of both direct engineering experience and collateral information on the basic processes involved. The extensive information and insight on subjects such as fuel atomization and vaporization, mixing, flammability limits, ignition energy, flame stabilization, flame propagation, smoke formation, and flame quenching are important aids in the analysis and design of high speed combustors.

Vol. II of this series, as well as Sec. B through G of this volume, have presented background information on combustion science in general. They also describe some of the engineering principles needed in combustor design. This section describes the kinds of combustion system that have been useful for gas turbines, the performance trends usually observed, and some of the main considerations that enter the design art. The relations of certain combustion principles to the applied problems are stressed. It is freely admitted that the result is not a design handbook. Rather, this and the preceding sections give the reader most of the background in the field of turbojet combustion. The references cited, while not a complete bibliography, provide a useful introduction into the scientific literature on the subject.

**H,2. Historical Development.** Historically, the successful application of combustion to the aircraft gas turbine was a prerequisite that both the British and German investigators had to meet in their independent and nearly concurrent development of workable turbojet engines. The basic technique that has evolved has already been sketched in Sec. B in order to permit the subsequent discussions of the appropriate processes. This article attempts to describe some particular combustor designs that are representative of the engineering experience and know-how in the field.

Four alternative arrangements of combustor on an engine are shown in Fig. H,2a; these are discussed here briefly to provide familiarity with terms and concepts discussed later. In Fig. H,2a(1) and H,2a(2) several tubular or can-type combustors are symmetrically arranged around the engine between the compressor and the turbine and, in Fig. H,2a(3) an annular combustor occupies the same space. Also, individual flame tubes can be contained in an annular housing, as seen in Fig. H,2a(4).

The can-type combustor may be either a reverse-flow type (Fig. H,2a(1)) or a straight-through type (Fig. H,2a(2)). The tubular combustors are connected by small cross-fire tubes to balance pressure and to spread flame among the several combustors. As the arrows show in Fig. H,2a, air that enters the combustor proper passes through holes distributed in the walls of a flame tube, or liner, that is concentric within the outer casing of the combustor. Some air may be admitted through the upstream end of the liner. Ordinarily a hydrocarbon fuel such as gasoline or kerosene, as discussed in Sec. G on fuels, is sprayed as atomized liquid (atomizing combustor) in the upstream end of the liner. Alternatively, fuel may be externally vaporized and admitted, or may be vaporized inside the combustor prior to being mixed with air and burned (vaporizing combustors). The fuel is ignited and burns with the air in the upstream end of the combustor, that is, with the primary air. The desired distribution of air in the combustor is achieved by design of the air openings in the liner, or flame tube. The fuel-air ratios in the zone of combustion, the so-called primary zone, are obviously richer than the over-all operating fuel-air ratios, as Sec. B stated to be necessary. The low initial velocities, turbulence, and gas flow reversals in this zone insure that fresh charge is brought to ignition temperatures and provide time for the combustion to be initiated and to occur. As the burned or burning gases pass downstream, more air, that is, the secondary air, enters the flame tube and dilutes and cools the gases to temperatures tolerable to the turbine. This dilution zone, or secondary zone, is not sharply demarcated from the primary zone.

In general the main combustors of aircraft gas turbines have been designed with the foregoing concepts.

Afterburners, or reheat systems, achieve mixture preparation, flame stabilization, and flame spreading by a somewhat different technique. As Fig. H,2b illustrates, fuel is injected into the stream leaving the turbine; correct mixtures at the flame holder depend on matching the fuel injection with the gas flow pattern. Flame is anchored by and spreads from an array of gutters or flame holders; the flame-holding principles are discussed in Sec. E. As Sec. B has discussed the main features of afterburners, and as design details are under military secrecy as of this writing,[1]

[1] Jan. 1957.

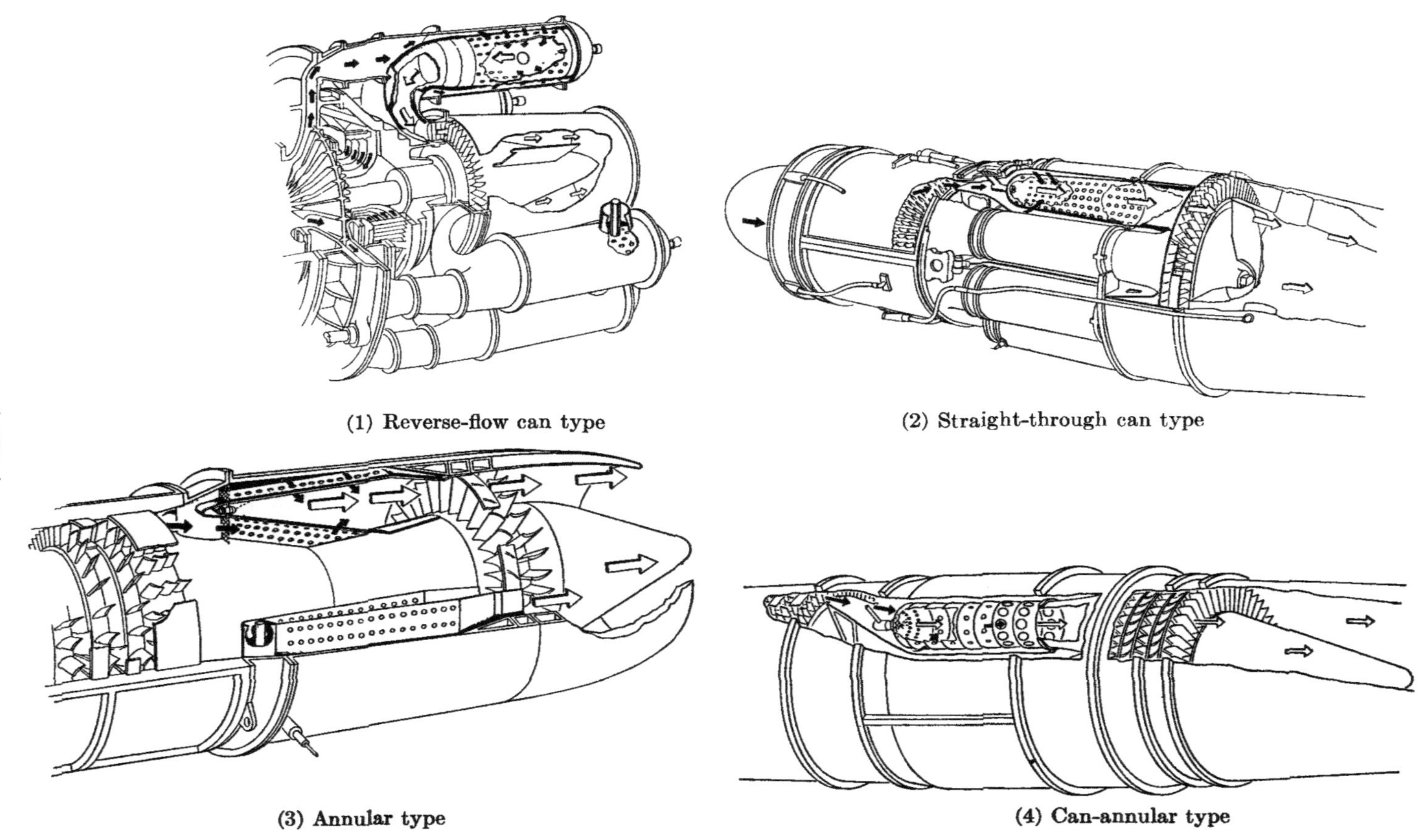


**Fig. H,2a.** Typical arrangements of combustors for aircraft gas turbine engines.

no further description of afterburner designs appears in this section. Only the main combustors for the aircraft gas turbine are discussed here.

The subsequent subdivisions are made only for convenience in presenting the material. Obviously there are many ways of accomplishing the necessary individual steps required of the combustor. Numerous combinations of the features to be described are possible and, in fact, have been tried.

*Fuel-atomizing, tubular combustors.* Combustion was found by Whittle [1] to be one of the major problems to be overcome in developing

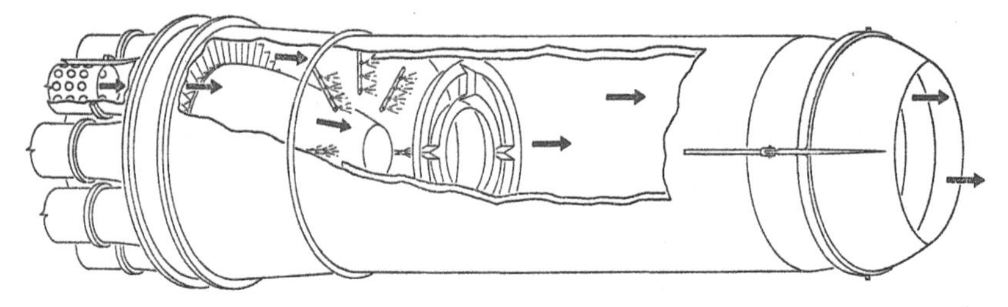

Fig. H,2b. Typical afterburner.

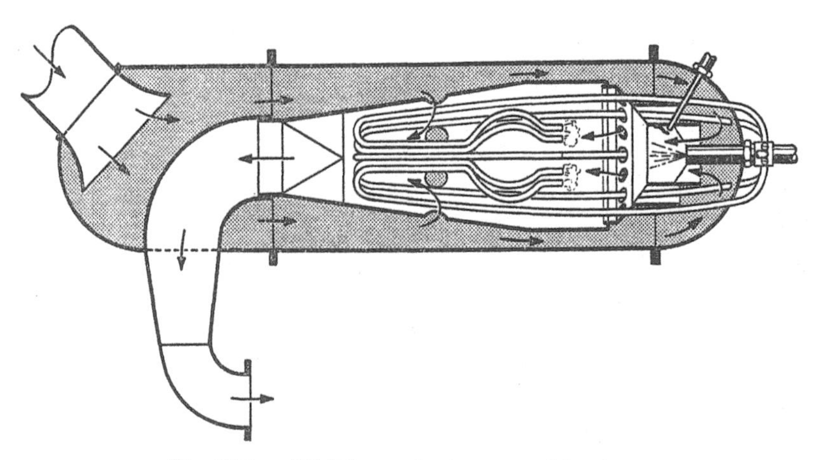

Fig. H,2c. Whittle combustor: vaporizing type.

an experimental turbojet engine. Although Whittle's very first bench engines used a single combustor, ten separate combustors were soon adopted to facilitate combustion testing. The reverse-flow arrangement was used to allow a short engine shaft. Whittle's early combustors involved attempts at combustion with low dispersion fuel sprays, and with fuel vaporizers, as shown in Fig. H,2c. A flame tube, or liner, divided the air flow in the combustor. These early attempts were not too successful. Whittle credits I. Lubbock of the Asiatic Petroleum Company with the innovation of the wide dispersion pressure-atomizing fuel nozzle and an improved flame tube to accompany it. In the new flame tube, most of the primary air was introduced to the combustion space through a large swirler located around the fuel nozzle, with the remainder of the air entering through short tubes that projected radially inward and through holes in the wall of the cylindrical flame tube. The back flow of hot gases caused by the air vortex created by the swirler served to anchor, or to stabilize the flame, and thus to make possible the high heat

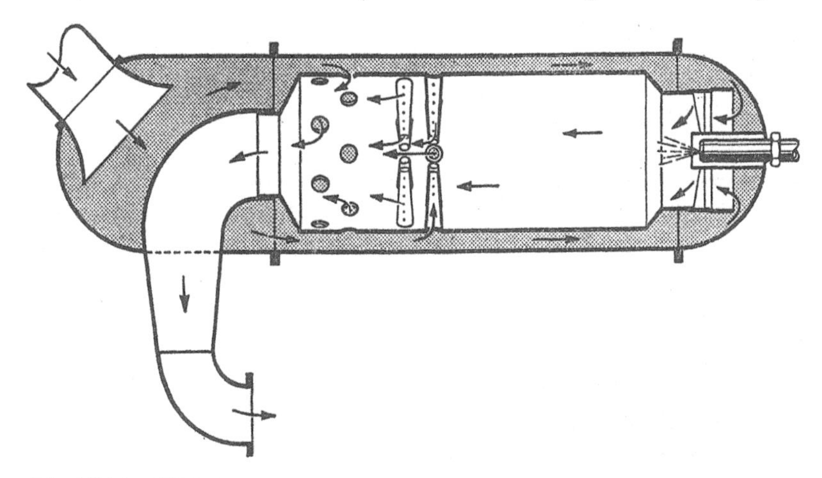

Fig. H,2d. Whittle combustor: atomizing type, as developed by Power Jets.

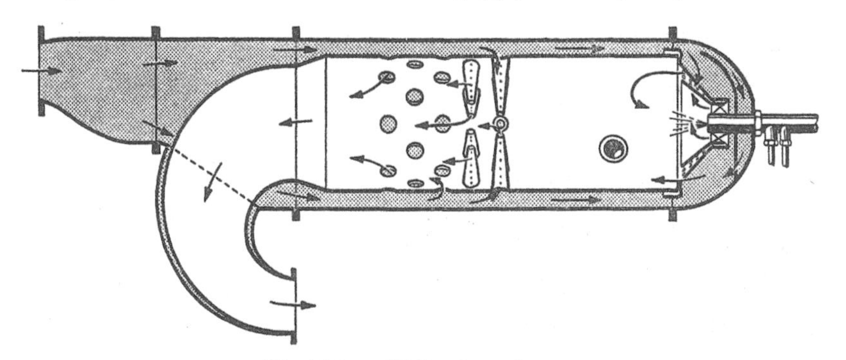

Fig. H,2e. Welland combustor.

release rates. After development, this combustor (Fig. H,2d) was incorporated in the Whittle W-1 engine that was built by Power Jets; that engine made the first British turbojet-powered flight in May 1941.

While the W-1 engine was being built, a larger Whittle engine, W-2, was started by Power Jets for production by the Rover Company. In 1940, Joseph Lucas, Limited, started production and design of the combustion and fuel systems for the Rover Company. The basic design evolved for the W-2 engine is shown in Fig. H,2e [*2*]. The swirler was much reduced in size, but still provided a flame-stabilizing vortex. Air entering the primary zone passed through a perforated dome, called a "colander," and intended to provide turbulence to produce high combustion rates. A shield inside the colander assisted in providing a quiescent zone in the upstream end of the flame tube to help stabilize combustion. The short tubes, or stub pipes, were retained in the liner to help mix the dilution air with the combustion gases. When Rolls Royce took over the Rover turbojet developments in early 1943, design features for the W-2 type engine from both Power Jets' version, W-2/500, and Rover's version, W2B/23, were incorporated in an engine which went on to production as the Welland, with the Lucas combustor of Fig. H,2e in it.

Straight-through combustor designs appeared in Great Britain on the Rolls Royce Derwent series, on the DeHavilland H-1, or Goblin, and on the Bristol Theseus turboprop. The combustor development work carried out by the Joseph Lucas Company for these engines and for their successors has evolved a type of combustor as shown in Fig. H,2f.

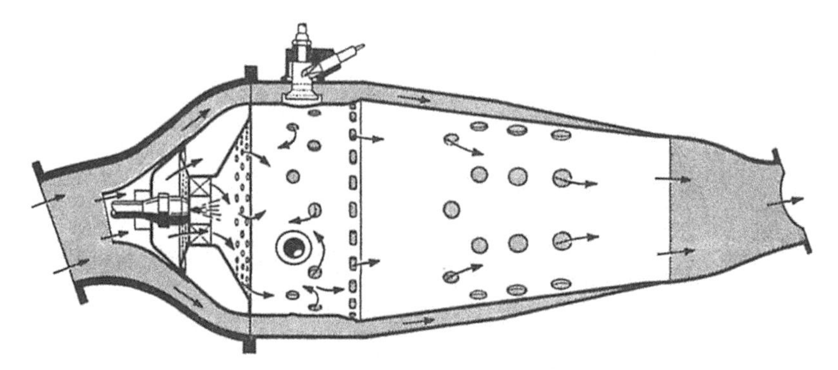

Fig. H,2f. Joseph Lucas Co. (England) combustor.

Part of the air entering the combustor from the compressor enters the upstream end of the flame tube through a snout whose size, location, and orientation in the inlet-air stream are designed to capture a desired fraction of the inlet air. Another orifice inside the snout assists in further metering the air that enters the snout. This air then flows through a punched plate into a calming zone to even the velocity profile, and thence into the combustion space both through a small set of swirl vanes around the fuel nozzle and through another punched plate, or colander. A ring of holes at the outside edge of the colander admits film-cooling air along the inside wall of the flame tube. Additional air for combustion at high fuel flow enters through holes in the wall of the upstream part of the liner. Cooling and dilution air enters through holes in a step, or shoulder, in the liner wall. Most of the dilution air is admitted through holes in the downstream part of the liner wall. Ignition is achieved by means of a separate supply of fuel and air provided at a spark plug during the starting cycle only. This torch ignition, as shown in Fig. H,2f, was later supplanted by a high energy spark for igniting the main fuel directly. Thorough discussions of the development of the Lucas combustion system may be found in [*2,3,4*]. Combustion systems of this general type have been incorporated in a number of important British engines. Among some of the early ones were Rolls Royce's Derwent, Dart, Nene, Avon; Bristol's Theseus, Proteus; DeHavilland's Goblin, Ghost; and Napier's Naiad, Eland. Lucas' combustor developments have had a strong influence on other British work in the field.

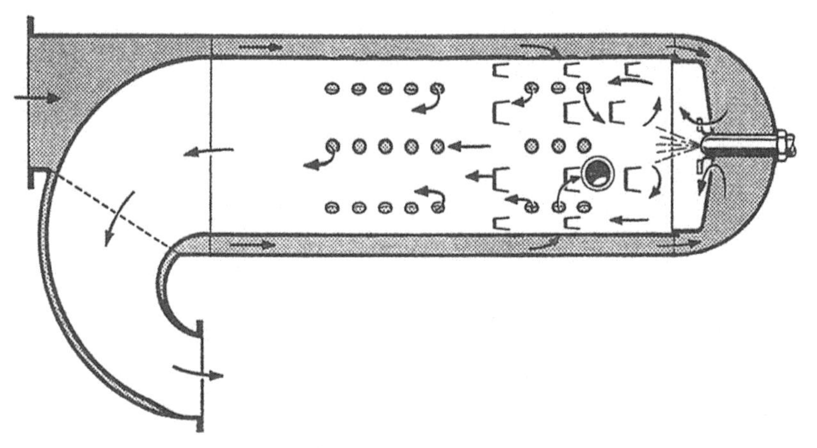

Fig. H,2g. J31 combustor.

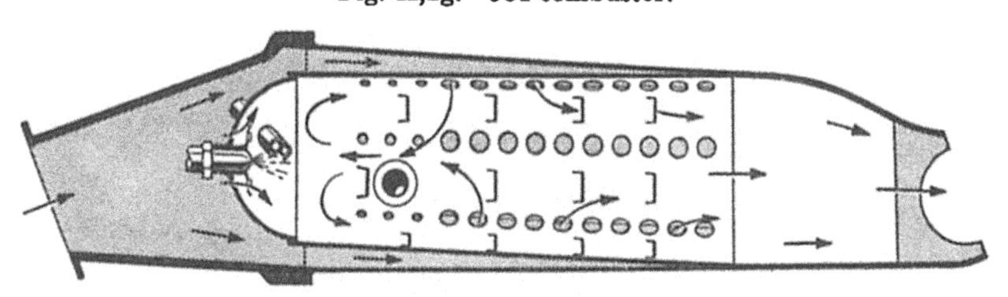

Fig. H,2h. J33 combustor.

In the United States, the reverse flow can-type combustor with atomizing nozzle appeared on the General Electric Company I-16, or J31 (Fig. H,2g), an engine largely copied from the Whittle W-2. General Electric independently adopted the straight-through version of the can combustor for their I-40, or J33, (Fig. H,2h [*5*]), and thereafter for subsequent engines such as J35 and J47.

An important variation of the tubular combustor comprises individual combustor liners of the tubular type contained in an annular housing. Examples are found on the Allison J71, General Electric J73 and J79, the Pratt and Whitney J57 and J75, and the Bristol Olympus engines. A sketch of the J57 combustor is presented in Fig. H,2i. It both illustrates the so-called "cannular" configuration and shows yet another type of tubular flame tube. Each of the eight cans has a short, perforated tube located down its central axis; in essence, then, each can is a small, annular combustor, fed by a ring of six fuel nozzles of the dual-flow type. Holes are provided in the can for primary and secondary air, while perforated steps in the can walls provide extensive film-cooling air. The individual cans are joined to a common annular duct by flaring transition pieces at their outlet.

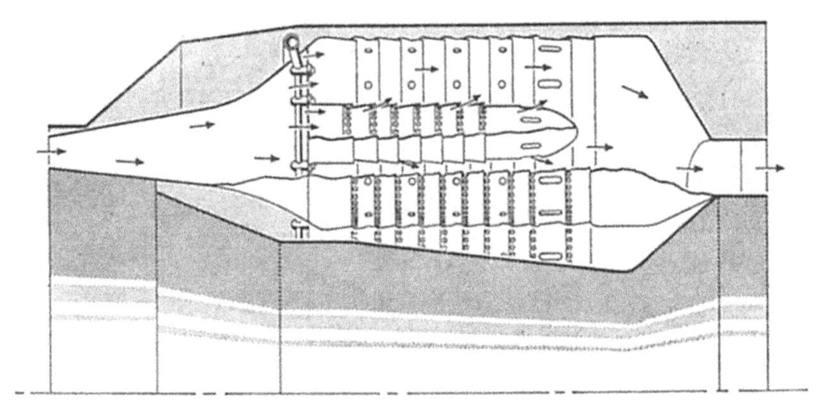

Fig. H,2i. J57 combustor.

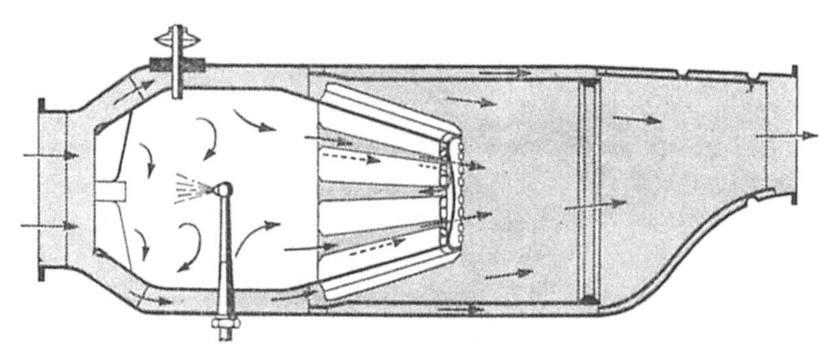

Fig. H,2j. Jumo 004 combustor.

Two other combustion systems should be mentioned. One of these is the combustor for the German Junker's Jumo 004 engine, the only German turbojet engine to reach mass production and extensive service use in World War II. This engine had six can-type combustors. One is illustrated in Fig. H,2j. Fuel similar to Diesel fuel is sprayed at injection pressures up to 750 psia upstream into the primary air that has entered the liner through six swirl vanes. The hot gases flow out of the combustion space between the stub pipes and mix with dilution air that has by-passed the combustion zone. A 4-in.-diameter baffle at the end of the stub pipes assists in diverting the hot gases to flow out between the stub pipes. The hollow pipes are cooled by dilution air flowing through them.

Another fuel-atomizing, can-type combustor is the so-called T-scheme [*6*] developed by the National Gas Turbine Establishment of Great Britain. In this combustor, fuel from a return-flow spray nozzle is sprayed upstream toward the apex of a cone-shaped baffle (Fig. H,2k). The spray angle is wide enough to clear the baffle. Some of the primary air enters through holes and slots in the baffle; the rest of the primary air enters through holes in the flame tube. Dilution air enters through holes in the downstream end of the flame tube.

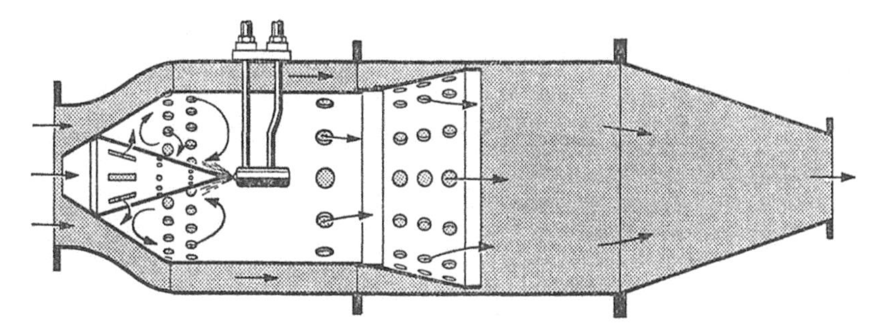

Fig. H,2k. NGTE T-Scheme combustor.

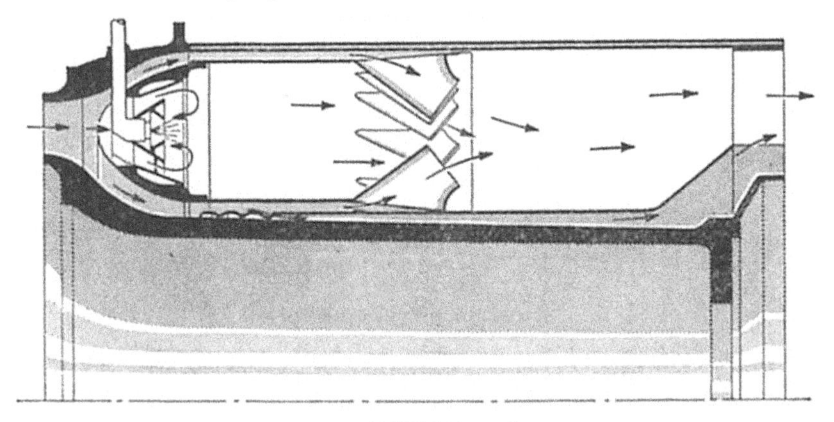

Fig. H,2l. BMW 003 combustor.

*Fuel-atomizing, annular combustors.* The annular combustor for atomized liquid fuel appeared on several of the early German engines. For example, the BMW 003, one of the two German turbojet engines to be developed to the production stage during World War II, used an annular combustor as shown in Fig. H,2l [*7*]; the combustor was developed with difficulty in the period from 1938 to 1943. There are 16 fuel nozzles around the annular liner. Each nozzle is surrounded by a baffle, and primary air flows in part through and in part around these baffles. The secondary, or dilution air enters through narrow scoops that lead the air into the hot combustion gases. The forty scoops on the outer liner wall alternate in circumferential position with forty scoops on the inner liner wall to form a so-called "sandwich" mixer.

Another example of an annular combustor developed by the Germans is the combustor for the Heinkel 011 engine. Development of this engine was under the direction of von Ohain, designer of the centrifugal engine HeS-3b that made the world's first turbojet-powered flight in August 1939 [*8*, pp. 377–379, 409]. The He 011 combustor is shown in Fig. H,2m [*9*]. Sixteen fuel injection nozzles are located uniformly in a ring at the upstream end of the annular combustor liner. Some of the air flowing past the nozzle ring is admitted as primary or combustion air through mixing fingers that extend into the combustion space just downstream of the

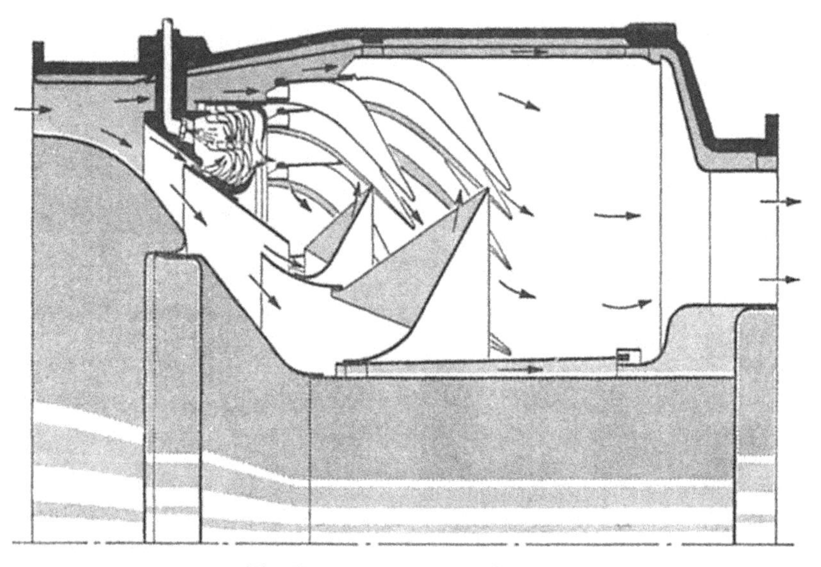

**Fig. H,2m. He 011 combustor.**

fuel sprays. The rest of the air enters the combustor through two sets of hollow vanes, or fingers, that extend into the combustion space from both the inside and outside walls of the annular liner; this air is mixed with the combustion gases as dilution, or secondary air. An annular shell between the combustor liner and the outer housing protects the outer housing from radiation.

The annular combustor was successfully used by the Westinghouse Company on one of the early turbojets of United States design, the 19A (later, J30). This type of combustor was developed and improved in later versions of Westinghouse engines by the combined efforts of the manufacturer and the NACA. Fig. H,2n shows the single annular liner, or "basket," as used in the J30 engine, and Fig. H,2o [*10*] the double annulus liner used in the J34 engine. Fuel is fed as hollow-cone sprays from a number of pressure-atomizing nozzles on a circular manifold at the upstream end of the basket. The air enters the combustion space through the many holes or slots shown in the inside wall and outside wall of the liner.

British use of the annular combustor appeared with Metropolitan-Vickers, Limited, in their early development of the F/2 engine, subsequently the Beryl. Primary air admitted to the combustion space is metered through holes in the upstream end plate of the combustor liner

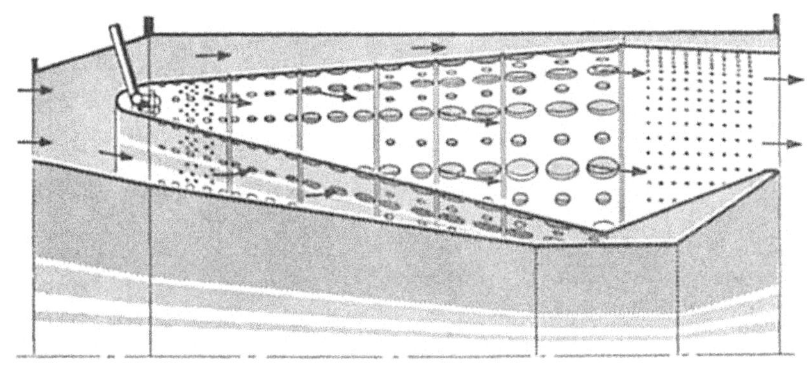

Fig. H,2n. J30 combustor.

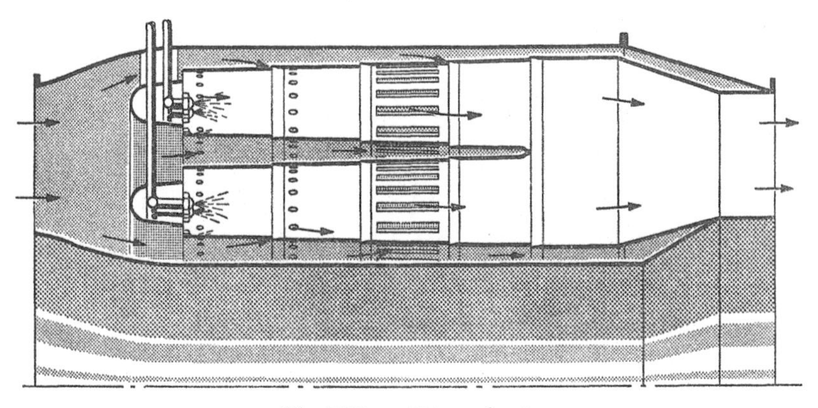

Fig. H,2o. J34 combustor.

(Fig. H,2p [*11*]). Fuel is sprayed upstream toward this perforated end plate from 20 atomizing-type nozzles. Dilution air is admitted through two rows of "sandwich" mixers as described for the German BMW 003 combustor.

Another British annular combustor development was an experimental unit based on the design principles evolved by the Joseph Lucas Company.

An unusual method of fuel injection was incorporated in a series of small gas turbine engines developed initially by Société Anonyme Turboméca of France, and subsequently, under license by Continental Aviation in the United States. These engines have been designed as air generators, shaft power producers, pure turbojets, and ducted fans. The Piméné turbojet engine, typical of the class, was rated at 220 pounds thrust with a specific fuel consumption of 1.10 pound per hour-pound thrust for take-off conditions. In these Turboméca engines, fuel is pumped and metered to the hollow compressor-turbine shaft, whence it flows outward

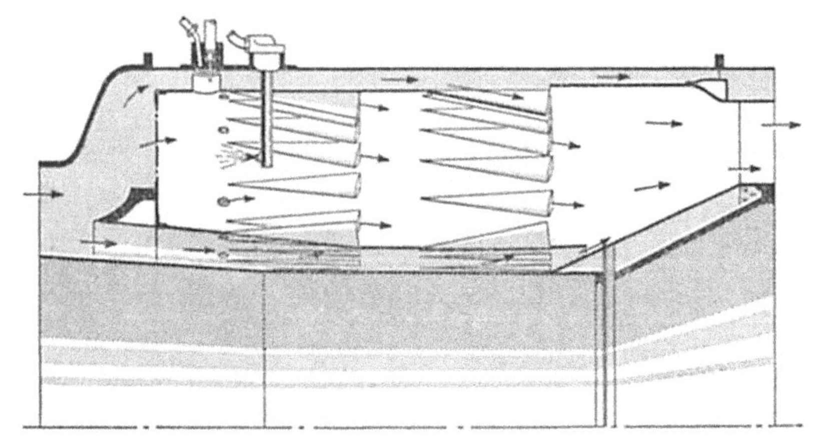

Fig. H,2p. Metrovick F/2 combustor.

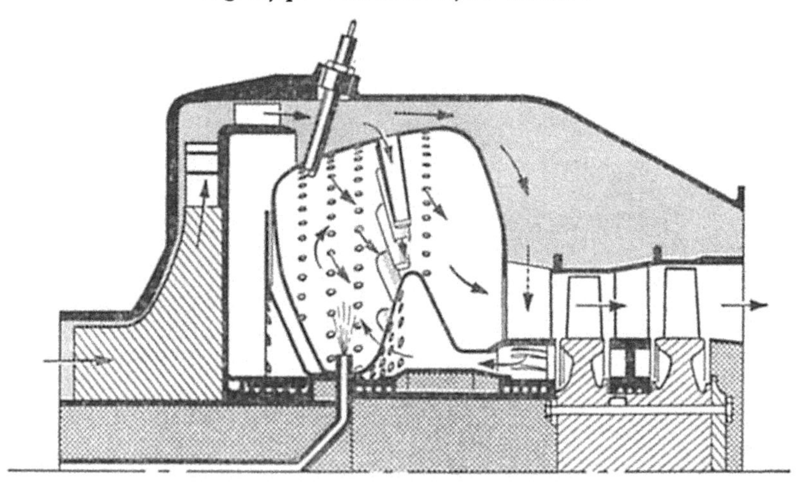

Fig. H,2q. Turboméca Piméné combustor.

in the fuel delivery ring attached to the shaft (Fig. H,2q). Centrifugal force sprays the fuel from the small holes in the periphery of the rotating fuel delivery ring into the combustion space. It may be estimated that for the fuel flows, fuel delivery ring sizes, and rotational speeds used at full power, drop sizes of the order of 100 microns are produced. A high pressure fuel pump is not needed. Most of the air flows directly through holes and tubes in the outside wall of the annular combustor liner. A part of the air flow is diverted around the combustor liner, through the hollow stator blades which are cooled thereby, and through holes in the inside wall of the annular liner into the primary zone. A labyrinth fed with compressed air seals the combustion space at the shaft.

*Vaporizing combustors.* Early attempts by Whittle in England [*1*] and by NACA in the United States [*12,13*] to vaporize fuel prior to injection by means of fuel tubes in the upstream end of the combustor (Fig. H,2c) met with varying success. The main problems were slugs of liquid fuel in the injectors, sudden increases in fuel flow, and coking in the fuel tubes at low fuel flows.

Armstrong-Siddeley successfully developed vaporizing-type combustors in which fuel is mixed and vaporized in air in the primary zone without resort to high pressure fuel atomization. In the Python engine, each of the eleven can-type combustors comprises a flame tube of 6-inch diameter and $16\frac{1}{2}$ in. long in a $7\frac{3}{4}$-in.-diameter outer casing (Fig. H,2r). Some air is admitted to the inside of the flame tube through a hole in its

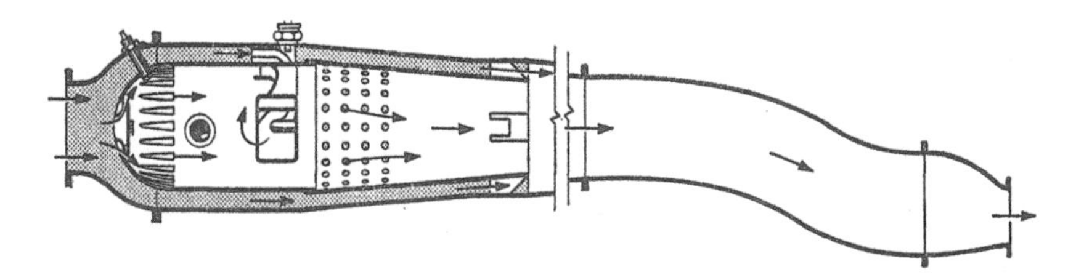

Fig. H,2r. Armstrong-Siddeley Python combustor.

dome and is directed along the inside wall of the dome and flame tube by a baffle located just inside the hole. More air enters flutes around the edge of the dome. A scoop that extends into the annulus between the outer casing and the flame tube carries primary air into a $3\frac{1}{2}$-in.-diameter cup located centrally in the flame tube. The cup is open on its upstream end, and the primary air enters the side of the cup tangentially, swirls, and spills out into the primary, or combustion zone. The fuel is fed at low injection pressure through a single tube having four equally spaced, radial holes, $\frac{3}{32}$-in. in diameter. This injector is located in the primary air scoop, and the fuel impinges on the walls of the scoop. An auxiliary injector for ignition only is located in the primary zone and upstream of the cup in one or two of the combustors. Holes in the flame tube add dilution air downstream. The very long, about 50-inch, combustion chamber is possible because the engine arrangement is such that air enters in the middle of the engine, turns 180°, flows in an axial compressor toward the front of the engine, and turns 180° again to pass through the combustor.

The same vaporizing principle has been applied to the Mamba combustor, Fig. H,2s [14]. The fuel injector is a four-pronged fork, each prong being a hollow tube for solid stream injection and arranged so that the tube discharges lie equally spaced on a circle about $1\frac{1}{2}$ in. in diameter. The prongs discharge into the upstream end of four similarly spaced tubes approximately $\frac{1}{2}$ in. in diameter that project into the combustion space. These tubes are 8 inches long and are attached at the upstream end of the flame tube. The tubes are shaped like a walking stick, or "candy cane," turning back upstream, and are arranged in such a way that they discharge on a circle about 3 inches in diameter into what becomes the primary zone. Primary air enters the tube and the liner dome, and the fuel is partially vaporized by the time it emerges from the

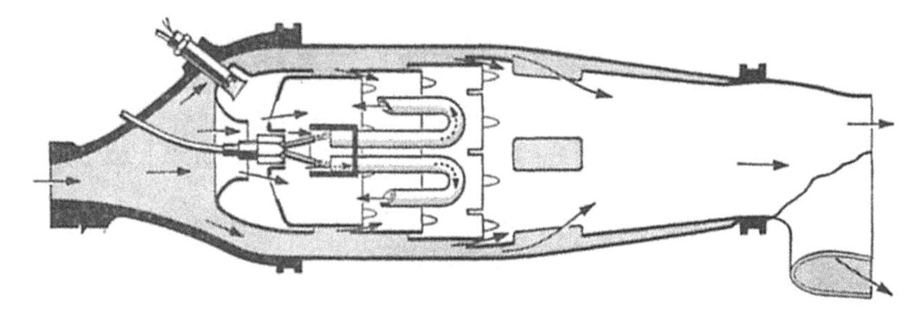

Fig. H,2s. Armstrong-Siddeley Mamba combustor.

tubes. Secondary air is introduced downstream in the combustor through holes in the liner wall. The type of vaporizing system developed for the Mamba was carried on into annular configurations by Armstrong-Siddeley in the Sapphire, in the Curtiss-Wright version of the Sapphire, J65, and by Westinghouse in the J46 and one model of the J40.

**H,3. Combustor Performance.** Extensive studies on combustor performance have been performed in flight tests and in tests in altitude rigs by numerous organizations and with many different combustors. The following discussion is intended to illustrate the effects of operating conditions on combustor performance. These effects both indicate the problems that must be solved by design and help to provide useful insight for their solution.

*Temperature rise and combustion efficiency.* Fig. H,3a from [15] indicates the performance characteristics of a turbojet combustor. The data of Fig. H,3a are for an early design of an annular combustor, but the general trends and phenomena are typical of other combustors. The data were obtained by operating the combustor in an altitude test rig that permitted independent control of the inlet-air pressure, temperature, and velocity. Methods used in making the temperature measurements are discussed in [16]. Fig. H,3a shows the measured values of the mean temperature rise through the combustor as a function of fuel-air ratio.

Curves that are included for 60, 80, and 100 per cent of the theoretical temperature rise permit interpolation of combustion efficiency.

Fig. H,3a(1) shows the effect of varying inlet-air pressure while maintaining constant inlet-air temperature and velocity. Also, the nature of the combustion is described on the figure with letter designations. A regular decrease in combustion efficiency with decreasing pressure is noted. Further, a maximum obtainable temperature rise was encountered at reduced pressures; increased fuel-air ratios beyond values giving this maximum temperature rise resulted in unsteady combustion and blowout. This so-called rich-limit blowout is ascribed to over-enrichment of the primary, or flame-stabilizing zone. The maximum obtainable temperature rise decreased as pressure decreased. At very low fuel-air ratios, and especially at low pressures, fuel flows were so low that the spray nozzles did not function well, and a marked loss of combustion efficiency resulted, even occasioned by lean-limit blowout due to an inadequate quantity of combustible mixture in the primary zone.

Similar data presenting the effect of inlet temperature on combustor performance are presented in Fig. H,3a(2). Comparison of Fig. H,3a(1) and H,3a(2) shows that a decrease in the inlet temperature produced the same general effects on performance as a decrease in inlet pressure. The broken-line curves in Fig. H,3a(2) represent unstable operation of the combustor when fuel flow was gradually increased from low values; rapid changes in fuel-flow rate caused the combustor to operate as indicated by the solid curves.

The effect of varying the reference velocity, the velocity based on inlet-air density, weight flow, and maximum cross-sectional area of the combustor, is shown in Fig. H,3a(3). An increase in reference velocity had the same general effects on performance as a decrease in inlet pressure or temperature.

Clearly, the deterioration of combustor performance with decreased inlet-air pressure and temperature and with increased velocity as reflected by reduced combustion efficiency and temperature-rise limits occurs because the processes converting chemical energy in the fuel to heat are too slow. With proper fuel-air ratios and sufficient residence time, complete combustion would occur; combustion efficiency would approach 100 per cent.

The conversion processes may be considered to be atomization of the fuel, vaporization of the fuel, mixing of fuel and air, ignition, and oxidation of the fuel to the final products of combustion. The combustion can be visualized as a competition between these conversion processes and the quenching that occurs when the reacting mixture is diluted with cold air and when the burning mixture contracts the relatively cool walls of the combustor liner. Reduced inlet-air pressure is adverse to combustion because: (1) it reduces the degree of fuel atomization thus increasing the time required for evaporation and combustion of fuel drops, (2) it decreases the turbulent mixing forces with respect to the viscous forces, (3) it increases ignition energy and ignition temperature requirements, (4) it enhances flame quenching, and (5) it decreases fuel-air reaction rates. Decreased inlet-air temperature is adverse to combustion because (1) it decreases fuel evaporation rate, (2) it increases the time required to achieve ignition, (3) it slows down combustion reactions, and (4) slows down flame propagation rates. Increased inlet-air velocity is adverse to combustion primarily because of reduced residence time for all conversion processes. Also, although velocity is beneficial in providing turbulent mixing, too rapid mixing of cold mixture with burning mixture may promote flame quenching.

If the rate of any one conversion process is substantially less than the rates of the others, then this one process will govern the over-all rate and hence will determine the combustion efficiency. It should then be possible to analyze the trends of combustion efficiency in terms of a simple model based on the rate-controlling step. Childs [*17*] found that the

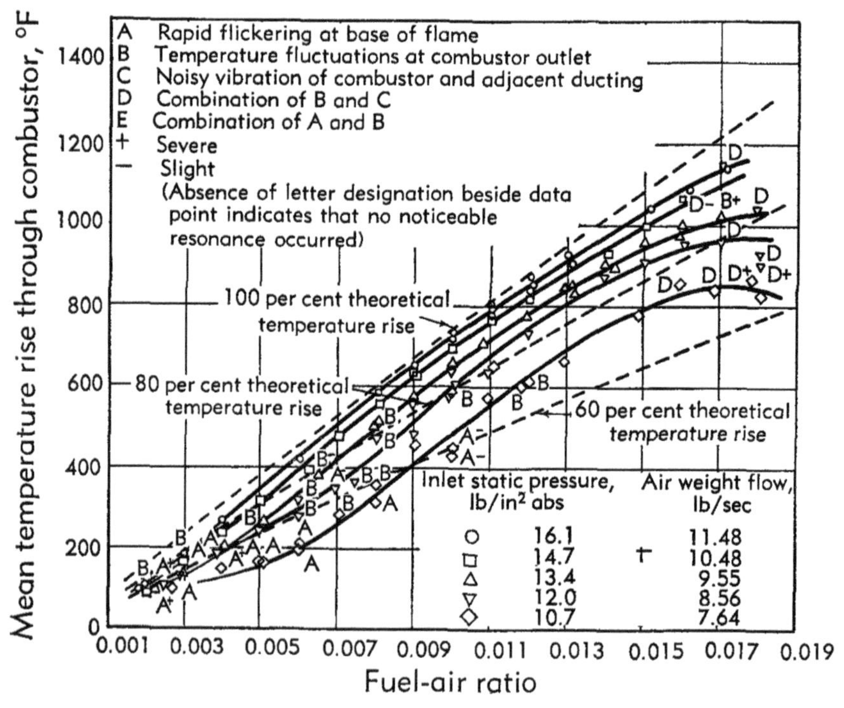


1. Effect of altering combustor-inlet static pressure. Inlet temperature = 65°F; reference velocity = 85.1 feet per second.

Fig. H,3a. Variation of mean temperature rise through combustor with fuel-air ratio for combustor-inlet conditions independently altered from values typical of those encountered in altitude operation of a gas turbine engine.

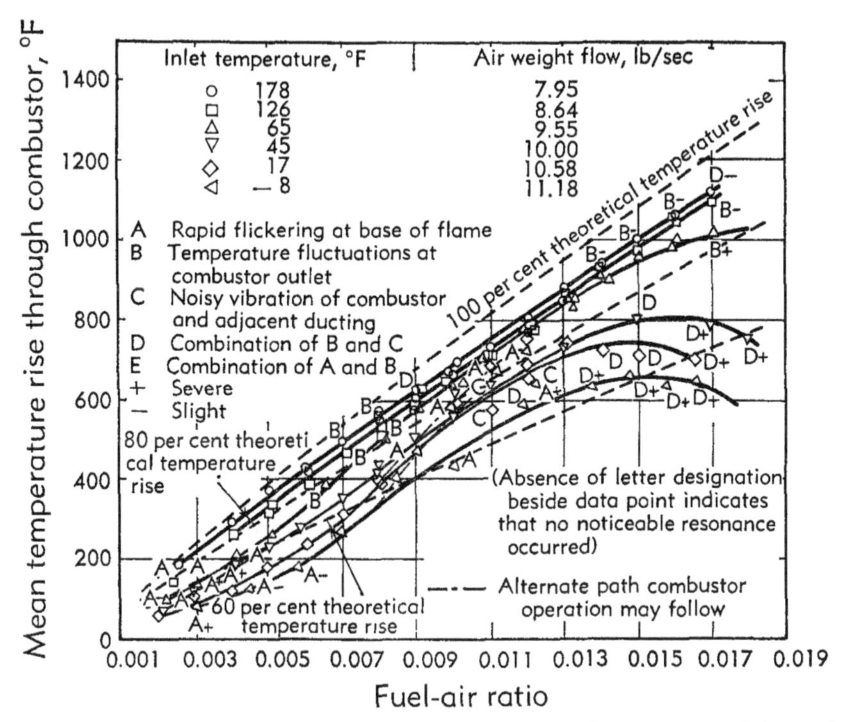


2. Effect of altering combustor-inlet temperature. Inlet static pressure = 13.4 pounds per square inch absolute; reference velocity = 85.4 feet per second.

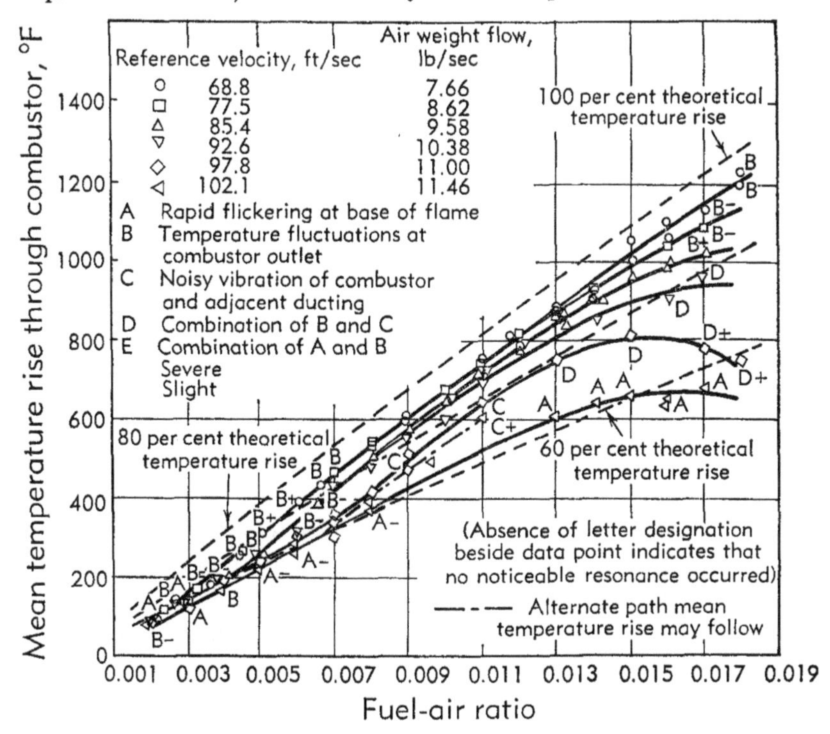


3. Effect of altering combustor reference velocity. Inlet temperature = 68°F; inlet static pressure = 13.4 pounds per square inch absolute.

change of combustion efficiency with inlet variables could quite often be correlated as

$$\eta_b = \varphi\left(\frac{p_i^0 T_i}{V_r}\right), \quad \text{or} \quad \varphi\left(\frac{(p_i^0)^2}{w_a}\right)$$

where $p_i^0$ is the inlet total pressure, $T_i$ the inlet temperature, and $V_r$ is the volume flow rate of inlet air divided by the maximum cross-sectional area of the combustor (e.g. Fig. H,3b). He then showed that this relation could be derived by assuming that a chemical process of second order was rate-controlling.

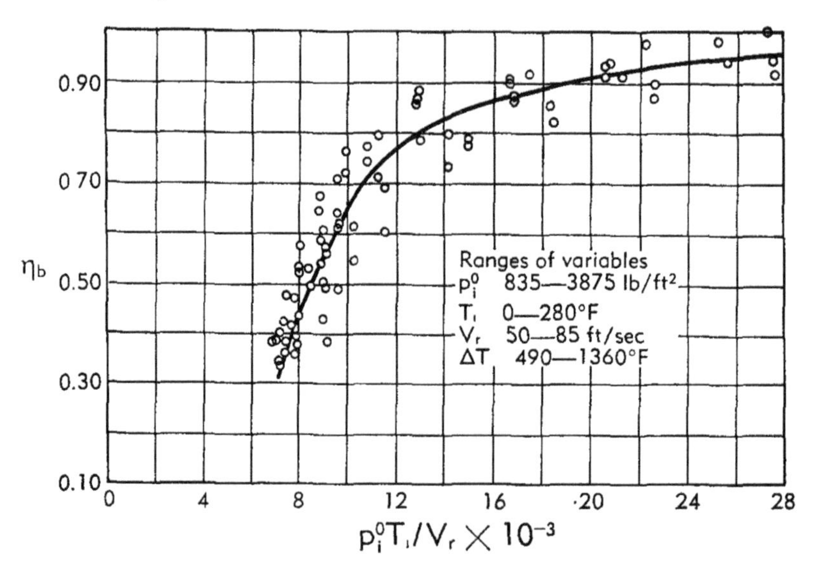


Fig. H,3b. Correlation of combustion efficiency with inlet-air conditions for an annular turbojet combustor.

Because detailed point-by-point sampling of gases in a turbojet combustor revealed considerable lack of homogeneity, the assumption was made that the reaction occurs in local, stoichiometric fuel-air ratio zones; rapid mixing to these zones is assumed. Similar correlation of efficiency with inlet parameters results if a homogeneous mixture is assumed, as Sec. B has discussed. Whether combustion efficiency should vary with $T_i$ as an exponent, or not, should be tested by experiment.

It is obviously important to the designer to know what process is limiting heat release rate and efficiency. It is exceedingly important to scaling and similitude studies to know the pressure exponent (see Way [*18*, pp. 296–327]). It is of equal importance to note, however, that such a convenient parameter does not apply to all combustor data. There have been some data that indicate a shift from one rate-controlling process to another as operating conditions vary through wide ranges. In one combustor, a parameter

$$\frac{(p_{\mathrm{i}}^{0})^{\frac{4}{3}}T_{\mathrm{i}}^{1.1}}{V_{\mathrm{r}}}, \quad \text{or} \quad \frac{(p_{\mathrm{i}}^{0})^{1.3}T_{\mathrm{i}}^{0.1}}{w_{\mathrm{a}}}$$

correlates data obtained at values of $p_{\mathrm{i}}^{0}$ and $w_{\mathrm{a}}$ higher than those where $\eta_{\mathrm{b}} = \varphi(p_{\mathrm{i}}^{0}T_{\mathrm{i}}/V_{\mathrm{r}})$ [*19,20*]. This parameter can be derived by assuming that the rate of flame spreading across flame surfaces of random size and shape establishes the over-all conversion rate. Also, neither parameter is adequate for correlating the data for some combustors. The evidence is that fuel vaporization and fuel-air mixing processes often play a dominant role. Grouping of inlet conditions to correlate with efficiency has been attempted by assuming fuel vaporization, or fuel-air turbulent mixing, or fuel droplet burning [*21*] to be controlling; the equations derived did not correlate experimental data, however.

The foregoing discussion, then, makes the point that the limit to conversion by combustion is probably chemistry for many combustors operating at low pressure; heat release rate varies with the square of the inlet-air pressure. In [*22*], however, it is shown that when the complete combustor volume is considered, heat release rates are only $\frac{1}{20}$ to $\frac{1}{100}$ of the maximums obtained in flames or highly stirred reactors. Much of the combustor volume is used for mixing processes of various kinds.

The effect of inlet-air pressure, temperature, and velocity on combustion efficiency may be translated into terms of flight altitude and engine speed. It is apparent that combustion efficiencies should be a maximum at low altitude and high engine speed where combustor inlet pressure and temperature are a maximum, and should decrease as altitude is decreased or as engine speed is decreased. The variation of combustion efficiency with altitude and engine speed for an engine with compressor pressure ratio of 4 is shown in Fig. H,3c. These data are for an early turbojet; they are considerably outdated. Combustor performance is vastly improved, with high efficiencies up to quite high flight altitudes. Nevertheless these performance trends are rather typical of other combustors, both tubular and annular, with both vapor injection and liquid injection, and with different kinds of fuel nozzles.

*Altitude operating limits.* As altitude is increased at a fixed engine speed, the appearance of the combustion changes regularly from a steady yellow flame, to a white flame tinged with blue to a blue flame tinged with white and flickering at rapid frequency, to a darker blue flame that flickers at lower frequencies and with greater amplitudes, and finally to a pulsating blue flame that may suddenly and unexpectedly become extinguished. Similar observations are made as engine speed is regularly reduced at a high altitude. These regular changes in the appearance of combustion are, of course, accompanied by a regular decrease in combustion efficiency until the altitude limit for combustion is reached. Quantitative significance is not attached to the color of the flames.

The altitude operational limit, or ceiling, imposed by the maximum obtainable temperature rise of the combustor is also shown in Fig. H,3c. The inlet-air conditions to the combustor determine whether or not

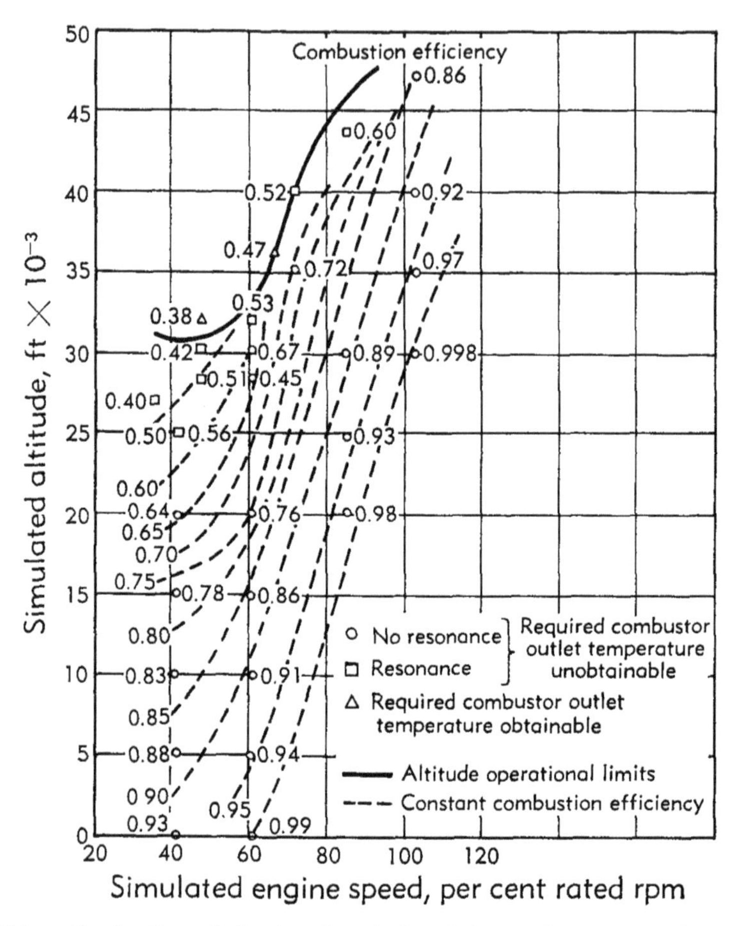


Fig. H,3c. Combustion efficiencies of typical turbojet combustor at various simulated flight conditions. Compressor pressure ratio = 4; flight speed = 0; fuel, aviation gasoline.

temperature rise sufficient to run the engine can be attained. Fig. H,3d, a hypothetical case, might represent the situation for an engine. As shown in the figure, a combustor outlet temperature of 1020°F is required for steady state operation of this engine at 90 per cent rated speed at altitudes above 35,400 feet. With combustor inlet conditions corresponding to 40,000 feet, a fuel-air ratio of 0.014 will supply the required temperature rise. At 47,000 feet, inlet-air pressures are lower and the combustor can supply just enough temperature rise to operate the engine, but no excess of temperature rise is available. This point would be on the curve of altitude limits against engine speed. At simulated combustor inlet conditions of 50,000 feet inlet-air pressure is still lower, and the combustor cannot supply the temperature required to run the engine. The engine would be inoperable at this altitude and rotational speed.

A further illustration of how combustor inlet conditions impose the altitude limits is in Fig. H,3e, which shows the combustor inlet conditions

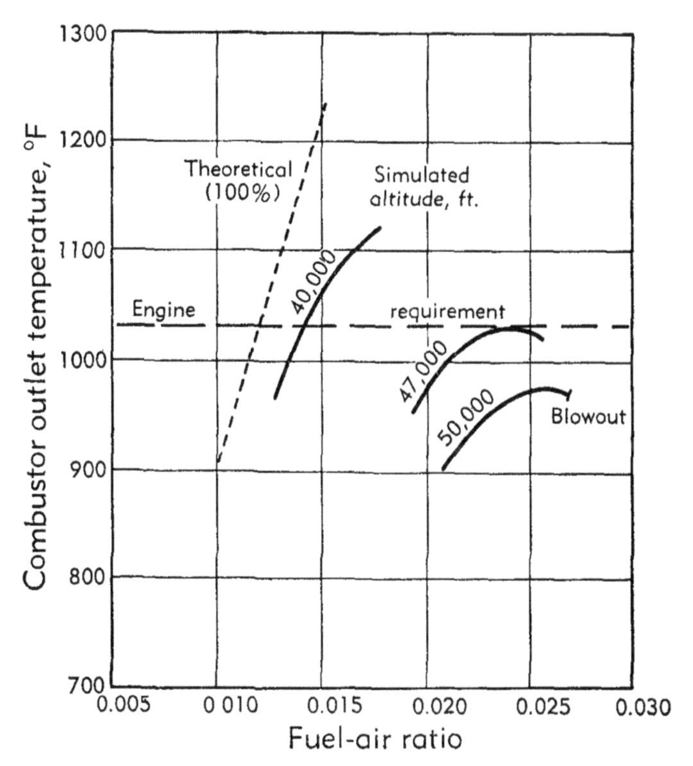


**Fig. H,3d. Temperature rise for hypothetical combustor at various altitudes, 90 per cent rated engine speed.**

for the different engine speeds at 32,000 feet (cf. Fig. H,3c). At engine speeds below 32 per cent inlet-air pressures and temperatures are low; but inlet-air velocities are sufficiently low to counteract the adverse pressure and temperature, and more than enough temperature rise can be attained with the combustor to run the engine. As simulated engine speed is increased through 32 per cent of rated rpm, although both combustor inlet-air pressure and temperature increase, inlet-air velocity increases so markedly as to offset the beneficial effect of increased pressures and temperatures, and the combustor either cannot burn at all or, even if it can burn, cannot produce the temperature rise required to operate the engine. An actual engine with the combustor performance illustrated in Fig. H,3c, and H,3e could not operate at 32,000 feet between 32 and 55 per cent engine speed. As the simulated engine speed is increased through the nonoperational region, the inlet-air pressure and temperature increase at accelerating rates, whereas the inlet-air velocity increases at a decelerating rate. At a simulated speed of about 40 or 50 per cent rated rpm, the favorable effects of inlet pressure and temperature become large enough to offset the adverse effect of the inlet velocity, and the temperature rise

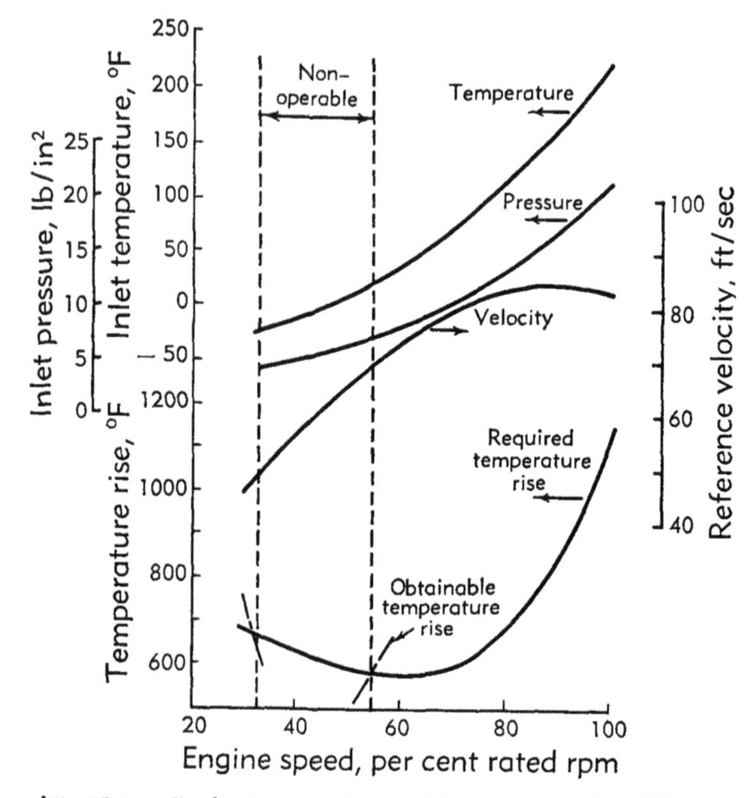


Fig. H,3e. Combustor operating conditions at 32,000 feet altitude.

attainable begins to increase with simulated engine speed. At 55 per cent rated speed and greater, the temperature rise attainable is equal to or greater than the temperature rise required, and the combustor is in the satisfactory operating range.

Flight and altitude wind tunnel tests of numerous turbojet engines have established the existence and nature of the altitude operational limit, or ceiling, imposed by combustion, as just discussed [*23*].

*Acceleration and ignition.* The same factors that impose decreased efficiencies and an altitude ceiling on the combustor also add to the difficulty of acceleration and ignition for start-ups at high altitudes.

Fig. H,3f shows the same altitude operating limits of Fig. H,3c, and a curve is also shown where a combustor temperature rise that is 100°F above the value required for nonaccelerating engine operation can be obtained. Energy corresponding to at least a 100°F rise in temperature is available only for altitudes and engine speeds below this curve. Similar

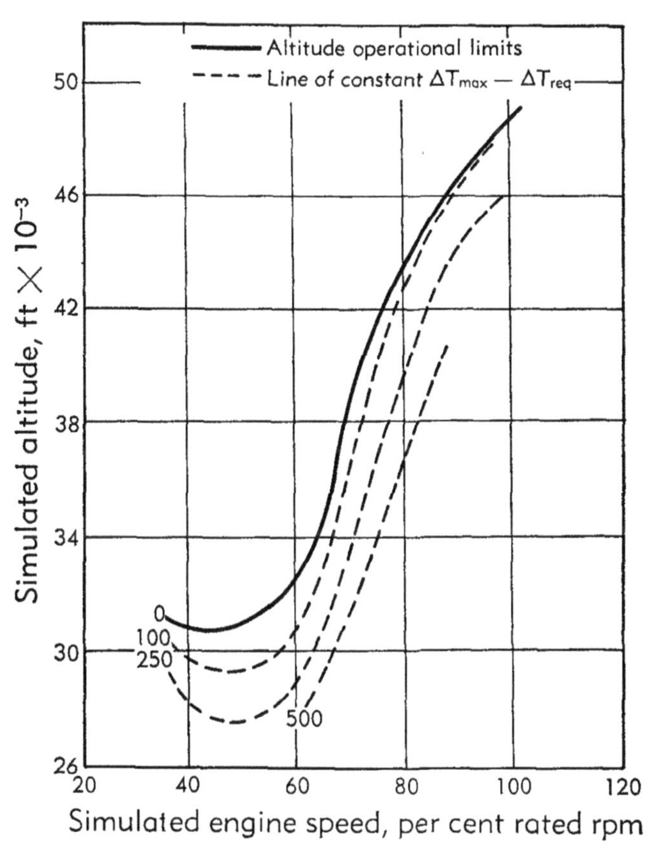


Fig. H,3f. Available acceleration of turbojet engine as indicated by difference between maximum temperature rise attainable from combustor and temperature rise required for nonaccelerating engine operation. Compressor pressure ratio = 4; flight speed = 0; fuel, aviation gasoline.

curves for 250°F and 500°F temperature rise available for acceleration are shown. These data are simply illustrative because, with an actual engine, engine speed would not remain fixed as the throttle is opened; whereas in the combustor research for Fig. H,3f, the simulated engine speed was unchanged while establishing each temperature rise limit. Attempts at rapid acceleration or deceleration of turbine engines at altitude can result in combustor blowout, because the lean or rich limit is momentarily exceeded. Compressor stall will greatly aggravate the problem. The rate of change of the fuel flow must be controlled in an attempt to avoid this problem.

Fig. H,3g for three turbojets illustrates the ignition problem encountered at altitude; ignition occurs only at flight speeds and altitudes below the curves. Lloyd and Mullins [*18*, pp. 405–425] have also discussed and illustrated the problem as shown in Fig. H,3h. These limits are the result of the inlet-air conditions to the combustor becoming adverse for ignition and combustion phenomena.

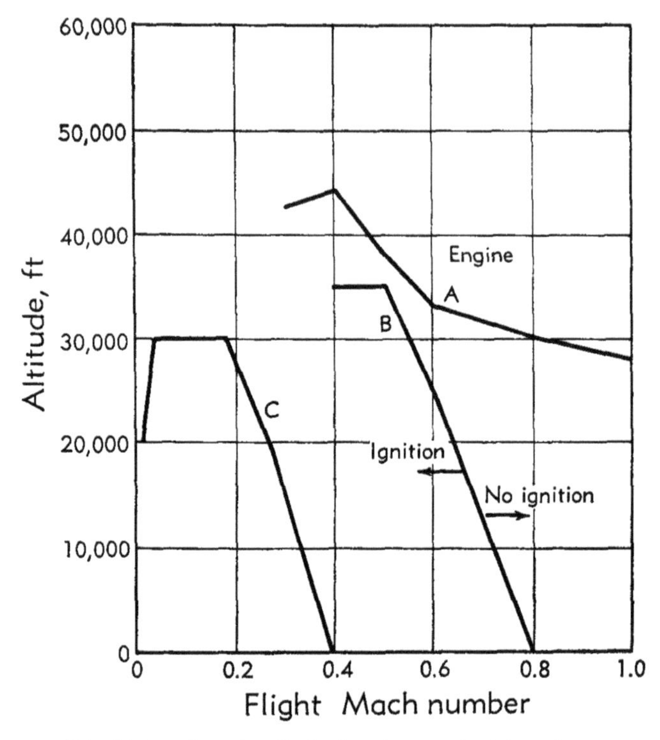


Fig. H,3g. Ignition limits for three turbojet engines.

In [*22*], it is noted that minimum ignition energies required to ignite turbojet combustors are many times those required to ignite ideal mixtures even in turbulent flow. This large difference may be attributed to the necessity for the spark to evaporate some fuel locally to provide a combustible mixture. Also, research has shown that the fuel and air in a combustor are seldom well mixed near the spark. The spark must ignite either a large pocket of mixture of correct composition, or several successive pockets, or slugs, of mixture, before flame will propagate through the combustor. High ignition energy (several joules) and detailed attention to location are the important design factors.

*Combustor pressure loss.* Two factors contribute to the loss of total pressure by the gases flowing through the combustor. One is the pressure change associated with acceleration of the fluid from its inlet velocity

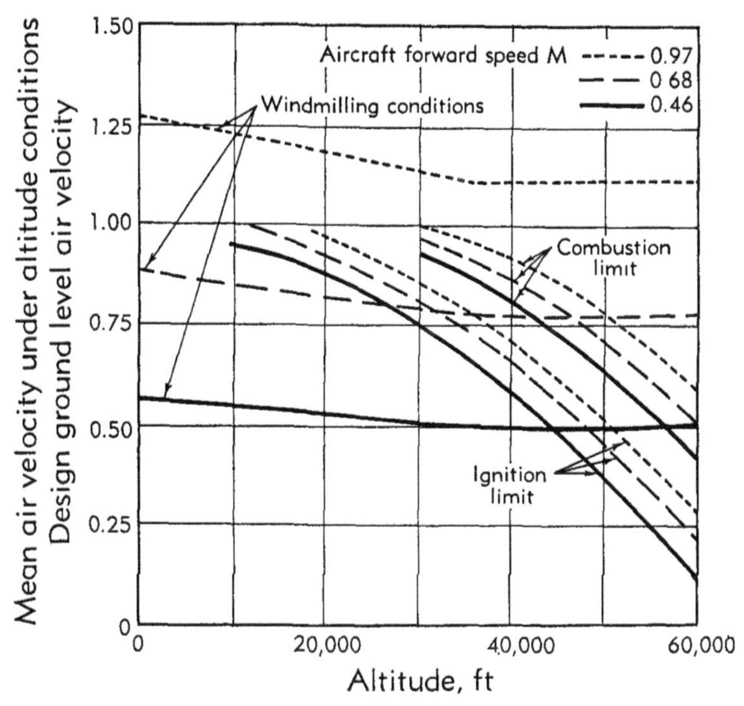


Fig. H,3h. Ignition and combustion limits for a combustion chamber; windmilling conditions corresponding to three flight Mach numbers [*18*, pp. 405–425].

to the outlet velocity. The other is the pressure drop due to aerodynamic drag, which includes friction in the passages and mixing losses of the air jets. From the simple momentum equation, assuming $A_3 = A_4$,

$$\frac{p_3^0 - p_4^0}{q_3} = \left(\frac{w_a + w_f}{w_a}\right)^2 \frac{\rho_3}{\rho_4} - 1 \tag{6-1}$$

while drag losses may be expressed as

$$p_3^0 - p_4^0 = f\frac{\rho_3 V_3^2}{2} \tag{6-2}$$

Thus the combined loss may be represented as

$$\frac{p_3^0 - p_4^0}{q_3} = \left(\frac{w_a + w_f}{w_a}\right)^2 \frac{\rho_3}{\rho_4} + f - 1 \tag{6-3}$$

where

$p^0$ = total pressure, lb/ft²
$w$ = weight flow, lb/sec
$\rho$ = density, slugs/ft³
$g$ = conversion factor, 32.2
$V$ = velocity, ft/sec
$q = \frac{1}{2}\rho V^2$
$f$ = const
$A$ = area, ft²

Subscripts:

a = air
f = fuel
3 = combustor inlet
4 = combustor outlet

Because it is practically impossible in a combustor to determine the correct $q_1$, $q_r$ is used instead, where

$$q_r = \frac{w_a^2}{2\rho_3 g^2 A_{max}^2}$$

When $w_f$ is small compared to $w_a$, a straight-line correlation results between $\Delta p^0/q_r$ and $\rho_3/\rho_4$; the value of $\Delta p^0/q_r$ corresponding to $\rho_3/\rho_4 = 1$, isothermal flow, is then the pressure loss due to drag. Typically, as shown

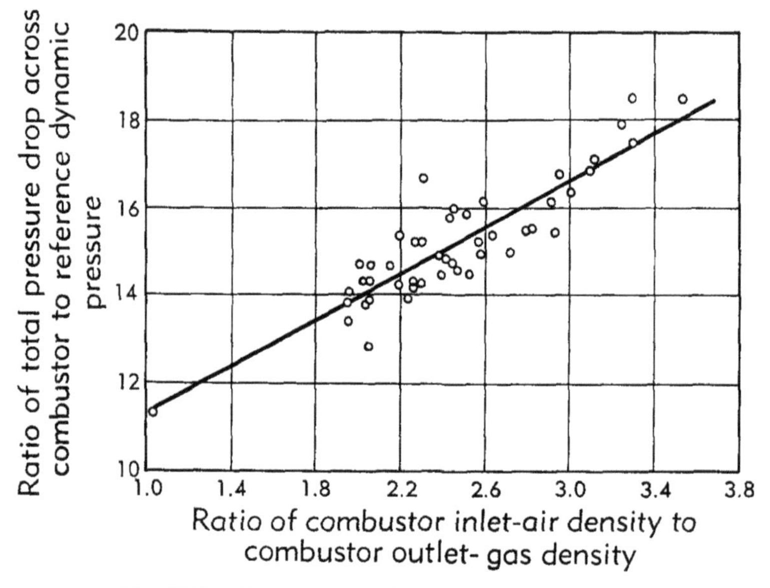


Fig. H,3i. Pressure losses in gas turbine combustor.

in Fig. H,3i, more than one half the total pressure loss is due to aerodynamic drag, rather than momentum change.

The loss due to aerodynamic drag of a given combustor increases with any change in engine operating conditions that increases the combustor-inlet dynamic pressure. The momentum pressure loss increases very nearly directly with any increase of the temperature ratio across the combustor. A chart is presented in [24], by means of which the pressure losses due to fluid friction and to momentum changes can be separately evaluated in terms of the combustor inlet-air pressure. By determining the total pressure loss without combustion at a known inlet-air condition and the total pressure loss with combustion at the same inlet-air condition and a known temperature rise, two constants can be determined from the chart. These constants then permit determination of the pressure drops at other combustion conditions.

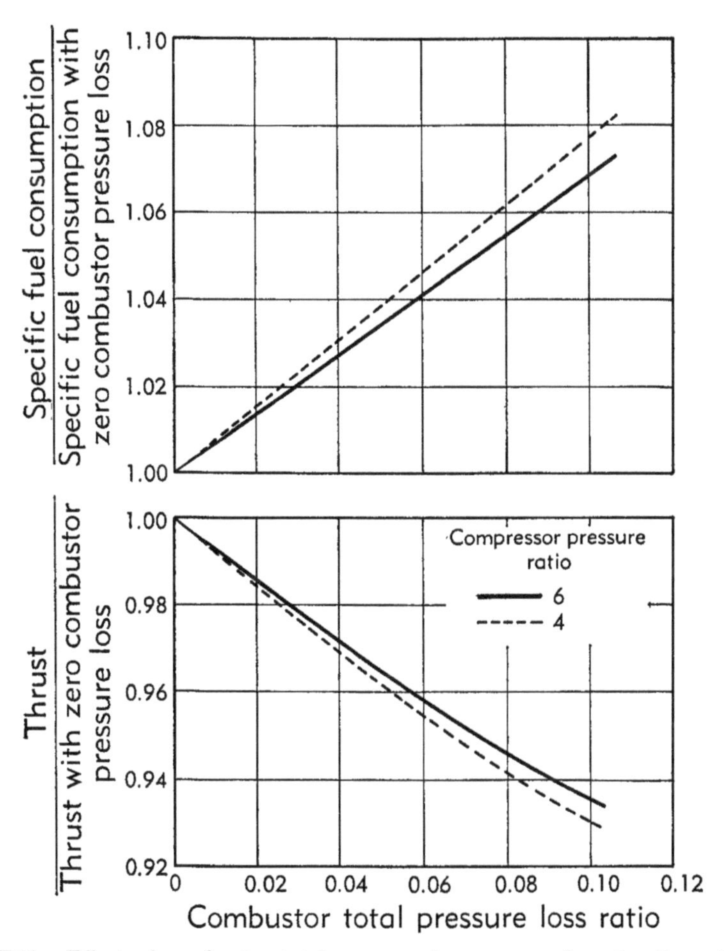


Fig. H,3j. Effect of combustor total pressure loss expressed as fraction of total pressure at combustor inlet on thrust and fuel consumption of typical turbojet engine. 100 per cent rated rpm; flight Mach number = 0.656; altitude = sea level; turbine-inlet temperature = 1500°F.

Combustor pressure loss reduces the cycle efficiency of the gas turbine power plant. Although the total pressure loss correlates for different combustor operating conditions as a function of combustor-inlet air density to combustor-outlet-air density when expressed in terms of an inlet dynamic pressure for the combustor, it is more meaningful for engine performance to express the pressure loss in terms of the fraction of the total pressure at the combustor inlet. In Fig. H,3j is shown the effect of combustor total pressure loss on the thrust and specific fuel consumption of two typical turbojet engines, one with a compressor pressure ratio of 6 and one with a compressor pressure ratio of 4, for a flight Mach number of 0.656 at sea level altitude with a turbine inlet temperature of 1500°F. The thrust and specific fuel consumption are shown as a ratio to the thrust and fuel consumption with no pressure loss in order to illustrate the order of magnitude of the effect of pressure loss on engine performance. Typical pressure loss ratios are of the order of 0.04 to 0.06 for turbojet combustors of the types described, corresponding to a 3 to

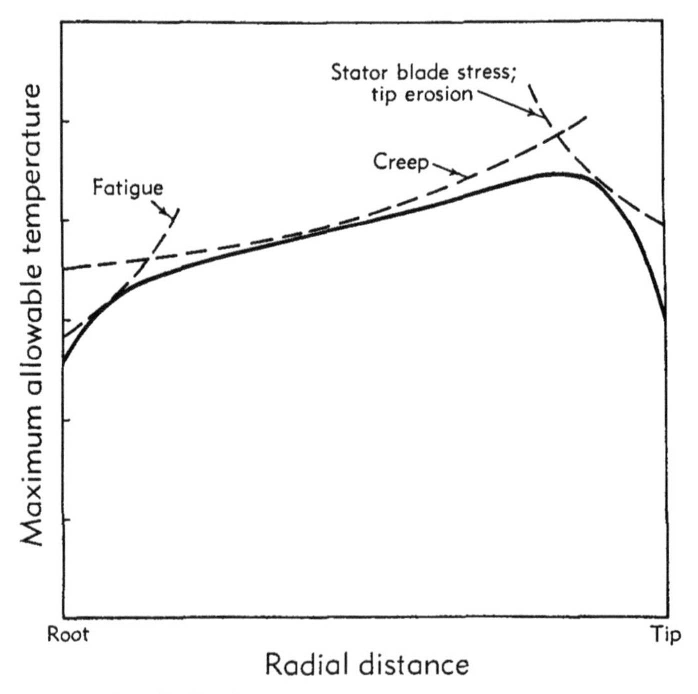


**Fig. H,3k. Limiting outlet temperature profile.**

4 per cent decrease in thrust and a 3 to 5 per cent increase in specific fuel consumption from the ideal. Higher pressure ratios and higher flight speeds reduce the cost of pressure drop to performance.

*Outlet temperature distribution.* Stress considerations of the turbine blading, especially on the rotor, determine the optimum distribution of gas temperatures at the combustor outlet. Usually the optimum temperature pattern is the envelope of several stress limit curves, as shown in Fig. H,3k, and has temperatures increasing radially from blade root to tip but circumferentially uniform.

If the optimum distribution of temperature is not achieved, either the turbine blade life will be shortened, or it will be necessary to operate the engine at lower temperature and thus at reduced power, or at reduced power and reduced efficiency.

Ordinarily the combustor inlet-air conditions of pressure and temperature do not affect the outlet-temperature distribution, unless these conditions are made so unfavorable for combustion that unsteady burning is encountered. Fig. H,31 shows that the nonuniformity of the combustor-outlet temperature distribution may be expected to increase as combustor temperature rise is increased, however. The data are for a typical combustor, and the mean deviation is the arithmetic mean of the deviations

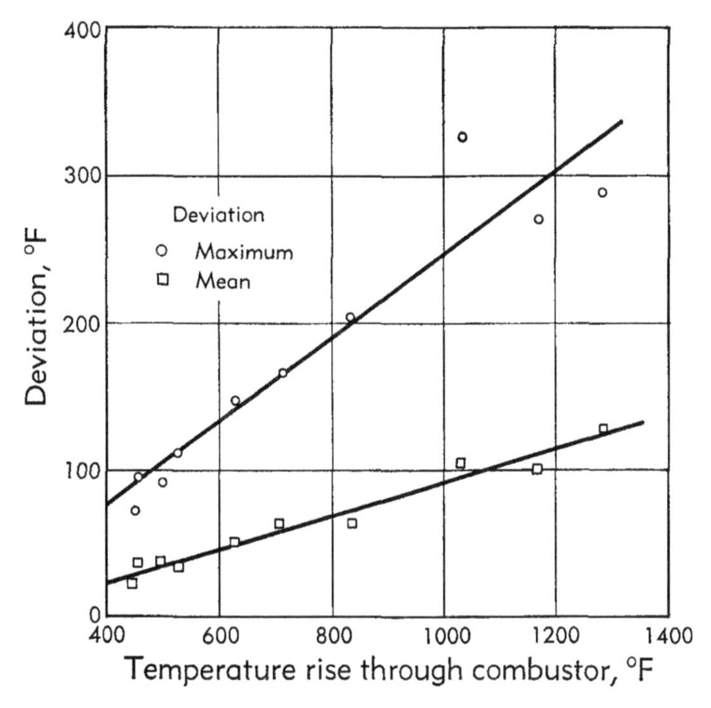


Fig. H,31. Combustor-outlet temperature uniformity expressed by deviations of point readings of outlet temperature from the mean combustor-outlet temperature.

of the individual thermocouple readings from the average combustor-outlet temperature; values for the maximum deviation of a single thermocouple reading from the average outlet temperature are also shown. The data points correspond to combustor operation at different inlet-air conditions.

The data in Table H,3 for two other typical combustors operated in an altitude rig show that the effects of altitude and engine speed are about as expected; in general, the mean temperature deviation from the average combustor-outlet temperature increases with increase in engine rotor speed and increase in altitude. Mean temperature deviation is greater at engine speeds requiring high combustor temperature rise. It is

*Table H,3.* Mean temperature deviation at various flight conditions.

| Altitude | Engine speed, per cent rated rpm | Mean temperature deviation at combustor outlet, °F | |
|---|---|---|---|
| | | Combustor 1 | Combustor 2 |
| 35,000 | 33 | 128 | --- |
| | 42 | 118 | 188 |
| | 67 | --- | 162 |
| | 92 | 165 | --- |
| | 100 | 200 | --- |
| 40,000 | 33 | --- | 215 |
| | 42 | --- | 197 |
| | 50 | 182 | --- |
| | 67 | 205 | 158 |
| | 92 | 201 | --- |
| | 100 | 202 | --- |
| 45,000 | 33 | --- | 224 |
| | 42 | --- | 196 |
| | 50 | 153 | 199 |
| | 58 | 115 | 188 |
| | 67 | 211 | 164 |
| | 83 | 230 | --- |
| | 92 | 240 | 255 |
| | 100 | 248 | --- |
| 50,000 | 67 | 225 | 180 |
| | 75 | 247 | --- |
| | 83 | 245 | --- |
| | 92 | 281 | 293 |
| | 100 | 275 | --- |
| 55,000 | 67 | --- | 195 |
| | 75 | 171 | --- |
| | 92 | 304 | 345 |
| | 100 | 355 | --- |
| 60,000 | 75 | --- | 255 |
| | 83 | --- | 308 |
| | 92 | --- | 335 |
| | 100 | 425 | --- |

greatest at the high engine speed, high altitude condition where not only a high temperature rise exists, but also combustor performance is depreciated to low efficiency and to unsteady burning because of the low inlet-air pressure.

Increased inlet-air velocity can generally be expected to make the outlet-temperature distribution more uniform. The effect is apparently the result of better gas mixing; the price is higher pressure loss.

Finally, any large variation in the velocity profile of the inlet air may be expected to influence combustor outlet temperature distribution.

**H,4. Combustor Arrangement.** A primary problem for the combustor designer is the initial arrangement, or layout, to be adopted for the combustion system. An initial choice might be made between annular and tubular combustors. There are advantages and disadvantages with either type.

The annular combustor will obviously use the space available for combustion more completely than tubular combustors. Large combustor

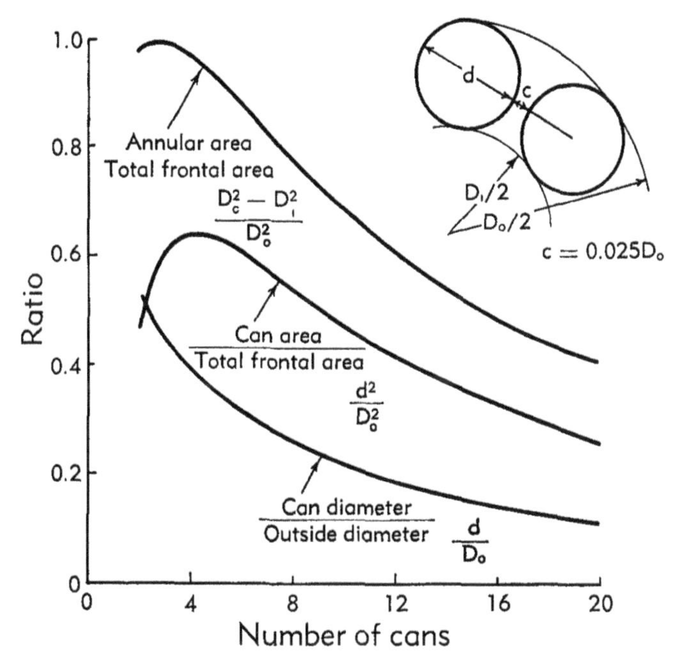


Fig. H,4a. Diameter, flow area, and occupied annular area for "can"-type combustors.

cross-sectional area is desired for efficient combustion and low pressure loss; on the other hand, high thrust from an engine of low frontal area and light weight requires that the combustor not be the element of largest diameter. Simple arithmetic shows that if the combustor outer dimension is kept within the limits of the engine frontal area, and if one-fourth of the engine diameter is allocated to engine shaft and engine frame, then an annular combustor space equivalent to 94 per cent of the engine frontal area is available; even if one half of the engine diameter is used for shaft and frame, 75 per cent of the engine frontal area can be used for the combustor.

Fig. H,4a shows the fraction of frontal area used by tubular or "can-type," combustors to be always less than about 65 per cent, and on the order of 50 per cent for 8 to 10 cans, a number of cans sufficiently large to avoid serious ducting problems at the compressor outlet. The clearance shown between the cans was assumed as $2\frac{1}{2}$ per cent of the outside diameter of the engine, and would correspond to 1 in. on a 40-in.-diameter engine. For from 4 to 20 cans, the flow area through the cans is about 65 to 70 per cent of the annular area that they occupy; in other words, reference velocities for tubular combustors will be about 50 per cent higher than if the same cross-sectional area had been occupied by an annular combustor.

As far as weight is concerned, calculations based on equal shell stress or on equal liner distortion indicate that tubular combustors should be lighter. However, the possibility of using the outer housing of the annular combustor to provide a part of the structural support of the engine may produce the lower midsection weight for the two combustor arrangements.

Because there is not the need to interconnect a number of combustors with small tubes as in the case of the can-type combustors, the annular combustor simplifies the ignition and starting problem. It also alleviates the blowout problem under certain critical operating conditions where one can might blow out and fail to reignite readily through the cross-fire tubes.

Because in the annular combustor there is less surface area of combustor liner exposed for a given flow cross section, it may be argued that cooling and elimination of carbon from the combustor walls is less of a problem [*25*].

An important reason for selecting tubular combustors is the ease of development of these units. Because considerable test-stand development time has been and will continue to be required for any new combustor, the smaller air flow requirements of the can combustor make possible extensive testing over a wide range of simulated operating conditions without exorbitant demands on air and exhaust facilities. Although segments of annular combustors can be used in test-stand work, the results will not approximate what will happen in an engine as closely as will results from tests of the full annulus.

Technically, it has appeared easier to develop tubular combustors because of the symmetry of air flow permitted around the fuel spray, which itself is most conveniently made symmetrical. Development of a strong ignition tore, or reverse flow zone, which will be discussed later, is easier in the can. The possibility of encountering air flow patterns that are axially skewed due to the merging of air streams from parallel axial slots or rows of holes in the annular configuration is described in [*25*]. The result would be a distorted outlet temperature profile.

It seems apparent that the small individual tubular combustors can be much more readily removed from inspection, serviced, and replaced. For removal, the annular combustor must either be sectioned longitudinally, or the entire rear assembly of the engine must be removed.

Fewer fuel nozzles are required for tubular combustors than for annular combustors. These nozzles are, therefore, larger, less sensitive to machining tolerances, and possibly easier to fabricate.

The foregoing, then, are the primary arguments for the selection of an annular or tubular combustor.

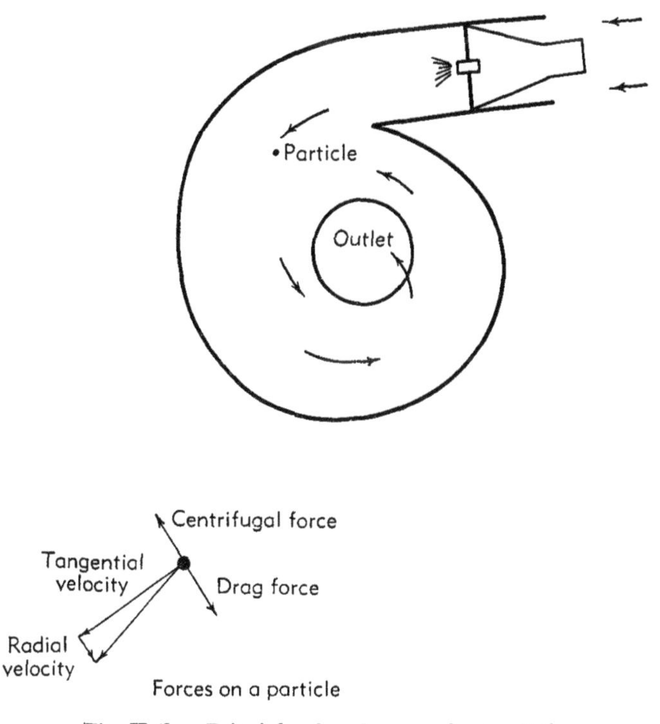


Fig. H,4b. Principle of cyclone combustor [*26*].

As to the selection between the reverse-flow tubular combustor and the straight-through tubular combustor, the pros and cons may be summed up as follows. The reverse-flow-type combustor allows a short engine shaft, prevents flame radiation from reaching the turbine stator and rotor, eliminates special joints for combustor expansion, and provides for a very convenient inspection and replacement of fuel nozzles and combustor liners. Only the first of these reasons has no strong counterargument. On the other hand, the straight-through combustor permits lower frontal areas for the same combustor flow area, has inherently less pressure loss by virtue of not having to turn the gas flow through 180° twice, and is easier to manufacture. For high performance engines, there is no argument; the straight-through arrangement is the choice.

The use of tubular flame tubes inside an annular housing has already been mentioned (Art. 2). These flame tubes may extend clear to the turbine stator blades, or may terminate at the end of the primary zone. This arrangement is an attempt to combine the simple, space-saving features of an annular chamber with certain desired features of the tubular chamber. The stiff walls of the flame tube are less subject to buckling, and it has symmetrical air-fuel patterns, as discussed above.

Lloyd and Probert [*26*] have described a cyclone, or vortex, burner for burning residual fuel oils in which the burning fuel droplets are subjected to both the tangential and inward-radial velocity components of the air flowing in the volute (Fig. H,4b). The centrifugal force of a fuel spray particle in the combustor is balanced by the inward component of drag force on the particle throughout the flow, and it can be shown that the size of the particle in equilibrium is proportional to the radius of its flow path. Thus the particles move in toward the discharge only as they burn down in size. Combustors of this cyclone type have not been incorporated in aircraft gas turbines, but may be worth exploring.

**H,5. Combustor Size.** The effect of the combustor cross-sectional area on combustor performance in a given engine is primarily the effect of combustor inlet-air velocity on pressure loss and combustion efficiency, as discussed in Art. 3. An aerodynamic limit exists for combustors that corresponds to choking at the combustor outlet. For example, for a temperature ratio of 3 at an inlet temperature of 200°F, an inlet velocity over 375 feet per second will choke a straight duct. If the designer tries to keep the combustor within the envelope of the engine, the compressor will set a theoretical maximum limit of 49 pounds of air per second per square foot of compressor inlet; practically, the limit will be somewhat less than this. The possibility of using even more cross-sectional area for combustion, if desired, exists for installations for supersonic flight speeds, where the inlet for the engine is larger than the compressor. At very high flight speeds, even the exhaust nozzle will be larger than the engine.

Pressure loss will become excessive even before choking is reached. The pressure loss coefficient, $\Delta p^0/q$, increases at inlet Mach numbers above about 0.15 and, as has been shown, increases with temperature ratio. Calculations [*27*] for tubular combustors having constant annulus and liner cross-sectional areas along the combustor axis and flush circular holes in the liner walls show the effect of liner area on pressure loss. The results may apply approximately to can-annular and annular combustors having equal velocities in the inner and outer annuli at all points along the combustor axis. Annulus wall friction represents 5 per cent of the combustor total pressure loss coefficient for a liner-to-total cross-sectional area ratio of 0.5, and 23 per cent at area ratio 0.7; these values are for a

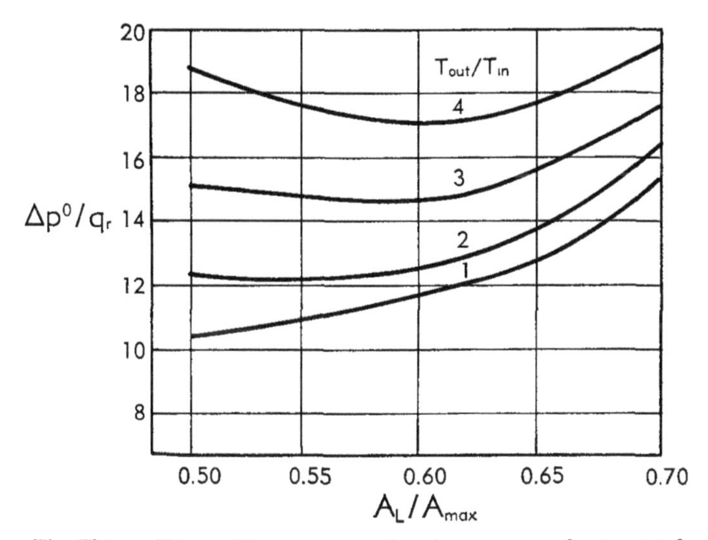


Fig. H,5a. Effect of liner cross-sectional area on combustor total pressure loss coefficient [27], $(A_h)_T/A_{max} = 0.6$.

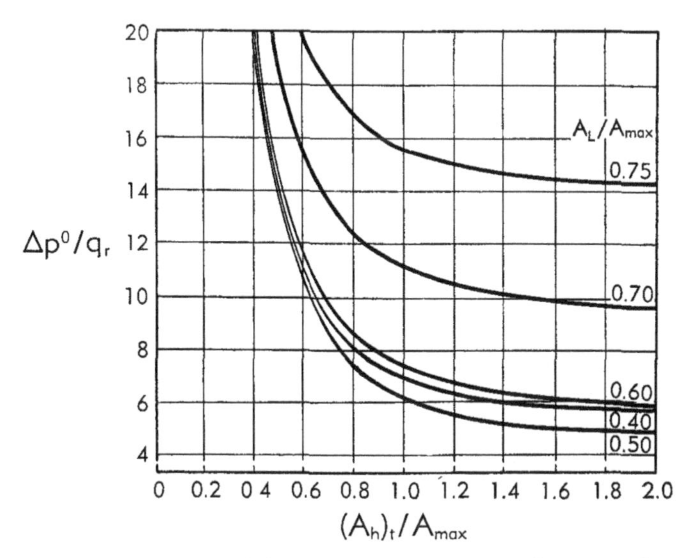


Fig. H,5b. Effect of liner open-hole area on combustor total pressure loss coefficient [27], $T_{out}/T_{in} = 1$.

total open-hole area in the liner walls equal to the reference, or maximum, cross-sectional area of the combustor. Fig. H,5a shows that a liner-to-total cross-sectional area of about 0.6 will minimize pressure loss.

The same calculations show the effect of liner open-hole area (Fig. H,5b). A total liner open-hole area greater than the combustor cross-sectional area minimizes pressure loss.

Fig. H,5c for a tubular combustor operating at sea level pressure and

620°R inlet-air temperature shows lines of constant combustion efficiency for a range of values of temperature rise and air weight flow per unit of cross-sectional area. At very low air flows, fuel flows became correspondingly low in these tests, and loss of combustion efficiency resulted from very poor fuel spray quality. At high inlet-air velocities, corresponding to small combustor cross-sectional area per weight flow of air, decreased combustion efficiency may be attributed to insufficient residence time for the combustion processes.

As far as larger combustors at the same values of inlet-air velocity are concerned, there are little direct performance data that can be presented. Other design features than size of combustor alone are invariably changed

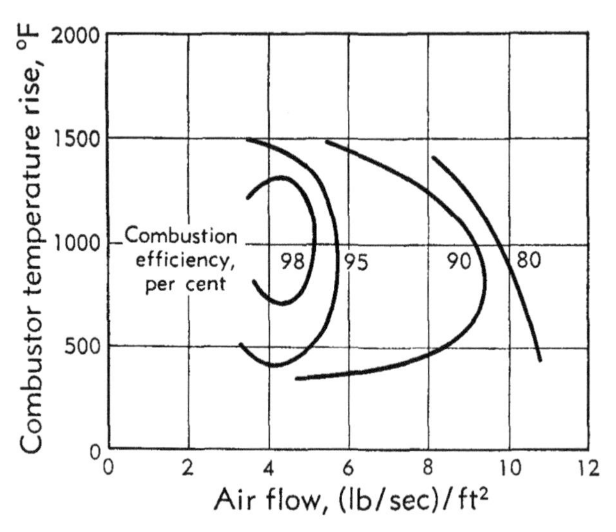


Fig. H,5c. Effect of air flow per maximum cross-sectional area on combustion efficiency of a turbojet combustor. Inlet-air pressure = 15 pounds per square inch; temperature = 620°R.

in any direct scaling of a combustor. In general, it has proved easier to design and develop large combustors than small ones. At least part of this easier development is simply because fabrication tolerances and small details do not influence the combustor performance as seriously in large combustors as in small ones, which is to be expected. Other factors will be discussed subsequently in Art. 6.

The other factor in combustor size, length, must next be considered. A compromise is required. Long chambers are desired in order to give the combustion processes time for completion with resulting high combustion efficiencies and in order to permit adequate mixing of the combustion gases to achieve desired temperature profiles at the turbine. But short chambers are desired in order to keep the rotor shaft short and to minimize engine weight. Conventional, successful practice has been with over-all combustor lengths of about four to five and one-half times the over-all diameter for tubular combustors or the distance across the annular space, for annular combustors. Similar proportions for flame tubes, or liners, are that they are between two and four times as long as wide.

**H,6. Primary Zone.** A most important design consideration for stable and efficient combustion is, of course, the primary combustion zone. This zone must provide the correct mixture compositions for ignition and combustion, continuous initiation of flame, and sufficient residence time for ignition and combustion, as Sec. B has pointed out.

*Mixture compositions.* Assume that a liner, or flame tube, is used to divide the air, permitting only part of it to burn with the fuel. Then the manner in which air and fuel are admitted to the upstream end of the combustor is exceedingly important.

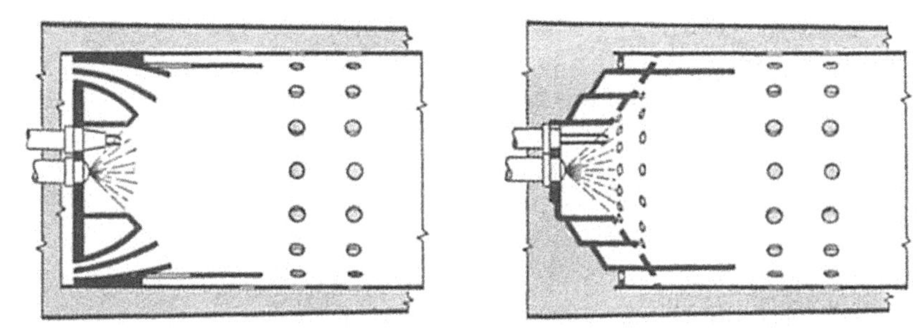

Fig. H,6a. Graduated air admission [*28*].

First, consider air admission. Research has been carried out on many combustor configurations, both tubular and annular, to show how the air for combustion should be admitted. The findings are that this air, the primary air, should be admitted gradually and in as much of the liner length as is reasonably possible. The quantity of this air depends on the combustor design. A graduated admission of primary air assists in the maintenance of local fuel-air ratios that are correct for combustion over wide ranges of fuel flow.

Mock [*28*] has illustrated two methods of achieving a graduated air velocity (see Fig. H,6a). In both forms shown in Fig. H,6a, the fuel spray and flame seat are contained in a central cup surrounded by annular sheaths of air. The flow openings for these annular air sheaths admit successively increasing velocities of the sheaths, for example, for an over-all velocity of 300 feet per second for the burner tube, 15 feet per second may be obtained for the inner sheath of air, 40 feet per second for the next outer cone, and so on.

Improvements in altitude operational limits and combustion efficiencies of engines with annular combustors have been made by modifications to the arrangement of the holes in the liner, or basket, wall. For example, the modified configuration of Fig. H,6b with the air-admission holes spaced to send more air downstream broadened the range of conditions at which combustion was maintained [*29*, pp. 289–290]. Fig. H,6c further illustrates this effect of air-passage distribution on altitude operational limits, and compares combustion efficiencies at four different operating conditions. Combustor pressure drop was essentially the same for each of the two configurations shown.

In a combustor design, the distribution of air entering a liner along its length can be established by prescribing the open-hole area in the liner walls. The fractional hole area for a fractional liner air flow depends on the discharge coefficient of the holes and the pressure drop for the jets through the holes.

Discharge coefficients for combustor liner air-entry holes have been measured for circular holes having flow parallel to the plane of the hole [*30*]. The discharge coefficient decreased markedly with an increase in parallel flow velocity up to 500 feet per second, the limit investigated. Discharge coefficient also decreased with a decrease in static pressure ratio across the hole. In the study, the effects of wall thickness and hole diameter were found to be small; the effects of duct height, boundary layer thickness, and static pressure level were negligible. Although details are available in [*30*] and in the further references cited there, curves such as Fig. H,6d and H,6e can be used to determine discharge coefficient provided the external flow passage walls are parallel and the jet velocity is greater than the internal parallel flow velocity. The corrected coefficient of Fig. H,6e is multiplied by the appropriate value from Fig. H,6d to obtain the coefficient to use. The total pressure $p_a^0$ and static pressure $p_a$ of the external approach stream between liner and housing and the jet static pressure $p_j$ are required. They may be obtained either by direct measurement, or by calculation from known values of combustor mass air flow, air density, and flow passage areas at pertinent stations. Jet static pressure is assumed equal to the static pressure in the liner at the station considered.

Thus, the pressure drop for the jet is the annulus total pressure minus the liner static pressure at the station. The liner static pressure depends on the density, and therefore the temperature, in the liner.

The calculation of combustor total pressure loss and air flow distribution involves trial and error methods. Flow and pressure drop must be balanced among the several stations. The effect of distribution of flow on the distribution of heat release, and vice versa, adds complication [*31*]. Alternatively, open-hole area distribution can be calculated for a given air flow distribution and total pressure loss.

Some generalized curves useful for estimates of the effects of geometry

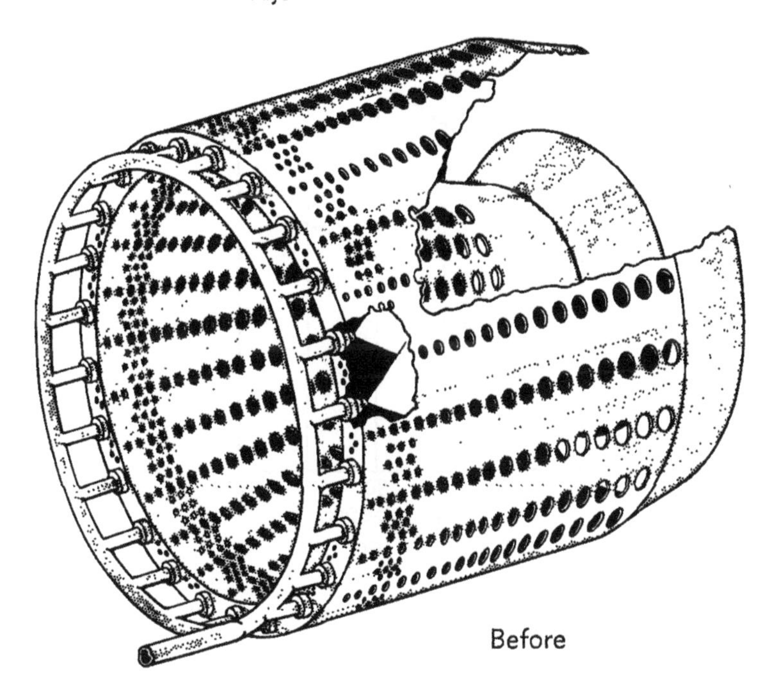


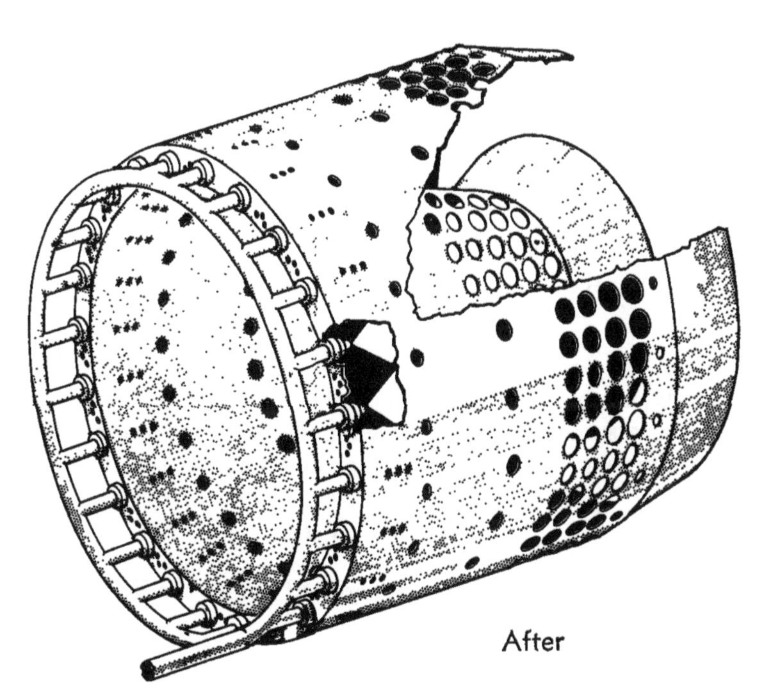


Fig. H,6b. Annular combustor liner, or basket, before and after modification to improve altitude performance [*29*].

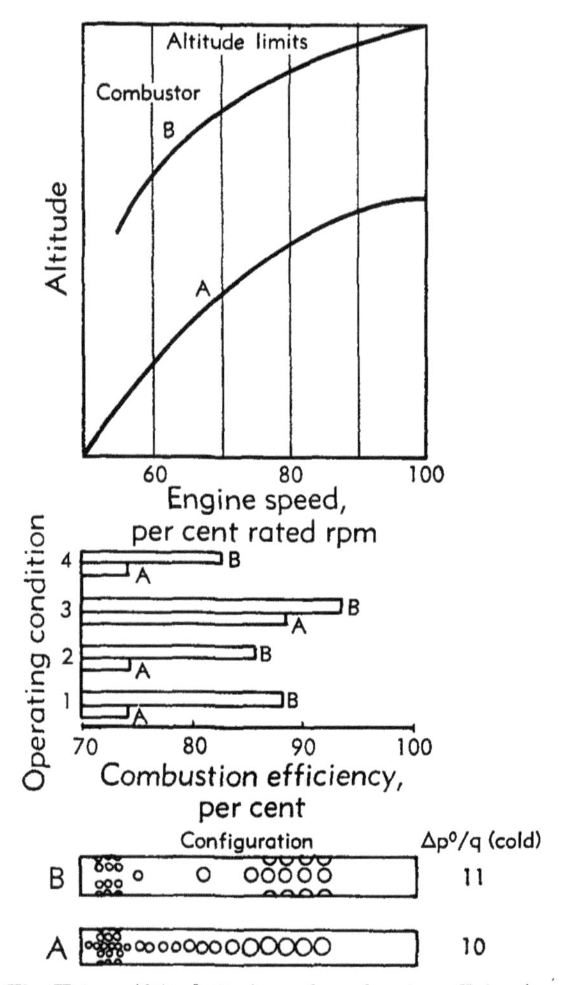


Fig. H,6c. Altitude limits and combustion efficiencies for two annular combustor configurations.

on pressure loss and air flow distribution have been presented in [27]. In this study, a tubular parallel wall combustor with continuous liner wall openings was assumed as the simplified model of a combustor. The orifice discharge equation for a liner wall opening was written in differential form,

$$d(w_a)_l = C\rho_j V_j dA_h$$

where $A_h$ is the open-hole area. Integrated across the liner length and made dimensionless, the resulting equation is

$$\int_1^0 \frac{d[(w_a)_l/(w_a)_t]}{C\sqrt{(p_a^0 - p_l)/q_r}} = \frac{(A_h)_t}{A_{max}}$$

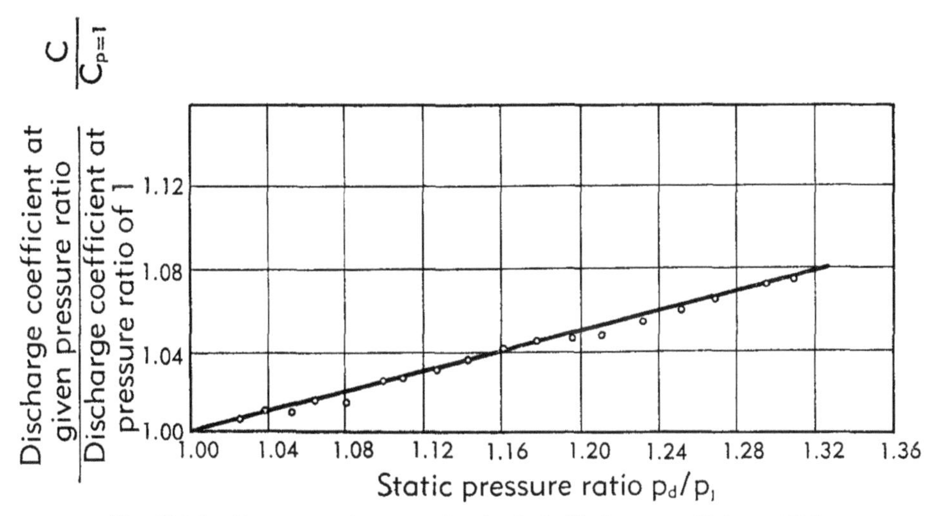


Fig. H,6d. Pressure ratio correction for hole discharge coefficients. Hole diameter = 0.75 inch; wall thickness at hole = 0.040 inch [30].

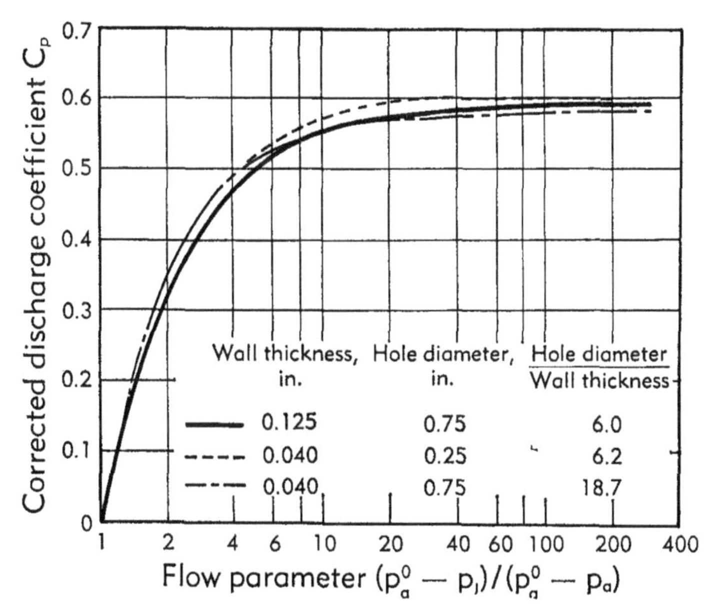


Fig. H,6e. Corrected discharge coefficient as function of ratio of dynamic pressure of jet to dynamic pressure in annulus. Parallel walls; circular holes [30].

for incompressible flow. The term $(p_a^0 - p_l)/q_r$ is related to the combustor total pressure loss coefficient, $\Delta p^0/q_r$, through the friction pressure drop along the annulus and through the momentum equation for the liner gas stream. This relation and the preceding equation are solved simultaneously, after some further simplifying assumptions. One of these assumptions is that the temperature in the liner corresponds to complete combustion of a stoichiometric mixture in the over-rich, upstream portion of the liner, and corresponds to complete combustion at a fuel-air ratio obtained by dividing the over-all fuel-air ratio by the fraction of total air that has entered the liner at stations where less than stoichiometric mixtures thereby result. Liner air flow distribution is given by

$$\frac{\displaystyle\int_0^{(w_a)_l/(w_a)_t} \frac{d[(w_a)_l/(w_a)_t]}{C\sqrt{(p_a^0 - p_l)/q_r}}}{\displaystyle\int_0^1 \frac{d[(w_a)_l/(w_a)_t]}{C\sqrt{(p_a^0 - p_l)/q_r}}} = \frac{A_h}{(A_h)_t}$$

From this equation, values of the fraction of total open-hole area, $A_h/(A_h)_t$, have been calculated for various air distributions, $(w_a)_l/(w_a)_t$, in [27]. For example, Fig. H,6f illustrates the effects of total open-hole

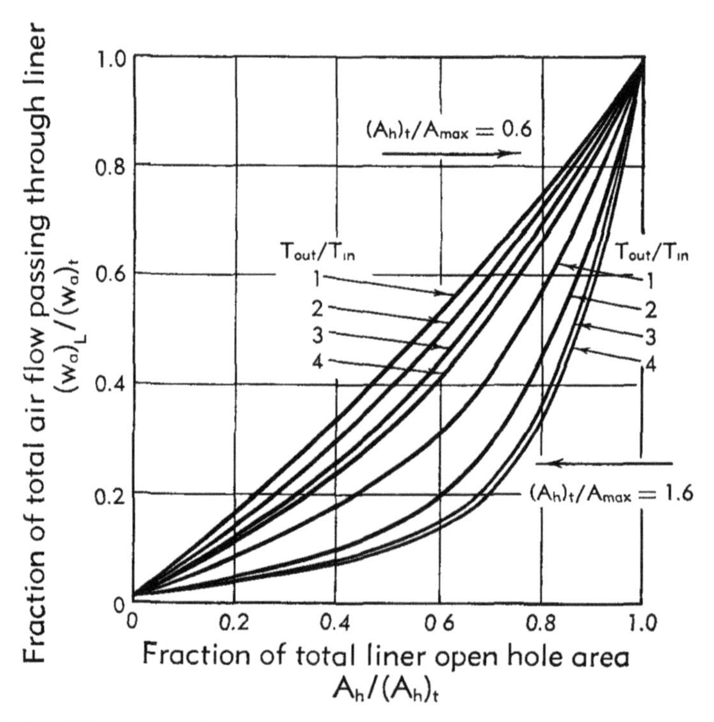


Fig. H,6f. Effects of total open-hole area and temperature ratio on fractional flow through liner wall. Ratio of liner area to reference area $A/A_{max} = 0.6$ [27].

area in the liner and temperature ratio on fraction of total air flow. The plots are for a liner to reference cross-sectional area, $A_l/A_{max}$ of 0.6. A straight line of unit slope would represent air flow distribution directly proportional to air-hole area distribution.

While the data and calculations from work such as found in [27,30] provide some useful design information, they also point out a difficult design problem. As total temperature is increased, the fraction of total air entering the upstream part of the liner decreases; static pressure rise from combustion reduces the static pressure drops for the jets and reduces the discharge coefficients. Just when more air is needed to keep flame lengths short, the supply decreases. And this effect is amplified by any increases in total liner open-hole area that might be made to decrease pressure loss.

Next, consider fuel admission. If satisfactory combustion requires localized fuel-air ratios that are at or near stoichiometric in the primary zone for all combinations of engine speed and altitude, then the manner in which fuel is admitted may be expected to be as important as the

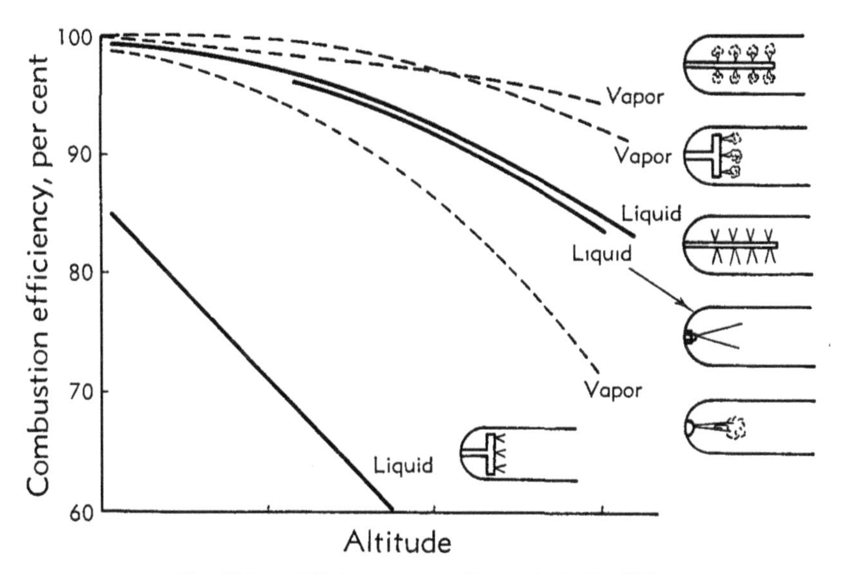


Fig. H,6g. Mixture preparation in turbojet [*32*].

manner in which air is admitted. Air admission and fuel admission must be tailored together to achieve correct mixtures in the primary zone. This consideration is illustrated in Fig. H,6g from [*32*]. Three methods of introducing vapor fuels (propane), and three different methods of introducing liquid fuel (JP-3) to the same tubular combustor are shown. Best performance was obtained from the liquid fuel when it was distributed axially from holes in a tube placed to match the fuel with the air. Almost as good performance resulted when the fuel was sprayed as a hollow cone; in this case molecular and eddy diffusion were counted on to complete mixing over the short distances between the many droplets in the disperse spray. The very low combustion efficiency that resulted when liquid was flowed uniformly across the upstream end was believed to have been caused by overenrichment.

The combustion efficiencies for vapor fuel were less affected by reduced combustor pressures. Even with vapor fuel, however, appropriate mixing of fuel and air must occur. Distribution of the vapor in the primary zone led to high combustion efficiencies; lowest efficiencies accompanied the high velocity, axially directed, solid jet of vapor.

When fuel is sprayed as a liquid, then, important variables will be the configuration of the spray, because it influences fuel distribution

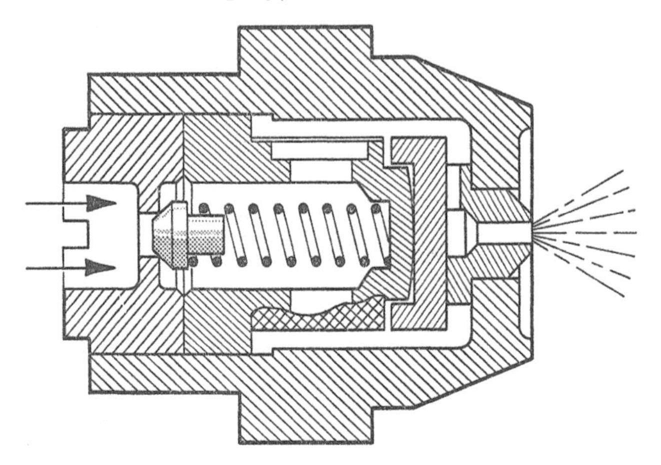

Fig. H,6h. Simplex fuel injection nozzle [*33*].

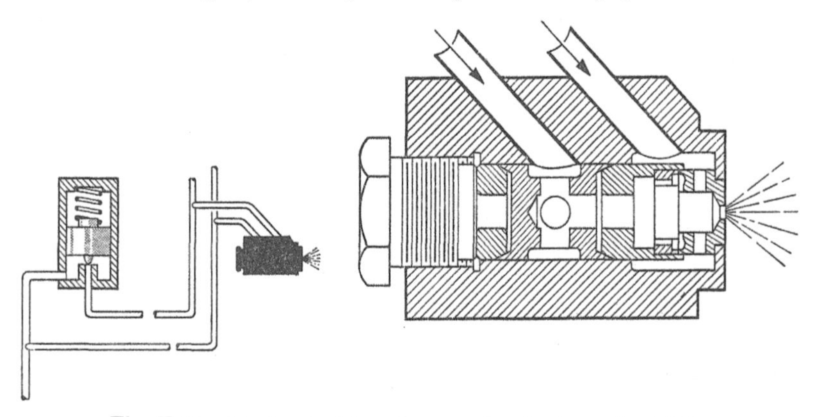

Fig. H,6i. Duplex fuel injection nozzle; dual manifold [*33*].

and mixing, and injection pressure, because it affects drop size and size distribution, as well as fuel distribution and mixing. If a fixed liner geometry is used, the characteristics of the fuel spray should not vary too widely over all operating conditions. This conclusion is a corollary to the concept that the fixed liner provides gradual admission of air so that correct mixtures can be maintained in the primary zone. Sec. G discusses the effects on performance of mismatching either the spray configuration or the injection pressure with the particular fuel and liner. Both Sec. B and D note the fundamental difficulty of operating over fuel flow ranges of up to 100 to 1 within satisfactory injection pressure ranges.

In Fig. H,6h, H,6i, H,6j, H,6k, and H,6l (cf. also [*33*]) are shown several types of liquid fuel injection nozzles that have been used success-

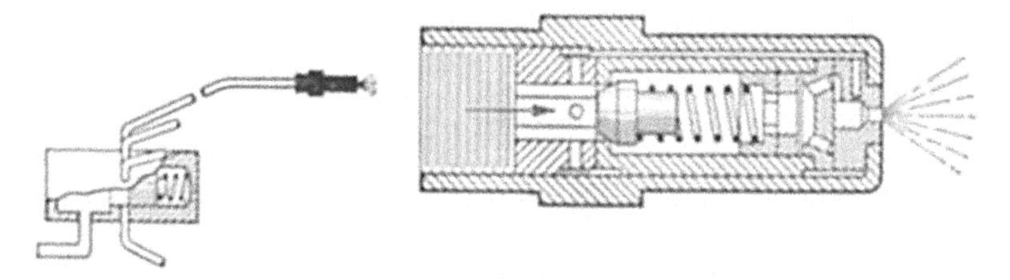

Fig. H,6j. Duplex fuel injection nozzle; single manifold [*33*].

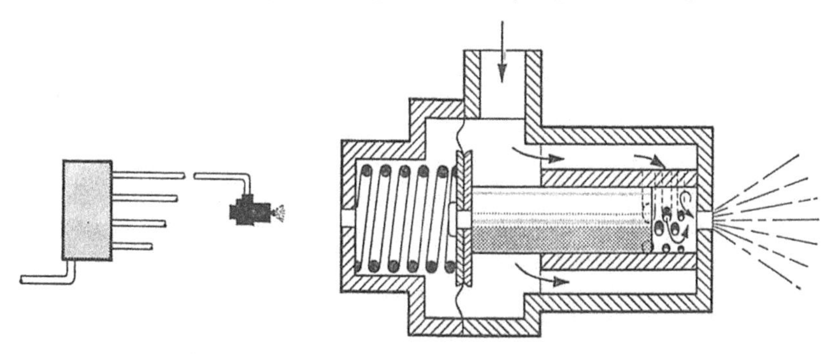

Fig. H,6k. Variable orifice fuel injection nozzle.

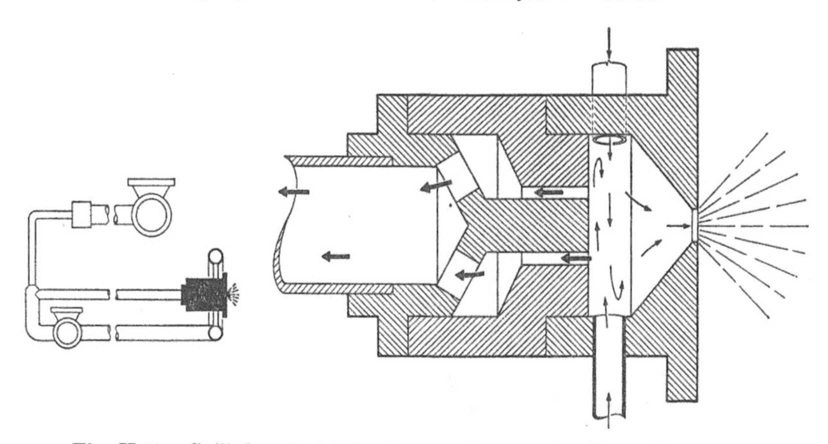

Fig. H,6l. Spill-flow fuel injection nozzle; two-circuit supply system.

fully on aircraft gas turbine engines. The centrifugal, or swirl-type atomizer (Fig. H,6h) meets the requirements that fine comminution of fuel be accomplished with high dispersion of the fuel in the primary zone, but without excessive penetration of the spray through the combustor into the secondary, or dilution zone. In this so-called simplex nozzle, fuel is fed to a swirl chamber by tangential slots or ports that give the fuel a high angular velocity, thus creating an air-cored vortex. The rotating liquid flows out under both axial and radial forces as a spinning sheet. This sheet rapidly disintegrates to ligaments that coalesce to droplets in the pattern of a well-defined hollow-cone spray. A flow-check-valve to prevent drooling of fuel after shut-down is shown incorporated in the nozzle.

Sec. D discusses the characteristics of sprays produced with this and similar atomizers.

The liability of the simplex nozzle is, of course, the wide range of injection pressure it requires.

The most direct attempt and currently the most widely used way to alleviate the problem imposed by the need for atomization over wide flow ranges is with the duplex nozzle. This nozzle uses two sets of tangential openings in the swirl chamber. The two sets of openings are separately fed and are of different size; at high flows the second, and larger set of openings becomes operative. The second set of openings may be operated from a secondary fuel manifold with a flow divider that opens proportionately with the application of fuel pressure, as shown in Fig. H,6i. Alternatively, the secondary orifices may be opened within the nozzle itself as in Fig. H,6j. In this latter case, a multiple flow divider is used to reduce the variation of flow among nozzles at the transfer point from primary to combined flow; the pressure drop at the divider is about equal to the pressure drop at the primary swirl ports.

The continuously variable orifice nozzle appears to offer a solution for a good spray over wide flow ranges without excessive injection pressures. In operation, an attempt to increase fuel injection pressure to increase fuel flow increases the flow area in the nozzle; thus the increased fuel flow occurs at a nearly constant injection pressure. Its liability is in the tolerances required and the resulting manufacturing difficulty and expense. A form of variable area nozzle shown in Fig. H,6k employs a piston to uncover more and successively larger tangential openings to the swirl chamber as the fuel rate is increased. A diaphragm or bellows backed by a spring is used to sense the fuel pressure and position the piston. Because a rather flat pressure-flow relation can result with variable orifice nozzles, an external fuel distribution control system is often required to equalize and control flow to the various nozzles. There are several schemes for achieving this flow control.

Fig. H,6m compares the pressure drops and useful flow ranges for representative examples of the three types of injection system just discussed. The maximum pressure drop across the control for the variable area nozzle was 125 pounds per square inch.

The spill-control nozzle, Fig. H,6l, recirculates fuel continuously at a high flow rate, insuring a high kinetic energy of fuel in the fuel nozzle.

The amount of fuel sprayed into the chamber is varied by varying the quantity of fuel in the return-flow part of the system. Alternatively, fuel can be circulated through the spill nozzle at a high constant rate and supply fuel can be metered into this circuit [*34*].

Air atomization, both with low pressure air, about 1 pound per square inch, and with high pressure air, about 75 pounds per square inch, is a well-known practice in the combustion of industrial fuel oils, and some work has been done to apply the principle to aircraft turbine engines.

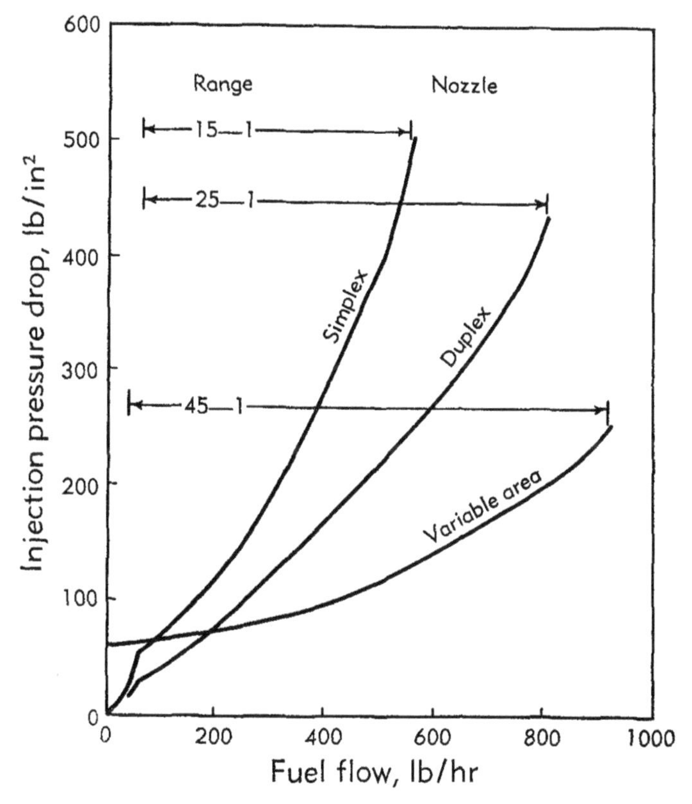


**Fig. H,6m. Comparison of pressure drops and operating ranges of three fuel-nozzle systems.**

The vaporizing systems of Armstrong-Siddeley have been described above (Art. 2), as have the fuel preheating experiments of Whittle. Experiments by the Germans at BMW and by others in which fuel is injected against either cold or hot surfaces to provide comminution and to aid evaporation have not been very successful. The fuel tends to agglomerate on the surface in sheets or as large droplets, and it is difficult to keep a surface hot when bathed with fuel because of the much more favorable heat transfer between liquid and solid than between hot gas and solid.

Fuel injection in the Turboméca engines discussed in Art. 3 is from a hollow ring on the compressor-turbine shaft. Centrifugal force of the rotating system forces the fuel through small ports in the periphery of the ring into the combustor. The need for a fuel pump is then eliminated. A bypass valve regulates fuel flow into the ring.

*Flame stabilization.* The obvious point to discuss after correct mixtures have been provided is flame stabilization. As Sec. E discusses, flame in high speed streams is stabilized, or anchored by means of a local low-velocity region in which burned and burning gases are circulated. This reversed flow carries air to the incoming fuel, carries flame gases back for continuous ignition, and helps to preheat the incoming air.

The basic characteristics of simple baffles or bluff objects, such as punched plates, as flame-stabilizing devices have been discussed in Sec. E in detail. The principle is seen in the combustors of Fig. H,2i, H,2l, and H,2n. With this type of flame stabilization for turbine engines, wide variations in fuel-air ratio without blowout are hard to achieve because it is difficult to maintain local fuel-air ratios within flammability limits in the recirculation region.

In the simples ttypes of tubular combustor the required backflow results from the merging of the air jets that enter radially from the first few rows of holes in the liner. Part of the air from the high pressure area of the merging jets flows toward the combustor outlet, but part of the air also flows toward the upstream end of the combustor (Fig. E,9b of Sec. E). This reverse flow has been observed visually by numerous investigators, and with tufts and with flame plumes from sodium-chloride-tipped probes [*25,35*]. The amount of reversed flow in the combustor can be varied with design, but about one fourth of the total air flow should be involved in the reversed flow for best results, according to [*36*]. The wide ranges of fuel-air ratio over which the simple, tubular turbojet combustor operates stably are attributed in [*36*] to the fact that the flame-stabilizing zone appears to shrink or expand to meet the changing fuel-air ratios. This variation in the quantity of air involved in the stabilization zone permits the maintenance of local fuel-air ratios within the flammability limits of fuel and air. The combustors of Fig. H,2g, H,2h, H,2i, H,2n, and H,2o are illustrative of the principle.

Reversed flow can be designed into combustors by the use of swirl vanes at the upstream end of the combustor. The vortex, or swirl, imparted to the air around the axis of the combustor by these vanes promotes a reversed flow back along the center of the combustor. This principle is seen in the combustors of Fig. H,2c, H,2d, H,2e, H,2f, and H,2j.

Reversed flow is imparted directly to some of the primary air by the design configurations of the Armstrong-Siddeley combustors of Fig. H,2r and H,2s.

*Flame propagation.* The fundamentals of combustion indicate that it is desirable to allow as much space or volume for the flame as possible if it is to propagate efficiently. For example, quenching by walls or cold air jets can render space unavailable for combustion, representing a loss in efficiency. This loss is greater for smaller combustors (i.e. smaller volume-surface ratio) and reduced pressure [*22*]. Experience has been in conformity with this idea. For example, Fig. H,6n shows the effect of volume-surface ratio on the combustion efficiency of a variety of combustors. The data are for operating conditions of equal severity; $p^0T/V$ was constant. The spread for any one size represents the effect of other design features. Either the mean curve or the upper envelope shows the effect of primary zone size.

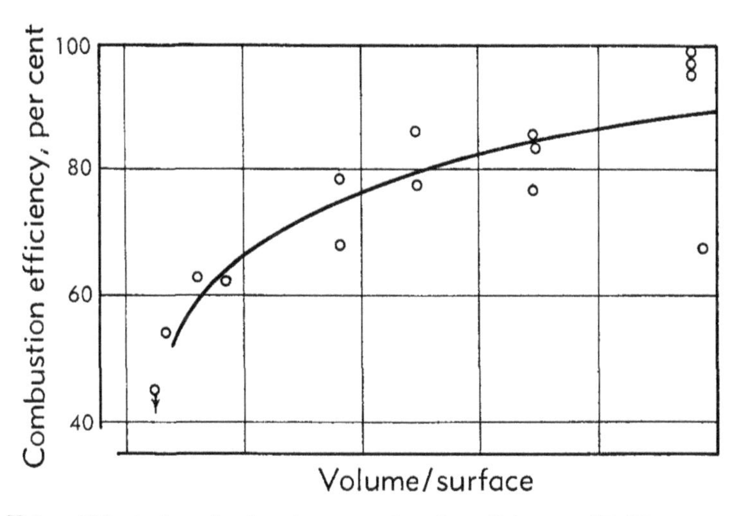


Fig. H,6n. Effect of combustor size on combustion efficiency. $p_i^0T_i/V_r$ constant [*22*].

In another study, Fig. H,6o compares the combustion efficiencies of two combustors that were identical in all respects except that the cross-sectional area for the primary zone of combustor D was 45 per cent greater than for combustor C. The increased width, or area, of the primary zone improved both the combustion efficiency and the altitude limits of the combustor.

Possible reasons for improvement with size are: (1) a smaller portion of the fuel spray impinges on the liner walls, (2) a larger quantity of material burns in the low velocity region of the combustor, and (3) quenching of chemical reactions by the cold walls and jets is reduced. Of course, making the combustion zone as large as possible must be done with a view to the total pressure loss involved, as discussed in Art. 5 and shown in Fig. H,5a.

Also, care should be taken that the cross-fire tubes of tubular com-

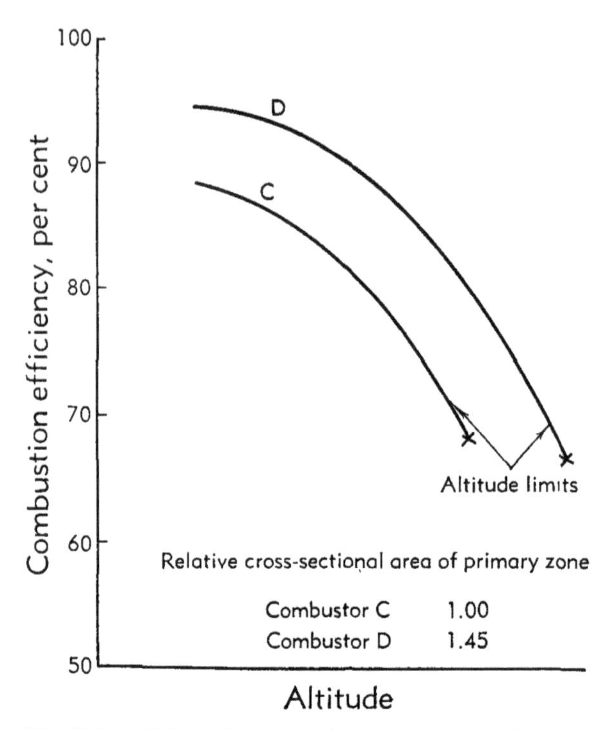


Fig. H,6o. Effect of size of primary zone on combustion efficiency and altitude limits of a turbojet combustor.

bustors exceed the minimum diameter set by quenching distances (Vol. II) for the lowest pressure to be encountered.

**H,7. Secondary Zone.** A preferred temperature distribution at the turbine section requires appropriate proportioning and mixing of the dilution air. The establishment of a temperature profile starts in the primary zone; the secondary air modifies the temperature profile at the end of the primary zone, and thus the flow patterns used in the primary zone have their influence.

Only a crude approximation to a temperature profile results from the use of basic data on flow-proportioning and air-jet mixing. A final temperature profile requires detailed adjustments in the holes and mixing devices in the liner wall. Also, it must be noted that the velocity profile from the compressor influences the temperature distribution, and, furthermore, this velocity profile may not remain fixed over ranges of engine operating conditions.

The simplest design for admitting and mixing the secondary, or dilution, air is the use of a number of holes of circular or other shape in the wall of the flame tube or liner. Proportioning of the air has already been discussed in Art. 6. Proper design of the flow paths requires consideration of both the flow coefficient and the pressure drop for each flow passage. The pressure inside the liner against which the quenching air must flow through a hole is decreased by a static pressure drop that can be calculated from momentum considerations and, strictly speaking, varies with the density ratio of the inlet air to the hot gas at the point where dilution occurs. Flow coefficients need to be known for the many kinds of opening that might be considered. Generalized curves such as those of [*27*] give a useful close approximation of air flow distribution.

The next most accurate approximation to air flow proportioning would be measurements of flow coefficients and static pressure in a liner with cold flow. Pavia [*37*] has computed the flows through the several flow passages of one kind of turbojet combustor based on measurements of static pressures and temperatures both inside and outside the liner and its parts at each of the several air-entry stations of the liner. Measurements made over a nearly fivefold range of air flows and at fuel-air ratios from 0 to 0.02 showed that the proportioning of the air affected was almost independent of the over-all air flow and varied only slightly with the fuel-air ratio. Departure from cold-flow distribution has been shown (Art. 6) to be greater in low pressure-loss combustors, however.

The least accurate, but easiest method is, of course, to assume that the flow proportions directly with the area. Results so calculated may be off by a factor of five or more.

Penetration of jets directed perpendicularly to a general stream have been shown by [*38*] to depend on the density ratio of the jet to the stream, on the velocity ratio of the jet to the stream, on the distance downstream from the jet, and on the product of the jet diameter and square root of flow coefficient. Penetration may be increased if rectangular slots with the long axis aligned with the direction of flow through the combustor are substituted for circular holes, as illustrated in Fig. H,7a [*39*].

In [*2*], rectangular slots located in a part of the flame tube wall where the flame tube diameter is increasing rapidly are described as minimizing the pressure loss. Reasons given are that this location permits a certain amount of reconversion of velocity head to pressure head, and that these slots lead the dilution air in at an angle to the combustor axis.

Another parameter of importance is the spacing between the holes or slots. For example, the effect of spacing on the temperature profile is illustrated in Fig. H,7b. The data are for annular combustors with slots for secondary air and with equal over-all pressure drop. A large number of narrow slots, closely spaced on the inner wall, produced a temperature profile that was cool at the turbine-blade root and hot toward the tip. The slots pushed a blanket of cold air in against the stream of hot combustion gases. The larger, more widely spaced slots produced a much flatter temperature profile. These slots permitted penetration and interleaving of the cold air with the hot gases.

Although some general principles such as the foregoing are helpful, the empirical approach is used in reaching the final design. A more exact, scientific approach is made difficult, or even impossible, by the complicated flow of combustion gases. Various methods and devices for mixing

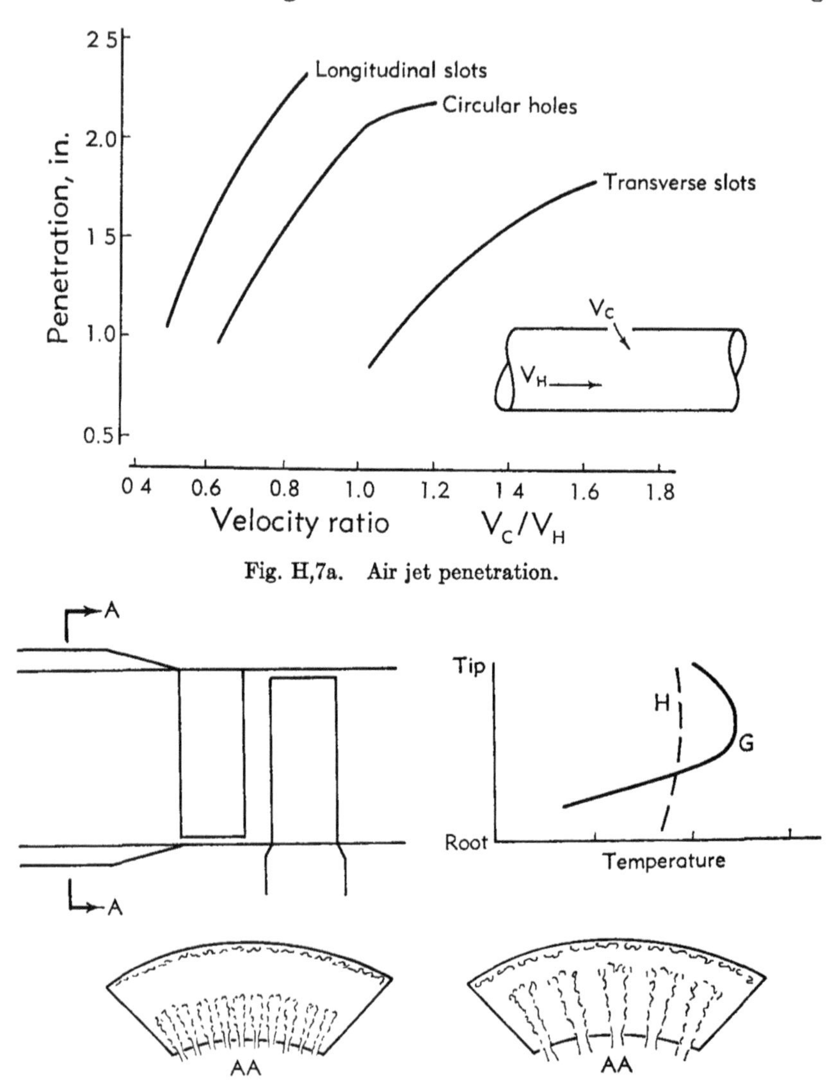


Fig. H,7a. Air jet penetration.



Fig. H,7b. Combustor-outlet gas temperature profiles [22].

or blending dilution air and combustion gases have been discussed in F,6. Some of these appear in combustors illustrated in this section, for example, stub tubes (Fig. H,2d), mixing fingers (Fig. H,2m), chutes or scoops (Fig. H,2l and H,2p), and a slot-baffle combination (Fig. H,2j).

Because it is difficult to cool adequately any mechanical mixers that protrude into the combustion gases, a series of orifices or a few long slots in the combustor wall is a preferred design.

## H,8. Construction and Durability.

*Materials.* The liners and inner parts for turbojet combustors, exposed as they are to the flame itself, must not only be capable of withstanding high temperatures without losing strength, but also must resist high temperature corrosion in the presence of air and combustion products. Also, all parts of the combustor are subjected to mechanical vibrations covering a wide range of frequencies, and often of quite large amplitude.

The most commonly used alloys for liners are Inconel in the United States engines, and Nimonic 75 in British engines. Typical analyses are shown in the following table:

| | Ni | Cr | Fe | Mn | Cu | Si | C | S | Ti |
|---|---|---|---|---|---|---|---|---|---|
| Inconel | 78.5 | 14.0 | 6.5 | 0.25 | 0.2 | 0.25 | 0.08 | 0.015 | --- |
| Nimonic 75 | 74.5 | 20.0 | 3.0 | 0.6 | 0.3 | 0.6 | 0.10 | --- | 0.4 |

Lack of high temperature alloys required the Germans to use ceramic coatings. The forward section of the Jumo 004 combustor liner (Fig. H,2j) was 22-gauge mild steel with an enamel coating, while the rear section and mixing tubes were made of aluminum-sprayed mild steel. Flame tube life was only 15 to 25 hours. The BMW-003 combustor liner was fabricated of an 8 per cent chromium steel with sicromal-10 mixing scoops (Cr, 13; Si and Al, 2.5; and C, 0.1).

Both 18-8 stainless steel, and aluminum-sprayed mild steel sheets serve as outer casings for combustors. Mild steel apparently has a greater ability to damp out vibration, thus is less liable to fatigue failure.

Light sheets, for example 22-gauge, are required to keep engine weight down. Any material selected should be readily drawn, swaged, formed, punched, and welded by any of the usual methods.

Typical service failures include cracks in the combustor liner. Cracks contribute to buckling which seriously distorts the air flow and depreciates the combustor performance. Cracking also may break metal from the liner, with the risk of turbine damage. Cracks frequently originate in fissures produced by punching operations, but may also originate in grain boundaries, according to a study in [*40*]. Scaling, or oxidation, both on the surface and in fissures, probably contributes to crack formation and spreading. The stresses imposed by thermal gradients and mechanical vibration promote cracks. Thus cracking is encountered in the vicinity of and between air-inlet holes, cooling louvres, and similar openings in flame tubes, where local temperature gradients may be several hundred degrees Fahrenheit in $\frac{1}{2}$ inch and where incipient cracks exist from punching or other forming operations.

To minimize cracking, proper deburring and treatment of the edges of holes, louvres, etc., such as by reaming, sanding, and vapor blasting, are effective.

Buckling of liners is caused by large thermal gradients. Small amounts of buckling can be caused by large thermal gradients produced at individual cooling louvres and at air-inlet holes; large amounts of buckling may be caused by overheating of an extensive portion of a flame tube. This overheating may result from inadequate cooling flows along the outside of the liner, either because of flow separation or because of a strong variation in the compressor-outlet velocity profile. It may also result from a skewed or distorted air flow or fuel flow pattern inside the liner, owing to carbon deposits on the liner or fuel nozzle that cause flame and hot gas to be directed strongly against the inside wall of the liner. Inadequate provision of clearances for expansion of the liner has been described by [*2*] as another cause of liner or flame tube buckling and distortion.

*Cooling.* The problem of cooling combustor liners adequately is aggravated by variations in radiant heat transfer from the flame as operating conditions vary, and by variations in the air flow profile from the compressor. Cooling by convective heat transfer to the air flowing in the annulus between the liner and housing is only partially effective at velocities commensurate with low pressure loss for the system. Further, nonuniform air distribution in the annulus interferes with proper cooling. The most successful method of cooling combustor liners has been to provide a sheath, or film, of cooling air along the inside wall of the flame tube. Thus, heat is removed by convection from both sides of the film-cooled wall and some further heat is lost by radiation, while heat transfer to the wall is by radiation from the flame, and increasingly by conduction and convection as the coolant film deteriorates.

Heat transfer calculations for a typical turbojet combustor reveal that more than half of the heat transferred to the liner is by radiation; the heat is transferred away mostly by convection [*41*]. The dominant terms governing the division of heat between radiation and convection are flame emissivity and mass velocity.

Nonluminous flames encountered with low pressure operation in combustors have low emissive power; their emissivity will be 0.1 or less. At operating pressures of one atmosphere and higher, radiation heat transfer to combustor walls increases markedly, both because the emissivities of the gaseous constituents increase with their pressures and because luminous flames are encountered at high pressures. Luminous flames may have emissivities approaching unity [*42*,*43*]. Increased fuel-air ratio and smoke-forming fuel types also promote luminous flames and, hence, increased radiant power.

Fig. H,8a shows the effects of some operating conditions on the wall temperature of a turbojet combustor. The data were obtained at a constant inlet-air velocity. The strong influence of inlet-air temperature is noted; this effect resulted in spite of a decreased convective heat transfer coefficient accompanying decreased mass flow. The effect of pressure illustrated was small; apparently the radiant heat transfer was offset by greater convective cooling. The flame was luminous at both pressures, and emissivity may not have changed much.

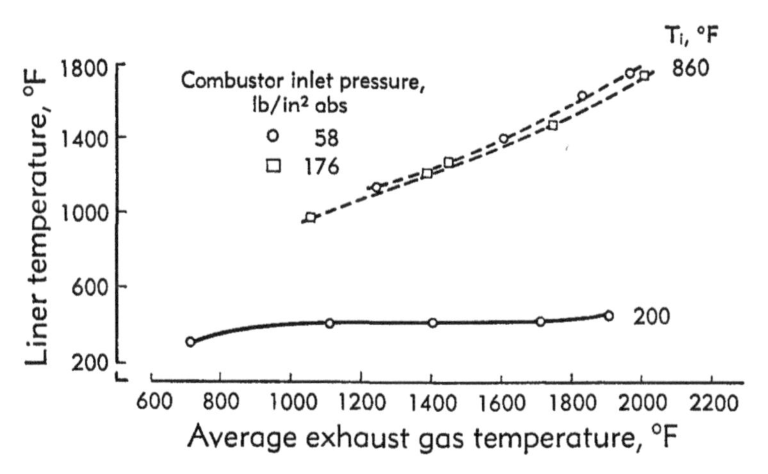


Fig. H,8a. Liner-wall temperatures for a single tubular combustor. Constant inlet-air velocity [*22*].

Techniques for introducing cooling-air films include louvres, shoulders perforated with many small holes, and continuous slots. Continuous slots are made by constructing the flame tube or liner of several sections, the downstream sections having successively larger diameters than the upstream sections, and joining these sections by spacers, such as corrugated strip. Too much air admitted with axial velocity will lengthen the flame region and depreciate altitude performance. Where individual louvres are used, care must be exercised in placing them to avoid high local temperature gradients in the vicinity of the louvers, for example, by staggering their locations.

Ceramic coatings appear attractive for aiding the cooling problem. Emissivity (absorptivity) appreciably less than the 0.5–0.6 for metal liners is required, however. The effective absorptivity will lie somewhere between the measured value and unity, depending on the transparency of the flame, because of internal reflection [*41*]. Several experiments have failed to demonstrate significantly lower temperatures on liner walls coated with ceramic. Another problem has been the tendency of ceramics to fail from thermal shock and mechanical vibration.

Corrosion and oxidation-resistant ceramics should be quite helpful in very high temperature (for example, 3500°F) devices, however. Here more than 90 per cent of the inlet air may be required to burn with fuel to reach these temperatures; little air is left for cooling.

*Mechanical vibrations.* Both the liners and the external housings and connecting ducts of combustors are subject to mechanical vibrations. Sources of these vibrations include pulsations from the compressor and turbine, buffeting from high velocity air, and flow separations as high velocity air negotiates turns and corners through passages, holes, and louvres. Studies of the frequencies of pressure pulsations in combustor inlets of turbojets have shown frequencies from very low to as high as 60 kilocycles, with some amplitudes as high as 70 per cent of the mean pressure [*2*]. Resonance of metal parts induced by these pulsations can promote stresses leading to fatigue failure. Thus a high damping capacity is desired of combustor alloys, as well as good resistance to fatigue failure.

A particular class of vibration known generally as combustion oscillations should be mentioned here. Combustion oscillations are essentially periodic fluctuations in combustor pressure that are maintained by the combustion process.

These oscillations may occur occasionally and unexpectedly in ducted burners of various kinds, although they are most often encountered in rockets [*44*, pp. 893–906; *45*], ramjets [*46*], and afterburners [*47*]. They are to be distinguished from the rough burning, or intermittent explosions, sometimes resulting from improper fuel-air mixtures.

The amplitudes of combustion oscillations are often quite high, equal to or greater than the operating pressure of the system. Frequencies, too, are often high, from hundreds to thousands of cycles per second, giving rise to terminology such as "chugging," "howling," "screaming," "screeching," and so forth. These high amplitudes and high frequencies, together with enhanced combustion temperatures and heat transfer rates, usually produce catastrophic failure of the combustion device.

Detailed studies have been made of the conditions producing combustion oscillations, and of the nature of the phenomena. Although the former is difficult to generalize, the latter resolves into a few possibilities.

Some combustion oscillations have been found to be a coupling between the flow of fuel or propellants into the combustor, and, after a time lag, the release of heat through combustion. The heat release rate varies with the flow rate; the flow rate varies with the pressure in the combustor; this pressure varies with the heat release rate. Frequencies may be in the hundreds of cycles per second, "chugging." Changes in some of the constants in the resonating system, for example, increased fuel pressure drop, usually provides a cure.

Other combustion oscillations have been shown to be periodic oscillations in the combustion gases themselves. High frequencies result,

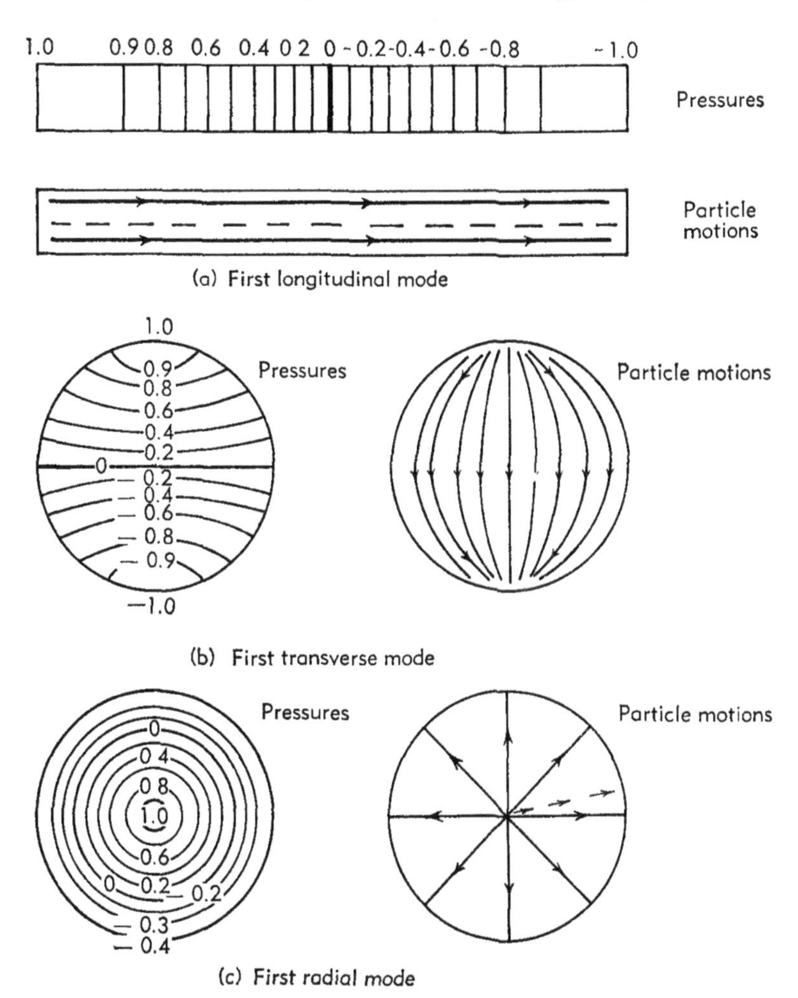


Fig. H,8b. Pressure contours and particle motions for fundamental modes of cylindrical duct.

"screeching" or "screaming." In one pertinent study, detailed measurements of frequency, relative amplitude, and phasing of the pressure oscillations and position of the flame were made in a 6-inch simulated afterburner [*48*]. Some of the possible modes of oscillation are shown in Fig. H,8b. Ionization gap and microphone probes established that when this burner "screeched," the oscillation was a standing transverse wave of the first order. While there were longitudinal components found ahead

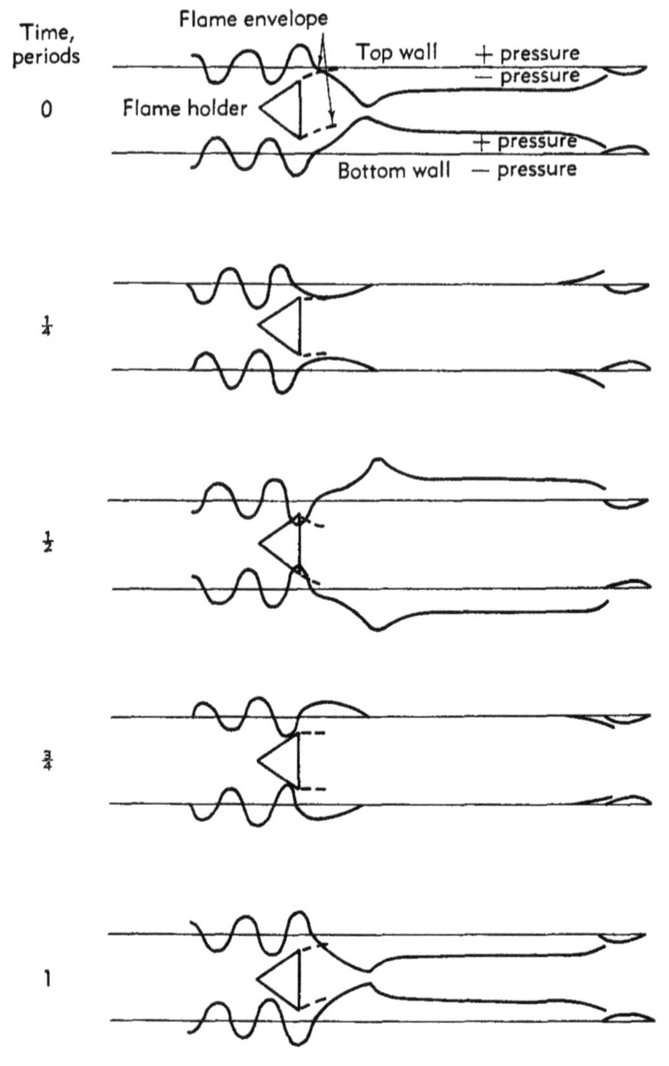


Fig. H,8c. Reconstruction of pressure and flame-displacement sequence.

of the flame holder, there were none in the hot gases. The flame front motion was in phase with the pressure wave. A construction of the corresponding events is shown in Fig. H,8c.

Other studies have also singled out the first transverse wave, either stationary or rotating, as a frequent mode for combustion oscillations [*44,45,47*]; longitudinal modes are also a problem in rockets.

Only speculation is possible about the nature of the coupling between heat release and acoustic oscillations that sustains these waves. If the heat release occurs at locations where the pressure varies and when the pressure is high, the oscillation may be sustained. Adiabatic compression alone may suffice to increase the rates of the reactions producing heat enough to maintain the oscillation. Alternatively, it is suggested in [*49*] that if the oscillations are driven by the kinetic energy of the stream, the heat release may intermittently release this kinetic energy to maintain the oscillation. Still another suggestion [*50*] is that vortices shed periodically from the alternate lips of the flame holder are responsible for oscillations in ramjet or afterburner-type combustors. The periodic transport of combustibles associated with the vortices into the hot wake of the flame holder excites the oscillation. The vortices, in turn, are generated by the fluctuating transverse velocity that accompanies the transverse pressure wave. The proper ignition time delay for the shed material permits amplification. Some modification of this suggestion will clearly be needed to apply it to rockets or other systems without flame holders.

The oscillations are damped by sound propagation, by mass transport downstream, by nonlinear and viscous effects accompanying strong sound waves, and by absorption and scattering at the wall. The usual lack of combustion oscillations in the main combustors of turbojets is probably due in part to the inherent large damping in the kinds of liners used.

*Carbon deposits.* The types of carbon encountered in turbojet combustors, soot and coke, and smoke, have been discussed from the standpoint of fuel characteristics in Sec. G by Gibbons. Soot produced in the gas phase from luminous flames may be deposited on liner walls or protuberances in the liner. Only if the wall is quite hot will the soot burn off. Coke may arise from cracking and partial oxidation of the heavy ends of liquid fuel striking or flowing along a liner wall. If the wall is always sufficiently cool, liquid fuel on it will not crack. If the wall is very hot, fuel boils away so rapidly that a vapor film prevents wetting of the wall and no coke forms. Also, coke may result from cracked and oxidized material, including soot, sticking to and baking on the liner walls. If the wall is hot enough, it burns off.

Operation with very hot liner walls is not desirable, and cures for the coke problem involve the use of fine atomization fuel sprays, avoidance of undue liquid impingement on walls, sweeping the walls with air films, and avoidance of protuberances or dead zones not swept with air. Similar considerations as for film cooling apply to the use of air films to sweep away potential carbon deposits.

**H,9. Future Trends for Design.** The main trends affecting the requirements for aircraft gas turbine combustors are larger engines, higher air mass flows per frontal area of compressor, higher pressure ratio engines, higher flight speeds, and cooled turbines and new materials that permit higher engine operating temperatures. Afterburning for thrust augmentation will continue in general use for some time. The effects on the combustor are to increase the heat release rate requirements, to widen the range of inlet-air pressures over which operation must be satisfactory, and to widen the fuel-air ratio range of operation.

Obviously the range of fuel flows to be accommodated will be greater in the future, both because of the wider ranges of mass flow of air and because of the wider fuel-air ratios. Attention will therefore need to be given to methods of introducing fuel to insure that for any fuel flow or for any over-all fuel-air ratio, local mixtures exist with composition and time sufficient for ignition and complete combustion under the prevailing local conditions.

The increased mass flows per frontal area impose the requirement of maintaining stable and efficient combustion at higher over-all velocities than those now in general use if the combustor is not to establish the engine frontal area.

High inlet-air pressures and temperatures, high heat release rates, and high temperature rise all increase the difficulty of cooling the combustion equipment. Radiant heat transfer is greater at high combustor pressures, for example. Improved applications of film and, possibly transpiration cooling, will be required. Some help may be expected from ceramics.

The high inlet-air velocities at high inlet-air pressures will impose high forces on the combustor liner. These, together with the higher temperatures, make liner durability a continuing, difficult problem.

Finally, there is always the demand for lighter weight, greater reliability and durability, more simplicity for inspection, maintenance and overhead, and for economy of fabrication inherent in any aircraft engine or its component.

## H,10. Cited References.


1. Whittle, F. The early history of the Whittle jet propulsion gas turbine. *The Aeroplane 69*, 445–448, 451–452, 503–507, 543–546 (1945). See also *Engineer 180 (1856)*, 288–290 (1945); *(1857)*, 210–212 (1945); *(1858)*, 328–329 (1945). Also *Proc. Inst. Mech. Engrs. 152*, 419–435 (1945).
2. Watson, E. A., and Clarke, J. S. Combustion and combustion equipment for aero gas turbines. *J. Inst. Fuel 21*, 2–34 (1947).
3. Watson, E. A. Fuel control and burning in aero-gas-turbine engines. *Chartered Mech. Engr. 3*, 91–126 (1956).
4. Clarke, J. S. A review of some combustion problems associated with the aero gas turbine. *J. Roy. Aeronaut. Soc. 60*, 221–240 (1956).
5. Streid, D. Design analysis of G. E. type I-40 jet engine. *Aviation 45*, 51–59 (1946).
6. Ashwood, P. F. The "T-Scheme." *Flight 52*, 630–631 (1947).
7. Schulte, R. C. Design analysis of BMW 003 turbojet. *Aviation 45*, 55–68 (1946).
8. Schlaifer, R., and Heron, S. D. *Development of Aircraft Engines and Fuels.* Andover Press, Andover, Mass., 1950.
9. Design and construction of German 109-011 A-O turbojet. *Aviation 45*, 63–64 (1946).
10. *Aviation Week 51*, 16 (1949).

11. Metro-Vick gas turbine. *Flight 49*, 423 (1946).
12. Nahigyan, K. K., and Manganiello, E. J. Combustion and static thrust investigations of a jet-propulsion power plant. *NACA Advance Restricted Rept. E6F19a*, July 1946.
13. Ellis, M. C., Jr., and Brown, C. E. NACA investigation of a jet propulsion system applicable to flight. *NACA Rept. 802*, 1944.
14. *Flight 59*, 370–371 (1951).
15. Childs, J. H., McCafferty, R. J., and Surine, O. W. Effect of combustor-inlet conditions on combustion in turbojet engines. *SAE Quart. Trans. 1*, 266–278, 344 (1947). Also *NACA Rept. 881*, 1947.
16. Olson, W. T., and Bernardo, E. Temperature measurements and combustion efficiency in combustors for gas turbine engines. *Trans. Am. Soc. Mech. Engrs. 70*, 329–334 (1948).
17. Childs, J. H. Preliminary correlation of efficiency of aircraft gas-turbine combustors for different operating conditions. *NACA Research Mem. E50F15*, 1950.
18. *Selected Combustion Problems, II.* Butterworths, London, 1956.
19. Childs, J. H., and Graves, C. C. Relation of turbine engine combustion efficiency to second-order reaction kinetics and fundamental flame speed. *NACA Research Mem. E54G23*, 1954.
20. Childs, J. H., and Graves, C. C. Correlation of turbine-engine combustion efficiency with theoretical equations. *Sixth Symposium on Combustion, Yale Univ.*, Aug. 19–24, 1956.
21. Graves, C. C. Combustion of single isooctane drops in various quiescent oxygen-nitrogen atmospheres. *Proc. Third Midwestern Conference on Fluid Mech., Univ. Minn.*, 759–778 (1935).
22. Olson, W. T. Combustion for aircraft engines. *Trans. Inst. Aeronaut. Sci. RAeS Conference*, June 20–23, 1955.
23. Silverstein, A. Research on aircraft propulsion systems. *12th Wright Bros. Lecture. J. Aeronaut. Sci. 16*, 197–227 (1949).
24. Pinkel, I. I., and Shames, H. Analysis of jet-propulsion-engine combustion-chamber pressure losses. *NACA Rept. 880*, 1947.
25. Way, S. Problems in the development of turbojet combustion chambers. *Westinghouse Research Labs. Sci. Paper 1421*, Mar. 1949.
26. Lloyd, P., and Probert, R. P. The problem of burning residual oils in gas turbines. *Proc. Inst. Mech. Eng. 163*, 206–220 (1950).
27. Grobman, J. S., and Dittrich, R. T. Pressure drop and air flow distribution in gas-turbine combustors. *Am. Soc. Mech. Engrs. Annual Meeting, New York*, Nov. 25–30, 1956.
28. Mock, F. C. Engineering development of the jet engine and gas turbine burner. *SAE J. (Trans.) 54*, 218–227 (1946).
29. Gray, G. W. *Frontiers of Flight.* Knopf, 1948.
30. Dittrich, R. T., and Graves, C. C. Discharge coefficients for combustor-liner air-entry holes. I: Circular holes with parallel flow. *NACA Tech. Note 3663*, 1956.
31. Knight, H. A. and Walker, R. B. The component pressure losses in combustion chambers. *Natl. Gas Turbine Establishment, England, Rept. R143*, Nov. 1953.
32. McCafferty, R. J. Effect of fuels and fuel-nozzle characteristics on performance of an annular combustor at simulated altitude conditions. *NACA Research Mem. E8C02a*, Sept. 1948.
33. Mock, F. C., and Ganger, D. R. Practical conclusions on gas turbine spray nozzles. *SAE Quart. Trans. 4*, 357–367 (1950).
34. Carey, F. H. The development of the spill flow burner and its control system for gas turbine engines. *J. Roy. Aeronaut. Soc. 58*, 737–753 (1954).
35. Olson, W. T., Childs, J. H., and Jonash, E. R. The combustion efficiency problem of the turbojet at high altitude. *Trans. Am. Soc. Mech. Engrs. 77*, 605–615 (1955).
36. Nerad, A. J. Some aspects of turbojet combustion. *Aeronaut. Eng. Rev. 8*, 24–26, 88 (1949).
37. Pavia, R. E. Static pressures and air flows in the flame tube of a Derwent V

combustor. *Aeronaut. Research Lab., Dept. of Supply, Australia, Engines Note 154*, Jan. 1951.

38. Ruggeri, R. S., Callaghan, E. E., and Bowden, D. T. Penetration of air jets issuing from circular, square, and elliptical orifices directed perpendicularly to an air stream. *NACA Tech. Note 2019*, 1950.
39. Hawthorne, W. R., Robers, G. F. C., and Zacek, B. Y. Mixing of gas streams—The penetration of a jet of cold air into a hot stream. *Roy. Air Establishment Tech. Note Eng. 271*, Mar. 1944.
40. Weeton, J. W. Mechanisms of failure of high nickel-alloy turbojet combustion liners. *NACA Tech. Note 1938*, 1949.
41. Winter, E. F. Heat transfer conditions at the flame tube walls of an aero gas turbine combustion chamber. *Fuel 34*, 409–428 (1955).
42. Topper, L. Radiant heat transfer from flames in a single tubular turbojet combustor. *NACA Research Mem. E52F23*, 1952. Also *Ind. Eng. Chem. 46*, 2551–2558 (1954).
43. Berlad, A. L., and Hibbard, R. R. Effect of radiant energy on vaporization and combustion of liquid fuels. *NACA Research Mem. E52I09*, 1952.
44. *Fourth Symposium on Combustion*. Williams & Wilkins, 1953.
45. Tischler, A. O., and Male, T. Oscillatory combustion in rocket propulsion engines. *Proc. Symposium on Aerothermochemistry, Northwestern Univ.*, Aug. 22–24, 71–82 (1955).
46. Kaskan, W. E., and Noreen, A. E. High-frequency oscillations of a flame held by a bluff body. *Trans. Am. Soc. Mech. Engrs. 77*, 885–895 (1955).
47. Truman, J. C., and Newton, R. T. Why do high-thrust engines screech? *Aviation Age 23*, 136–143 (1955).
48. Blackshear, P. L., Jr., Rayle, W. P., and Tower, L. K. Study of screeching combustion in a 6-inch simulated afterburner. *NACA Tech. Note 3567*, 1955.
49. Blackshear, P. L., Jr. Driving standing waves by heat addition. *NACA Tech. Note 2772*, 1952.
50. Rogers, D. E., and Marble, F. E. A mechanism for high frequency oscillation in ram-jet combustors and after-burners. *Jet Propulsion 26*, 456–462 (1956).

# PART THREE

# *MECHANICAL AND METALLURGICAL ASPECTS*

# SECTION I

# *MECHANICS OF MATERIALS FOR GAS TURBINE APPLICATIONS*

EGON OROWAN
C. RICHARD SODERBERG

**I,1. Introduction.** The gas turbine, more than the internal combustion engine and the airplane itself, is a typical instance of a device that was relatively easy to invent but difficult to make work. Since its net output is the difference between the turbine output and the compressor input, both components have to reach high degrees of efficiency before their combination can deliver power. This requires a considerable knowledge of the aerodynamics and thermodynamics involved, and the necessarily high temperatures and stresses demand materials, and a knowledge of their behavior, far beyond what was sufficient at earlier stages of engineering.

The growing demands of the engineer have been met by the development of new materials and by the increasing understanding of their mechanical behavior. Until recently, materials for high temperature service have been developed largely by trial-and-error methods involving the increasing use of less common chemical elements. At present, however, the Periodic System is nearly exhausted as a source of new alloying elements available in sufficient quantities, and further progress can only be hoped from a more rational use of those already being utilized. Both the design of new materials, and their effective application in engineering, therefore, call for a deeper fundamental knowledge of the physics of strength and plasticity. Much of what the engineer needs in this field has been made available to him by the tremendous development of the physics of solids since 1912 (the date of Laue's discovery of X-ray diffraction); more could be obtained relatively easily once the needs of the designer are clearly formulated.

The present article attempts to give a brief summary of the fundamental aspects of the mechanical properties of solids as far as they are of direct or indirect interest to the gas turbine designer. It will be concluded by a brief survey of the most important materials available; this, however—being of relatively ephemeral value—will occupy less space than the discussion of the more permanent fundamental points of view.

**I,2. Influence of Temperature on Mechanical Behavior of Metals and Other Crystalline Materials. General Survey.** The influence of the temperature on the mechanical behavior of a purely elastic or (in addition to its elasticity) purely viscous material is relatively simple: it is known if the temperature dependence of the elastic constants or of the coefficient of viscosity has been measured. The only complication occurs in the "softening range" of supercooled liquids (e.g. between 400 and 600°C for the common glasses) in which the change of the viscosity with the temperature involves a considerable inertia effect: hours or days at a constant temperature may be required to obliterate the molecular structure developed at a preceding temperature [*1,2,3*]. Below the softening range the structure is practically frozen, and viscosity is largely replaced by delayed elasticity [*4*].

By comparison, the influence of the temperature on the mechanical behavior of crystalline materials (in particular, of metals) is extremely complex. The fundamental cause of this difference is clear enough. Elastic deformation (apart from rubber elasticity and certain types of delayed elasticity) consists in the straining of atomic bonds without atomic rearrangements; likewise, viscous deformation in amorphous materials does not change the atomic structure in the sense that the degree of disorder which corresponds to (relative) thermodynamic equilibrium at the given temperature is preserved. In contrast to this, crystalline materials suffer various structural injuries in the course of nonelastic deformation; strain hardening is the most common manifestation of these. These injuries can be removed by thermal agitation if the temperature is high enough ("thermal softening" or "recovery"). This process may occur during the deformation itself and can lead to deformation progressing at a constant stress, in the course of which new injuries are produced at the same rate at which existing ones are removed ("recovery creep"). An entirely different type of creep is due to the local superposition of thermal stress fluctuations upon the applied stress; this is the creep observed after the application of a constant load to a specimen of steel, copper, brass, or aluminum in the tension or compression test at room temperature. Its name is "transient creep"; it disappears in the course of time because the increase of strain hardening makes it increasingly difficult for thermal stress fluctuations to produce further slip.

Polycrystalline metals contain in the grain boundaries an element of atomic disorder, a "two-dimensional glass," which can give rise to viscous flow consisting in the sliding of the grains upon their neighbors if the temperature is above the "softening point" of the grain boundary (the "equicohesive temperature"). Thus, the grain boundaries which are sources of strength and hardness at lower temperatures can become sources of weakness at higher temperatures. However, the grain boundaries are usually of irregular shape, so that sliding along them cannot continue unhindered by geometrical nonconformities. If the grains remained rigid, boundary sliding would come to a rest almost as soon as it had started, or else it would lead to the opening up of cracks between the grains. The latter process does in fact occur after considerable sliding; it is the cause of the characteristic form of creep fracture ("stress rupture"). However, a considerable amount of sliding may precede cracking because the grains can readjust themselves to the movements of their neighbors by means of slip, recrystallization, or "polygonization" at the most highly stressed prominences along their boundaries. The velocity of creep due to boundary sliding, therefore, is usually determined not by the viscosity of the boundary but by the rate at which readjustments of the shapes of the grains can remove the hindering influence of geometrical nonconformities.

A wide and technologically very important range of structural changes at high temperatures, strongly influencing plastic deformation and influenced by it, embraces various precipitation and re-dissolution phenomena. The highly creep-resisting alloys upon which modern gas turbines and jet engines are based owe their performance usually to high-melting microscopic or submicroscopic precipitates, frequently carbides of titanium or niobium. Deposited in the grain boundaries, they act like sand on the rails and hinder sliding. Even more important, however, is their action inside the grains. Strain hardening can increase the stress carrying capacity, but it is currently removed by thermal agitation already well below the melting point of the metal, and recovery creep (creep due to thermal softening) results. In many materials, such as refractory oxides or carbides, diamond, etc., recovery does not set in until very high temperature are reached; this, however, is due to the very strong atomic binding forces which at the same time make plastic deformation very difficult and render the material brittle. If now such brittle refractory materials are present in the form of small particles dispersed in a ductile matrix, they harden the material just like the lattice injuries produced by cold working; unlike these, however, they cannot be wiped out by thermal fluctuations until temperatures are reached that may have to be far higher than the recrystallization temperature of the matrix. At the same time, moderate though highly effective amounts of brittle refractory precipitates may still leave the material sufficiently ductile. In fact, most of the highly creep-resisting alloys can be considered as "cermets" (metal-ceramics) containing relatively small amounts of hard refractory materials finely dispersed in a metallic matrix. A simple calculation (see Art. 3 and 5) shows that such highly dispersed "cermets" can be more effective in preventing creep than coarse mechanically mixed and sintered ones, unless these contain much of the hard refractory material—in which case they may be too brittle.

The precipitation of refractories can be nucleated by injuries due to plastic deformation (for example, dislocations). Plastic deformation of a supersaturated solid solution, therefore, may lead to precipitation prevalently at the weakest points where the deformation has been particularly strong; in this way, the precipitate may automatically reinforce the regions of lowest resistance to creep. This seems to be the basis of the "warm-working" treatment which can raise the creep resistance of some alloys very considerably.

It is apparent from the foregoing sketch that the mechanical behavior of crystalline materials is influenced by many factors, representing fundamentally different processes which come into play in different temperature ranges depending on the individuality of the material. The hope of a general mathematical treatment based on the knowledge of a few simple functions of the temperature (such as are the coefficient of viscosity and the elastic constants), therefore, belongs to the realm of utopia. Only two realistic approaches seem possible at present. One is a purely empirical one, based on tests imitating service conditions with materials of practical interest. Useful and indispensable as such an approach is, it is not sufficient as a guide for long range development, and it does not eliminate the possibility of unpleasant surprises if the tests used do not imitate the service conditions accurately and exhaustively enough. The alternative approach requires a scientific acquaintance with the molecular mechanisms of the various processes that may occur in materials in different temperature ranges before, during, or after plastic deformation. Such an acquaintance is indispensable to the engineer to whom the high temperature mechanical behavior of materials is of importance.

**I,3. Yield Stress. Dislocations and Plastic Deformation.** At temperatures too low for recovery to occur (for pure metals, usually below about one third of the absolute melting point), the deformation of metals beyond the elastic limit is of the plastic type, governed by an approximate functional relationship between stress and strain (the stress-strain curve). The relationship is only approximate because transient creep ("cold creep"; see below, Art. 6) exists at all temperatures. When a stress exceeding the elastic limit is applied, initially very rapid but rapidly decelerating creep occurs as indicated in Fig. I,3a, curve $L$. Since at constant stress the transient creep rate dies away rapidly and the strain "settles down" to a fairly definite value, a stress-strain curve can be constructed which does not depend strongly upon the rate of loading, i.e. upon the time elapsed between the application of the stress and the strain reading. At higher temperatures, the strain-time curve is of type $H$ (Fig. I,3a); the decelerating transient creep is followed by a constant-rate viscous creep.

The ordinate of the stress-strain curve is the yield stress, i.e. the stress needed for further plastic deformation after a given amount of preceding plastic strain. In some metals, such as annealed low-carbon steel, plastic deformation starts almost abruptly, often with a drop of stress. In such cases, the stress at which plastic deformation begins is called the (upper) "yield point"; the level to which the stress drops after the first traces of plastic deformation is the "lower yield point."

With most engineering metals, the stress vs. plastic strain curve joins the stress axis tangentially: there is no sharp stress value at which plastic deformation begins, no sharp elastic limit. However, single crystals usually have a fairly distinct initial yield stress; a few per cent below this value plastic strains cannot be recognized even with the most sensitive methods available which could reveal deformations almost of

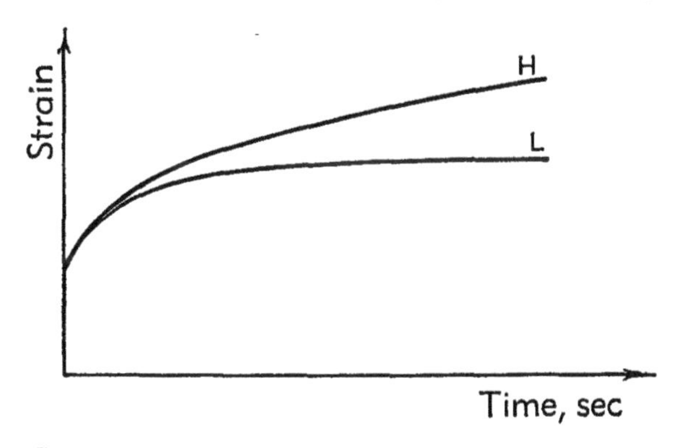


Fig. I,3a. Creep under constant stress at low ($L$) and high ($H$) temperatures.

the order of magnitude of the atomic spacings. The most frequent reason for the lack of a sufficiently sharp elastic limit in engineering tests on polycrystalline metals is that crystal grains of different orientation begin to yield at different loads.

The plastic behavior of metals in the low temperature range as just described raises the following physical questions:

1. What determines the value of the initial yield stress of an undeformed crystal or polycrystalline material?
2. What is the cause of strain hardening, and what determines the rate of increase of the yield stress with the plastic strain?
3. What are the molecular mechanisms of transient and of viscous creep?

The first two questions are briefly treated in this article; the third in Art. 6 and 7.

There is a remarkably simple way of estimating the order of magnitude of the stress at which slip would be expected to start in a perfectly regular crystal. Let Fig. I,3b (left) represent a metal crystal upon which a shear stress $\tau$ acts; the circles stand for atoms, and one of the possible slip planes should be the horizontal plane *S-S*. In the elastic region shear stress $\tau$ and shear strain $\gamma$ are related by the line *AB* in Fig. I,3c. If the parts 1 and 2 of the crystal above and below *S-S* have been displaced horizontally by one identity spacing (Fig. I,3b, right), the interatomic forces are again in equilibrium without any external force. Thus at the strain AC (Fig. I,3c), corresponding to this amount of displacement between neighboring atomic slip planes, the shear stress is zero, and in the neighborhood of *C* again a Hookean straight line *DCE* represents the

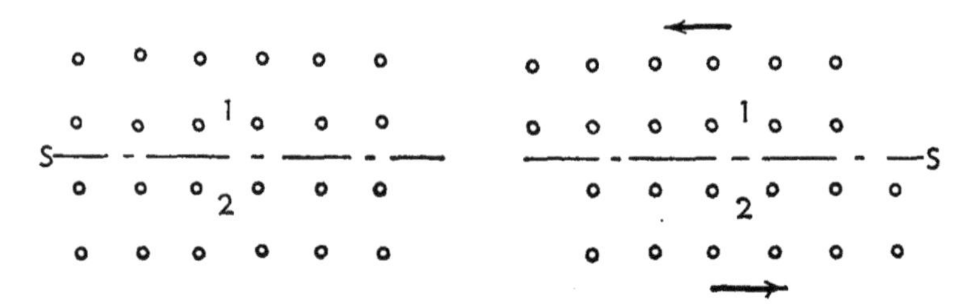


Fig. I,3b. Slip in a crystal.

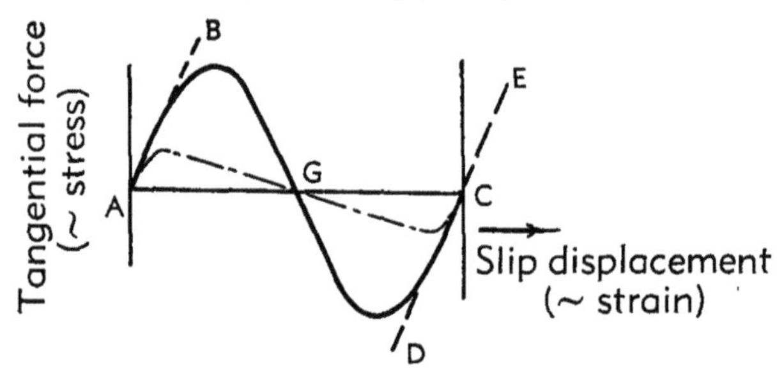


Fig. I,3c. Variation of tangential force with shear displacement during slip.

stress-strain relationship. At *G*, halfway between *A* and *C*, there is an unstable equilibrium between parts 1 and 2 of the crystal; the stress is again zero, but the sign of the gradient of the stress-strain line around this point is opposite to that at *A* and *D*. The simplest stress-strain relationship that satisfies the local requirements at *A*, *G*, and *C* is the sine:

$$\tau = \tau_m \sin 2\pi \frac{\gamma}{\gamma_0} \tag{3-1}$$

where $\gamma_0$ is the strain that corresponds to a tangential displacement of adjacent atomic planes by one identity spacing. If $\gamma$ is small,

$$\tau \cong \tau_m 2\pi \frac{\gamma}{\gamma_0} \tag{3-2}$$

and this must be identical with Hooke's law $\tau = G\gamma$ where $G$ is the modulus of shear; consequently,

$$\tau_m \cong \frac{G}{2\pi}\gamma_0 \tag{3-3}$$

The exact value of the identity shear strain depends on the structure of the crystal; however, it is always of order unity for simple crystals such as those of the metals. Consequently, the maximum shear stress beyond which irreversible deformation starts must be of the order

$$\tau_m \cong \frac{G}{2\pi} \tag{3-4}$$

A closer consideration shows that the sine curve is far from representing the real conditions; the true relationship must be more of the type indicated by the dash-dotted curve in Fig. I,3c. However, it can be shown that the difference is not likely to change the order of magnitude of the estimate (Eq. 3-4); for copper, the probable, more accurate value of $\tau_m$ cannot be less than some $\frac{1}{3}$ or $\frac{1}{4}$ of that given by Eq. 3-4 [*5*]. The shear stress at which plastic deformation starts in a perfect and undisturbed crystal, therefore, must have an order of magnitude between, say, $\frac{1}{10}$ and $\frac{1}{50}$ of the shear modulus. For the common metals, the order of magnitude of the shear modulus is $10^{12}$ dynes/cm$^2$ = $10^6$ bars $\cong$ 15,000,000 psi; the theoretical expectation of the elastic limit of the perfect crystal, therefore, would be $10^4$ to $10^5$ bars (150,000 to 1,500,000 psi). In contrast to this, the observed yield stresses of single shear of crystals of the softer metals (Zn, Cd, Mg, Pb, Cu, Ag, Au), i.e. the shear stress in the slip plane resolved in the slip direction at which plastic deformation starts, is between 5 and 10 bars (70 and 140 psi). The first problem of the physics of plasticity is to explain this discrepancy.

A long series of investigations has shown that the crude estimate just given cannot be wrong in the order of magnitude; consequently, some deviation of the crystal from ideal perfection must be responsible for the discrepancy. For several reasons, it cannot be due to stress concentrations at microscopic or submicroscopic cracks which are responsible for the similar discrepancy between the observed value and the theoretical estimate of the tensile strength of brittle materials like glass. This is apparent already from the fact that the depth (length) of a crack of atomic sharpness that would have a stress concentration factor of several thousands would be a multiple of the thickness of many single crystal specimens for which such discrepancies were observed. Another suggestion was that thermal stress waves, superposed to the applied stress, would cause yielding at extremely low values of the latter. A more detailed analysis, however, has shown that under usual conditions thermal stress waves cannot reduce the yield stress of the common metals more than by a factor of the order of unity. This does not mean that their influence is negligible, because transient creep must be due entirely to such thermal stress fluctuations; but they cannot be responsible for the large discrepancy between the observed and the estimated values of the initial yield stress of crystals.

What seems to be the most frequent cause of the discrepancy between the observed and the theoretically expected value of the yield stress can be recognized from Fig. I,3d. In reality, slip cannot take place simultaneously at all points of a slip plane, as would be the case if parts 1 and 2 in Fig. I,3b moved as rigid units. It is bound to start at some point and spread out from it. In Fig. I,3d it is assumed that slip has started at point $P$ and progressed halfway across the crystal along the slip plane $S$-$S$ to the point $Q$. The atoms missing from the dotted strip above $P$ have moved in the direction of the upper arrow, each displacing its neighbor to the left, and so on; obviously, the same number of atoms as are missing above $P$ can be found now in the dotted strip above $Q$, at the boundary of the slipped region, forming an extra half-sheet of atoms inserted above the slip plane. As a consequence, the atoms facing each other across the slip plane at $Q$ are no longer in register: they have been lifted more or less completely out of the potential troughs in which they were originally resting between the atoms on the opposite side of the slip plane. A relatively low shear stress is sufficient, therefore, to move the atoms above the slip plane at $Q$ past those below it; in other words, if a sufficiently large patch of slip is already present in a slip plane, a relatively low stress can extend its boundaries and increase the amount of slip. This, however, does not explain yet why the yield stress is lower than the theoretical estimate for the perfect crystal: for a full explanation, it must be shown that easily moved slip boundaries can arise without very high stresses. The salient point in this respect is that slip boundaries of the type seen in Fig. I,3 can arise without any slip; for instance, they can be present as a consequence of accidents of growth [*6,7*]. A simple possibility is illustrated in Fig. I,3e. If two small crystals grow together at the moment when an atomic plane on one of the adhering faces is half-completed, the result is a crystal with an extra half plane of atoms inserted, just as in Fig. I,3d. If the edge of the half plane lies in a potential slip plane, its surroundings are indistinguishable from a slip boundary and it can give rise to slip at stresses far below the theoretical estimate for the perfect crystal.

A slip boundary as that at $Q$ in Fig. I,3d is called a dislocation. In the course of the last 20 years it has become clear that dislocations represent a basic element in the structure of deformed crystals and that they are present in small numbers in most undeformed crystals. Their role in the plastic behavior of crystalline materials is as fundamental as is that of dipoles in the dielectric behavior of matter. It is appropriate, therefore, to give some consideration here to the structure and properties of dislocations.

Fig. I,3f shows a schematic view of an atomic plane just above the slip plane of a crystal after slip has occurred over a square patch at the

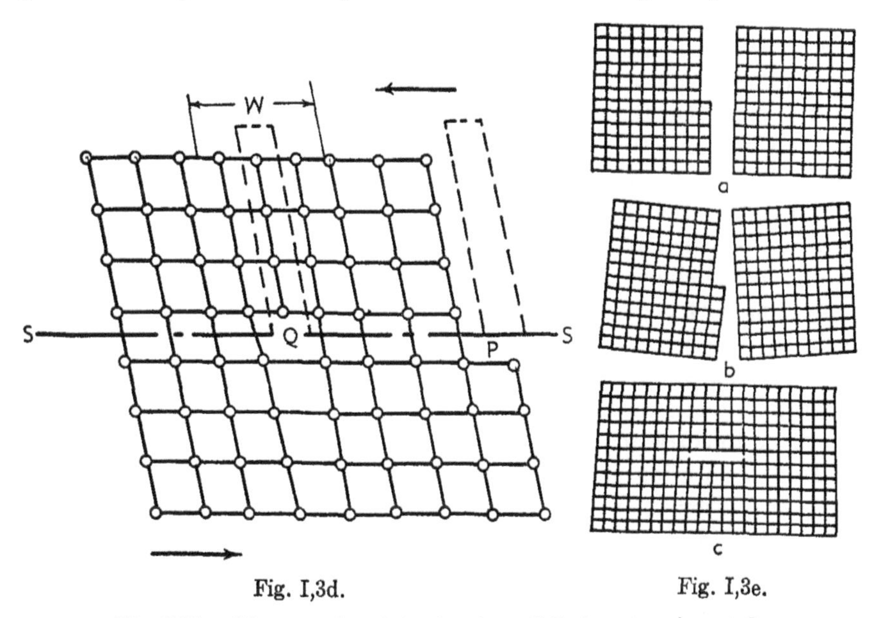


Fig. I,3d. Fig. I,3e.

Fig. I,3d. Slip started at $P$ in the plane $S$-$S$ after stopping at $Q$.
Fig. I,3e. Formation of an edge dislocation by the fusion of two cyrstals.

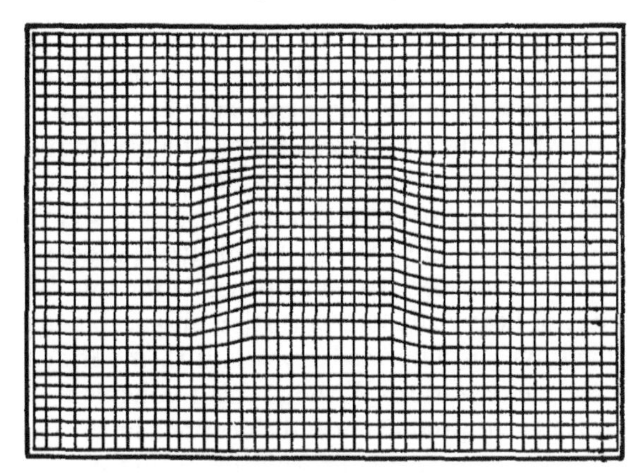

Fig. I,3f. Lattice plane with square area of slip.

center, vertically upward. The top of the square slip boundary is compressed; this corresponds to the insertion of the extra plane above the slip plane in Fig. I,3d. The bottom side is extended; the corresponding extension is seen in Fig. I,3d below the extra plane at $Q$. Naturally, there is an extension along the top side of the square in the atomic plane below the active slip plane (not shown in the figure), and a compression along the bottom side. Top and bottom of the square represent "edge dislocations" (they run along the edges of the extra plane); Fig. I,3d shows the structure of an edge dislocation in cross section. Quite different is the structure of the vertical sides of the square in Fig. I,3f. The upward motion, which creates tension or compression normally to the top and bottom sides, gives rise to shear in the atomic plane itself, along the vertical sides. On the right-hand side the shear is clockwise; on the left, counterclockwise. Its directions are reversed in the atomic plane underneath. This type of slip boundary is a "screw dislocation."[1] Obviously, a slip boundary that is perpendicular to the slip direction is an edge dislocation; one that is parallel is a screw dislocation. Boundaries that are neither perpendicular nor parallel are dislocations of the mixed type, containing both edge and screw components. The relatively easy mobility under a shear stress is a property common to all types of dislocations if they represent slip boundaries for potential slip planes (some dislocations, generated by the insertion of half planes of atoms into a crystal, have no slip plane to correspond to).

If a dislocation loop (closed slip front) lies entirely in one active slip plane, it expands or contracts under a shear stress (according to the direction of the stress), and finally disappears if the specimen is a single crystal. In a polycrystalline material, an expanding slip front is ultimately jammed against the grain boundary and may become immobilized unless the direction of the stress changes. Such a dislocation loop, therefore, can only produce slip by one identity period before it becomes ineffective. However, if only a part of the loop lies in the active slip plane while its continuations go off the slip plane and run into the crystal at an angle, the end or ends of the part in the slip plane may be more or less immobilized at the points where the dislocation leaves the plane. In this case, the active part of the dislocation loop can sweep over the slip plane again and again without disappearing or becoming immobilized, and it can produce potentially any amount of slip. Such "Frank-Read dislocation sources" or "dislocation mills" [*8*] seem to be responsible for the "slip bands" or "slip zones," sites of heavy slip concentration observable microscopically or with the naked eye on deformed crystals if their surface was smooth enough before the deformation. Without this mechanism, or an equivalent mechanism that can produce new dislocations in the course of deformation, a small amount of plastic deformation would use up the existing dislocations in the crystal, which would then attain the extremely high theoretical yield stress.

The theoretical conclusion that the yield stress of a perfect crystal

[1] The name "dislocation," borrowed from geology, was originally introduced into the theory of elasticity by A. E. H. Love. The term "edge" and "screw" are due to J. M. Burgers.

must be extremely high has recently been confirmed experimentally by the observation that electrolytically grown, extremely thin "whiskers" of tin and other metals when bent under the microscope show an extremely high yield stress, fairly close to the theoretically expected value [*9*].

The realization that dislocations capable of acting as slip fronts can arise without any slip, by accidents of growth and perhaps also in other ways, explains at least qualitatively the discrepancy between the observed and the theoretically estimated values of the yield stress. It has to be complemented, however, by considerations that may lead to a quantitative calculation of the yield stress. If a crystal contains dislocations that can act as slip fronts, the yield stress may be determined by two factors: (1) the shear stress required to move the dislocation in its slip plane even if this contains no significant obstacles; or (2) the shear stress needed for pressing a slip front across a field of obstacles in the slip plane, the obstacles being dislocations or other lattice injuries produced by preceding slip or by accidents of growth, precipitated particles, regions of adverse internal stress, etc.

Without any theoretical consideration it is obvious that both factors must be of importance [*10*, p. 69]. The initial yield stress of soft metals is very low compared with the level to which the yield stress can rise by strain hardening; in the strongly hardened state, therefore, the yield stress is determined mainly by lattice injuries acting as obstacles to slip. On the other hand, hard metals or nonmetallic crystals have a very high initial yield stress which, as a rule, is due to the high value of the "driving stress," (Peierls-Nabarro force) required for moving the dislocation in the absence of obstacles. That it cannot be due to injuries present before deformation (except in special cases) is indicated by the fact that a soft carborundum, diamond, or tungsten has never been found; consequently, their hardness is unlikely to be due to avoidable imperfections. A direct proof of this exists for diamond which, according to X-ray evidence, can be one of the most perfect crystals available. There are many highly imperfect diamonds also, but practically no difference is observed between the hardness of the "perfect" and imperfect crystals. It can be said, therefore, that inherently hard crystals are hard because their dislocation driving stress is high. The dislocation is difficult to move if the atoms on opposite sides of the slip plane are crosslinked by strong directional covalent bonds, or if the lattice energy and the intrinsic pressure are so high that the atoms in the dislocation are "jammed together" in the direction perpendicular to the slip plane. Both factors are present in metals with a partially filled *d* shell (the transition metals), since the *d* electrons produce strong directional bonds. Metals with nearly filled *d* shells (e.g. Ni, Pd) are somewhat softer because the electric charge distribution approaches spherical symmetry. A glance into the Periodic Table shows that, in fact, as far as the mechanical properties are known, all strong metals, with the exception of beryllium, have a partially filled *d*-shell [*10*, p. 69; *11*, p. 415].

The next question is: When the yield stress is determined mainly by the difficulty of forcing slip through a field of obstacles, how is its magnitude to be calculated? Since there are numerous types of obstacles, and usually several alternative ways in which slip may occur in spite of them, no simple general answer can be given. However, a few typical and apparently particularly important cases may be mentioned.

*Accumulation of dislocations at fixed obstacles* (*slip confined to areas between the obstacles*). Suppose that the material is subdivided into cells by walls impermeable to slip. This case is approximately realized in polycrystalline metals, particularly if their lattice is hexagonal so that only one favorable plane of slip exists and the slip cannot penetrate easily from one grain into the next. The first dislocation moves relatively easily to the grain boundary; the next, however, has to overcome the opposing shear stress created by the presence of the first dislocation, and so on. As dislocations accumulate at the grain boundary, it becomes increasingly difficult to move another dislocation toward it, and the material strain hardens without necessarily suffering additional lattice injuries, merely by the accumulation of dislocations which impede the movement of following dislocations. This mechanism was first considered by G. I. Taylor [*12*], with the assumption of closely spaced hypothetical mosaic boundaries within the individual crystals. If the geometry of the obstacles is known, the rise of the yield stress can be calculated as a function of the strain. The relative importance of this effect in the observed strain hardening can be recognized from the magnitude of the Bauschinger effect: since the accumulation of dislocations would make reverse slip easier, the difference between the yield stress for continuing straining and that for reverse straining is an indication of the Taylor component in the observed strain hardening. Usually, it is relatively small; however, if hard inclusions act as fixed obstacles, they give rise to a distinct additional hardening of the Taylor type [*13*].

*Forcing of dislocations through regions of adverse internal stress.* In hardening of the Taylor type, the yield stress is due to the increase of the adverse internal stress due to the accumulation of dislocations at obstacles; the initial yield stress is nearly zero unless other sources of a yield stress are present (e.g. a finite driving stress). On the other hand, the crystal may contain already in the undeformed state regions of adverse internal stress which, of course, give rise to a finite initial yield stress. Thus, Mott and Nabarro [*14*] have treated the case of precipitated particles larger or smaller in the unstrained state than would be the cavity they occupy in the matrix if the matrix were unstrained. Such particles exert a (positive or negative) pressure upon the surrounding matrix and produce stresses in it which, in certain regions, involve a shear stress in the slip plane opposite in direction to the applied stress. A slip front cannot progress, then, unless the applied stress is large enough to overcome the adverse local internal stress.

The magnitude of the yield stress arising from this cause would be easy to calculate in a two-dimensional case if, in addition, each slip front had to overcome the same maximum internal stress. In the three-dimensional case, however, the situation is entirely different. If the regions of adverse stress are patches much smaller in diameter than the distance between their centers, the slip front is suspended on them like a garland on a row of pegs (Fig. I,3g). From the consideration leading to

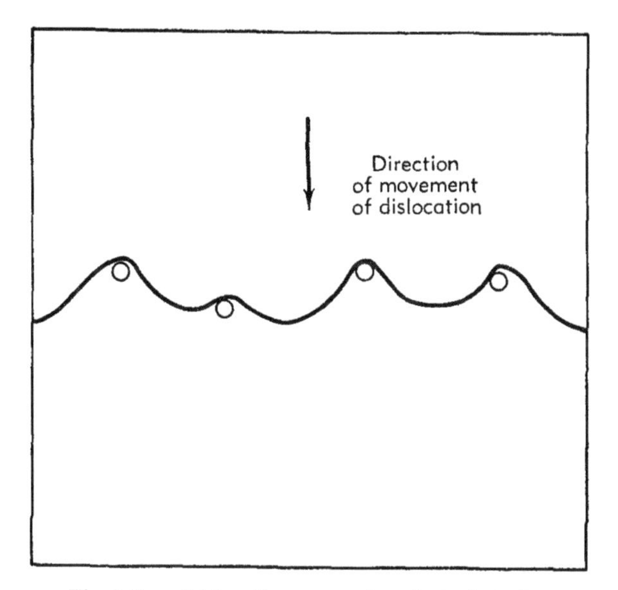


Fig. I,3g. Dislocation pressed against obstacles.

Eq. 3-6 (further below) it is easy to recognize that the applied stress required for pulling the dislocation across the "pegs" may be much lower than the adverse internal stress acting upon the dislocation at the pegs; the ratio between the two is approximately the ratio between the radius of curvature on the dislocation at the pegs and that between them.

*Extrusion of dislocations between obstacles.* If the adverse stresses in Fig. I,3g are high and the distances between the patches much greater than both the diameters of the patches and the effective depth of the slip front (the "width" of the dislocation, denoted by $W$ in Fig. I,3d), slip may extend over the slip plane in a peculiar way, by engulfing the obstacles without shearing through them [*15*]. Like the processes considered above, this is a case of slip being confined to the areas between the obstacles; an important characteristic point is introduced, however, by the circumstance that it requires a definite magnitude of the applied stress to squeeze even the first slip front through between two obstacles.

This process is illustrated by Fig. I,3h. The obstacles may be centers of high adverse stress, or hard inclusions, or even points at which dislocations go off the slip plane at an angle to it. The common feature is that the dislocation is held back at certain fixed points in the slip plane. Let the shaded patches in Fig. I,3h represent, for instance, cross sections of hard inclusions; the plane of the figure is the slip plane. A dislocation 1 coming from above is held up at a number of obstacles, and the applied stress makes it bulge between them as seen in the position 2. With increasing stress, the bulges increase until they come into contact with one another (position 3) and fuse together to form a smoother line 4 moving past the row of obstacles and leaving each obstacle encircled with a dislocation

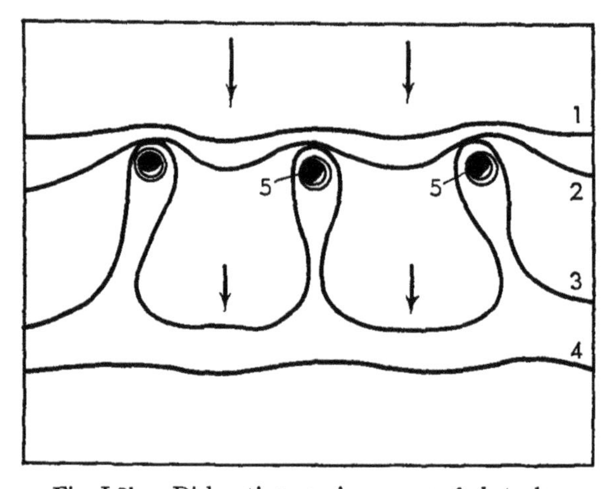


Fig. I,3h. Dislocation passing a row of obstacles.

ring 5. In this way, the dislocation can sweep over the entire slip plane with the exception of the small patches within the rings encircling the obstacles, just as two steel plates joined by rivets can slip on each other by a certain amount without immediately shearing through the rivets. This process of extrusion of dislocation bulges between anchoring points can repeat itself; if the obstacles are inclusions, every wave of slip adds one dislocation ring to those already encircling them. The accumulation of dislocation rings at the obstacles is quite analogous to the accumulation of dislocations at the impermeable walls in Taylor's model; it will be seen that this process is accompanied by increasing strain hardening. However, in contrast to Taylor's model, the slip process does not start at zero stress; a finite characteristic shear stress must be applied to start it. In view of the importance of the process, a brief sketch of the calculation of this initial yield stress is worth giving.

The elastic stresses in a slip front, and the disregister between the atoms facing each other across the slip plane (see Fig. I,3d) give rise to a characteristic specific energy $\Gamma$ per unit length of the dislocation. The elastic part of this energy can be calculated approximately [*16,17*] and the disregister part can be estimated. The result is that the energy of a dislocation, per interatomic spacing along its length, is of the order of one electron volt for the common engineering metals. Since 1 ev = 1.6 $\times 10^{-12}$ erg, and the atomic spacings are of the order of 1 Angstrom unit = $10^{-8}$ cm, the dislocation energy $\Gamma$ is of the order of $10^{-4}$ erg/cm. If a dislocation line is lengthened by $dx$, the stretching force $F$ does the work $Fdx$ which must be equal to the increase of the energy $\Gamma dx$ if the stretching is under a tension numerically equal to its energy $\Gamma$ per unit length, just as the surface of a liquid is under a tension numerically equal to the surface energy per unit area.

If a dislocation bulges forward between anchoring points, as in Fig. I,3h, its tension which tends to straighten out the bulge must be balanced by a "shear pressure" exerted by the applied stress, just as the tendency

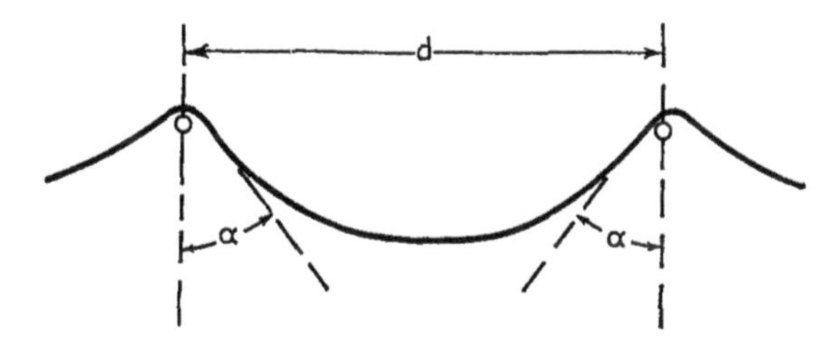


Fig. I,3i. Calculation of the extrusion stress of a dislocation.

of the surface tension to contract a soap bubble formed at the nozzle of a pipe must be balanced by the air pressure. If, as in Fig. I,3i, the dislocation bulge makes angles $\alpha$ with the slip direction at the anchoring points, the dislocation tension at the ends of the arc exerts a resultant force $2\Gamma \cos \alpha$ opposite to the slip direction, which is here assumed to be normal to the chord of the arc. This force must be balanced by the applied stress. If a slip plane is assumed to be of unit area, and the identity spacing in the slip direction is $b$, the work of the applied shear stress $\tau$ for one atomic step of slip is $\tau b$. If a dislocation line of length $y$ advances by $dx$ normal to itself, slip extends by an area $ydx$ and the mean displacement between the parts of the crystal above and below the slip plane is $bydx$; the work done is $\tau bydx$. This can be interpreted, with Mott, by saying that a force $\tau b$ acts upon the dislocation per unit of its length.

In Fig. I,3i, therefore, the bulge is pulled back by the tensions at its ends, and at the same time it is forced outward by the "shear pressure" $\tau b$. This is quite analogous to the equilibrium of a soap bubble blown from a pipe: the air pressure $p$ has to balance the surface tension acting at the orifice of the pipe. The resultant force exerted on the bubble by the pressure $p$ in the direction of the axis of the pipe is $p$ times the area of the orifice; similarly, the resultant force exerted by the shear stress upon the dislocation bulge, in the slip direction, is the shear pressure $\tau b$ times the distance $d$ between the ends of the arc. Thus, the equilibrium condition is

$$\tau b d = 2\Gamma \cos \alpha \tag{3-5}$$

with the assumption that the chord of the bulge is normal to the slip direction.

The maximum value of $\tau$ is needed when $\cos \alpha = 1$, i.e. when the bulge becomes semicircular; its amount is

$$\tau_e = \frac{2\Gamma}{bd} \tag{3-6}$$

$\tau_e$ is the shear stress required for extruding a dislocation line between obstacles a distance $d$ apart. If it is lower than the shear stress needed for shearing across the obstacles, it represents the contribution of the obstacles to the yield stress $\tau_y$. The yield stress itself is the sum of the extrusion stress $\tau_e$ and the driving stress $\tau_d$ (dislocation friction), which is the shear stress required for moving the dislocation over a slip plane not containing any obstacles:

$$\tau_y = \tau_d + \tau_e \tag{3-7}$$

Since, as mentioned, the ends of the bulge may be fixed not only by obstacles, but also by the dislocation going off the slip plane, Eq. 3-6 represents a quantity that may be a significant contribution to the yield stress under a variety of conditions.

**I,4. Strain Hardening.** Among the various ways in which slip can occur in spite of the presence of obstacles in the slip plane, that of slip remaining confined to the areas between the obstacles has been mentioned on p. 364. In this case, the yield stress may be very small initially, but it rises with the amount of slip, i.e. with the accumulation of dislocations at the obstacles. This is the simplest mechanism of strain hardening; one of its characteristics is that, in its pure form, the hardening ought to disappear when the plastic strain is reversed and the accumulated dislocations are withdrawn again. Such behavior is not known in reality, but the Bauschinger effect is its blurred realization.

The principal cause of strain hardening is the formation of lattice injuries, acting as obstacles to slip, in the course of plastic deformation. The process of strain hardening, then, consists of two components: the formation of obstacles and the accumulation of dislocations at them. Many types of lattice injuries capable of producing hardening have been considered; which of them are practically important can hardly be recognized at present. However, two important facts about the mechanism of strain hardening must be accounted for by any detailed theory. Polycrystalline zinc, cadmium, or magnesium have far higher initial yield stresses and a far higher rate of strain hardening than the single crystals, because such metals have only one plane of easy slip (the 0001 plane), and so the grains cannot accommodate themselves to the directions of slip possible in their neighbors. In such polycrystalline metals, therefore, most of the strain hardening is due to the presence of obstacles represented by the grain boundaries; the hardening within the crystal grains is insignificant by comparison. In cubic metals, on the other hand, there are enough slip possibilities for any grain to undergo any kind of homogeneous distortion (apart from the infinite multitude of nonhomogeneous deformations); for this reason, the grain boundaries are relatively unimportant as sources of strain hardening. On the other side, the rate of strain hardening of cubic crystals is much higher than that of comparable hexagonal crystals. This shows that the possibility of slip on intersecting slip planes, and "slip collisions" arising from this, are the main sources of strain hardening in cubic single crystals.

When a strain-hardened (cold-worked) metal is subjected to increasing temperature, first of all the microscopic internal stresses due to the accumulation of dislocations at obstacles are removed by the recession and recombination of dislocations. During this process, the dislocations move mainly in their slip planes; it can be followed by measurements of the electric resistivity, of the X-ray line breadth, and of the yield stress which diminishes without quite reaching the initial value of the undeformed material. When the temperature is raised further, dislocations begin to move out of their slip planes. A simple example of this [*18,19*] is for the atoms at the edge of the extra half plane of an edge dislocation (at $Q$ in Fig. I,3d) to diffuse into vacant lattice sites; if the edge is being nibbled away in this manner, the dislocation moves towards its compressed side, out of its slip plane. Once the dislocations are free to leave their slip planes and to go around obstacles, the network of dislocations in the distorted metal can shorten its total length just as a foam gradually reduces its total area [*20*]; in addition, it can rearrange itself in various ways so as to reduce its total energy. One of the important types of rearrangements is "polygonization" which consists of the straightening of bent slip lamellas; in the course of this the profile of the lamella changes from a more or less continuously curved arc to a polygon with straight sides. This can be observed in X-ray photographs taken with "white" radiation: the continuous asterism streaks break up into rows of dots, each dot being reflected from one side of the polygon.

If the temperature is raised sufficiently, and if the preceding deformation was strong enough, the dislocation network bursts up and disappears completely; this is recrystallization. Since the process starts at the points of intense local distortion and spreads by the growth of the distortion-free lattice arising at such points, the recrystallized regions may have orientations quite different from the mean orientation of the crystal before recrystallization.

**I,5. Precipitation Hardening.** If a solid solution becomes supersaturated below a certain temperature, particles of the solute, or of a phase containing solvent and solute, tend to segregate. If the solution is quenched rapidly to a low temperature, there may be no time for the segregation to occur. If it is now held at this or at a somewhat higher temperature for a longer time, segregation may take place with the formation of extremely small, initially submicroscopic, particles. The high dispersion is due on the one hand to the ease of nucleation at high degrees of supercooling when the difference between the free energies of the stable and the unstable phase is very high [*21*]; on the other hand, to the very low velocity of diffusion and so to the slow growth of the numerous nuclei.

With continued aging at a suitable temperature, the size of the precipitated particles increases, not only because further segregation takes place, but also because the finest grains having the highest "vapor pressure" may distill over to the coarser ones. As the precipitated particles grow in size from close-to-atomic to nearly microscopic dimensions, the yield stress rises; later, when the particle size becomes microscopic, the yield stress may decrease again ("overaging"). It reaches a maximum at a critical dispersion.

The precipitated particles cannot be regarded simply as pegs immobilizing the slip planes: if the total amount of precipitated material is given, the stress needed for shearing through the pegs in a slip plane is determined by their total cross section, independently of whether there are many thin or a few thick pegs present. Precipitated particles may also act by the internal stresses they produce in the matrix; if the natural (stressless) size of the particle differs from that of the hole it occupies, it generates a state of tension or compression accompanied by shear stresses in the slip planes, and these may counteract the movement of dislocations. However, here again the stress fields around the particles would remain geometrically similar to themselves as the precipitated particles grew in size, provided that their shape remained the same; if the magnitude of the internal stresses alone were to determine the yield stress, there could be no influence of the degree of dispersion. The simple peg theory and the internal stress theory, therefore, cannot explain the phenomenon of age hardening. What seems to be the basic mechanism of age hardening was discovered by Nabarro [*17*]. The internal stress field around very finely dispersed particles varies on such a small scale that the highest probable adverse balance of the positive and negative internal stresses acting on a sufficiently large segment of a dislocation is very small. The dislocation can move relatively easily over such a stress field which is almost "smooth," although the highest local stresses may be quite high. If the same amount of segregated material is distributed on a coarser scale, the possibility of higher accidental adverse balances of internal stress acting upon a dislocation segment is higher, and the yield stress is increased.

If the yield stress decreases again in the course of aging, this may have two causes. The natural size of the precipitate may be smaller than the hole occupied in the matrix; in this case, the particle may break away from the matrix in the course of aging and the state of internal stress around it may disappear [*22*]. The possibility of this process is supported by X-ray observations. But even if no such breakaway occurs, the coarsening of the dispersion is bound to lead to a decrease of the yield stress when the spacings between the particles become large enough for the dislocations to be extruded between them. As soon as dislocation extrusion becomes possible, the yield stress is determined primarily by the extrusion stress

$$\tau_e = \frac{2\Gamma}{bd} \tag{3-6}$$

(see Art. 3), where $d$ is the spacing between the particles [*15,23*]. Thus the yield stress increases with the spacing between the particles in the Nabarro region of high dispersions, but decreases according to Eq. 3-6 when the extrusion process becomes feasible. It has a maximum between the Nabarro region and the extrusion region, at a critical dispersion, as postulated by Merica, Scott, and Waltenberg in 1920.

When slip by dislocation-extrusion progresses in a precipitation-hardened material, the precipitated particles become surrounded with an increasing number of dislocation rings, according to Fig. I,3h and the attached consideration. This has two consequences. The accumulating dislocation rings produce increasing adverse internal stresses in the area between the obstacles, and a decrease of the free space $d$ (see Eq. 3-6) between them; thus the yield stress required for sending new dislocations over the slip plane increases. This component of strain hardening is substantially identical with that considered in G. I. Taylor's theory [*13*].

Secondly, the accumulating rings give rise to an increasing shear stress in their interior (around the segregated particle). If this reaches a critical value, the particle (or a patch of high internal stress) is sheared through and the dislocation rings disappear. If this process takes place in a large fraction of the precipitated particles, the extrusion mechanism may cease to be the main source of the yield stress; soon after this phase, shear fracture is likely to occur.

**I,6. Transient Creep (Cold Creep).** If a metal undergoes progressive deformation at a constant stress, it is said to show creep. If the stress-strain curve represented an exact law of plastic deformation, no creep would exist.

If a specimen of a ductile metal is loaded with a constant stress, and the strain is plotted as a function of time, a creep curve of type 1, Fig. I,6a,

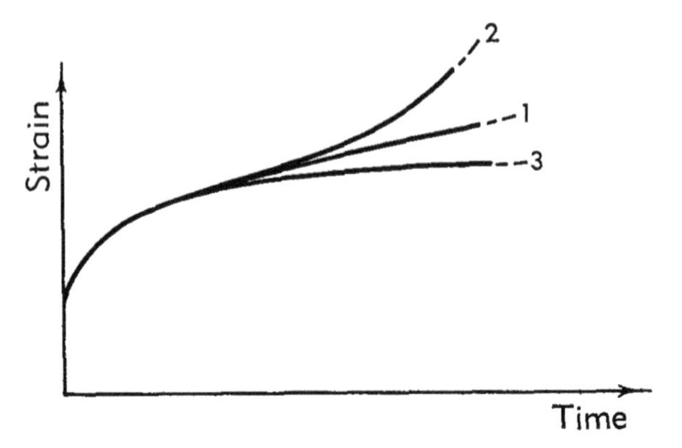


Fig. I,6a. Creep curves of a metal at constant stress (1), constant tensile (2), and compressive (3) load.

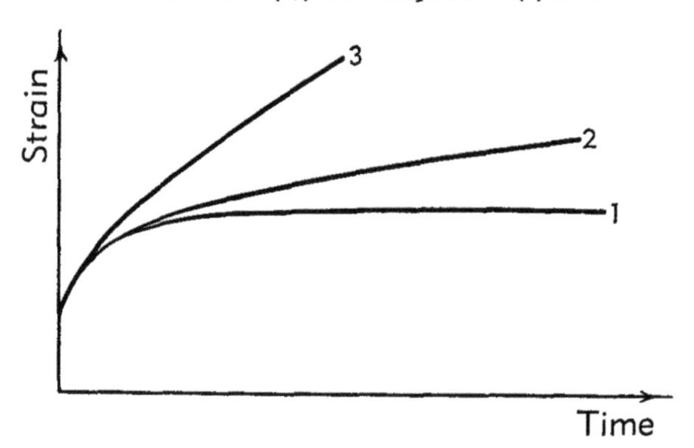


Fig. I,6b. Creep curves at constant stress, at low (1), intermediate (2), and high (3) temperature.

is obtained. More complicated curves are observed if structural or metallurgical changes occur during creep. If a specimen is subjected to constant tensile load, the tensile stress rises as the cross section of the specimen decreases by the extension. Consequently, the creep rate speeds up as compared with the constant stress test, and curves of type 2, Fig. I,6a, result. A constant load, compression creep test with the same material would lead to a curve of type 3.

Fig. I,6b shows schematically constant stress creep curves such as would be obtained at different temperatures. Curve 1 corresponds to a very low temperature (e.g. room temperature for iron or the temperature of liquid air for lead). Curve 2 would be observed at a moderately high temperature, e.g. at room temperature for lead or around 750°C (about 1400°F) for iron; finally, curve 3 would apply at a still higher temperature. As the curves indicate, the application of a stress above the yield stress is followed by a very rapid, but rapidly decelerating, creep which asymptotically tends towards a constant rate. At low temperatures (below the temperature at which the first traces of removal of strain hardening appear) the asymptotic creep rate is zero; the initial decelerating part of the creep, however, is always present, even in the neighborhood of the absolute zero [*24*]. (Strain-aged metals represent apparent exceptions to this type of behavior.)

From these characteristics of the creep process, Andrade [*25*], who was the pioneer of creep research, drew the conclusion that in general two physically different components of creep exist: a decelerating component which is present at all temperatures, and a steady rate component which disappears below a certain temperature range. Andrade denoted these components as $\beta$ creep and $K$ creep; one of the writers of this section introduced the terms "transient" and "viscous" creep which are currently in use. In the lowest temperature range (e.g. below some 250°K for pure copper) transient creep is of the type

$$\gamma = \gamma_0 + B \log (t - 1) \tag{6-1}$$

above this range, but before viscous creep become significant, it is usually well represented by

$$\gamma = \gamma_0 + Bt^{1/n} \tag{6-2}$$

with $n$ mostly around 3. The latter type of transient creep was observed by Andrade in his early experiments.

At high temperatures viscous creep is usually overwhelming in engineering applications, and the transient component is frequently of little practical importance. In the low temperature range, however, transient creep alone is significant; this is the case, e.g. in creep problems connected with prestressed reinforced concrete. Particularly important is transient creep in structures subjected to rapid heating (e.g. aerodynamic heating of ultrasonic aircraft). Until the temperature reaches the region of hot creep, the creep under such conditions is prevalently of the transient type, and calculations based on the viscous creep rate derived from conventional creep tests may be grossly misleading in the dangerous direction.

Before dealing with the physical background of transient creep, a remark about the possibility of finding general functional expressions for the creep strain seems appropriate. It would appear plausible to generalize the deformation laws of pure viscosity $(d\gamma/dt) = f(\tau)$ and of pure plasticity ($\gamma = f(\tau)$) for crystalline materials showing both strain hardening and creep in the manner

$$\frac{d\gamma}{dt} = f(\tau, \gamma, t) \tag{6-3}$$

by assuming that the creep rate would be a function of stress, strain, and temperature. Such a relationship is called a "mechanical equation of state." However, experiments of various types have shown that such a relationship does not exist [*23,26*]. The simple reason for this is that the mechanical response of a crystalline material to a certain stress at a certain temperature depends on the structural state it has attained in the course of preceding straining. This state is not determined by the magnitude of the preceding strain alone but depends very essentially on the temperatures and the rates at which the straining has been carried out. Thus the rate of creep under a given stress at a given temperature is not a function of the current values of $\tau$, $\gamma$, and $t$ alone but depends on the entire mechanical-thermal history of the specimen. For limited practical purposes, however, it is often possible to approximate the rate of a creep process by an expression of type (Eq. 6-3) in sufficiently narrow ranges of the variables. In particular, at very high temperatures where recovery and recrystallization rapidly wipe out the effects of preceding strain history, metals often behave approximately as purely viscous materials obeying the viscous deformation law.

Concerning the origin of transient creep (already briefly indicated in Art. 2), there seems to be no doubt that it is due to thermal stress waves locally superposed upon the applied stress [*23,27,28*]. Although thermal stresses are far too weak to bridge the gap between the observed and the theoretically estimated values of the yield stress in the sense of Becker's original theory (see, for example, [*10*, p. 69]) they can well initiate slip processes if the applied stress is almost high enough to do so. A property of transient creep that is very unusual among thermally activated reactions is that it can take place with considerable velocity in the immediate neighborhood of absolute zero [*24*]. This is easy to understand because the slightest thermal stress can cause creep if the applied stress is raised to the level at which it is nearly capable of producing slip without any thermal help.

The logarithmic creep expression (Eq. 6-1) can be derived simply from this picture by assuming that the activation energies for local slip processes, at a given level of the applied stress, are initially distributed with a constant density in the low energy fringe of the distribution curve in which activation is feasible at the given temperature. The Andrade-type of creep (Eq. 6-2) is due to the possibility of recovery processes with very low activation energies; it is not, however, a simple superposition of some recovery creep upon pure transient logarithmic creep.

**I,7. Viscous Creep (Hot Creep).** All nonelastic deformation involves movements of atoms or molecules from one equilibrium position to another; in slip, for instance, atoms of one lattice plane move from one potential well to another between the atoms of an adjacent atomic plane. The consecutive equilibrium positions are separated by potential barriers; if the deformation takes place with a velocity that is low compared with that of the relevant elastic waves, its rate must be determined by the frequency of thermal fluctuations powerful enough to help the atoms over the potential barrier into the adjacent equilibrium position. This frequency $f$ is given, with a good approximation, by the Boltzmann exponential as

$$f = Ce^{-A/kT} \tag{7-1}$$

where $A$ is the height of the potential barrier (the "activation energy"), $k$ the Boltzmann constant $= 1.38 \times 10^{-16}$ erg/°C, and $T$ the absolute temperature in °K. $C$ is not strictly constant, but its dependence on the temperature and on the variables influencing $A$ is usually small compared with that of the exponential. The Boltzmann exponential $e^{-A/kT}$ represents that fraction of particles (or complexes of particles) which has an energy around the value $A$ at a given moment. If the velocity distribution of the particles changes, on the average, $\nu$ times per second (e.g. by $\nu$ collisions per second suffered by each molecule of a gas), and if $N$ particles are present in the assembly considered, $\nu Ne^{-A/kT}$ will be a measure of the number of particles per second which obtain enough energy to fly over the potential barrier. Since not only the magnitude but also the direction of the velocity matters, and for other reasons, the constant $C$ in Eq. 7-1 contains other factors in addition to $\nu$ and $N$.

Since the value of $kT$ at room temperature is about $\frac{1}{40}$ electron volt, an activation energy of 1 electron volt gives a value of $e^{-40} \cong 10^{-17}$ to the Boltzmann exponential. $C$ contains as a factor the number of the relevant particles, and the frequency with which they change their dynamic states; it can therefore be large enough to make the frequency $f$ of successful activations sufficiently high for the process to progress with an easily observable velocity at room temperature. However, a value $A = 10$ electron volts, for instance, would make the exponential equal to $\cong 10^{-170}$, and no reaction could have a constant $C$ high enough to counterbalance this: the process would be completely frozen in. Most reactions—in particular, the processes considered in this section—can progress with a measurable speed at room temperature only if the activation energy does not exceed 1 or 2 electron volts. (At higher temperatures, the upper limit of the feasible activation energies increases in proportion with the absolute temperature.)

This circumstance is of crucial importance in that it determines the character of plastic as contrasted with viscous deformation. It will be seen further below that it is also the basis for the understanding of the properties of highly creep-resistant materials.

It was mentioned above (Art. 3) that thermal stresses cannot initiate slip in a faultless crystal. The energy barrier for moving one atom into the next equilibrium position among the atoms of the adjacent atomic plane is only of the order of 1 electron volt; however, a single atom cannot slip because the surrounding equilibrium positions are occupied. It can be shown that many thousands, or millions, of atoms have to slip simultaneously in order to be firmly enough anchored in their new equilibrium positions not to be pushed back into the old ones by the elastic reaction of the surroundings (by the stresses around the dislocation surrounding the slipped patch). This would require an activation energy of many thousands of electron volts, quite apart from the energy of the dislocation loop produced at the same time; and it is easy to prove that applied stresses of the usual magnitude could supply only a small fraction of this. Consequently, thermal activation of local slip in a faultless crystal at the usual values of the yield stress is utterly impossible. Suppose now that dislocations are already present, and that the process of slip involves extruding them through a field of obstacles (see Fig. I,3g, I,3h, and I,3i). The stress required for this is given by Eq. 3-6; and the question is now, can a substantial part of this stress be replaced by thermal activation? In other words, can a thermal stress fluctuation complete the extrusion process long before the applied stress can bend the dislocation to a semicircle? Again, the answer is in the negative. To account for the observed values of the yield stress, the obstacles must have spacings of the order of hundreds of interatomic spacings; since the energy of the dislocation is of the order of 1 electron volt per interatomic spacing (see p. 367), the energy of the dislocation between two obstacles amounts to hundreds of electron volts before bulging begins, and it has to increase by a factor of about $((\pi/2) - 1)$, i.e. more than 50 per cent, in order to form a semicircular bulge. It would then require an activation energy of tens or hundreds of electron volts to extrude a dislocation loop between the obstacles at a stress that is significantly lower than the extrusion stress; thermal activation of the extrusion process becomes possible only when the applied stress is so close to the extrusion stress that the length of the bulge is only a few interatomic spacings less than that of the semicircle.

In the particular case of dislocation extrusion, therefore, it is clear why thermal activation can replace only a very small part of the stress required for plastic deformation. Slip is not possible until the stress has come close to the critical value at which it would cause slip without any thermal help. Similar conditions are present if the yield stress is determined by other factors (see Art. 3). This is the reason why crystalline materials (except at very high temperatures; see below) are *plastic*, i.e. why they have fairly definite yield stresses below which no deformation occurs, in contrast to Newtonian *viscous* materials which can undergo slow deformation even at lowest stresses.

Viscous materials are either amorphous, i.e. of irregular atomic arrangement, or at least they contain amorphous (noncrystalline) regions. Fig. I,7a shows a simple example of how, in such materials, atomic rearrangements involving only two or three atoms can occur: the atoms shown can switch into the alternative equilibrium positions indicated by dotted circles without having to displace any of the surrounding atoms from their equilibrium positions. Such a rearrangement requires only one or two electron volts or less, even if no stress is acting. If a

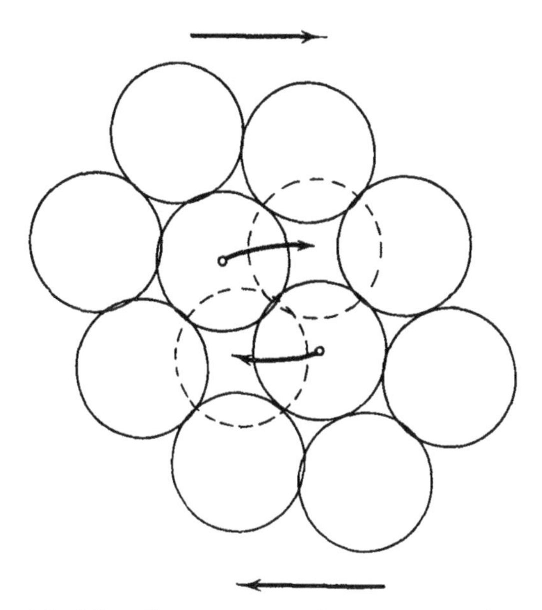

Fig. I,7a. Rearrangement of two atoms causing shear distortion indicated by arrows.

stress is present, it gives only a slight preference to those rearrangements which lead to a small distortion of the body as a whole in the direction of the applied stress; the repetition of such rearrangement processes with the bias supplied by the stress results in viscous flow. Viscous flow is possible only if the temperature is high enough to provide the activation energy necessary for the atomic rearrangement; once this condition is satisfied, however, there is no lower limit to the stress that can give a preferential direction to the rearrangements, unless some kind of structural order is present, so that the viscous behavior is non-Newtonian. Plastic deformation, on the other hand, is possible at all temperatures because its main driving force is the applied stress; it ceases, however, if the stress drops below the critical value of the yield stress. The fundamental cause of the difference is that the elementary processes of viscosity involve only a few atoms or molecules, and therefore have relatively low activation energies; in a crystal, however, only concerted actions of large numbers of atoms can take place, and the necessary activation energy is extremely high unless the stress is close to the level at which it could carry out the process alone, without thermal help.

Since viscous flow is possible only if some kind of deviation from perfect crystalline order is present, the question arises, what kind of amorphous disorder capable of giving rise to viscous creep can exist in a substantially crystalline metal? Two types of disorder seem to be of particular importance: that represented by the grain boundaries in a polycrystalline metal, and the disorder produced in the crystal lattice by cold work and strain hardening. In addition, the presence of vacancies and of interstitial atoms, as well as of atoms adsorbed on the surface, may give rise to viscous creep; whether these possible sources of creep are of practical importance, however, is not sufficiently known.

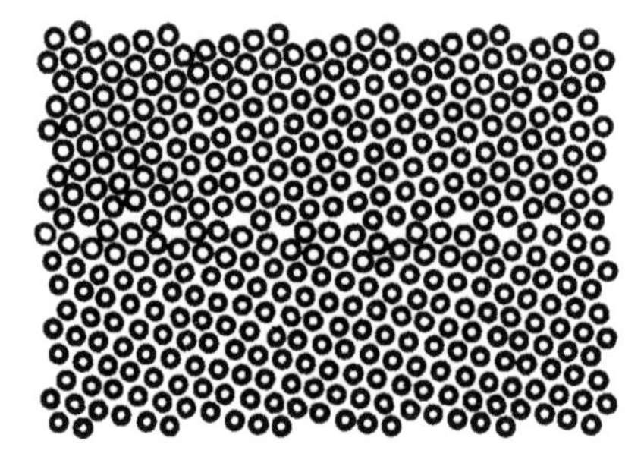

Fig. I,7b. Grain boundary in the Bragg-Nye soap bubble model of a metal.

*Creep due to grain boundaries (boundary sliding).* The two-dimensional soap bubble model of a crystal studied by Bragg and Nye [*29*], Bragg and Lomer [*5*], Lomer [*30*], and Lomer and Nye [*31*], which consists of numerous soap bubbles of equal size floating on a soap solution and arranging themselves in a two-dimensional poly- or monocrystalline array, shows clearly the structure of the grain boundary of a simple metal (Fig. I,7b). If the neighboring grains differ more than 5 or 10 degrees in orientation, the typical configuration repeated along the grain boundary is that shown in Fig. I,7a, representing a small group within which thermally activated rearrangements, e.g. the transition indicated by arrows and dotted circles may be possible at temperatures below the melting point. Each such rearrangement changes the direction and amount of pressure exerted by the rearranged atoms upon their surroundings; consequently, it causes a change of strain which, of course, is not limited to the surroundings but affects the whole body and produces a small strain in it.[2] Without an applied stress, the rearrangements

[2] It is easily seen that the order of magnitude of the average strain produced in a body of volume $V$ by a rearrangement of the type shown in Fig. I,7a, involving atoms in a small volume $v$ is $v/V$.

of various groups of atoms take place in random directions and the resulting mean strain is zero. However, if a stress is present, it gives a preferential direction to the possible rearrangements and the result is a progressive flow, i.e. sliding of the grains along their boundaries. If the number of the atomic groups that are mobile enough to be activated at the given temperature is small, so that they are separated by purely elastic regions, application of a stress produces a drift into a new distribution of the "orientations" of the groups with respect to the stress. As this distribution is approached, the deformation comes to a standstill: it is a "delayed elastic" deformation ("elastic creep"), not a viscous one, and it disappears gradually when the stress is removed. However, if the temperature is high enough to make atomic rearrangements possible at most points, the mobile groups are no longer constrained by immobile surroundings to act as two- or three-position switches; progressive rearrangements, i.e. viscous flow, become possible.

Since the applied stress is small compared with the interatomic forces (the "intrinsic pressure"), it has only a small influence upon the potential energies of the possible equilibrium positions of the mobile atom groups. Calculation shows [*32,33*] that the rate of flow is then proportional to the applied stress because the higher terms in the power series representing the influence of the stress are negligible; in other words, Newton's law of viscosity

$$\tau = \mu \frac{d\gamma}{dt} \tag{7-2}$$

results ($\tau$ is the shear stress, $\gamma$ the shear strain, and $\mu$ the coefficient of viscosity). This seems to disagree with the fact that the rate of grain boundary creep is far from being proportional to the applied stress; it is usually given by a curve of the character shown in Fig. I,7c. There is a fairly definite "creep strength" below which the creep rate soon becomes extremely small; above it, the rate rises very rapidly. The usual cause of this behavior, however, is a simple one. Almost as soon as sliding begins between the grains of a polycrystalline metal at a higher temperature, it is hindered by geometrical locking between the grains. When this sets in, creep cannot continue unless either the geometrical resistance is overcome by plastic deformation of those parts of the grains which are in the way of sliding, by grain boundary migration, or similar changes; or else cracks and cavities open up between the grains. In the first instance, the rate of creep is determined by the rate at which the geometrical obstacles are overcome; it is usually of a lower order of magnitude than the rate of sliding that would occur between grains separated by a perfectly plane boundary. If the stress is alternating, and the creep strains are small, the sliding is reversed before it would be strongly hindered by geometrical locking; in accordance with this, the rate of boundary sliding deduced from measurements of vibration damping is far higher than that under steady load. The case of separation between the grains with crack formation is discussed in Art. 8.

In most of the modern highly creep-resistant alloys, boundary sliding is not the main source of hot creep. This is due partly to the high hot-hardness of the grains which makes the removal of geometrical locking difficult, and partly to the precipitation of hard particles (e.g. of refractory carbides) in the grain boundaries which create additional geometrical locking. In many metals and alloys, however, grain boundary sliding is the main cause of hot creep. This was the case in the tungsten filaments of incandescent lamps before single crystal filaments were available.

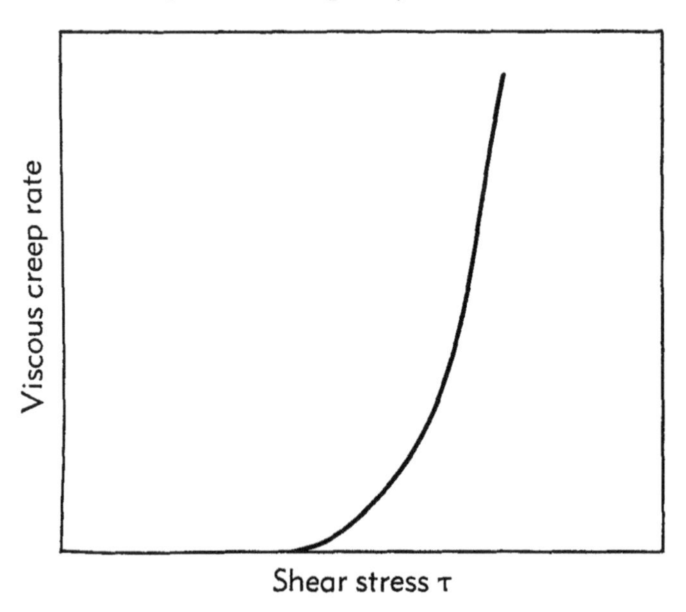


Fig. I,7c. Typical dependence of the viscous ("hot") creep rate upon the shear stress.

The creep was reduced first by the addition of small amounts of silica or thoria which produced the very jagged grain boundaries typical of "secondary recrystallization"; and ultimately by the introduction of single crystal filaments. It is a remarkable fact, easily understood from the above considerations, that while polycrystalline metals are harder than single crystals of average orientation at low temperatures, they usually become softer at temperatures high enough for grain boundary sliding to take place (above the so-called "equicohesive temperature").

*Creep due to thermal softening ("recovery").* Fig. I,7d shows schematically the origin of a type of viscous creep that can occur even in single crystals where no grain boundaries are present; it is frequently more important in polycrystalline metals than grain boundary sliding. *OA* is the yield stress-strain curve at the temperature in question; if a stress

$\tau = OS$ is applied, a corresponding plastic deformation $SB$ takes place as fast as inertia effects and transient creep allow. After this, the material has acquired the strain hardening needed to resist the applied stress, and no further deformation would take place in the absence of thermal softening. However, if the temperature is high enough for thermal softening (loss of strain hardening, recovery of yield stress) to occur, the yield stress diminishes by a certain amount $d\tau$ in the time element $dt$. In order to carry the load, therefore, the material has to undergo during

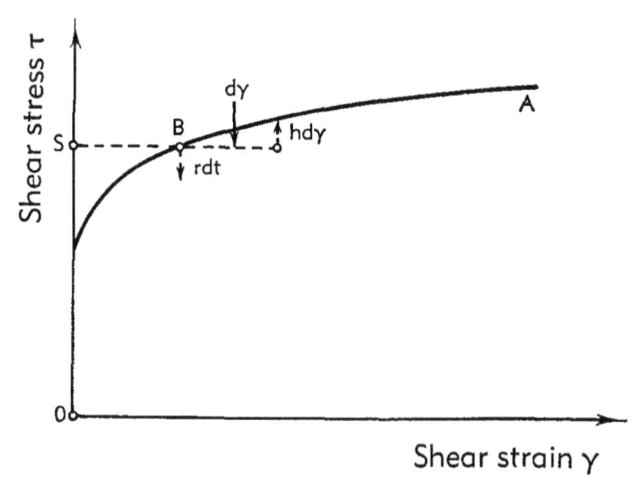


Fig. I,7d. Illustration of recovery creep.

the time $dt$ an additional deformation $d\gamma$ just large enough to provide the strain hardening $\delta\tau$ lost by recovery. In a stationary state, both $d\tau$ and $\delta\tau$ are proportional to the time element:

$$d\tau = rdt \tag{7-3}$$

and

$$\delta\tau = hd\gamma = h\frac{d\gamma}{dt}dt \tag{7-4}$$

although no simple proportionality exists between finite increments of hardening, recovery, and time. Hence, the rate of "recovery creep" is

$$\frac{d\gamma}{dt} = \frac{r}{h} \tag{7-5}$$

This equation, however, is little more than a mathematical illustration because the "rate of thermal softening (recovery)" $r$ is not a constant but depends on the preceding thermal and mechanical history of the material.

The interpretation of recovery creep in terms of molecular concepts follows from what was said about strain hardening and recovery in Art. 4.

In the course of plastic deformation, various types of obstacles to slip arise (among them, dislocations in intersecting slip planes which hinder the movement of other dislocations in slip planes they penetrate). The result is the formation of a three-dimensional network of entangled dislocations suspended on obstacles; plastic deformation is possible only if dislocations are extruded between anchoring points, or if the anchoring obstacles are sheared through. If such a cold-worked metal is exposed to a temperature at which atomic rearrangements (self-diffusion) are possible, obstacles may be dissolved by local ordering processes, and dislocations may gain additional freedom of movement out of their slip planes by the diffusion process mentioned in Art. 4. In these ways, the dislocation network and its anchoring points become less dense and the yield stress decreases. This process is retarded, and increased resistance to creep is given to the metal, if particles of high melting point precipitate at the anchoring points (e.g. where dislocations in different slip planes cross) and protect them against being obliterated or weakened by thermal rearrangements. Similarly, if such precipitation occurs along dislocations, in a way similar to Cottrell's picture of strain-aging [*34*], the dislocation loses much of its mobility and the progress of thermal softening with the resulting creep is retarded.

This simple picture suggests much of the general strategy used, consciously or unconsciously, in producing creep-resistant materials. In metals of high melting point and high elastic modulus (i.e. high binding energy) self-diffusion starts at a high temperature; consequently, thermal softening and recovery creep are suppressed until a relatively high temperature is reached. Unfortunately, some of the best metals from this point of view (e.g. molybdenum or tungsten) have low corrosion resistance, high density, and high price. This makes it necessary in most practical cases to use metals less resistant to thermal softening and to reinforce them by precipitated particles of high lattice energy (e.g. refractory carbides). The latest development is to get rid of the limitations imposed by the chemical requirements of the feasible precipitation processes by introducing the refractory particles mechanically, by mixing and sintering (or by sintering oxidized metal particles).

Whether produced chemically or mechanically, the effectiveness of refractory particles in reducing creep must depend upon their spacing $d$. According to Eq. 3-6, the shear stress required for extruding a dislocation between obstacles spaced at $d$ is

$$\tau_e = \frac{2\Gamma}{bd} \tag{3-6}$$

Embedded hard particles cannot be expected to be very effective in increasing creep resistance unless they are so close to one another that the extrusion stress $\tau_e$ exceeds the "creep strength" of the matrix metal, i.e. the shear stress at which the creep rate reaches a permissible value.

In some cases, as in that of oxidized sintered aluminum, the refractory component (e.g. $Al_2O_3$) is present as an envelope of metal grains, not as granular inclusion. The hard envelope may then effectively immobilize dislocations at grain diameters much above the value $d$ given by Eq. 3-6.

Since nucleation of precipitated particles is particularly easy at lattice defects, such as dislocations or some types of obstacles to slip, plastic deformation may induce precipitation at the weakest points where slip occurs abundantly during plastic working. This seems to be the basis

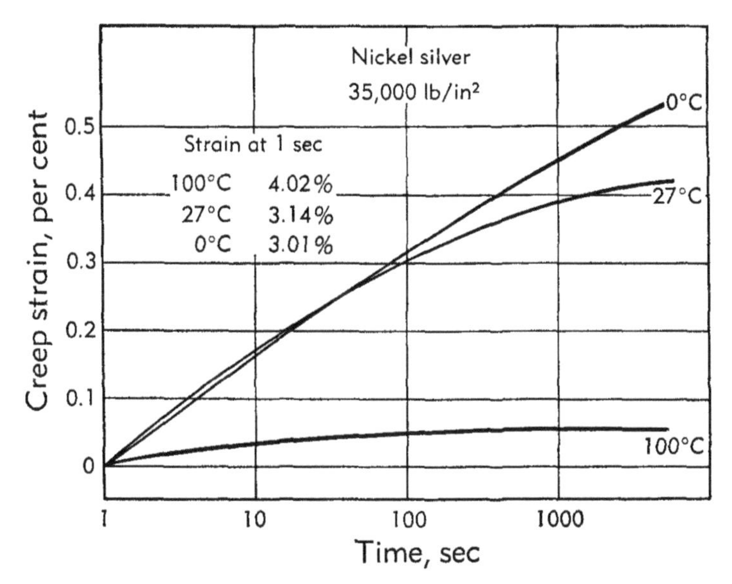


Fig. I,7e. Creep of nickel silver at various temperatures. Effect of strain-aging. From [*36*].

of the beneficial effect of "warm-working" upon the creep resistance of many alloys (see Harris and Bailey [*35*, p. 60]). A striking example of the effect of precipitation is shown by Fig. I,7e [*36*], in the case of nickel silver. Creep is rapid at 0°C; at 27°, it decelerates after two or three minutes. At the same stress at 100°C, the initial creep rate is already much lower, and creep stops altogether after little more than 1 minute. The case of nickel silver is not quite relevant to hot creep because the curves in Fig. I,7e represent almost pure transient creep; however, they demonstrate the drastic effect of strain-aging (probably precipitation along the dislocations) in reducing creep.

*Creep due to individual migration of atoms.* If there is a vacant lattice site available, one of the neighboring atoms may step into it and leave its former site vacant. By a continuation of this process, the vacancy may migrate to the surface and disappear, leaving the specimen "densified." If a tensile stress acts upon a rod, vacancies may arise at its ends and leave at the side surfaces more frequently than the other way around; obviously, the effect is then an extension accompanied by the corresponding transverse contraction. If there is a steady production and removal of vacancies, the result is a kind of viscous creep. Whether such creep is practically important cannot be recognized with certainty.

There is another possibility of viscous creep by the individual migration of atoms. Some atoms of a solid may form an adsorbed layer on its surface; since they are not fully incorporated in the lattice, they are highly mobile. When such an atom arrives at a grain boundary, it can easily enter it because, as seen in Fig. I,7b, grain boundaries are loosely packed and offer easy accommodations for additional atoms. If a tensile stress is present, the equilibrium between atoms leaving the surface to enter a grain boundary, and those leaving the boundaries to go to the surface, is disturbed: more atoms go into the boundaries than away from them, and the result is a gradual increase of the length and decrease of the thickness of the rod, i.e. tensile creep.

**I,8. Creep Fracture.** If sufficiently prolonged, a hot creep test ends with fracture which is usually of a characteristic type. As a rule, no neck formation occurs, and the fracture appears brittle; it is due to the separation of grains along their boundaries. Sometimes this is mainly a stress-corrosion process due to the simultaneous action of the stress and of a corrosive medium, e.g. oxygen, capable of penetrating into the boundaries. However, typical creep fracture can occur without corrosion effects, by the process of boundary sliding mentioned in the preceding section. The grain boundaries can be considered as a viscous medium, a two-dimensional glass. If the temperature is only moderately high, sliding between the grains at not too low speed requires an applied stress that is high enough to overcome geometrical locking by deforming the protuberances of the grains that are in the way of sliding. In this way continuity is preserved. However, at higher temperatures and low strain rates the stress required for sliding is too low for removing protuberances; the grains slide on their neighbors almost like rigid bodies, and cracks and cavities open up between them. After a few per cent extension, this leads to fracture by separation between the grains.

One of the remarkable properties of creep fracture is that, in the temperature and stress range in which the grains behave as nearly rigid, the strain required for their complete separation, i.e. for fracture, is almost a matter of geometry alone and depends little upon stress or temperature. To measure the time required for fracture at a given stress is then equivalent to the measurement of the mean creep rate. Since the fracture process itself serves as a crude extensometer by indicating a fracture strain that is approximately constant for a given material, the creep fracture test has the advantage of not requiring any strain measurement. In addition to serving as a crude creep rate test, however, it fulfills an important practical need. Occasionally a material fractures under creep conditions at strains that would not be objectionable in themselves for the particular application; this, however, may not be revealed by an abridged creep test in which the fracture strain is not approached. Whenever this possibility exists, therefore, creep fracture tests must be carried out in addition to creep rate measurements. In particular, this has to be done if considerable stress concentrations (by notches and the like) within small volumes are present where high local creep strains could be tolerated but the formation of a crack would be dangerous (cf. [*37*]).

**I,9. Creep in Engineering Design.** Up to the end of the nineteenth century, the procedures of engineering design with materials were very simple. A structural part was dimensioned so that the highest stress in it, calculated from the service load on the assumption that the material would remain purely elastic right to the point of failure, should not exceed a critical value which was substantially identified with the ultimate strength, i.e. the maximum load in the tensile test divided by the initial cross-sectional area of the specimen. With ductile metals, this procedure has no rational basis since excessive plastic deformation takes place long before the maximum load is reached in the tensile test; moreover, the maximum load usually occurs long before fracture. Whichever of these two modes of failure is of importance, the quantity that determines the resistance of the material is certainly not the ultimate tensile strength. In cyclic loading, the ultimate strength is equally unconnected with the occurrence of fatigue fracture. However, by choosing fictitious safety factors derived from practical experience for individual types of materials and often for individual structural parts under special loading conditions, the illusion of a scientific calculation of structural dimensions could be more or less maintained, although in reality the working stress was far from being a material property. It had to be adapted to the individuality of the structural part and to the service conditions; in fact, the application of the theory of elasticity was usually little more than a method of scaling up or down the dimensions of a structural part for different service loads.

Under the impact of more exacting demands to materials and to their utilization this simple way of life for the designer collapsed in the first quarter of the present century. One of the main reasons for this was the increasing importance of creep in materials subjected to high temperatures. That the resistance of a material to creep cannot be described by a single physical quantity, even if this is made dependent on the temperature, is immediately evident from Andrade's analysis of the creep process. As mentioned in Art. 6, Andrade [*25*] recognized long before the importance of creep became clear in engineering that two fundamentally different processes are involved in creep. The initial decelerating part of the creep curve (transient creep) is present at all temperatures down to absolute zero. In addition, if the temperature is high enough, viscous creep appears, the rate of which is independent of time unless it is

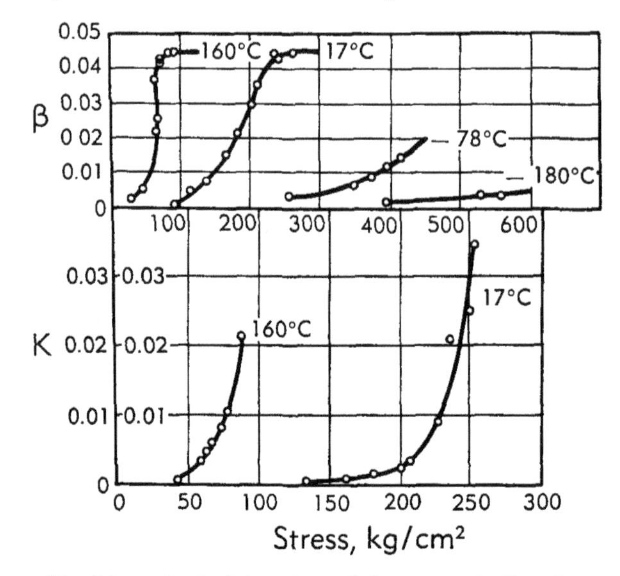


Fig. I,9a. Andrade's values of the constants B and K in Eq. 9-1 and 9-2 for lead.

influenced by structural changes. In Andrade's experiments, the transient creep rate of pure metals with not too high melting points could be represented by

$$\frac{d\gamma}{dt} = \frac{B}{3} t^{-\frac{2}{3}} \tag{9-1}$$

obtained by differentiating Eq. 6-2 with $n = 3$; on the other hand, the viscous creep rate was constant:

$$\frac{d\gamma}{dt} = K \tag{9-2}$$

Thus, for the metals in question, all transient creep curves belong to the same family, and, of course, all viscous creep curves also. The resulting creep curves, however, could be correlated only if there were a simple relationship between the dependence of $B$ and that of $K$ upon stress and temperature. Fig. I,9a shows Andrade's original curves giving this dependence for commercially pure lead. Obviously, there is no simple relationship between $B(\tau,T)$ and $K(\tau,T)$, which is evident already from the fact that $K$ vanishes if $T$ decreases below the range of thermal softening, whereas $B$ remains finite provided that the stress is adjusted to the level of the yield stress.

A graver complication is that Eq. 9-1 or, for that matter, any alternative expression for transient creep, such as Eq. 6-1, applies only for a constant stress. Any change of stress changes the rate of transient (as well as that of viscous) creep in a way that cannot be obtained by taking into account, e.g., the stress dependence of $B$ as given in Fig. I,9a and introducing it into Eq. 6-1; Eq. 9-1 is valid only for creep under a fixed stress. In fact, the considerations attached to Eq. 6-3 show that a general creep equation giving the rate of creep as a function of time, stress, and temperature cannot exist because the creep rate is influenced by the entire mechanical-thermal history of the specimen.

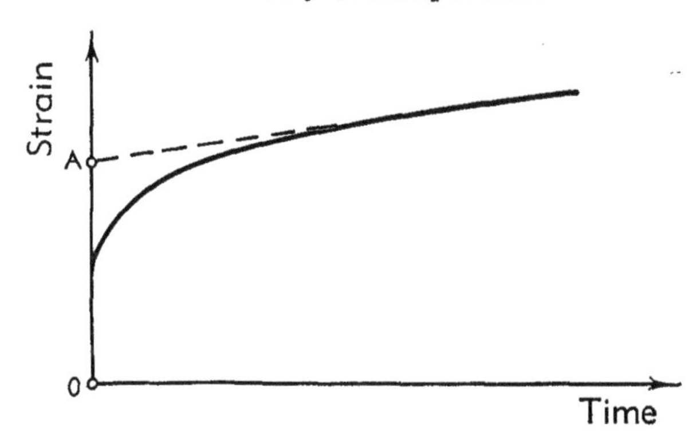


Fig. I,9b. Odquist's method of adding the total transient creep strain to the plastic strain.

These complications largely disappear if the designer is interested only in creep at a high temperature and a constant low stress, extending over a time so long that, within a fraction of the life of the structure, transient creep practically disappears. In this case, the creep curve can be extrapolated backward and the entire transient creep strain $OA$ (Fig. I,9b) can be regarded as a part of the sudden strain that follows the application of the stress; transient creep as a time-dependent process is then disregarded [*38*]. If, in addition, no structural changes occur during creep, only the dependence of the viscous creep rate $K$ upon stress and temperature remains to be taken into account. Its general character is shown in Fig. I,9a; simple mathematical expressions to approximate the stress dependence are, e.g.,

$$K = \frac{d\gamma}{dt} = \text{const } \tau^n \quad [39,40,41,42,43] \tag{9-3}$$

$$K = \text{const } (e^{a\tau} - 1) \quad [44] \tag{9-4}$$

$$K = \text{const } \sinh (a\tau) \quad [45] \tag{9-5}$$

where the constant still contains the dependence upon the temperature. Since viscous creep is the consequence of a self-diffusion process (see Art. 7), its temperature dependence must be given approximately by a Boltzmann exponential (Eq. 7-1) in which the activation energy $A$ is very nearly that of ordinary self-diffusion. Thus, the rate of viscous creep is

$$K = \frac{d\gamma}{dt} = C\tau^n e^{-A/kT} \tag{9-6}$$

if the stress dependence is assumed to be of the power type (Eq. 9-3).

As mentioned in Art. 8, the temperature and stress dependence of $K$ can be obtained experimentally in a crude approximation without any elaborate strain measurement from creep fracture tests [*37*]. In sufficiently limited temperature ranges, at low stress and high temperature, the creep fracture strain is determined largely by grain-geometrical factors; in such ranges, therefore, it is little dependent on stress and temperature. Since the fracture strain $F$ is the product of the mean creep rate $\overline{K}$ and the fracture time $\vartheta$,

$$\overline{K} = \frac{F}{\vartheta} \tag{9-7}$$

If the temperature is lower or the stress higher, the deformation within the grains is no longer negligible and the fracture strain increases. This puts an end to the applicability of creep fracture tests for the approximate study of the temperature and stress dependence of the viscous creep rate.

According to Larson and Miller [*46*], the Boltzmann exponential (Eq. 7-1 and 9-6) describes the temperature dependence of viscous creep accurately enough for practical purposes in many cases. The logarithm of the constant of proportionality (e.g. of $C\tau^n$ in Eq. 9-6) is remarkably constant; the physical reason for this has been given by Orowan [*47*].

In all preceding considerations, the "engineering" (constant load tensile) creep curve (curve 2 in Fig. I,6a) was regarded as being due to the superposition of transient and of viscous creep to the initial sudden deformation which follows the application of a stress and which represents the purely plastic (plus elastic) component of the total deformation. In the engineering literature, on the other hand, the creep curve is subdivided into an initial decelerating part ("primary" creep), an intermediate straight portion ("secondary" or "steady state" creep), and a final accelerating part which ends with fracture ("tertiary" creep). The tertiary stage is due either to the increase of stress due to the decrease of cross-sectional area in the ordinary constant load tensile test, or to structural changes (e.g. recrystallization, internal crack formation) which reduce the resistance of the specimen; it does not reflect an essential feature of the creep process. The only relevant question is, then: are the terms "primary" and "secondary" synonymous with "transient" and "viscous"? The answer was mainly of academic importance a few years ago; recently, however, it has become interesting even from the designer's point of view.

Obviously, the rate of secondary creep (the minimum creep rate) can be equal to that of viscous creep only if the transient component has completely disappeared. This is the case in most engineering creep tests serving the design of structures that have to stand up at a high temperature for a long time (e.g. years). The stresses of practical importance are then relatively low, so that the test extends over a long time and transient creep has plenty of time to disappear long before the end of the secondary stage.

Recently, however, structures that have to withstand high stresses at elevated temperatures for a short time (for example, expendable aircraft) have become important. Under such conditions, the whole creep process up to the final fracture (if the stress is tensile) may take place in hours or even minutes, and then the secondary stage may contain very substantial amounts of transient creep. An extreme case of this is present if a tensile specimen of a ductile metal (say, copper) is loaded at room temperature only slightly below the ultimate (maximum) load. In the course of the initial decelerating creep necking develops, and the consequent increase of stress at constant load leads to the acceleration of the third stage and finally to fracture. Between the first and the third stages, of course, there is a point of inflexion with an approximately straight portion in the creep curve which satisfies entirely the definition of the secondary stage; nevertheless, under the conditions just assumed (copper at room temperature) there is no trace of viscous creep in the entire test, and the minimum creep rate is that of a purely transient creep. Obviously, its uncritical use for design purposes would be quite mistaken. The minimum creep rate, therefore, can be identified with the viscous creep rate and used for design purposes in the traditional way only if the stress level is low enough for the secondary stage to extend far beyond the time required for the transient component to disappear.

For the reason just mentioned, the traditional methods of creep testing and design cannot be used for structures loaded at high stress and high temperature. Since transient creep is essentially involved in such cases, particularly if the temperature rises and therefore the yield stress falls in the course of service, design calculations based on simple creep tests made at constant temperature and stress or load are usually not feasible at present. However, the short duration of service at high stress removes most of the characteristic difficulties of creep testing; full-size or model tests on structural parts under realistic loading conditions become possible and often necessary.

Transient creep arises not only when a stress is applied, but also whenever it is increased, or reversed. Design based on the viscous creep rate alone, therefore, is possible only if the stress changes slowly as compared with the time required for transient creep to become insignificant. Very little research work has been carried out on incremental transient creep, so that for the present any design in such cases has to be based on ad hoc tests. If, however, the load remains constant long enough

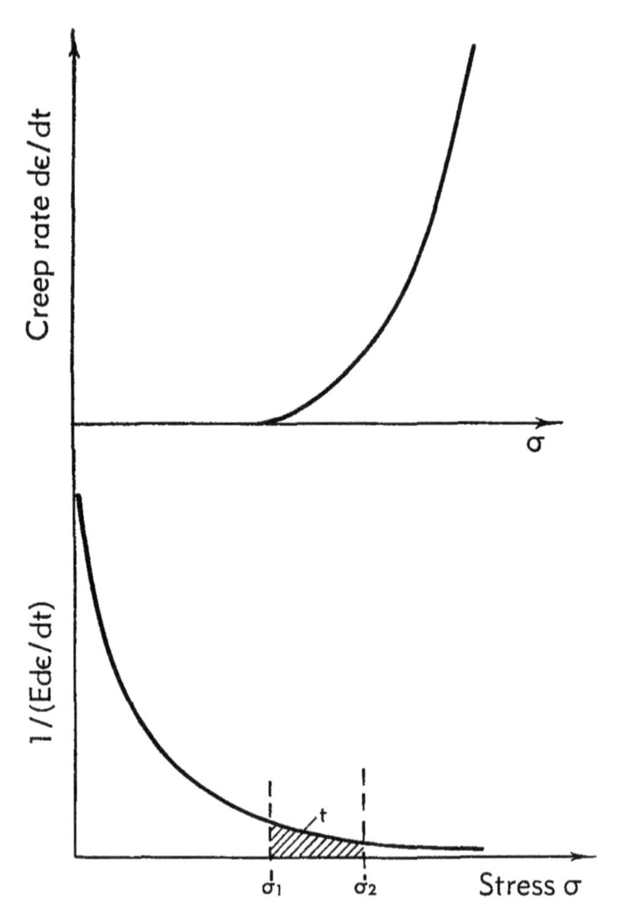


Fig. I,9c. Time $t$ during which the stress relaxes from $\sigma_0$ to $\sigma$, as determined by Eq. 9-8 and the $d\epsilon/dt$ vs. $\sigma$ curve.

for transient creep to disappear, and if the viscous creep strain between changes of stress is much greater than the transient creep strain following a stress change, then (and only then) the calculation can be based on the viscous creep rate as obtained from ordinary creep tests.

An important creep effect is *relaxation*, i.e. the decrease of stress in a body held between rigidly fixed grips. Stress relaxation in bolted joints is a familiar example. Let $\sigma_0$ be the initial stress in the bolt, $\sigma$ the stress, and $\epsilon$ the plastic strain after time $t$. If the ends are held rigidly,

$$\frac{\sigma}{E} + \epsilon = \frac{\sigma_0}{E} \tag{9-8}$$

where $E$ is Young's modulus. Hence,

$$\sigma_0 - \sigma = E\epsilon \tag{9-9}$$

where $\epsilon$ is the total creep strain up to time $t$.

From Eq. 9-8 or 9-9 it follows that

$$\frac{d\sigma}{dt} = -E\frac{d\epsilon}{dt} \tag{9-8a}$$

If, therefore, the creep rate $\frac{d\epsilon}{dt}$ is a known function of the stress, the time during which the stress relaxes from $\sigma_0$ to $\sigma$ is obtained as

$$t = \int_{\sigma_0}^{\sigma} \frac{d\sigma}{E\frac{d\epsilon}{dt}} \tag{9-8b}$$

This is illustrated in Fig. I,9c which implies the assumption that transient creep is negligible and that viscous creep depends on the stress according to Eq. 9-3. Since the viscous creep rate decreases very rapidly with decreasing stress, the drop of stress is very rapid initially and slows down soon.

In the general case when transient creep is not negligible, relaxation in metals cannot be calculated from constant-stress or constant load creep data on the basis of the available knowledge.

**I,10. Creep Tests and Their Evaluation.** Engineers became interested in creep between 1920 and 1930; during this period, creep testing was concerned mainly with the measurement of the time after which, in a constant load tensile test, fracture took place. The immediate results of such creep fracture ("stress rupture") tests are usually plotted as curves showing the lifetime of the specimen as a function of stress, at a given temperature; it is convenient to plot both stress and lifetime, or the lifetime alone, logarithmically. (See Fig. I,15 in Art. 15.)

Soon it became clear that failure by creep would usually occur by excessive deformation long before fracture took place [*48,49,50,51*], and creep tests in the narrower sense of the word (i.e. creep rate measurements) increasingly replaced the creep fracture test, although the latter is still necessary for special purposes (see Art. 8) and also useful as a cheap approximate substitute of the creep rate measurement. The main result of the standard creep test was the secondary (minimum, steady state) creep rate which, at low stress levels, practically coincides with that of the viscous creep; it is the main basis for the design calculation of longevous structures. In such cases the complications due to transient creep largely disappear, and the main problem is to estimate with sufficient accuracy the creep that occurs during the intended lifetime of the structure (which may amount to 20 or 30 years) from a creep test of necessarily much shorter duration.

If the creep rate remained constant during the service life, it could be obtained directly from relatively short time tests at the service temperature and the service stress. Of course, the test would have to extend over a period long enough for transient creep to disappear; but the third stage would not have to be approached. However, structural and metallurgical changes (recrystallization, thermal softening, precipitation, redissolution, cracking, corrosion) that may not appear in a test of 10,000 hours duration may radically change the creep rate later in the course of the service life. One may think, therefore, of embracing the full range of creep strains up to fracture within a test of 10,000 or even 1000 hours duration by applying a stress, or a temperature, higher than the values corresponding to the intended service conditions. By recording several creep curves for various stresses, all higher than the service stress, and extrapolating the viscous creep rate to the permissible design value, the corresponding design stress would be obtained. Or, creep curves might be obtained with stresses in the range of the service stress but at higher temperatures; by extrapolation to the service temperature, the stress corresponding to a desired creep rate could be obtained.

Unfortunately, neither the stress-accelerated nor the temperature-accelerated test is always reliable. At higher stresses, structural changes may occur that would never take place at the service stress (e.g. recrystallization); at a higher temperature, phase transformations, precipitation and dissolution phenomena may make the result inapplicable to the service conditions. There is no foolproof formal way of extrapolating creep test results: unless the long-time behavior of the material is reliably known from experience, only close acquaintance with the physical and metallurgical factors involved can reduce the danger of a serious error.

**I,11. Design for Temperature Rise during Service.** The aerodynamic heating of structural parts in highly supersonic aircraft raises the question of how in such cases creep can be taken into account by the designer. Very little experimental work has been published on this problem; however, the available general knowledge of plastic deformation and creep gives fairly definite indications about the main points involved.

If the highest service temperature is below the range of hot creep, a stress-strain curve of the material at this temperature can be obtained without difficulties. The strains represented in the curve include transient creep strains: these are lower if the loading rate is high, higher if it is low. At room temperature the loading rate does not considerably affect the stress-strain curves of the common strong metals; at higher temperatures, however, its effect may be great (i.e. the creep strain contained in the measured strain may be large, and it may increase rapidly with the duration of loading). Consequently, the design cannot be based on a yield stress-strain curve obtained with some conventional loading time; the loading rate under service conditions has to be used.

Equally misleading, in the dangerous direction, would be the uncritical use of the minimum creep rate measured in a conventional creep test. In this test, it may take hours or days before the transient creep disappears and the rate of the viscous creep can be measured. This is usually a small fraction of the average creep rate during the first minutes or hours of loading, during which the creep may be predominantly transient; its use for designing a structure that may be in service only a few minutes or hours, therefore, may be unsafe.

The correct procedure is to base the design on a test in which the intended service conditions (duration of loading and variation of the temperature during loading) are imitated as accurately as possible, and the entire deformation, from the time of the application of the load, is measured. If the periods of service are relatively short (minutes, hours, or a few days), such imitative tests are far less time-consuming than the shortest conventional hot creep tests which must extend at least over 1000 hours.

It need hardly be mentioned that, whenever a metal is used at an elevated temperature, the question of structural and chemical stability must be considered. Strain hardening, grain boundaries, precipitates, martensitic components, which can give a high yield stress to a metal at a lower temperature, may become completely ineffective at higher temperatures if the strain hardening is removed by recrystallization or thermal softening, if grain boundaries soften, precipitates are redissolved, and martensitic inclusions decompose into relatively soft phases. In addition, oxidation or other types of chemical attack may occur, possibly combined with the effect of the applied stress to stress corrosion. If a new material is used for long-time high temperature service, no amount of care can eliminate completely the possibility of unpleasant surprises. Whenever the duration of service is short enough to be duplicated in a practically feasible test, the design must be based on a closely imitative test.

**I,12. Plastic Instability; Ultimate Strength. Instability of Non-Newtonian Viscous Materials.** A plastic instability is present if, after a certain amount of plastic deformation which may be zero, the load required for further deformation begins to decrease. If the load is not reduced quickly enough after the point of instability has been reached, rapidly accelerating deformation occurs which may end with complete failure.

Appropriately one should distinguish plastic instabilities due to physical causes from those due to geometrical causes. If the yield stress decreases with the plastic strain, the load required for continuing the deformation decreases even if the cross-sectional area of the specimen remains constant. The best-known type of such a physical instability is the yield phenomenon of annealed mild steels which manifests itself in the appearance of an upper and a lower yield point and in the development of Lüders bands on the specimen. In the hot creep range many materials suffer structural changes which reduce their resistance to stress and increase the creep rate even if the stress and not the load is kept constant.

A plastic instability in the narrower sense of the word is one due to purely geometrical causes. As in the elastic range, so in the plastic range also there is the possibility of buckling; creep buckling has been studied particularly by Hoff and his collaborators, Kempner and Libove [*52,53,54,55,56,57,58*]. The most characteristic geometrical type of plastic instability, however, is one to which there is no parallel in elasticity (if rubberlike materials are excluded). It is due to the decrease of the load-carrying cross section of the loaded body, caused by the transverse contraction in tensile loading. In the initial part of the stress-strain curve the rise of the yield stress with the plastic strain is usually rapid enough to overcompensate the effect of diminishing cross section; it leads to a steady increase of the load required for further deformation. In many cases, however, the rate of strain hardening diminishes so rapidly that, before fracture would take place, the yield load begins to decrease with increasing extension; i.e. a plastic instability develops. The point on the ("true") stress-strain curve at which this occurs in the ordinary tensile test is easily determined by means of a graphical construction first given by Considère [*59*]. If $x$ is the current and $x_0$ the initial length of the specimen, the linear strain is

$$\epsilon = \frac{x - x_0}{x_0} \tag{12-1}$$

If $A$ is the current and $A_0$ the initial cross-sectional area, and $V$ the volume of the specimen which remains constant during plastic deformation,

$$V = Ax = A_0 x_0 \tag{12-2}$$

and

$$A dx = -x dA$$

The tensile force is

$$F = \sigma A \tag{12-3}$$

It reaches a maximum when

$$dF = \sigma dA + A d\sigma = 0 \quad \text{or} \quad \frac{d\sigma}{\sigma} = -\frac{dA}{A} = \frac{dx}{x} \tag{12-4}$$

In view of Eq. 12-1,

$$\frac{dx}{x} = \left(\frac{dx}{x_0}\right)\left(\frac{x_0}{x}\right) = \frac{d\epsilon}{1+\epsilon}$$

Hence the condition of a load maximum is

$$\frac{d\sigma}{d\epsilon} = \frac{\sigma}{1+\epsilon} \tag{12-5}$$

This condition finds a simple graphical expression in the (true) yield stress-strain diagram (Fig. I,12a). If point $P$ is plotted at unit distance from the origin on the negative strain axis, the gradient of the line connecting it with any point $(\sigma,\epsilon)$ of the stress-strain curve $OY$ is $\sigma/(1+\epsilon)$. At the point of contact of the tangent to the curve drawn from $P$, there-

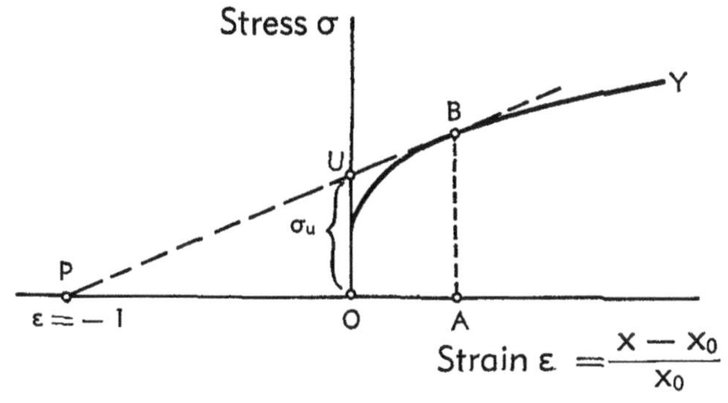


Fig. I,12a. Considère's construction of the ultimate strength $\sigma_u$, the maximum load point $B$, and the uniform extension $OA$ from the true stress-strain curve.

fore, the condition of the load-maximum equation (Eq. 12-5) is satisfied. $OA$ is the strain $\epsilon_m$ at the load maximum, and $AB = \sigma_m$ the (true) maximum load stress. The ultimate tensile strength $\sigma_u$ is defined as the maximum load divided by the initial cross-sectional area; in view of Eq. 12-1,

$$\sigma_u = \frac{F_m}{A_0} = \sigma_m \frac{A_m}{A_0} = \sigma_m \frac{x_0}{x_m} = \frac{\sigma_m}{1+\epsilon_m}$$

or

$$\frac{\sigma_u}{\sigma_m} = \frac{1}{1+\epsilon_m} \tag{12-6}$$

The similarity of the triangles $PAB$ and $POU$ in Fig. I,12a shows that the intercept $OU$ between the abscissa axis and the tangent $PB$ is the ultimate strength.

So long as the yield load rises with the extension, any weak cross section of the specimen, in which most of the deformation is concentrated at a certain moment, soon hardens in spite of the decrease of its cross section and ceases to be the weakest point; the deformation continues elsewhere. At the maximum load, however, any further deformation reduces the yield load; the cross section that was the weakest when the load maximum was first reached in it remains the weakest point, and all further deformation is concentrated in its neighborhood and ceases elsewhere. In other words, a neck develops in the specimen.

The ultimate strength has nothing to do with fracture; it is simply the load per unit of the initial cross-sectional area at which plastic instability sets in in uniaxial tension. Whether or not it has any significance to the designer depends on whether the structure contains a tensile member in which the load can conceivably reach the instability value and whether the consequences of failure (usually disruption) in this event are graver than those of the milder failure which precedes it, consisting of excessive plastic deformation.

It is of interest that the construction of Considère gives the point of plastic instability on the stress-strain curve only for a specimen under uniaxial tensile stress. A cylindrical or spherical vessel under internal pressure, or a plate under equal tensions in two mutually perpendicular directions, also reach points of instability if the yield stress-strain curve of the material is of the usual character, and if fracture does not occur before the point of instability is reached. It is easy to see, however, that plastic instability occurs in these cases at different points of the stress-strain curve, and that the highest conventional stress (maximum load divided by the initial cross-sectional area) does not coincide with the ultimate tensile strength (for details, see Sachs and Lubahn [*60*], Swift [*61*], and Orowan [*47*]).

The preceding discussion of plastic instability applies only to purely plastic materials, i.e. to materials that have a mechanical response to stresses completely described by a yield stress-strain curve not depending on the rate of straining. Since metals and other crystalline materials always show at least transient creep, they are not purely plastic; in the low temperature range (below the hot creep range), however, transient creep has only a minor influence upon the conditions of plastic instability. In the hot creep range, on the other hand, metals behave at not too high stresses as non-Newtonian viscous, rather than as plastic, materials, and the question must be investigated whether they then still show an instability phenomenon.

Fig. I,12b gives a picture of the mechanical behavior of a metal at relatively low stresses and high temperatures; the real behavior is represented by curve $OA$ which is approximated by the two straight lines of curve $OYB$. In the simplified diagram, the material has a sharp yield point below which only elastic deformation occurs; above the yield point, the strain rate is a linear function of the stress. It is given by

$$\frac{d\epsilon}{dt} = C(\sigma - Y) \tag{12-7}$$

where $\sigma$ is the (uniaxial) tensile stress acting upon the rod-shaped specimen, $Y$ the yield stress, and $\epsilon$ the tensile strain. $C$ is the gradient of the slope of the inclined line $YB$. Since

$$\frac{d\epsilon}{dt} = \frac{dx/x}{dt} = -\frac{dA/A}{dt}$$

where $x$ is the current length and $A$ the current cross-sectional area (see Eq. 12-1 and 12-2), Eq. 12-7 can be written as

$$\frac{dA}{dt} = -CA(\sigma - Y)$$

or, if the axial force $F = A\sigma$ is introduced,

$$\frac{dA}{dt} = -C(F - AY) \tag{12-8}$$

In the special case of Newtonian viscosity, the yield stress $Y$ is zero. If the specimen contains a slightly thinner cross section, the rate of

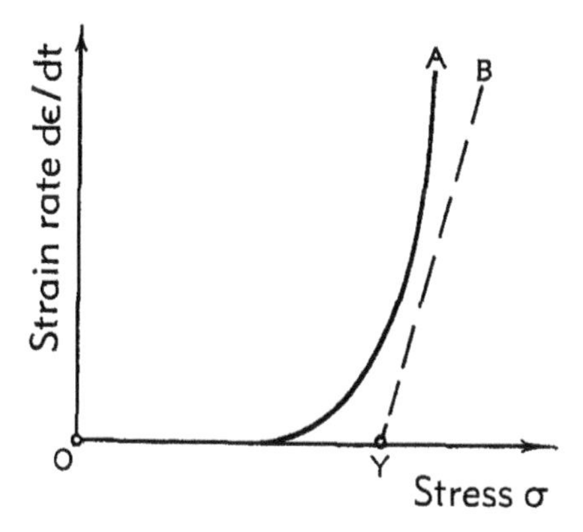


Fig. I,12b. Comparison of the $d\epsilon/dt$ vs. $\sigma$ curves of viscous creep in metals ($OA$) and of Bingham flow ($OYB$).

decrease of its area will then be the same as that of any other cross section, since the right-hand side of Eq. 12-8 does not depend on the cross-sectional area if $Y = 0$. This means that there is no instability and necking if the material is Newtonian—hot glass can be blown and drawn, in sharp contrast with metals. However, if the yield point is not zero, the conditions are quite different. Suppose that the tensile force $F$ is slightly lower than the yield force $AY$ for the normal cross sections, but slightly higher than the yield force $A'Y$ for a subnormal cross section ($A' < A$). The normal sections will then not extend at all, while the subnormal section undergoes creep which is initially slow but accelerates as the cross-sectional area decreases. The result is necking followed by rupture at the weak point. If the tensile force is considerably higher than the yield force, the instability is less extreme; all parts of the specimen extend, and it requires some extension before necking at the weakest point becomes marked. If the tensile force is very high compared with the yield force, the behavior of the material is approximately Newtonian and there is no significant tendency to instability [*47*; *52*; *62*, p. 156].

In accordance with these conclusions, many viscous materials with a yield point can be extended considerably if the force is high and the extension rapid; but they soon neck and rupture if the force is small as shown by Nadai and Manjoine [*63*,*64*]. Metal single crystals, particularly those of the hexagonal metals such as zinc and cadmium, break with very little extension by highly localized slip involving only a few planes if the temperature is high and the strain rate is very low; at the same temperature, however, they can be extended considerably and more or less uniformly if the straining is rapid. Creep fracture (see Art. 8) can be regarded as an extreme case of such an instability: the specimen breaks in the weakest cross section by viscous sliding of the grains along their neighbors before the extension of the specimen as a whole exceeds a few per cent.

### I,13. Plastic Deformation and Creep under Triaxial Stress.

Plastic Deformation. As stated in Art. 3 a material is called purely plastic if the irreversible (nonelastic) part of its deformation is

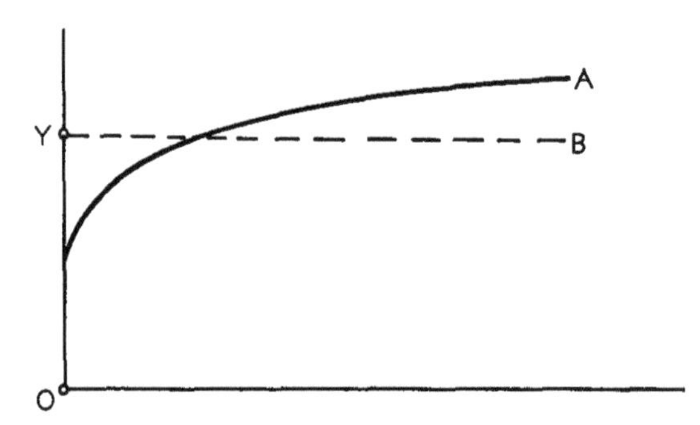


Fig. I,13a. Typical plastic yield stress-strain curve *OA*, compared with the curve of the "ideally plastic" material *OYB*.

governed by a functional relationship between stress and strain, i.e. a "stress-strain curve," which is not influenced by time. If the stress has to be increased with increasing strain, the material is said to show strain hardening (curve *OA* in Fig. I,13a); for simplicity in the mathematical treatment of the distribution of stress and strain in plastically deformed bodies, strain hardening is usually disregarded, so that the stress-strain curve assumes the shape seen in curve *OYB* in Fig. I,13a. An idealized material of such type is called "ideally plastic."

A stress-strain curve can describe the behavior of a plastic material only in simple special states of stress such as uniaxial tension or compression, or pure shear. The relationship between stress and strain in the general case of plastic deformation under a triaxial state of stress is unknown: it is bound to be very complicated because plastic deformation produces directional internal stresses which make the material anisotropic. The laws of plastic deformation in a material made anisotropic by internal stresses due to cold working are unknown; they are fundamentally different from those describing the plastic behavior in cases of anisotropy due to a preferred orientation of the grains (about the latter, see Hill [*65*]). Consequently, the treatment of plastic deformation under triaxial states of stress is usually confined to an ideally plastic material which shows no strain hardening and develops no anisotropy. Since in such a material the stress need not change during deformation, there is no relationship between stress and strain (see curve *OYB*, Fig. I,13a); once the stress is high enough for yielding, it can produce any amount of plastic strain. The plastic stress-strain relationship which would govern the deformation of real materials, therefore, degenerates into: (1) a *yield condition*, determining those combinations of the stress components for which yielding takes place; and (2) a *stress-strain relationship* in which, however, the strain components contain an indeterminate common factor: only their ratios are determined by the components of the applied stress.

Any yield condition has to satisfy the requirement that the superposition of a hydrostatic (isotropic) state of stress (one with three equal principal stresses) must not influence the occurrence of yielding. Only very high pressures (higher than those commonly occurring in engineering) have any influence. This demands that only shear stress components (differences of principal stresses) can enter into the yield condition. The two yield conditions that are widely used at present are: (1) The *von Mises condition:* yielding occurs when the principal stresses $\sigma_1$, $\sigma_2$, $\sigma_3$ satisfy the equation

$$\frac{1}{\sqrt{2}}\sqrt{(\sigma_1-\sigma_2)^2+(\sigma_2-\sigma_3)^2+(\sigma_3-\sigma_1)^2}=Y \qquad (13\text{-}1)$$

where $Y$ is the yield stress in uniaxial tension or compression; and (2) the *Tresca* or *maximum shear stress condition:*

$$\sigma_1-\sigma_3=Y \qquad (13\text{-}2)$$

where $\sigma_1$ is the algebraically greatest and $\sigma_3$ the smallest principal stress. The characteristic difference between the two conditions is that the intermediate principal stress $\sigma_2$ has no effect upon yielding in the Tresca condition while it has a significant influence in the Mises condition.

Experiments carried out at Nadai's suggestion by Lode [*66*], by Ros and Eichinger [67], and by Taylor and Quinney [68] have shown that the behavior of most common polycrystalline metals in the annealed isotropic state is intermediate between that corresponding to the Tresca and the Mises conditions, but it is much closer to the Mises condition. Morrison's careful investigations [69] have revealed that the upper yield point of annealed low carbon steels shows a very erratic behavior which cannot be approximated by any simple yield condition. For instance, the upper yield point increases with decreasing diameter of rod-shaped tensile specimens but decreases with decreasing wall thickness of thin-walled tubular tensile specimens. More important to the engineer than this capricious quantity is the lower yield point. In the lower yield region the material cannot obey the Mises condition because, as soon as a thin Lüders band is formed in a plane close to that of the maximum shear stress, the intermediate principal stress cannot influence its widening because the band is constrained between nonplastic regions of the specimen which shield it from any influence of the intermediate principal stress. This is verified by experience, according to which an annealed low carbon steel (1) does not obey any simple yield condition insofar as the upper yield point is concerned; (2) obeys approximately the Tresca condition in the region of yield (at the lower yield point); and (3) obeys approximately the Mises condition in the region of strain hardening which follows that of the yield.

In regard to the plastic stress-strain relationship, the strain is obviously a simple shear in the Lüders band, and probably generally whenever the Tresca condition applies. If the Mises condition applies, it is usually assumed that the corresponding stress-strain relationship is given by the St. Venant-Lévy-Mises equations:

$$
\begin{aligned}
\delta\epsilon_1 &= \delta\lambda[\sigma_1 - \tfrac{1}{2}(\sigma_2 + \sigma_3)] \\
\delta\epsilon_2 &= \delta\lambda[\sigma_2 - \tfrac{1}{2}(\sigma_3 + \sigma_1)] \\
\delta\epsilon_3 &= \delta\lambda[\sigma_3 - \tfrac{1}{2}(\sigma_1 + \sigma_2)]
\end{aligned}
\tag{13-3}
$$

where $\delta\epsilon_1$, $\delta\epsilon_2$, $\delta\epsilon_3$ are small increments of the principal (plastic) strains. $\delta\lambda$ is an arbitrary quantity which may be called the *extent* of the deformation; it depends on how far the deformation is carried on before the state of stress is changed. These equations are formally identical with the equations of Hookean elasticity; $\delta\lambda$ has taken the place of the elastic constant $1/E$, and Poisson's ratio is 1/2 since there is no significant volume change in plastic deformation.

The Lévy-Mises, stress-strain relationship has been tested very carefully by Taylor and Quinney [68], who have found that annealed polycrystalline metals show considerable departures from it. However, the simplicity of Eq. 13-3 have led to their widespread use in the analysis of plastic deformation; the errors involved are within the permissible limits for most engineering applications so far made.

NEWTONIAN VISCOSITY. If a viscous material obeys Newton's law

$$\tau = \eta \frac{d\gamma}{dt} \tag{7-2}$$

where $\tau$ is the shear stress, $\gamma$ the shear strain, and $\eta$ the coefficient of viscosity, the strain rate $d\epsilon/dt$ of a tensile specimen under the tensile stress $\sigma$ is

$$\frac{d\epsilon}{dt} = \frac{1}{3\eta}\sigma \qquad \text{(see below)} \tag{13-4}$$

Under a triaxial state of stress with principal stresses $\sigma_1$, $\sigma_2$, and $\sigma_3$, the principal strains are given by the Navier-Stokes equations

$$\begin{aligned} \frac{d\epsilon_1}{dt} &= \frac{1}{3\eta}[\sigma_1 - \tfrac{1}{2}(\sigma_2 + \sigma_3)] \\ \frac{d\epsilon_2}{dt} &= \frac{1}{3\eta}[\sigma_2 - \tfrac{1}{2}(\sigma_3 + \sigma_1)] \\ \frac{d\epsilon_3}{dt} &= \frac{1}{3\eta}[\sigma_3 - \tfrac{1}{2}(\sigma_1 + \sigma_2)] \end{aligned} \tag{13-5}$$

These follow from the corresponding equations of elasticity if the principal strains are replaced by the corresponding strain rates, Poisson's ratio by $\frac{1}{2}$ (constancy of volume), and Young's modulus

$$E = 2G(1 + \nu)$$

by $3\eta$, since $\eta$ is the quantity that corresponds to the shear modulus $G$ in the case of viscosity (see Eq. 7-2), and $\nu = \frac{1}{2}$. With $\sigma_2 = \sigma_3 = 0$, the first equation of Eq. 13-5 leads to Eq. 13-4.

In spite of the superficial resemblance between the Navier-Stokes equations and those of Lévy and Mises, they refer to different types of material and fundamentally different processes of deformation. In the Lévy-Mises equations, the strain increments $\delta\epsilon_1$, $\delta\epsilon_2$, and $\delta\epsilon_3$ are sometimes written as $(d\epsilon_1/dt)dt$, $(d\epsilon_2/dt)dt$, and $(d\epsilon_3/dt)dt$; the appearance of time, however, has nothing to do with the nature of the deformation process. $dt$ is merely the time during which a given state of stress is maintained, and $d\epsilon_1/dt$ the (arbitrary) deformation rate during this time. Written in this form, the Lévy-Mises equations merely say that if one of the principal strain rates has the arbitrary value $(d\epsilon_1/dt)$, the others are determined in the manner given.

VISCOUS CREEP OF METALS. The plot of the viscous creep rate $K$ as a function of the stress in Fig. I,9a shows that the high temperature viscosity of metals is far from being Newtonian. Instead of increasing proportionally to the stress, the strain rate is vanishingly small until the stress has reached a certain level, and then it rises very rapidly, as reflected by the empirical formulas (Eq. 9-3, 9-4, and 9-5). Viscous creep under a triaxial state of stress cannot be treated, therefore, by the Navier-Stokes equations (Eq. 13-5); the Lévy-Mises equations, of course, are out of question since they refer to a type of deformation entirely unrelated to time. This circumstance was not clearly realized for some time; the strain increments in a plastic material (which are not determined by the applied stress and the time increment but depend on the entirely arbitrary rate of straining) were confused with the strain differentials in a viscous material which are fully determined by the applied stress and the differential of time.

The first treatment of non-Newtonian viscosity under a triaxial state of stress was due to Hencky [*70*]. It referred to the particularly simple idealized case of non-Newtonian behavior shown in Fig. I,12b, curve $OYB$. The material was assumed to have a sharp yield stress $OY$ (e.g. in tension or compression) below which no viscous deformation is present; above it, the rate of strain is proportional to the excess of the applied stress over the yield stress:

$$\frac{d\epsilon}{dt} = \frac{1}{3\eta}(\sigma - Y) \tag{13-6}$$

Such non-Newtonian behavior is said to be of the Bingham type.

Hencky obtained a mathematically plausible generalization of the Bingham equation (Eq. 13-6) to the three-dimensional case. For simplicity and clarity, the Hencky equations will be written in what follows for the principal stresses $\sigma_1$, $\sigma_2$, and $\sigma_3$ and the principal strain rates

$$\frac{d\epsilon_1}{dt}, \quad \frac{d\epsilon_2}{dt}, \quad \frac{d\epsilon_3}{dt}$$

In this case, they are

$$\begin{aligned} \frac{d\epsilon_1}{dt} &= \frac{1}{3\eta}\left(1 - \frac{Y}{\bar{\sigma}}\right)[\sigma_1 - \tfrac{1}{2}(\sigma_2 + \sigma_3)] \\ \frac{d\epsilon_2}{dt} &= \frac{1}{3\eta}\left(1 - \frac{Y}{\bar{\sigma}}\right)[\sigma_2 - \tfrac{1}{2}(\sigma_3 + \sigma_1)] \\ \frac{d\epsilon_3}{dt} &= \frac{1}{3\eta}\left(1 - \frac{Y}{\bar{\sigma}}\right)[\sigma_3 - \tfrac{1}{2}(\sigma_1 + \sigma_2)] \end{aligned} \tag{13-7}$$

Here $Y$ is the yield stress in uniaxial tension or compression, and $\bar{\sigma}$ the Mises expression (the "equivalent stress")

$$\bar{\sigma} = \frac{1}{\sqrt{2}}\sqrt{(\sigma_1 - \sigma_2)^2 + (\sigma_2 + \sigma_3)^2 + (\sigma_3 - \sigma_1)^2} \tag{13-8}$$

Obviously, the Hencky equations differ from those of Navier-Stokes only in that they contain an additional factor (the first bracket on the right-hand side); together with the factor $1/\eta$, this may be regarded as representing the reciprocal of a variable coefficient of viscosity. At the yield point, the Mises expression is equal to $Y$ and so the first bracket vanishes; this amounts to an infinitely high viscosity. If the shear stresses are extremely high, the bracket is practically equal to unity, and the Hencky equations coincide with the Navier-Stokes ones for a Newtonian material of viscosity $\eta$.

The Hencky equations were later independently re-obtained and applied to various problems by Oldroyd [*71*]; a generalized theory has been given by Prager [*72*].

It may be asked how it comes about that the Mises expression, first met with in the theory of plasticity, appears again in the fundamentally

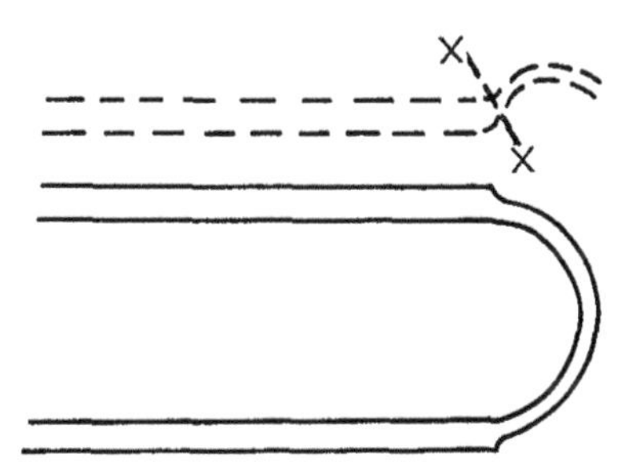


Fig. I,13b. Discontinuity of creep strains at junction between cylindrical tube and hemispherical cap leading to fracture at $x$-$x$.

different field of viscosity. The reason is that the plasticity and non-Newtonian viscosity have one point of contact: a Bingham material with vanishing coefficient of viscosity is identical with an ideally plastic material. With both, any strain and any strain rate can be reached with a fixed value of the stress. Consequently, the Hencky equations must go over into the Mises yield condition (if this is assumed to apply to the plastic case) for $\eta = 0$. Obviously, this requirement is satisfied: to give a finite strain rate at finite shear stresses in spite of $\eta = 0$, the first bracket in Eq. 13-7 must become zero, and equating it to zero gives the Mises yield condition (Eq. 13-1).

The solution of the problem of the Bingham material is not sufficient for the treatment of viscous creep under triaxial stress, because the creep rate vs. stress curve (Fig. I,9a) is very different from the Bingham curve, particularly in the yield region (at very low strain rates) which is of the greatest practical importance (cf. Fig. I,12b). If a set of stress vs. strain rate equations for non-Newtonian flow goes over with vanishing yield stress into the equations of Newtonian viscosity, it can certainly not be used for dealing with creep in metals. The first treatment of triaxial viscous creep in metals was carried out by Bailey [*41*,*42*]. In spite of its somewhat complicated nature, his set of equations has been used and is still being used successfully for design under creep conditions. Bailey's strain rate vs. stress relationships are so constructed that, for the uniaxial case, they go over into the power-law expression (Eq. 9-3); to be sufficiently flexible, they contain two exponents, both empirical constants.

With no reference to the viscous creep of metals, Fromm [*73*] suggested a relationship which did not merge into Newtonian viscosity with vanishing yield stress; however, he did not seem to have aimed at treating non-Newtonian viscosity of the creep type, and his equations do not appear to represent the behavior of any important type of material. A general treatment of viscous creep in metals, independent of empirical formulas like those given in Eq. 9-3, 9-4, and 9-5, and at the same time simple enough for practical use in engineering, was given by Soderberg in 1936 [*44*]. As indicated by the Hencky equations (Eq. 13-7), the equations of any non-Newtonian-type of viscosity are likely to have the form of the Navier-Stokes equations with a variable coefficient of viscosity, defined as the ratio of the stress to the strain rate in the corresponding uniaxial case or in that of simple shear. The only question is how to find, in the case of a triaxial stress, the equivalent uniaxial stress at which the apparent coefficient of viscosity, obtained from the uniaxial strain rate-stress relationship (see the lower curves in Fig. I,9a), has the value to be used in the triaxial strain rate-stress equations. Soderberg's answer is a straight generalization of that of Hencky. Since Bingham viscosity is included as a special case in the general one, the equations of non-Newtonian viscosity must degenerate for $\eta = 0$ to the Mises yield condition (Eq. 13-1). In this case, three principal stresses $\sigma_1$, $\sigma_2$, and $\sigma_3$ must be equivalent to a uniaxial stress

$$\bar{\sigma} = \frac{1}{\sqrt{2}} \sqrt{(\sigma_1 - \sigma_2)^2 + (\sigma_2 - \sigma_3)^2 + (\sigma_3 - \sigma_1)^2} \qquad (13\text{-}9)$$

It is plausible to assume that the equivalent uniaxial stress is given by Eq. 13-9 generally, not only at the "yield surface" of a Bingham material. This leads immediately to the Soderberg solution of the general problem: The strain rates are obtained from the Navier-Stokes equations (Eq. 13-5) in which the effective coefficient of viscosity is given by

$$3\eta_{\text{eff}} = \frac{\bar{\sigma}}{(d\epsilon/dt)_{\bar{\sigma}}} \qquad (13\text{-}10)$$

$(d\epsilon/dt)_{\bar{\sigma}}$ being the creep rate in uniaxial tensile or compressive loading for a stress equal to the Mises expression. Under the following subheading, a few practical examples of creep rate calculation under triaxial states of stress are given.

A valuable contribution to the calculation of the distribution of stresses and strain rates in a body undergoing pure viscous creep under any load distribution has been made recently by Hoff [*52*]. He pointed out that, if the creep rate-stress relationship can be approximated by the expression

$$\frac{d\epsilon}{dt} = C\sigma^n \tag{9-3'}$$

the distribution of stress in the body is the same as if the material obeyed the nonlinear law of elasticity

$$\epsilon = C\sigma^n \tag{13-11}$$

with the same constants $C$ and $n$, provided that the initial and boundary conditions are equivalent. The surface tractions must be the same in both cases, and the prescribed surface displacements in the elastic problem must be numerically equal to the surface velocities of the creep problem.

It seems that the Soderberg treatment of the general problem of non-Newtonian viscosity is equivalent to the formal theory developed by Prager [*72*]. The latter has not been applied to the problem of creep in metals.

The above treatment of non-Newtonian creep can be correlated with the Mises-Geiringer theorem of the "plastic potential" [*74*; *75*, pp. 17–26] in the following way. The simplest type of non-Newtonian viscosity of an isotropic material is one governed by a relationship that differs from Eq. 13-5 only in that the first factor on the right-hand side ($1/3\eta$) is not a constant but a function of the stress components:

$$\begin{aligned}
\frac{d\epsilon_1}{dt} &= F(\sigma_1, \sigma_2, \sigma_3)[\sigma_1 - \tfrac{1}{2}(\sigma_2 + \sigma_3)] \\
\frac{d\epsilon_2}{dt} &= F(\sigma_1. \sigma_2, \sigma_3)[\sigma_2 - \tfrac{1}{2}(\sigma_3 + \sigma_1)] \\
\frac{d\epsilon_3}{dt} &= F(\sigma_1, \sigma_2, \sigma_3)[\sigma_3 - \tfrac{1}{2}(\sigma_1 + \sigma_2)]
\end{aligned} \tag{13-5a}$$

The factor $1/3\eta$ in the Navier-Stokes equations is a material constant which may be called the "coefficient of tensile fluidity." If the corresponding quantity $F$ in the non-Newtonian case is stress-dependent (i.e. a function of the general stress components $\sigma_x$, $\sigma_y$, $\sigma_z$, $\tau_x$, $\tau_y$, and $\tau_z$), it cannot depend in an isotropic material on the coordinate axes to which the stress components are referred; i.e. it must be a function of the invariants of the stress tensor. Moreover, it may be assumed that it is not influenced significantly by the hydrostatic pressure. The simplest function satisfying these requirements is one that depends on the Mises equivalent stress

$$\bar{\sigma} = \frac{1}{\sqrt{2}} \sqrt{(\sigma_1 - \sigma_2)^2 + (\sigma_2 - \sigma_3)^2 + (\sigma_3 - \sigma_1)^2} \tag{13-8}$$

alone. For brevity, let principal axes be used. The derivative of any function $f(\bar{\sigma})$ with respect to the principal stress $\sigma_1$ is

$$\frac{df(\bar{\sigma})}{d\sigma_1} = \frac{f'(\bar{\sigma})}{\bar{\sigma}} 2[\sigma_1 - \tfrac{1}{2}(\sigma_2 + \sigma_3)] \tag{13-8a}$$

If, therefore, the function $f(\bar{\sigma})$ is determined so that

$$\frac{2f'(\bar{\sigma})}{\bar{\sigma}} = F(\bar{\sigma}) \tag{13-8b}$$

where $F$ is the stress-dependent tensile fluidity of the material, comparison of Eq. 13-8a and the corresponding equations for the other derivatives with Eq. 13-5a shows that

$$\begin{aligned} \frac{d\epsilon_1}{dt} &= \frac{df(\bar{\sigma})}{d\sigma_1} \\ \frac{d\epsilon_2}{dt} &= \frac{df(\bar{\sigma})}{d\sigma_2} \\ \frac{d\epsilon_3}{dt} &= \frac{df(\bar{\sigma})}{d\sigma_3} \end{aligned} \tag{13-8c}$$

In words, the principal creep rates are the derivatives of a function $f(\bar{\sigma})$ of the Mises invariant with respect to the corresponding principal stress. It is easy to see that, if stress and creep rate are not referred to principal axes, the creep rate components are still the derivatives of $f(\bar{\sigma})$ with respect to the stress component with the same subscript.

After Von Mises, this situation is described by saying that the creep-rate tensor, as a function of the stress tensor, is derived from the tensor potential $f(\bar{\sigma})$. If $f(\bar{\sigma}) = C\bar{\sigma}^2$, (where $C$ does not depend on the stress components, although it may vary from point to point in the deformed body), and if the creep rates are replaced by strain increments $\delta\epsilon$, etc. (see Eq. 13-3), Eq. 13-8c lead to the St. Venant-Lévy-Mises equations (Eq. 13-3). The equivalent stress $\bar{\sigma}$, therefore, is not only the quantity that determines the occurrence of plastic yielding; it is also a "plastic potential" in that its derivatives with respect to the stress components are proportional to the strain increment components. In ideal plasticity, any function of $\bar{\sigma}$ is a plastic potential since $\bar{\sigma}$ and so $f(\bar{\sigma})$ assume fixed values when plastic deformation takes place. In non-Newtonian viscosity there is in general no sharp yield point and no yield condition; but there may exist a "creep potential" of which the derivatives are the creep-rate components. In particular, if the creep-rate components are proportional to the "reduced stresses" (the square brackets on the right-hand side of the Navier-Stokes and of the Lévy-Mises equations) the creep potential is a function of the Mises equivalent stress alone. In the particular case where the stress dependence of the creep rate is represented by a power function $A\bar{\sigma}^n$,

$$f(\bar{\sigma}) = \frac{A}{2(n+1)} \bar{\sigma}^{n+1} \tag{13-8d}$$

The mathematical elegance of the Mises-Geiringer theorem of the plastic potential, of course, is no proof of its physical truth. Experiments of Ros and Eichinger [*67*] and of Taylor and Quinney [*68*] have shown that the behavior of ductile metals deviates considerably from the Lévy-Mises equations. This may be due to a slight anisotropy of the tubular specimens used in the experiments; or to the anisotropy developed during the experiment itself; or it may be a consequence of the fact that the Mises yield condition, while a good approximation, is not exactly true; or it may indicate a limit to the validity of the theorem of the plastic potential. As far as creep is concerned, experiments of Norton [*76*] and Soderberg [*77*] indicate that the creep rate-stress relationship obtained from a creep potential of the Soderberg type is at least a good enough approximation for most engineering applications.

It is interesting that the plastic potential based on the maximum shear stress (Tresca) yield condition

$$\sigma_1 - \sigma_3 = Y \tag{13-2}$$

leads to a stress-strain relationship that represents simple shear in the plane of the maximum shear stress (Geiringer [*75*]). This agrees with the fact, discussed at the beginning of this article, that the progress of yielding in annealed low carbon steels is governed by the Tresca, rather than the Mises, yield condition, and that the plastic deformation here is a simple shear in the Luders bands. In single crystals the Mises condition is obviously inapplicable; yielding takes place when the shear stress in the slip plane resolved in the slip direction reaches a critical value, and the strain is a simple shear along the slip plane. Here again, the theorem of plastic potential leads to a correct conclusion.

Transient Creep under Triaxial States of Stress. The preceding considerations apply only to purely viscous creep. No experimental or theoretical work appears to have been done on transient creep. It would seem plausible to assume as a point of departure that, at least at low stress levels, the principal stresses $\sigma_1$, $\sigma_2$, and $\sigma_3$ would produce transient principal creep strains given by

$$\begin{aligned}
\epsilon_1 &= \epsilon_{01} + (B \ln t)[\sigma_1 - \tfrac{1}{2}(\sigma_2 + \sigma_3)] \\
\epsilon_2 &= \epsilon_{02} + (B \ln t)[\sigma_2 - \tfrac{1}{2}(\sigma_3 + \sigma_1)] \\
\epsilon_3 &= \epsilon_{03} + (B \ln t)[\sigma_3 - \tfrac{1}{2}(\sigma_1 + \sigma_2)]
\end{aligned} \tag{13-12}$$

if the creep is of the logarithmic type, and by

$$\begin{aligned}\epsilon_1 &= \epsilon_{01} + (Bt^{1/n})[\sigma_1 - \tfrac{1}{2}(\sigma_2 + \sigma_3)] \\ \epsilon_2 &= \epsilon_{02} + (Bt^{1/n})[\sigma_2 - \tfrac{1}{2}(\sigma_3 + \sigma_1)] \\ \epsilon_3 &= \epsilon_{03} + (Bt^{1/n})[\sigma_3 - \tfrac{1}{2}(\sigma_1 + \sigma_2)]\end{aligned} \tag{13-13}$$

if it is of the Andrade type. The constants $\epsilon_{01}$, $\epsilon_{02}$, and $\epsilon_{03}$ represent purely plastic strains; the creep constants $B$ might be related to the corresponding quantities in the uniaxial tensile or compressive test in the following way. The uniaxial test gives $B$ as a function of the stress. As in the procedure of Soderberg, the appropriate value of $B$ for the triaxial stress case might be that corresponding to a value of the uniaxial stress equal to the "effective" stress in the triaxial case, i.e. to the Mises quantity

$$\bar{\sigma} = \frac{1}{\sqrt{2}} \sqrt{(\sigma_1 - \sigma_2)^2 + (\sigma_2 - \sigma_3)^2 + (\sigma_3 - \sigma_1)^2} \tag{13-9}$$

It should be noted that the left-hand sides of Eq. 13-12 and 13-13 represent *strains*, not strain rates or strain increments as in the Navier-Stokes or Lévy-Mises equations. Eq. 13-12 and 13-13 apply only when both the principal stresses and the directions of the principal axes remain constant. If the magnitude of the stress changes, transient creep is no longer represented by the simple logarithmic or Andrade-type expressions. If the principal stress axes, or the signs of the principal stresses, change, the structural (microscopic) residual stresses produced by the preceding stage of creep make the material anisotropic, so that the simple isotropic stress-strain relationships considered here no longer apply. This complication is largely absent in the case of viscous creep where the relatively high temperature does not permit the development of considerable micro-residual stresses.

Practical Applications. Practical problems of creep in high temperature machinery usually require the long-time prediction of the viscous deformation. It is in general permissible to neglect the elastic deformation and often also the primary component of the creep. For uniaxial states of stress, the time to fracture can be obtained from creep fracture tests; the condition of creep fracture under triaxial stress, however, has not been investigated so far. It is very unlikely to resemble either the Mises or the Tresca yield condition because a superposed hydrostatic pressure or tension does not affect these while it has a profound influence upon the occurrence of fracture. The superposition of a moderate hydrostatic pressure, for instance, may prevent fracture whereas it would have no significant effect upon plastic or creep deformation. In all probability, superposition of an equitriaxial tension would greatly accelerate creep fracture, so that reliance on creep fracture tests in uniaxial tension might be dangerous, particularly if the experimentally obtained creep fracture strength were used in connection with a yield condition (e.g. in place of the uniaxial yield stress $Y$ in the Mises condition, Eq. 13-1).

In what follows a few examples of the calculation of creep rates in simple engineering structures will be given. Naturally, the results cannot claim the same degree of accuracy which is possible when the deformation is purely elastic.

The treatment is based upon the availability of creep data in tension, in which under the stress $s$ at the temperature $T$ the creep rate $e$ may be expressed in the form

$$\dot{e} = f(s)e^{-b/T} = \dot{e}_{\mathrm{m}} \left(\frac{s}{s_{\mathrm{m}}}\right)^{n} e^{(T-T_{\mathrm{m}})/B} \tag{13-14}$$

In the second expression a power function has been used for $f(s)$ (see Eq. 9-3), and the temperature function has been approximated by a more tractable type of exponential function. $s_{\mathrm{m}}$ and $T_{\mathrm{m}}$ are arbitrary reference values; $\dot{e}_{\mathrm{m}}$ the creep rate at $s_{\mathrm{m}}$; and $T_{\mathrm{m}}$, $n$, and $B$ are material constants. The influence of the temperature has been simplified by restricting the range of temperature to values that are small in comparison with the absolute temperature $T_{\mathrm{m}}$. It is easily seen that the constant $B$ has the value $T_{\mathrm{m}}^2/b$. Typical values of $n$ vary from 5 to 7; the value of $B$ is of the order of 20 to 40°F.

In the present discussion, the Mises condition is used in its usual simple form which is valid only for isotropic materials (see further below).

In the triaxial state of stress, $\sigma_1$, $\sigma_2$, $\sigma_3$, the three principal creep rates then have the values

$$\begin{aligned} \dot{\epsilon}_1 &= \frac{\dot{e}}{s}\left[\sigma_1 - \tfrac{1}{2}(\sigma_2 + \sigma_3)\right] \\ \dot{\epsilon}_2 &= \frac{\dot{e}}{s}\left[\sigma_2 - \tfrac{1}{2}(\sigma_3 + \sigma_1)\right] \\ \dot{\epsilon}_3 &= \frac{\dot{e}}{s}\left[\sigma_3 - \tfrac{1}{2}(\sigma_1 + \sigma_2)\right] \end{aligned} \tag{13-15}$$

Here $\dot{e}$ is the creep rate in the uniaxial tension test under a tensile stress $s$ which is numerically equal to the equivalent stress $\bar{\sigma}$, i.e.,

$$s = \bar{\sigma} = \frac{1}{\sqrt{2}} \sqrt{(\sigma_1 - \sigma_2)^2 + (\sigma_2 - \sigma_3)^2 + (\sigma_3 - \sigma_1)^2} \tag{13-16}$$

The stress-strain relationships in triaxial viscous creep are formally similar to those of elasticity: Eq. 13-15 take the place of Hooke's law. The apparent "coefficient of viscous traction" plays the role of the modulus of elasticity; however, it now depends on stress and temperature. Poisson's ratio is 0.5. The stresses must satisfy the equations of equilibrium as well as the boundary conditions; the strains must satisfy the compatibility relations. The equations of compatibility are sometimes simplified because the volume remains constant.

Eq. 13-15 can be used only if the transient creep strains are negligible compared with the viscous (steady state) creep strains. If this condition is satisfied, the stress distribution does not alter in the course of deformation at constant load and temperature, except when the strains become large. Very little work has been done so far on the calculation of stress distributions under conditions of transient creep.

As in the elastic case, most practical problems must be reduced if possible to two-dimensional ones in order to avoid excessive complications. Two important cases exist:

The case of *plane strain*, usually the simpler for calculation, is characterized by zero creep rate in one direction. With $\dot{\epsilon}_3 = 0, \sigma_3 = (\sigma_1 + \sigma_2)/2$ (see the third equation of Eq. 13-15). Then 13-15 and 13-16 become

$$\dot{\epsilon}_1 = \frac{\sqrt{3}}{2}\dot{e}, \quad \dot{\epsilon}_2 = -\frac{\sqrt{3}}{2}\dot{e}, \quad \dot{\epsilon}_3 = 0$$
$$s = \frac{\sqrt{3}}{2}(\sigma_1 - \sigma_2) \qquad (\sigma_1 > \sigma_2) \tag{13-17}$$

The case of *plane stress* is characterized by zero stress in one direction. With $\sigma_3 = 0$, Eq. 13-15 and 13-16 become

$$\dot{\epsilon}_1 = \frac{\dot{e}}{s}[\sigma_1 - \tfrac{1}{2}\sigma_2]$$
$$\dot{\epsilon}_2 = \frac{\dot{e}}{s}[\sigma_2 - \tfrac{1}{2}\sigma_1] \tag{13-18}$$
$$\dot{\epsilon}_3 = -\frac{\dot{e}}{s}\frac{1}{2}[\sigma_1 + \sigma_2]$$
$$s = \sqrt{\sigma_1^2 + \sigma_2^2 - \sigma_1\sigma_2}$$

The application of these relationships will now be illustrated by a few examples.

*Thin cylinder and sphere under internal pressure.* For thin shells in general the stress may be assumed to be constant throughout the wall, at least when there is no temperature gradient across the thickness $t$. It is then possible to evaluate the creep equations for the mean radius $r$. The stresses are statically determinate. Denoting by $\sigma_1$, $\sigma_2$, and $\sigma_3$ the tangential, radial, and axial stresses respectively, we have, for the cylinder,

$$\sigma_1 = \frac{pr}{t} - \frac{p}{2}$$
$$\sigma_2 = -\frac{p}{2} \tag{13-19}$$
$$\sigma_3 = \frac{pr}{2t} - \frac{p}{2}$$

where $p$ is the pressure, and for the sphere

$$\sigma_1 = \sigma_3 = \frac{pr}{2t} - \frac{p}{2}$$
$$\sigma_2 = -\frac{p}{2} \tag{13-20}$$

Since the quantity $p/2$ is common to all the stresses, it can be dropped, and both cases become two-dimensional. Obviously, the cylinder is at the same time a plane strain problem, for which the results are

$$\dot{\epsilon}_1 = \frac{\sqrt{3}}{2}\dot{e} = -\dot{\epsilon}_2, \quad \dot{\epsilon}_3 = 0$$
$$s = \frac{\sqrt{3}}{2}\frac{pr}{t} \tag{13-21}$$

The sphere represents a case of plane stress; the results are

$$\dot{\epsilon}_1 = \dot{\epsilon}_3 = \tfrac{1}{2}\dot{e}, \quad \dot{\epsilon}_2 = -\dot{e}$$
$$s = \frac{pr}{2t} \tag{13-22}$$

Norton [*76*] and Soderberg [*77*] have published the results of experiments on superheater tubes which agree with these results within the experimental errors. The axial creep rate was found to be zero in all cases except one in which there was reason to doubt the isotropy of the material, but even here it was very small.

The above result shows that the cylinder and the sphere, with the same mean radius and under the same pressure, will expand at the same rate if the thicknesses are in the ratio

$$\frac{t_{\mathrm{cyl}}}{t_{\mathrm{sph}}} = 3^{\frac{n+1}{2n}} \tag{13-23}$$

For instance, for $n = 6$, $t_{\mathrm{sph}}/t_{\mathrm{cyl}} = 0.528$. In the Norton-Soderberg experiments, this situation was not fully appreciated, and the ends of some cylindrical tube specimens were closed with spheres that were thinner than was required by this rule and expanded faster than the cylinders. This led to strain concentrations (discontinuity of strain rate) and fracture at the region $x$-$x$ as shown in Fig. I,13b. It is an important consideration in high temperature design that discontinuities of creep rate which may give rise to internal stresses must be avoided.

*Thick cylinder under internal pressure and variable temperatures.* The long thick tube represents a case of plane strain, the axial creep rate being zero. This situation is not immediately obvious, but Bailey [*40*] found in experiments with thick tubes of lead that the axial creep rates were zero within the experimental error. With the notations of Fig. I,13c the creep rates are thus given by Eq. 13-17.

The compatibility relations may be developed as follows: Denoting by $\rho$ the radial displacement and by $\dot{\rho}$ its time derivative, the tangential

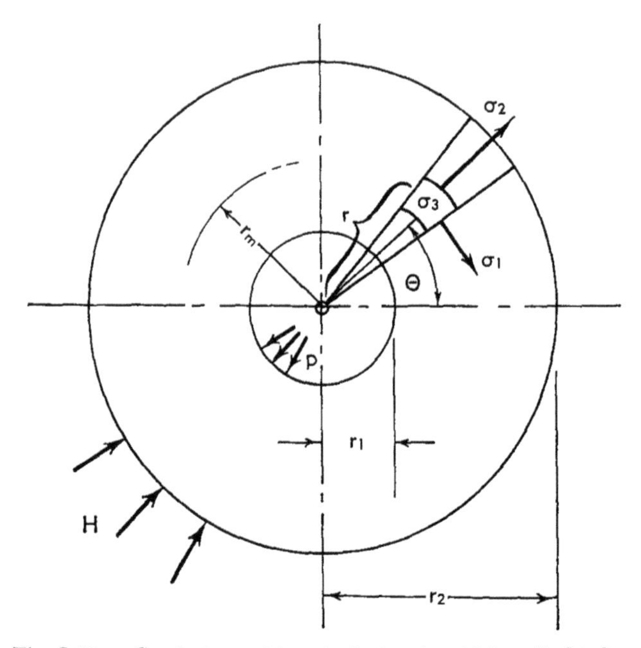


Fig. I,13c. Symbols used in calculation for thick-walled tube.

and radial creep rates are $\dot{\rho}/r$ and $d\dot{\rho}/dr$, respectively. Since the axial creep rate is zero, constancy of volume gives

$$\frac{\dot{\rho}}{r} + \frac{d\dot{\rho}}{dr} = 0, \qquad \dot{\rho} = \frac{\text{const}}{r}$$

and hence, by Eq. 13-17,

$$\dot{\epsilon}_1 = \frac{\sqrt{3}}{2}\dot{e} = \frac{\text{const}}{r^2} = -\dot{\epsilon}_2 \tag{13-24}$$

These equations indicate that $\dot{e}$ must vary as $1/r^2$, i.e.,

$$\frac{\dot{e}}{\dot{e}_{\text{m}}} = \left(\frac{r}{r_{\text{m}}}\right)^{-2} \tag{13-25}$$

The variations of temperature are assumed to arise from heat flow from the outside surface into the tube. For this case, the temperature is determined by

$$T = T_{\text{m}} + \frac{Hr_2}{K}\ln\frac{r}{r_{\text{m}}} \tag{13-26}$$

where $H$ is the flow of heat per unit area of the outer surface and $K$ is the thermal conductivity of the tube material. With the notations

$$\eta = \frac{Hr_2}{KB} \qquad m = \frac{\eta + 2}{n} \tag{13-27}$$

the creep equation (Eq. 13-14) now becomes

$$\frac{s}{s_m} = \left(\frac{r}{r_m}\right)^{-m} \tag{13-28}$$

The stresses $\sigma_1$ and $\sigma_2$ obey the equations of equilibrium

$$r\frac{d\sigma_2}{dr} = \sigma_1 - \sigma_2 = \frac{2}{\sqrt{3}}\,s = \frac{2}{\sqrt{3}}\,s_m\left(\frac{r}{r_m}\right)^{-m} \tag{13-29}$$

The last two expressions are the direct results of Eq. 13-17 and 13-28. In this form the differential equation in $\sigma_2$ is directly integrable. With the boundary conditions $r = r_1$, $\sigma_2 = -p$; $r = r_2$, $\sigma_2 = 0$, the result is

$$s_m = \frac{\sqrt{3}}{2}\,mp\,\frac{1}{\left(\frac{r_m}{r_1}\right)^m - \left(\frac{r_m}{r_2}\right)^m} \tag{13-30}$$

$$\begin{aligned}
\sigma_1 &= p\,\frac{(m-1)\left(\frac{r_m}{r}\right)^m + \left(\frac{r_m}{r_2}\right)^m}{\left(\frac{r_m}{r_1}\right)^m - \left(\frac{r_m}{r_2}\right)^m} \\
\sigma_2 &= -p\,\frac{\left(\frac{r_m}{r}\right)^m - \left(\frac{r_m}{r_2}\right)^m}{\left(\frac{r_m}{r_1}\right)^m - \left(\frac{r_m}{r_2}\right)^m} \\
\sigma_3 &= p\,\frac{\frac{m-2}{2}\left(\frac{r_m}{r}\right)^m + \left(\frac{r_m}{r_2}\right)^m}{\left(\frac{r_m}{r_1}\right)^m - \left(\frac{r_m}{r_2}\right)^m}
\end{aligned} \tag{13-31}$$

Note that the axial stress is not constant as in the Lamé solution, except when $m = 2$, or $2n - \eta = 2$. If $\eta = 0$, this corresponds to $n = 1$, the linear case.

The character of the stress distribution is clearly dependent on the values of the exponent $m$. For typical superheater tubes, this may have values ranging from 2 to 8. Negative values would be encountered for heat flow in the reverse direction.

The latest results on creep in thick tubes were published by Bailey [78].

*Rotating disks.* This is a typical plane stress problem for which Eq. 13-18 are directly applicable. The problem does not lend itself to explicit solutions, but step-by-step methods [*78,79,80*] can be used. Such methods also permit the treatment of disks of variable thickness.

Wahl [*81*] has compared the theoretical and experimental results obtained for the creep of disks made of gas turbine materials and has made the observation that the Mises condition underestimates the creep of the disk if the analysis is based on tension creep tests with specimens taken in tangential direction from disks. The main reason for this discrepancy seems to be the lack of isotropy of the disks.

The proper procedure would be to use the Mises condition in the form developed by Hill [*65*] for anisotropic materials and to obtain the necessary constants for (tensile) creep tests on specimens machined from disks in the appropriate directions. This, of course, would be a cumbersome procedure; it is of practical interest, therefore, that Wahl could obtain reasonable numerical agreement by basing the calculation on the Tresca condition and neglecting the anisotropy.

**I,14. Fatigue.** The word "fatigue" is used to denote several fundamentally different types of failure in materials. In static fatigue, fracture occurs after prolonged action of a stress; whether the stress is interrupted or steady is not of primary importance. Creep fracture is a static fatigue phenomenon; another, entirely different one, is the delayed fracture of glass. Chemical attack may play a role; this is the case in stress corrosion which may be called a static fatigue due to combined chemical and mechanical factors.

Sharply contrasting to static fatigue is cyclic fatigue in which the integrated duration of the stress is almost irrelevant; fracture occurs after a certain number of reversals or repetitions of the load almost independently of the frequency. In chemically active surroundings cyclic fatigue may occur at stress cyclic amplitudes much lower than in an inert medium ("corrosion fatigue").

The existence of purely mechanical cyclic fatigue raises several fundamental questions. First, how can progressive damage occur under cycles of constant stress amplitude? Purely elastic deformation cannot lead to irreversible structural changes; progressive plastic deformation, on the other hand, could not occur under cycles of constant stress amplitude in a material whose behavior would be completely described by a unique stress-strain curve covering both tension and compression. The first cycle would strain-harden the material to the level of the highest stress of the cycle, and after this no further plastic deformation could take place without applying a higher stress.

The second question is, what type of structural damage is responsible for fatigue failures? And why do safe ranges of stress exist, i.e. stress amplitudes below which the life of the specimen (measured in cycles) increases extremely rapidly with a slight decrease of the stress amplitude?

The answer to the second question is not clear yet, although much important work has been done on the matter in the course of the last few years by Forsyth [*82*, p. 20], Thompson [*82*, p. 43] and Wood [*82*, p. 1]. The answers to the first and third questions, however, have been known for some time [*33*,*83*]. The salient point is that all crystalline materials either contain inhomogeneities (such as stress concentrations due to cracks, or grains of different orientation in a polycrystalline metal), or (if they are nearly perfect single crystals) they become inhomogeneous when plastic slip takes place. From the present point of view, the important inhomogeneities are regions in which plastic deformation begins

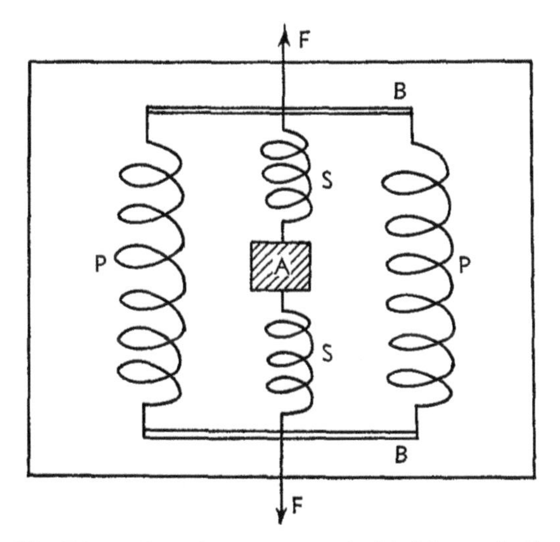


Fig. I,14a. Plastic region $A$ embedded in an elastic matrix represented by springs $P$ and $S$.

before the bulk of the body reaches its so-called elastic limit; such inhomogeneities may be called for brevity "soft spots." The shaded area $A$ in Fig. I,14a represents such a soft spot. The external forces $F$ do not act upon it directly: since $A$ is embedded in a substantially elastic matrix, the stresses and strains in it depend on the elastic deformation of its immediate surroundings as well as on the applied forces (i.e. on the mean stress in the body). It is easy to see that the interaction between $A$ and its elastic surroundings can be represented schematically by springs in series ($S$) and parallel ($P$) to it; the rigid bars $B$ stand for regions in the elastic matrix so far from $A$ that their displacements are not influenced significantly by the deformation of $A$ but are determined by the applied force $F$ alone. Clearly, the plastic deformation of $A$ under a given force $F$ is a maximum if $A$ is perfectly soft (i.e. a liquid); in this case, the stress in it is zero. Conversely, if $A$ is perfectly rigid (i.e. if its yield stress is extremely high, so that the plastic deformation is zero), the stress in it has the highest possible valuc because the entire displacement between the bars $B$ has to be taken up by the springs $S$ (for simplicity, the elasticity of $A$ itself may be regarded as negligible, or it may be represented by a part of the length of the springs $S$). If now an alternating force $F$ is applied, $A$ undergoes strain hardening; as it hardens, the stress that has to be exerted on it by the spring $S$ in order to produce further plastic deformation increases. Thus the stress amplitude in $A$ rises while the plastic strain amplitude decreases, because the increase of the stress amplitude causes an increase of the strain amplitude in the springs $S$; a simple calculation or graphical construction allows one to follow this process quantitatively (see Orowan [*4,33,83*]). In general, $A$ suffers both progressive strain hardening, causing an increase of the stress amplitude, and progressive damage (reduction of fracture stress). Strain hardening alone may raise the stress to the level of the strength (fracture stress) and thus cause local fracture initiating a fatigue crack; vice versa, a decrease of strength due to structural damage may cause local fracture even if the stress remains constant. In reality, a rise of local stress due to strain hardening and reduction of local strength due to structural damage will cooperate in producing or extending a fatigue crack (in the latter case, $A$ is the region of stress concentration at the tip of the crack).

The calculation just referred to shows that the stress amplitude in $A$ converges towards a limit as the strain amplitude decreases towards zero; the limit, of course, increases with the amplitude of the applied force $F$. If the fracture stress, although possibly reduced by structural damage due to alternating straining, always remains higher than the limit toward which the local stress amplitude converges, the stress range is a safe one and fracture will never occur. If, however, progressive strain hardening raises the stress in $A$ above the strain-impaired value of the cohesion, cracking will take place after a certain number of cycles.

The situation illustrated by Fig. I,14a is of importance not only in the physical theory of fatigue, but also in the engineering use of fatigue data in the case of nonuniform distribution of stress in a structure subjected to cycles of constant load (or deformation) amplitude [*83*]. The region $A$, which in the context of the physical theory is usually microscopically small, may also represent the area of stress concentration around a notch or a fillet. In this case, the considerations attached to Fig. I,14a show that the critical region of high stress in which fatigue failure may occur is subjected to cycles of increasing amplitude, rising towards the stress amplitude that would be present if the material were purely elastic. This amplitude would be reached ultimately if strain hardening progressed in the region of stress concentration until any trace of plastic deformation was suppressed. For fairly plausible reasons [*83*], however, the progress of strain hardening ceases when the strain amplitude drops to a small but finite limiting value; this is the reason why the stress-strain hysteresis loop in cyclic stressing never closes completely. This being so, the stress amplitude at $A$ cannot rise to the value it would reach there if the material were perfectly elastic: the "notch sensitivity" of a ductile metal in cyclic loading is less than that of a perfectly brittle material, and it can be very much less [*84*].

The physical background of cyclic fatigue indicates what to expect at higher temperatures. Both the structural damage, and the strain hardening leading to a rise of the local stress, are consequences of plastic deformation; their extent at a given stress amplitude, therefore, is determined by the level of the yield stress. At least approximately, then, the fatigue strength would be expected to decrease with increasing temperature roughly in proportion to the yield stress. This seems to be the case with many materials, up to the temperature region of recovery, recrystallization, and hot creep. Recovery should currently wipe out a part of the local structural damage and strain hardening; however, the experimental data available are too scanty to give definite hints about this, particularly, because the appearance of hot creep introduces another complication: creep fracture. At first sight, it would seem that creep fracture could not occur if the mean stress of the stress cycle were zero: the creep and the intergranular cracking produced by the positive half of the cycle would be undone by the negative half. However, closer consideration shows, and experience confirms, that this is by no means so. If a crack opens up between grains during one half of the cycle, it will hardly be closed completely by the second half; plastic and viscous deformation in an extremely complex structure like a metal are not fully reversible. For this reason, cracks may be produced, and creep fracture may occur, under purely alternating stress, although in general much more slowly than under repeated tensile stress without reversal. If the mean stress of the cycle is not zero but a tension, of course, it can cause creep fracture in the usual manner, apparently somewhat accelerated by the superposed alternating stress; at the same time, the alternating component may give rise to cyclic fatigue. The separation of the two effects has not been carried out fully, but experiments by Lazan and Westberg [*85*] give a very clear picture of how they superpose (Fig. I,14b). If the mean stress is zero, the failure is due to creep-influenced cyclic fatigue; if the alternating component of the cycle is zero, it is pure creep fracture. The curves rise vertically from the abscissa axis; this shows that the superposition of a relatively small alternating stress does not have a significant influence upon the creep fracture process. Similarly, the curves for short lifetimes leave the abscissa axis nearly horizontally, indicating that the superposed creep, if it is small, has little influence on the cyclic fatigue process. If cyclic and creep fatigue occurred independently without influencing each other, each curve would consist of a straight horizontal and a straight vertical branch. That there is a curved transition between the horizontal and the vertical ends shows that the two types of fatigue cooperate in creating damage.

Fig. I,14b shows how little the fatigue strength for purely alternating stressing, and how much the creep fracture strength, depend on the time to fracture. This corresponds to a similar difference between the temperature dependence of the cyclic fatigue strength and of the creep fracture stress for a given lifetime of the specimen. As a rule, the fatigue strength decreases with increasing temperature much more slowly than the creep fracture stress or the stress required for a specified creep strain in a given time. Thus the stress that causes fracture in 300 hours in Nimonic 80

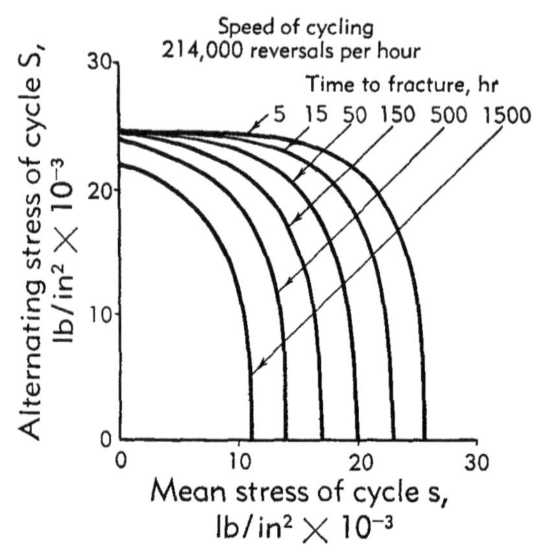


**Fig. I,14b. Fracture under simultaneous hot creep and cyclic fatigue. After Lazan and Westberg [*85*].**

(1942–1943 type) decreases from 29.5 tons/in.² at 600°C to 6.1 tons/in.² at 800°C. The fatigue strength of the same material, at 2000-2200 cycles/min., decreases in the same temperature range from 19 to 9 tons/in.² (see Tapsell [*35*, p. 169]).

An interesting case of fatigue is that due to cyclically changing thermal stresses. The rapid change of temperature in a gunbarrel during and after firing, or in a water-cooled hot-roll, give rise to the well-known phenomenon of heat-cracking (crazing); thermal expansion and contraction in a mechanically constrained pipe or rod, if many times repeated, can produce fatigue fracture. Obviously there is no difference whatever between a stress produced mechanically and one due to thermal expansion; cyclic mechanical fatigue and fatigue due to thermal cycling, therefore, are identical phenomena. The only complicating factor in thermal fatigue is that the temperature changes during the stress cycle. It is plausible to expect that a thermally produced stress cycle will have the same effect as the same stress cycle produced mechanically at some constant temperature between the extreme temperatures of the thermal cycle. Since thermal cycling is slow unless the specimen is very thin, the amount of experimental information available on this point is very limited: it consists mainly of measurements made by Coffin [*86*,*87*] on an 18Cr-8Ni steel containing about 1 per cent columbium. As far as can be recognized, the fatigue effect of thermal cycling in this case was about the same as would have been the effect of the same stress cycle at a constant temperature not far from the mean temperature of the cycle.

**I,15. Materials for High Temperature Service.** A material may be endowed with high creep resistance at high temperatures by a variety of circumstances:

1. It may have a high melting point. This is usually accompanied by (a) a high "driving stress" of the dislocations, i.e. a high value of the shear stress required to extend slip over the slip plane even if no obstacles are present (see Art. 3); (b) a low rate of self-diffusion at a given temperature, i.e. a low rate of thermal removal of structural injuries acting as obstacles to slip.
2. It may contain in solid solution an element that forms strong bonds with atoms of the matrix and thus gives rise to effects (a) and (b) mentioned above.
3. It may contain an alloying element that precipitates particles of a temperature-resistant phase acting as obstacles to slip, or to the sliding of grains upon their neighbors. The precipitation may take place (a) at grain or subgrain boundaries; (b) more or less at random within the grains; (c) along dislocations.
4. It may be compounded mechanically (by pressing and sintering) from a ductile metallic component and a hard refractory material satisfying the requirements given under (1) above.

Although a tremendous amount of testing has been carried out on creep-resistant materials, very little is known about the specific mechanisms of their behavior. This is due partly to the short time passed since interest in creep arose; partly to the very complex nature of the practically important alloys and the lack of fundamental work on simple alloys capable of showing clearly the salient points involved; and partly to the circumstance that most of the testing has been done under pressure of practical needs without adequate guidance by the available scientific knowledge. For these reasons, only a few instances are known in which the creep resistance of a material is clearly connected with some of its atomic or structural features. The simplest and most important fact that stands out is shown strikingly by Fig. I,15 compiled by Cross and Simmons [88]. It shows the stress at which fracture occurs after 1000 hours as a function of the temperature for the most important light alloys, alloys of titanium, iron, and of molybdenum (including the highly creep-resistant alloys of the metals of the iron group). The regions representing aluminum and magnesium, titanium, and molybdenum alloys are so widely separated that it is obvious at once that the main factor in creep resistance is the melting point of the dominant component of the alloy. For the three groups mentioned, the absolute temperature at which

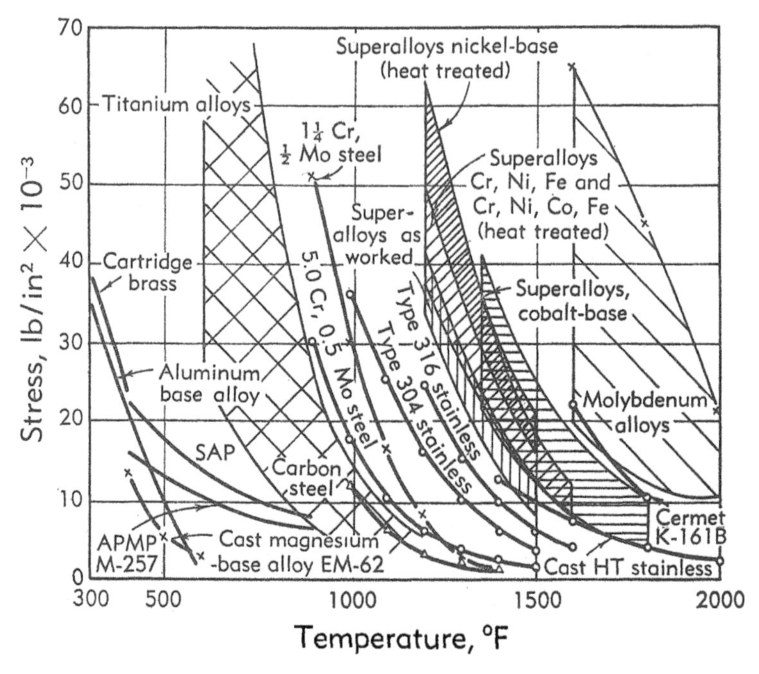


Fig. I,15. Creep fracture stress for fracture in 1000 hours of various types of materials. After Cross and Simmons [88].

the material breaks after 1000 hours under a stress of 20,000 psi is roughly 40 ($\pm$10) per cent of the absolute melting point of the dominant metal; the width of the bands obtained by alloying is relatively small. Plain carbon and low alloy ferritic steels are near the upper limit of the band for titanium alloys, corresponding to the small difference between the melting points of iron and titanium. However, the peculiar weakness of titanium as compared with its somewhat higher melting point is conspicuous. It is no doubt a consequence of the fact that titanium has only 2 d-electrons while iron has 6; that titanium is intrinsically a weaker metal is shown by its higher compressibility. In fact, titanium is so much lighter than iron mainly because its cohesive forces are weaker; its atomic weight is not much lower than that of iron (47.90 versus 55.85).

The curves SAP and APMP refer to sintered products manufactured from aluminum powder with superficially oxidized grains. That the creep resistance of such products at higher temperatures is much higher than that of purely metallic alloys is not surprising.

Between the bands of the titanium and of the molybdenum alloys lies the domain of the so-called "superalloys" containing metals of the first transition group (Fe, Cr, Mn, Co, Ni); in these alloys there is either no clearly dominating element, or at least the total of the alloying elements added to the dominant exceeds 20 per cent. In the majority of the superalloys (except in the cobalt-based ones) there is a small percentage of columbium, titanium, or vanadium. These elements (together with zirconium, hafnium, and tantalum) form very stable carbides of approximate composition $MC$ ($M$ = metal) which crystallize in the sodium chloride lattice. They can precipitate at very low concentrations, and the precipitate is stable up to high temperatures. Thus they are particularly suitable for increasing the creep resistance at high temperatures (see Art. 4). The use of these elements for "stabilizing" stainless steels against grain boundary oxidation is closely connected with the high bonding energy of their carbides: by binding carbon, they prevent the formation of chromium carbides at the grain boundaries and the depletion of the boundary regions in chromium. Physical reasons for the sodium chloride structure and the high stability of the carbides of Ti, V, Zr, Cb, and Hf have been suggested by Rundle [*89*] and by Hume-Rothery [*90*, pp. 218ff.].

Between the bands of the cobalt-based and the molybdenum-based alloys, an example of a "cermet" is shown in Fig. I,15. Cermets are metal-ceramic mixtures in which grains of a nonmetal (e.g. $Al_2O_3$) or a half-metal (e.g. TiC) are embedded in a matrix of a metal capable of wetting them (e.g. cobalt or nickel). The ceramic component alone would be too brittle and relatively weak. Embedded in a ductile metal, however, even if this amounts to only 20 or 30 per cent of the weight of the product, its performance is greatly improved. Cracks in the brittle component are arrested at the metal envelope of the grain, and the metal gives a certain amount of ductility which, however modest, can afford effective protection from moderate mechanical shock loads. In addition, the high thermal conductivity of the metal increases the resistance of the product to thermal shock.

An important point about cermets is that their creep resistance can be much higher than that of the matrix metal. If the ceramic grains are fairly closely packed, they form a rigid framework with the metal acting as a kind of adhesive. Because the extremely thin layers of metal between the points of contact of the ceramic grains do not contain significant numbers of mobile dislocations, and because the free lengths of slip planes at such points are too small to give easy passage to dislocations coming from bulkier parts of the metal envelope, the effective yield stress of the metal around the points of contact may be far higher than its ordinary bulk yield stress, just as the yield stress of metal "whiskers" that are too thin to contain effective dislocations is extremely high [*9*].

For general engineering use, molybdenum seems to be the ultimate metal that can be used as the base of an alloy. At the very highest temperatures to which metals are exposed in present technology—in electric incandescent lamps—both the rate of evaporation and the creep rate of molybdenum would be too high; consequently, tungsten alone is used. Unfortunately, the high density of tungsten (almost twice that of molybdenum) and its very low resistance to oxidation at high temperatures excludes its use in most cases. The only other plausible candidate, rhenium, is eliminated by its rarity and high density.

There are numerous nonmetallic crystals whose ductility comes close to that of ductile metals; silver and thallium halides, naphthaline, gypsum are among the most familiar examples. Can one hope to find a nonmetallic material that would combine the creep resistance of a hard refractory with at least a moderate ductility? The outlook is not very bright, for three reasons. Many hard refractories become quite ductile at a high temperature; $Al_2O_3$ is an example. However, as they gain in ductility they lose in yield stress; it is doubtful whether the combination of high yield stress and high ductility found in the strong engineering metals can be approached by a nonmetallic material. Secondly, only metals are known to remain ductile over a wide temperature interval; the known refractories may acquire some ductility at the service temperature but they need extreme care if they have to be exposed to stress before they reach this temperature. Thirdly, the high thermal conductivity of metals is unique; during heating, nonmetals develop much greater temperature gradients and thus much higher thermal stresses for a given value of the coefficient of thermal expansion. Those nonmetals which have a low thermal expansion are glasses (silica glass) with a relatively low softening range and vanishing yield point (creep limit) above the softening range. In addition, the maximum safe design stress for glasses below the softening range is very low because, unlike a polycrystalline material in which the grain boundaries provide effective obstacles to crack propagation, a glass breaks when the first crack begins to run.

## I,16. Cited References.

1. Lillie, H. R. *J. Am. Ceram. Soc. 16*, 619 (1933).
2. Jones, G. O. 1944. Rept. Progr. Phys. (Phys. Soc. London) *12*, 1948–1949.
3. Jones, G. O. *Glass*. Methuen, 1956.
4. Orowan, E. *First U. S. Natl. Congress Appl. Mech.*, 453 (1951).
5. Bragg, W. L., and Lomer, W. M. *Proc. Roy. Soc. London A196*, 171 (1949).
6. Burgers, J. M. *Proc. Koninkl. Med. Akad. Wetenschap. 42*, 292, 378 (1939).
7. Bragg, W. L. *Proc. Phys. Soc. London 52*, 54, 105 (1940).

8. Frank, F. C., and Read, W. T. *Phys. Rev. 79*, 722 (1950).
9. Galt, J. K., and Herring, C. *Phys. Rev. 86*, (*II*), 656 (1952).
10. Orowan, E. In *Dislocations in Metals.* (M. Cohen, ed.), AIME, 1954.
11. Pauling, L. *The Nature of the Chemical Bond*, 2nd ed. Cornell Univ. Press, 1948.
12. Taylor, G. I. *Proc. Roy. Soc. London A145*, 362 (1934).
13. Fischer, J. C., Hart, E. W., and Pry, R. H. *Acta Metall. 1*, 336 (1953).
14. Mott, N. F., and Nabarro, F. R. N. *Proc. Phys. Soc. London 52*, 86 (1940).
15. Orowan, E. Symposium on internal stresses. *Inst. Metals London*, 451 (1947).
16. Peierls, R. *Proc. Phys. Soc. London 52*, 34 (1940).
17. Nabarro, F. R. N. *Proc. Phys. Soc. London 58*, 669 (1946); *59*, 256 (1947).
18. Mott, N. F. *Proc. Phil. Soc. B64*, 729 (1951).
19. Mott, N. F. *Rept. Ninth Solvay Conference*, 1952.
20. Smith, C. S. *Trans. Am. Inst. Mining Met. Engrs. 175*, 15 (1948). Also *Am. Soc. Mech. Engrs. 45*, 533 (1948).
21. Volmer, M. *Kinetik der Phasenbildung.* Steinkopff, Dresden, 1939.
22. Mehl, R. F., and Jetter, L. K. In *Symposium on Age Hardening.* Am. Soc. Metals, Cleveland, 1940.
23. Orowan, E. *Trans. West of Scotland Iron Steel Inst.*, 45 (1947).
24. Meissner, K., Polanyi, M., and Schmid, E. *Z. Phys. 66*, 477 (1930).
25. Andrade, E. N. da C. *Proc. Roy. Soc. London A84*, 1 (1911); *A90*, 329 (1914).
26. Dorn, J. E., Goldberg, A., and Tietz, T. E. *Trans. Am. Inst. Mining Engrs. 180*, 205 (1949).
27. Becker, R. *Physik. Z. 26*, 919 (1925).
28. Becker, R. *Z. tech. Phys. 7*, 547 (1926).
29. Bragg, W. L., and Nye, J. F. *Proc. Roy. Soc. London A190*, 474 (1947).
30. Lomer, W. M. *Proc. Roy. Soc. London A196*, 182 (1949).
31. Lomer, W. M., and Nye, J. F. *Proc. Roy. Soc. London A212*, 576 (1952).
32. Eyring, H. *J. Chem. Phys. 4*, 283 (1936).
33. Orowan, E. *Proc. Roy. Soc. London A168*, 307 (1938).
34. Cottrell, A. H. Report on strength of solids. *Proc. Phys. Soc. London*, 30 (1948).
35. Harris, S. T., and Bailey, W. H. *Symposium on High Temperature Steels and Alloys for Gas Turbines.* Iron & Steel Inst., London, 1952.
36. Orowan, E., and Torti, M. L. To be published.
37. Grant, N. J., and Bucklin, A. G. *Trans. Am. Soc. Metals 42*, 720 (1950).
38. Odquist, F. K. G. *Trans. Roy. Inst. Technol. Stockholm 66*, 1953. Also *Proc. Eighth Intern. Congress Theoret. Appl. Mech.*, 1953.
39. Norton, F. H. *Creep of Steel at High Temperatures.* McGraw-Hill, 1929.
40. Bailey, R. W. *World Power Conference*, Tokyo, 1929.
41. Bailey, R. W. *Engineering (London) 129*, 265, 327, 772 (1930).
42. Bailey, R. W. *J. Appl. Mech. 3*, A1 (1936).
43. Bailey, R. W. *J. West of Scotland Iron Steel Inst. 45*, 11 (1937).
44. Soderberg, C. R. *Trans. Am. Soc. Mech. Engrs. 58*, 733 (1936).
45. Nadai, A. *S. Timoshenko Sixtieth Anniversary Volume.* Macmillan, 1938.
46. Larson, F. R., and Miller, J. *Trans. Am. Soc. Mech. Engrs. 74*, 765 (1952).
47. Orowan, E. In *Design of Piping Systems* (The M. W. Kellogg Co.). Wiley, 1956.
48. French, H. J. *Natl. Bur. Standards Tech. Paper 219*, 679 (1921).
49. French, H. J. *Chem. & Met. Eng. 33*, 591 (1926).
50. French, H. J. *Proc. ASTM 26, Part 2*, 7 (1926).
51. French, H. J., Cross, H. C., and Patterson, A. A. *Natl. Bur. Standards Tech. Paper 362*, 1927.
52. Hoff, N. J. *J. Appl. Mech. 20*, 105 (1953).
53. Hoff, N. J. *Quart. Appl. Math. 12*, 49 (1954).
54. Hoff, N. J. *J. Roy. Aeronaut. Soc. 58*, 1 (1954).
55. Kempner, J. *NACA Tech. Note 3136*, 1954; *3137*, 1954.
56. Kempner, J., and Patch, S. A. *NACA Tech. Note 3138*, 1954.
57. Libove, C. *J. Aeronaut. Sci. 19*, 459 (1952).
58. Libove, C. *NACA Tech. Note 2956*, 1953.
59. Considère, M. *Ann. Ponts et Chausées Sixth Series 9*, 574 (1885).

60. Sachs, G., and Lubahn, J. D. *Trans. Am. Soc. Mech. Engrs. 68*, 277 (1946).
61. Swift, H. W. *J. Mech. Phys. Solids 1*, 1 (1952).
62. Orowan, E. In *The Principles of Rheological Measurement.* Nelson & Sons, London, 1946.
63. Nadai, A., and Manjoine, M. J. *Proc. ASTM 40*, 822 (1940).
64. Nadai, A., and Manjoine, M. J. *J. Appl. Mech. 8*, A77, A91 (1941).
65. Hill, R. *The Mathematical Theory of Plasticity.* Clarendon Press, Oxford, 1950.
66. Lode, W. *Z. Physik 36*, 913 (1926).
67. Ros, M., and Eichinger, A. *Proc. Second Intern. Congress Appl. Mech., Zurich*, 315 (1926).
68. Taylor, G. I., and Quinney, H. *Trans. Roy. Soc. London A230*, 323 (1931).
69. Morrison, J. L. M. *Proc. Inst. Mech. Engrs. London 142*, 193 (1940).
70. Hencky, H. *Z. angew. Math. u. Mech. 5*, 115 (1925).
71. Oldroyd, J. G. *Proc. Cambridge Phil. Soc. 43, Part 1*, 100 (1946).
72. Prager, W. *Mécanique des solides isotropes au-delà du domaine élastique.* Gauthier-Villars, Paris, 1937.
73. Fromm, H. *Ing. Arch. 4*, 432 (1933).
74. von Mises, R. *Z. angew. Math. u. Mech. 8*, 161 (1928).
75. Geiringer, H. *Fondements mathématiques de la théorie des corps plastiques isotropes.* Gauthier-Villars, Paris, 1937.
76. Norton, F. H. *Trans. Am. Soc. Mech. Engrs. 61*, 239 (1939).
77. Soderberg, C. R. *Trans. Am. Soc. Mech. Engrs. 63*, 735 (1941).
78. Bailey, R. W. *Proc. Inst. Mech. Engrs. London 164*, 425 (1951).
79. Johnson, A. E. *Proc. Inst. Mech. Engrs. London 164*, 433 (1951).
80. Popov, E. P. *J. Franklin Inst. 243*, 365 (1947).
81. Wahl, A. M. *J. Appl. Mech. Paper 55A46*, 1955.
82. Freudenthal, A. M., ed. *Fatigue in Aircraft Structures.* Academic Press, 1956.
83. Orowan, E. *Welding J. Research Suppl. 273*, June 1952.
84. Cazaud, R. *La Fatigue des Métaux.* Dunod, Paris, 1948. English transl. *Fatigue of Metals.* Chapman & Hall, 1953.
85. Lazan, B. J., and Westberg, E. *Proc. ASTM 52*, 837 (1953).
86. Coffin, L. F. *Trans. Am. Soc. Mech. Engrs. 76*, 931 (1954).
87. Coffin, L. F. *Am. Soc. Mech. Engrs. Paper 56-A-178*, 1956.
88. Cross, H. C., and Simmons, W. F. Utilization of heat-resisting alloys. *ASM*, 55 (1956).
89. Rundle, R. E. *Acta Cryst. 1*, 180 (1948).
90. Hume-Rothery, W., and Raynor, G. V. *The Structure of Metals and Alloys*, 3rd ed. Inst. of Metals, London, 1954.

# SECTION J

# *FLUTTER PROBLEMS IN GAS TURBINES*

JAN R. SCHNITTGER

**J,1. Introduction.** Despite half a century of successful turbomachinery design, there still exist blade vibrations which lead to fatigue failures. The occurrence of these has been more common in the aircraft turboengine, which is natural, since the aircraft turbine has been developed close to its design limits in a process of testing up to the point of destruction.

The main origin of blade vibration is aerodynamic excitation, which is opposed by aerodynamic and mechanical damping. The design of large steam turbines has depended for many years on the correct determination of natural frequencies of the blades, a problem which is now well under control, both theoretically and experimentally, and no space will be devoted here to this problem. Steam turbine engineering practice also works with the concept of a simple, direct relation between the steady state, gas-bending stresses and the aerodynamic excitation forces, but this relation does not appear to hold throughout for the slender blading of axial compressors.

In order to materialize the situation, we consider the cascaded blades in Fig. J,1a, immersed in an air flow with the upstream velocity $V$. Suppose this flow to be entirely uniform, so that no external disturbances are active on the blades. In this ideal case any continued blade vibration must be self-sustained, the exciting forces arising as a result of the motion of the blades. This is what has been called flutter, a phenomenon known from airplane wings. When the airfoil is stalled at large angles of attack, the situation is aggravated and the *classical flutter* [*1,2*] in potential flow changes to *stalled flutter* in separated flow. Classical flutter has never been observed in turbomachine practice, though attempts were made to grasp the problem by slight modifications of classical flutter theory to corroborate the results of stalled flutter in compressor blades [*3*]. It must still be freely admitted that the complex problem of blade vibrations in axial compressors is not completely understood.

As distinguished from classical flutter of airplane wings, the flutter of turboblades is always connected with discontinuities in the flow. Stalled flutter with boundary layer separation at high angles of attack was first studied by Studer [*4*] for a single airfoil. However, very active research is known to be taking place and important contributions are to be expected in the near future.

In annular cascades the stall flutter is complicated by the stall propagation. The most stable flow pattern at increased angle of attack is not a uniform increase in the flow angle. Instead there appear one or several isolated waves of stalled flow, traveling in the $U$ direction at absolute speeds less than $U$ in Fig. J,1a [*5,6,7,8,9*]. Because stalled flow

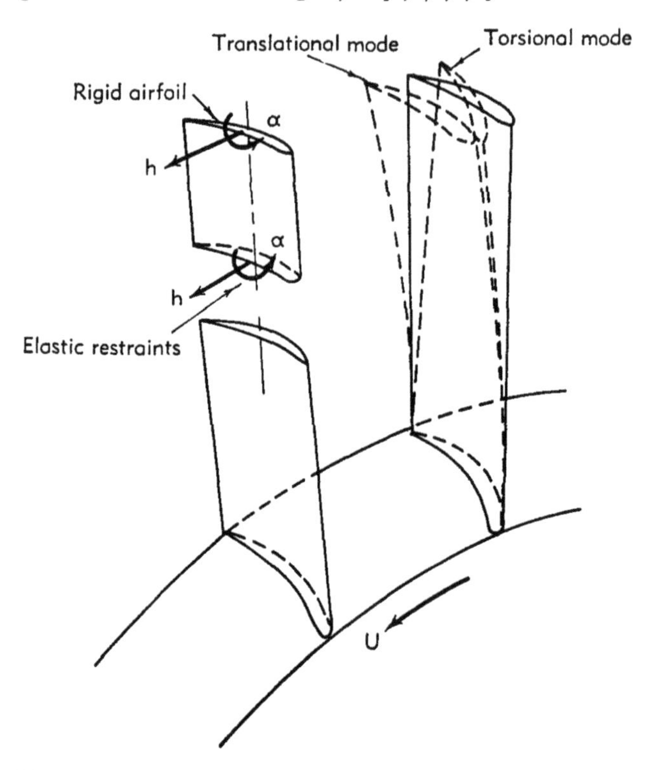


**Fig. J,1a. Turbomachine blades and their fundamental modes.**

does not easily arise in turbines, the approach here is toward compressor blading, although many remarks, except for those concerning stalled flow, are also applicable to turbine blades.

Shock flutter or choked flow flutter is another type of self-excitation which occurs in a very narrow range of flow velocities, slightly above that free stream Mach number at which the maximum local velocity first becomes supersonic [*10,11*]. It is likely that here one encounters a case of shock wave-boundary layer interaction. The position of the shock wave is extremely sensitive to Mach number variations and moves for small changes of the velocity from the position of maximum local velocity to the trailing edge. Wake excitation is present when blade rows of turborotors pass through wakes of upstream objects, causing variations with respect to time of the magnitude and direction of approaching flow velocity. Thus a distinct case of forced vibration arises.

In contrast, stalled flutter is not a clearcut case of self-excitation on account of the stall propagation.

A general phenomenon observed, the mechanism of which is not yet satisfactorily explained, is the fact that most recorded blade vibrations in actual turbomachines show amplitudes varying in time and sometimes the amplitudes change rather spasmodically. This may be connected

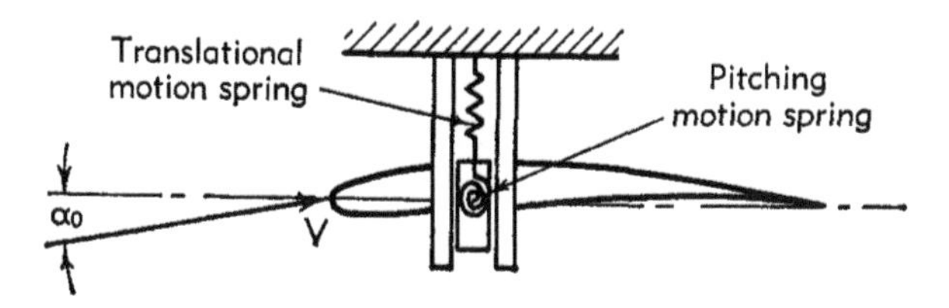


**Fig. J,1b. The two-dimensional equivalent blade system.**

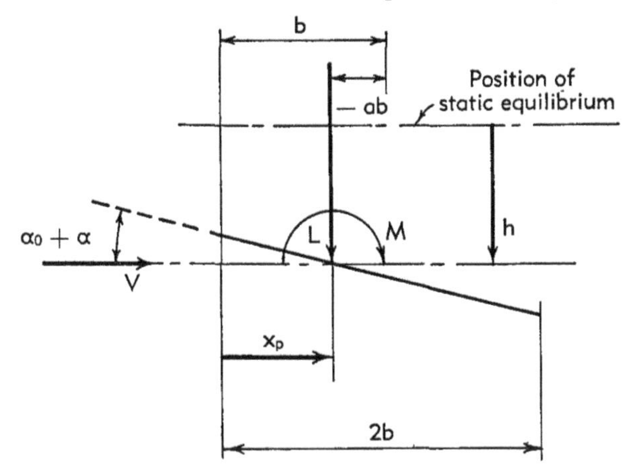


Fig. J,1c. Pitching and blending deflection in two-dimensional equivalent blade system.

with stall propagation. Another possibility is the cascade effect, studied in stalled flow by Sisto [*12*]. He also initiated the study of unsteady aerodynamic reactions on airfoils in cascade due to the influence of vibrating, adjacent blades in potential flow [*13*]. This method has been followed lately by others [*14,15*].

Theoretically the individual cantilever blade has an infinite number of degrees of freedom. The cascade arrangement increases the complexity even more. In order to arrive at some general conclusions, simplification is necessary. One common way to study the aerodynamics is to isolate the tip section of a cantilever, which part exhibits the largest amplitudes. This outermost part of the blade is replaced by an equivalent stiff airfoil, connected to the rest of the similarly stiff blade with two elastic springs (Fig. J,1a and J,1b). These springs allow for a pitching and a bending deflection according to Fig. J,1c. In the following the equations of motion are discussed within this model for forced and self-sustained vibration, and the aerodynamics thus derived are applied in Art. 6 to the design problem of actual blades.

**J,2. Equations of Motion for Equivalent Airfoil.** Introducing the following nomenclature:

- $b$ semichord of airfoil
- $C_h = m\omega_h^2$ spring constant of translational motion
- $C_\alpha = I_\alpha\omega_\alpha^2$ spring constant of pitching motion
- $c = 2b$ chord of airfoil
- $F_L$ aerodynamic reaction in equations of motion due to lift
- $F_M$ aerodynamic reaction in equations of motion due to moment
- $h$ translational deflection of center of torsion, positive downward
- $I_\alpha$ moment of inertia of airfoil about center of torsion per unit span
- $L$ aerodynamic lift in center of torsion per unit span of airfoil
- $l$ length of airfoil in spanwise direction
- $M$ aerodynamic moment about center of torsion per unit span of airfoil
- $m$ mass of airfoil per unit span
- $x_g$ chordwise distance from leading edge of center of gravity of airfoil
- $x_p$ chordwise distance from leading edge of center of torsion of airfoil
- $\alpha$ angle of attack, counted from angle of static equilibrium $\alpha_0$
- $\alpha_0$ initial angle of attack at static equilibrium (flow direction − chord direction)
- $\kappa = \frac{\rho\pi b^2}{m}$ ratio of the mass of air contained in a cylinder with the chord as diameter per unit span to the mass of airfoil per unit span
- $\omega$ frequency, rad/sec

then for the two-dimensional blade system of Fig. J,1b and J,1c the equations of motion may be written

$$
\begin{aligned}
m\ddot{h} + (x_g - x_p)m\ddot{\alpha} + C_h h &= L \\
(x_g - x_p)m\ddot{h} + I_\alpha\ddot{\alpha} + C_\alpha\alpha &= M
\end{aligned}
\tag{2-1}
$$

The terms on the left-hand side of Eq. 2-1 contain the inertia and elastic forces of the system in translational ($h$) and pitching ($\alpha$) motion. $L$ and

$M$ are the aerodynamic reactions on the airfoil, referred to the center of torsion at $x_p$. No structural damping is considered so far. Introducing

$$\left.\begin{aligned} \frac{C_h}{m} &= \omega_h^2 & x_g - x_p &= x_\alpha \\ \frac{C_\alpha}{I_\alpha} &= \omega_\alpha^2 & I_\alpha &= m \cdot r_\alpha^2 \\ \kappa &= \frac{\rho\pi b^2}{m} & &\text{mass ratio} \\ \kappa &= \frac{\omega b}{v} & &\text{dimensionless frequency} \\ \alpha_0 &= & &\text{initial angle of attack} \end{aligned}\right\} \tag{2-2}$$

one obtains

$$\begin{aligned} \ddot{h} + x_\alpha\ddot{\alpha} + \omega_h^2 h &= \kappa F_L \\ \frac{x_\alpha}{r_\alpha^2}\ddot{h} + \ddot{\alpha} + \omega_\alpha^2\alpha &= \kappa F_M \end{aligned} \tag{2-3}$$

In aircraft turboengine design the mass ratio $\kappa$ equals about 0.002 (static sea level conditions of axial flow compressors). The aerodynamic reactions $\kappa F_L$ and $\kappa F_M$ are extremely small in comparison to the inertial and elastic forces of the equivalent blade under consideration. Therefore the modes of vibration and corresponding frequencies may be determined with a sufficient degree of accuracy by solving the homogeneous part (left-hand side) of the system (Eq. 2-3). Assuming a solution

$$\begin{aligned} \alpha &= (\Delta\alpha)_0 e^{i\omega t} \\ h &= (\Delta h)_0 e^{i\omega t} \end{aligned} \tag{2-4}$$

one obtains a frequency equation of the second order, giving two natural frequencies $\omega_1$ and $\omega_2$ and two amplitude ratios $\{[(\Delta h)_0/r_\alpha]/(\Delta\alpha)_0\}_1$ and $\{[(\Delta h)_0/r_\alpha]/(\Delta\alpha)_0\}_2$. Some results [*16*] are given in Table J,2.

*Table J,2.* Frequency and amplitude ratio of the normal modes of a two-dimensional blade system. Data are given for three ratios of $\omega_\alpha/\omega_h$.

| | $\omega_\alpha/\omega_h$ | | |
|---|---|---|---|
| | 4 | 6 | 8 |
| $\omega_1/\omega_h$ for $x_\alpha/r_\alpha = 0.4$ | 0.995 | 0.997 | 1.000 |
| $\omega_1/\omega_h$ for $x_\alpha/r_\alpha = 0.8$ | 0.980 | 0.993 | 0.993 |
| $\omega_2/\omega_\alpha$ for $x_\alpha/r_\alpha = 0.4$ | 1.097 | 1.094 | 1.093 |
| $\omega_2/\omega_\alpha$ for $x_\alpha/r_\alpha = 0.8$ | 1.70 | 1.68 | 1.67 |
| $\{[(\Delta h_0)/r_\alpha]/(\Delta\alpha)_0\}_1$ for $x_\alpha/r_\alpha = 0.4$ | 37.6 | 60.7 | 746.0 |
| $\{[(\Delta h)_0/r_\alpha]/(\Delta\alpha)_0\}_1$ for $x_\alpha/r_\alpha = 0.8$ | 19.2 | 56.8 | 56.8 |
| $\{[(\Delta h)_0/r_\alpha]/(\Delta\alpha)_0\}_2$ for $x_\alpha/r_\alpha = 0.4$ | −0.42 | −0.41 | −0.40 |
| $\{[(\Delta h)_0/r_\alpha]/(\Delta\alpha)_0\}_2$ for $x_\alpha/r_\alpha = 0.8$ | −0.82 | −0.81 | −0.80 |

The following conclusions may be drawn:

1. In general the free modes of the system consist of both translational and angular displacements due to inertia coupling (which exists as soon as $x_\alpha$ is different from zero). A linear transformation defines the relation between these *normal* modes and those of the $h$ and $\alpha$ motion.
2. There exists a pitching motion around a point located between $x_p$ and $x_g$ and a slightly distorted translational mode. For solid blades of the NACA 4-digit series the center of torsion is located at 33 to 35 per cent chord and the center of gravity at 42 per cent chord. These figures correspond to about 90 per cent of the lower degree of coupling in Table J,2.

It is seen that this analysis justifies a consideration of two different cases, namely that of a pure translation and one case of pitching around a point at 35 to 40 per cent of the chord. Eq. 2-3 may thus be written approximately as two independent equations:

$$\ddot{h} + \omega_h^2 h = \kappa F_L \tag{2-5}$$

$$\ddot{\alpha} + \omega_\alpha^2 \alpha = \kappa F_M \tag{2-6}$$

Some cases where this concept does not hold are discussed in Art. 6. By and large, however, these equations represent the actual conditions of turboblades, since almost all available measurements indicate that either a pitching or an essentially translational motion takes place in the case of cantilever blades, like those of Fig. J,1a.

## J,3. Aerodynamic Forces in Potential Flow.

*Dynamic forces on a single airfoil in translational motion.* In undisturbed flow the aerodynamic reaction on an airfoil in translational motion is [1]

$$L = -\pi\rho b^2\ddot{h} - 2\pi\rho Vb\dot{h}C(k_h) \tag{3-1}$$

where $\rho$ = density
$V$ = flow velocity
$k_h = \dfrac{\omega_h b}{V}$ = reduced bending frequency

The first term represents the inertial force of the air surrounding the airfoil and the second is essentially the aerodynamic damping. The function $C$ is the well-known complex Theodorsen function [2]. The real part $F(k_h)$ which determines the magnitude of the aerodynamic damping varies between $\frac{1}{2}$ for reduced frequency $k_h = \infty$ and is close to 1 for actual values of $k_h = \omega_h b/V$.

Consider next a disturbance in the form of a variation of the free stream velocity $V$ (horizontal gusts):

*Case 1.*
$$V = V_0(1 + \sigma e^{i\omega t}) \tag{3-2}$$

and

*Case 2.* Variation of the angle of attack of the free stream velocity (vertical gusts):

$$\Delta\alpha = (\Delta\alpha)_0 e^{i[\omega t - k(x-b)]} \tag{3-3}$$

When these disturbances are introduced in the flow, Eq. 3-1 is completed by additional terms, due to Greenberg [*17,18*].

These periodic, sinusoidal variations of the magnitude and direction of the free stream velocity around the blade are very similar in action to wakes of upstream objects, when a rotating blade row passes by. However, it must be recalled that the premises of incompressible potential flow around a single airfoil, executing a small harmonic motion and leaving behind plane, infinitely extended wakes, are rather limited.

*Cascade effect in potential flow.* With the introduction of cascaded airfoils the flutter problem assumes a new complexity, characterized by an additional length ratio. The dimensionless ratio that fixes the cascade geometry is taken by Sisto [*13*] to be

$$P = \left(\frac{\pi b}{s}\right) e^{i\beta} \tag{3-4}$$

where $b$ is the semichord, $s$ is the spacing, and $\beta$ is the cascade stagger angle.

Sisto makes the assumption that all the blades are oscillating in the same mode, at the same frequency, and with some constant phase $m\pi$ between the motions of adjacent blades. He restricts the development to a cascade of thin, slightly cambered blades and assumes small displacements, small stream turning, and plane wakes. He formulates and solves the pertinent integral equation for the amplitude of the vorticity distribution. Fig. J,3 from the rather complicated analysis of Sisto shows the resulting lift coefficient, due to pure bending for zero stagger angle $\beta$. The result corresponds to a modification of lift equation [*13*] above for the case of cascaded airfoils. No radical change appears and the lift coefficient, which is equal to $L/4bq$, is changed mainly in the out-of-phase component which is in phase with inertial and spring force. Sisto concludes that the radically lowered critical speeds encountered in axial flow compressors are not caused by interference effects in classical flutter aerodynamics. Calculations and experiments of Wang, Lane, and Vaccaro [*15*] seem to confirm this conclusion. Schnittger [*19*] arrived at the same

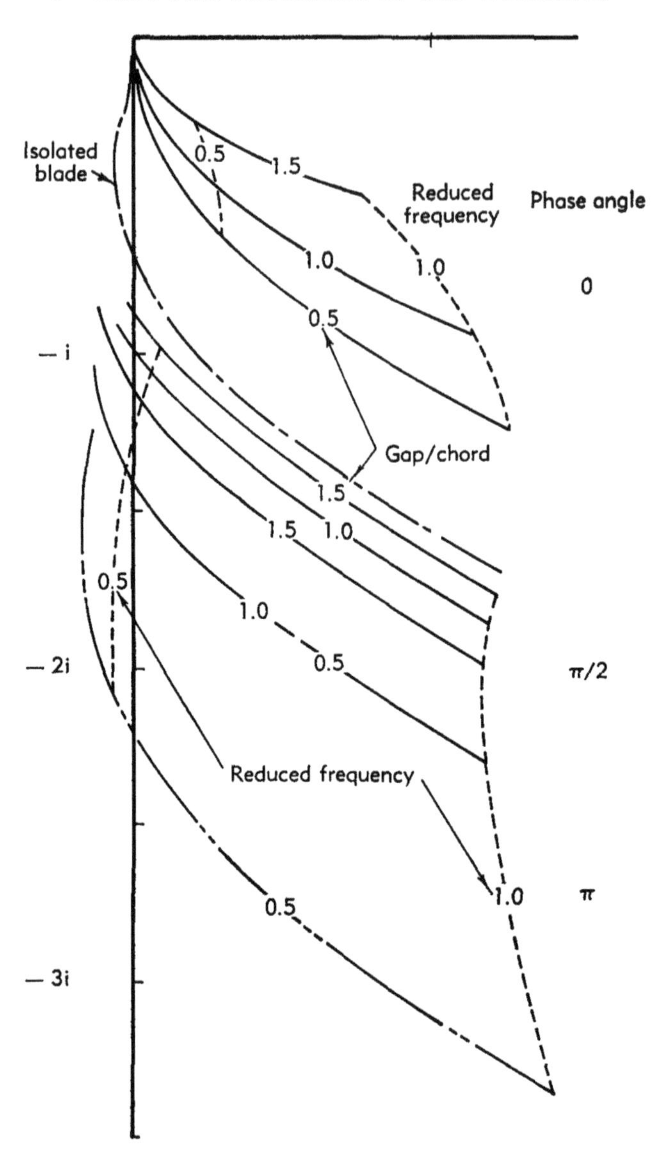


Fig. J,3. Lift coefficient due to pure bending, $E_{L_h}$, for zero stagger, several phase angles, and several gap-chord ratios [*13*].

conclusion regarding cascade effect, due to steady state forces imposed by fixed neighbor blades in an infinite cascade.

With an increasing trend toward transonic compressors and very slender blading it may be necessary in the future, however, to return to the effect calculated by Sisto.

**J,4. Forced Vibrations of Two-Dimensional Airfoils in Potential Flow.** Introducing the expression for the lift according to Eq. 3-1 (but

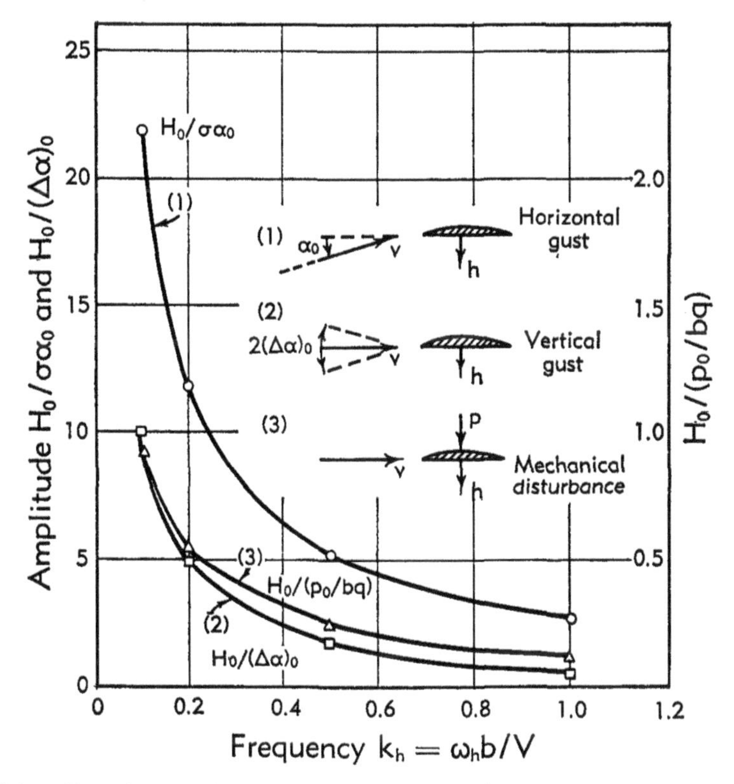


Fig. J,4a. Forced translational motion. Variation of the steady state amplitude $H = h_0/b$ with the frequency $k_h$. (1) Horizontal velocity disturbance: $V = V_0(1 + \sigma e^{i\omega t})$; (2) vertical velocity disturbance (angle-of-attack variation): $\beta = \beta_0 e^{i[\omega t - k(x-b)]}$ and (3) mechanical disturbance: $p = p_0 e^{i\omega t}$.

modified according to the development by Greenberg) into the equation of translational motion (Eq. 2-5), one obtains the amplitude response of the system due to external disturbances.

The amplitudes may be expressed in the following forms [*16*]:

*Case 1.* $V = V_0(1 + \sigma e^{i\omega t})$

$$\frac{H_0}{\sigma\alpha_0} = f_\sigma(k_h, \kappa, \nu) \tag{4-1}$$

*Case 2.* $(\Delta\alpha) = (\Delta\alpha)_0 e^{i[\omega t - k(x-b)]}$

$$\frac{H_0}{(\Delta\alpha)_0} = f_{\Delta\alpha}(k_h, \kappa, \nu) \tag{4-2}$$

*Case 3.* Mechanical force per unit span of airfoil $p = p_0 e^{i\omega t}$

$$\frac{H_0}{p_0/bq} = f_p(k_h, \kappa, \nu) \tag{4-3}$$

To the left in Eq. 4-1, 4-2, and 4-3 the dimensionless amplitude $H_0 = h_0/b$ is related to the magnitude of the amplitude of the actual disturbance, i.e. the velocity variation $\pm\sigma V_0$ at the angle of attack $\alpha_0$, the angle-of-attack variation $\pm(\Delta\alpha)_0$, and the mechanical force $\pm p_0$ per unit span per unit "dynamic pressure force" $bq = b\frac{1}{2}\rho V^2$, respectively. The amplitude responses are functions of the natural dimensionless frequency $k_h = (\omega_h b/V)$, the mass ratio $\kappa = (\rho\pi b^2/m)$, and the ratio $\nu = k/k_h$ of the exciting frequency $k = (\omega b/V)$ to the natural frequency $k_h$.

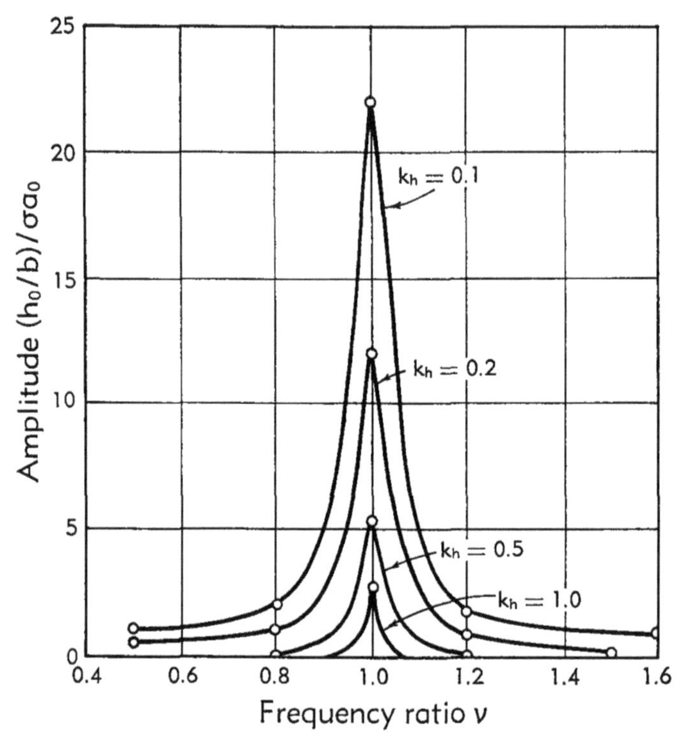


**Fig. J,4b.** Forced translational motion. Variation of the steady state amplitude $(h_0/b)/\sigma\alpha_0$ with the frequency ratio $\nu = k/k_h$ for given $k_h$ values. Horizontal velocity disturbance. Mass ratio $\kappa = 0.002$.

The maximum amplitude of any system is obtained when the exciting frequency is tuned to the system, that is $\nu = 1 + \epsilon$ where $\epsilon$ is a small number $\ll 1$. The maximum responses of the above three cases are given in Fig. J,4a as functions of $k_h$. It is seen that there is a general increase in the amplitude response with decreasing $k_h$ (that is, for increasing velocity at a given system or decreasing elastic stiffness for a given velocity).

When the natural frequency is different from the exciting frequency, the amplitude response becomes strongly influenced by the mass ratio $\kappa$, the resonance peak being considerably sharper for the heavier blade (small $\kappa$). With larger $\kappa$ the relative magnitude of the reaction $\kappa F_L$ grows and the disturbances are able to force a considerable amplitude on the system at the exciting frequency, even when there is no exact resonance. There is an upper limit of $\kappa$, above which the previously demonstrated method of obtaining the natural modes and frequencies of the system becomes unsatisfactory. This is the range of classical flutter which is considered in Art. 5.

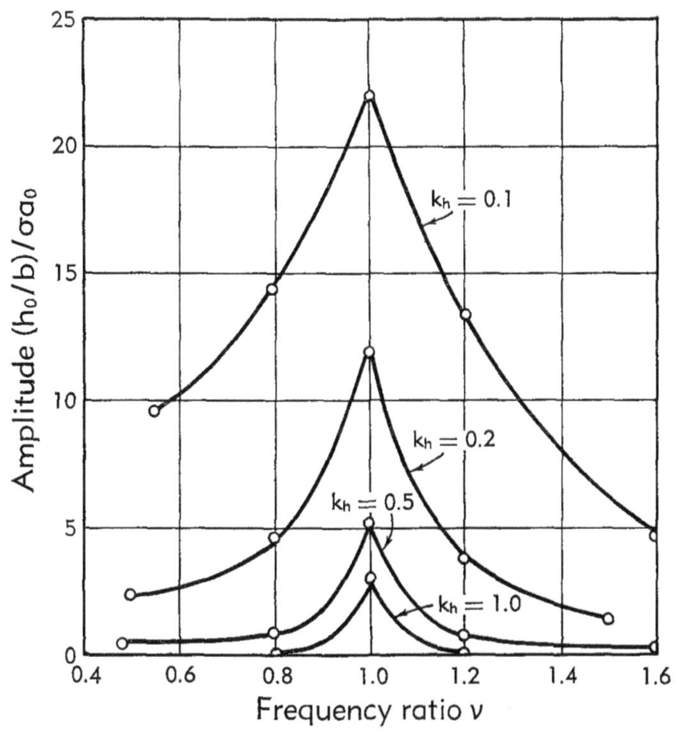


Fig. J,4c. Forced translational motion. Variation of the steady state amplitude $(h_0/b)/\sigma\alpha_0$ with the frequency ratio $\nu = k/k_h$ for given $k_h$ values. Horizontal velocity disturbance. Mass ratio $\kappa = 0.02$.

The resonance curves of case 1, Eq. 4-1, are given in Fig. J,4b and Fig. J,4c for $\kappa = 0.002$ and $\kappa = 0.02$. They are very similar for the two other cases. The above results constitute the familiar problem of forced vibrations. The resonance is effected whenever any integer multiple of the rotor speed times the number of upstream objects checks the natural frequency of the blade system. The most important features (which are very similar for pitching motions) may be summarized as follows:

1. The correct aeroelastic parameter is the natural dimensionless frequency $k_h = (\omega_h b)/V$ or $k_\alpha = (\omega_\alpha b)/V$.

2. The amplitude of the system is proportional to the magnitude of the disturbance.
3. The influence of the mass ratio $\kappa$ is essentially felt when no exact resonance is at hand. It is important to note that the smaller the mass of the airfoil is (thin blades), the larger are the out-of-resonance stresses and the greater is the danger of an increased stress level over a broad range of flows and rotational speeds (forced excitation at integer multiples of rotor angular rotation).

No systematic tests are at present known to the author, which could verify these results. Experimental research is necessary to obtain further knowledge.

## J,5. Aerodynamic Forces in Stalled Flow.

*Difference between classical flutter and stalled flutter of turbomachine blades.* According to the classical theory a pitching motion around a fixed center is aerodynamically damped in smooth, potential flow, and the same is true for a pure translational motion. At sufficiently large velocities the natural modes are changed into one combined motion at a common frequency by the influence of the aerodynamic forces and the motion may eventually become self-sustained through an interchange of energy between the components of translation and pitching in this new mode, a fact qualitatively shown by den Hartog [*20*] and thoroughly demonstrated in a paper by Halfman [*21*]. It can be demonstrated that this mechanism is a very distant possibility in the case of turboblades. A crossplot (Fig. J,5a) was made of a graph in a report by Theodorsen and Garrick [*2*, Graph I-A]. In Fig. J,5a the reduced critical velocity $V/(\omega_\alpha b)$, at which classical flutter suddenly starts, is plotted as a function of the mass ratio $\kappa$ for two values of $x_\alpha$ ($x_\alpha$ represents the distance between the center of torsion and the center of gravity). It is seen that the very small values of $\kappa$ in the actual application correspond to exceedingly large velocities. The change of modes does not take place and the frequency ratio $\omega_\alpha/\omega_h$ remains of the order of 6:1 for most actual blades. It follows that the basic mechanism of turboblade flutter is a single degree-of-freedom vibration.

*Aerodynamic mechanism in stalled flow of single airfoil in pitching motion. The concept of limit cycle.* Consider a two-dimensional airfoil elastically restrained, so that a pitching motion around a fixed center $x_p$ may take place. An estimate of possible motions may be obtained by investigating the work per cycle transferred from the flow to the airfoil during a cycle of vibration. The work per cycle is obtained by evaluating the integral:

$$W_{MR} = \oint M d\alpha \tag{5-1}$$

This expression represents work fed into the vibrating blade, when a closed area in the $M$, $\alpha$ plane is circumvented in a clockwise sense in the course of a cycle. In unstalled flow it was shown first by Halfman, et al. [22] that the aerodynamic moment $M$ and aerodynamic lift $L$ form elliptical traces when plotted versus the deflection in harmonic motion, as in Fig. J,5b. Work is withdrawn from the system and such a motion is damped. At large angles of attack, however, the boundary layer separates and the trace deviates from the elliptical path. This was thoroughly

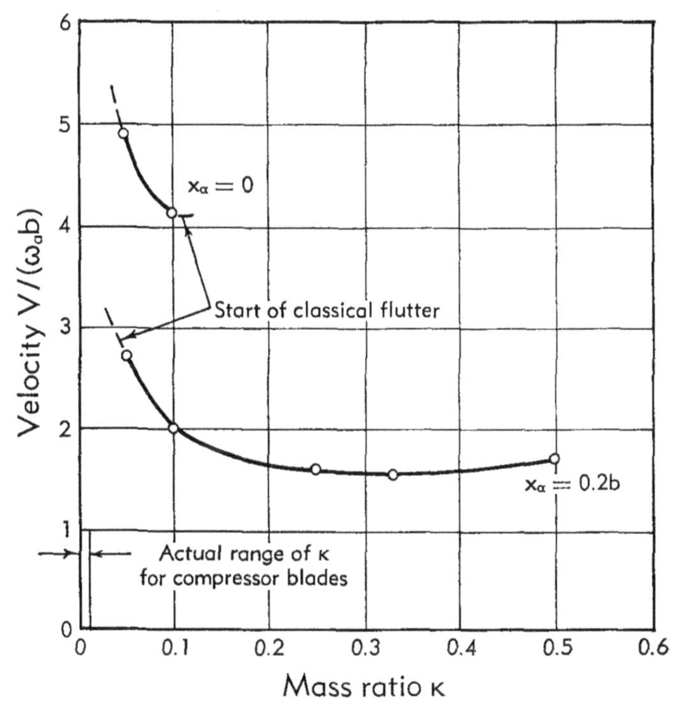


Fig. J,5a. Two degree-of-freedom classical flutter. Reduced velocity $V/\omega_\alpha b$ is plotted as function of the mass ratio $\kappa = (\rho\pi b^2/m)$. The fixed variables of the diagram are $\omega_\alpha/\omega_h = 5$, center of torsion $x_p$ at 30 per cent chord, radius of gyration $r_\alpha = 0.5$. Two different values of the location of center of gravity $x_g$ are shown.

demonstrated from records of deflections by the above research group, and Plate J,5 from [19] shows in the form of an interferogram how the boundary layer separation on the upper surface takes place, causing a great change in the aerodynamic moment in comparison to the unstalled case. The trace of the aerodynamic moment may be calculated for airfoils investigated in [22], under the assumption that the deviation from a quasi-steady state follows in the same manner, both in stalled and unstalled flow. Such calculations were undertaken by Schnittger [19] for a blunt airfoil in the above test material and two examples are presented in Fig. J,5c for reduced angular frequency $k_\alpha = 0.1$ and in Fig.

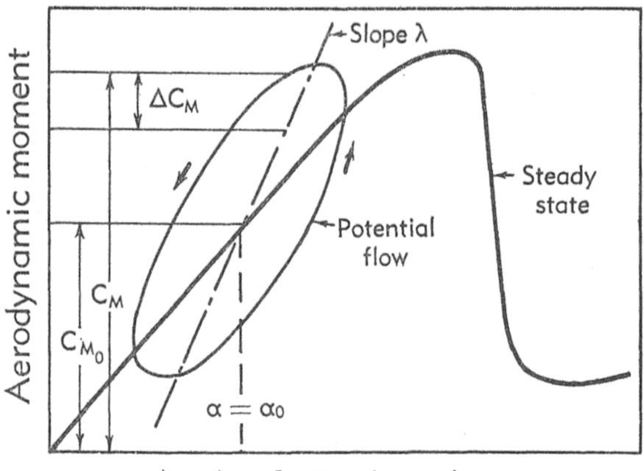


Fig. J,5b. Notations for aerodynamic moment due to pitch at low initial angle of attack. From [21].

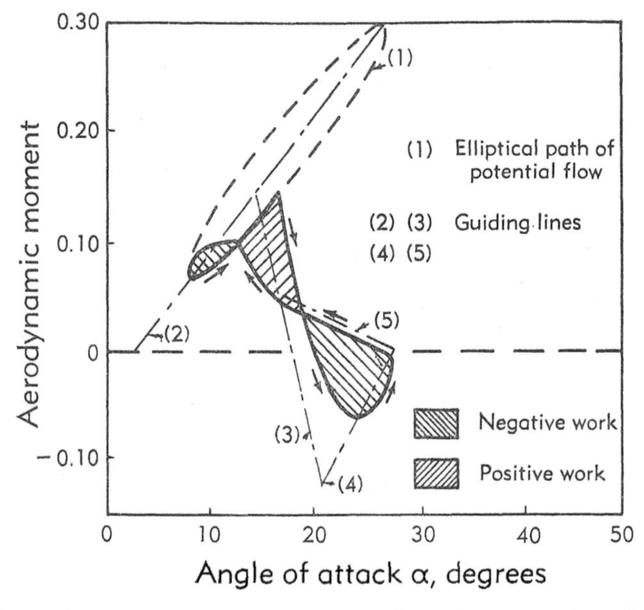


Fig. J,5c. Calculated hysteresis loop of aerodynamic moment in stalled flow. Initial angle of attack $\alpha_0 = 18°$. Amplitude $(\Delta\alpha)_0 = 10°$. Frequency $k_\alpha = 0.10$.

J,5d for $k_\alpha = 1.0$. The quasi-steady state in the stall region ($\alpha_0 + \alpha$ larger than 15°) is indicated by a number of lines (3, 4, and 5 in Fig. J,5c), which, according to the experimental results of [22], replace the line of slope λ. It is found that increasing $k_\alpha$ increases the deviation from the quasi-steady state. At $k_\alpha = 1.0$, again, the tendency from the classical elliptical path has almost disappeared, so that the lines of quasi-steady stall will be gradually replaced by the quasi-steady state along the line of slope $\lambda$. This means that the time for boundary layer separation becomes too short. There exists an upper limit of $k_\alpha$, beyond which no separation is to be expected.

It is seen in Fig. J,5c and J,5d that areas of both positive and negative work are formed. The work is negative in the unstalled region and positive in the stall region and eventually becomes negative again at very high angles of attack. Unless the integral $W_{MR}$ is not exactly equal to zero, it is evident that the cycle will not represent a steady state motion. The

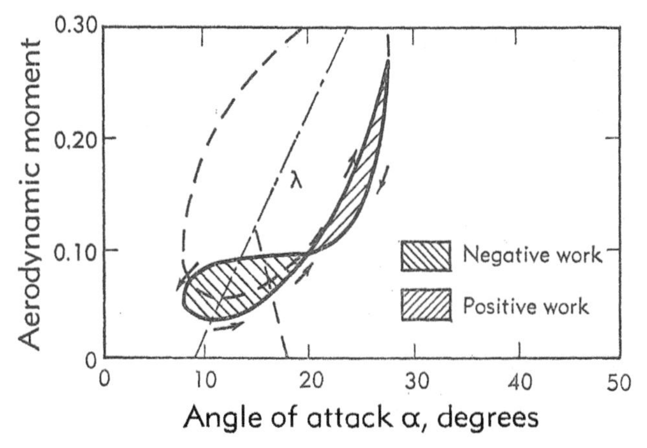


Fig. J,5d. Calculated hysteresis loop of aerodynamic moment in stalled flow. Initial angle of attack $\alpha_0 = 18°$. Amplitude $(\Delta\alpha)_0 = 10°$. Frequency $k = 1.0$.

excess negative or positive work gradually changes the amplitude $(\Delta\alpha)_0$ of the cycle until

$$W_{MR} = \oint M d\alpha \equiv 0 \tag{5-2}$$

Eq. 5-2 defines the *limit cycle* of the motion, that is, it determines an amplitude $(\Delta\alpha)_0$, other conditions being given. The simplest possible case of pitching would be one where one assumes that the Mach number, the Reynolds number, and cascade geometry are fixed quantities and Eq. 5-2 determines the amplitude of the limit cycles as functions of the angle of attack $\alpha_0$ and the frequency $k_\alpha$ (based on the natural frequency $\omega_\alpha$), that is,

$$(\Delta\alpha)_0 = (\Delta\alpha)_0(\alpha_0, k_\alpha) \tag{5-3}$$

The author made calculations for the above case to determine the function (Eq. 5-3) for one value of $\alpha_0$ (Fig. J,5e). At small $k_\alpha$, the deviations from the quasi-steady state are small and the arising quantities of work are small with the effect that the limit cycles have moderate $(\Delta\alpha)_0$ values. At large $k_\alpha$ values the trend of vanishing separation effects has a similar result. Concerning the influence of $\alpha_0$, which is not directly demonstrated in Fig. J,5e, a somewhat similar behavior is found. If the initial angle of attack $\alpha_0$ is very different from that static value of $\alpha_0$ at which stall takes place, it is clear that the flow will be essentially unstalled or completely stalled during the main part of a cycle, giving a predominance for negative, damping work. In flow at low Mach number, when no shock flutter distorts the picture, one therefore obtains a function $(\Delta\alpha)_0 = (\Delta\alpha)_0(\alpha_0, k_\alpha)$, which goes toward zero or very small values outside a limited range of $\alpha_0$ and $k_\alpha$. Outside a very limited range of $\alpha_0$ values, an initial impulse is necessary to start vibrations (hard flutter).

It should be pointed out that the time lag of boundary layer separation is essential in order to explain self-excited torsional flutter in compressors.

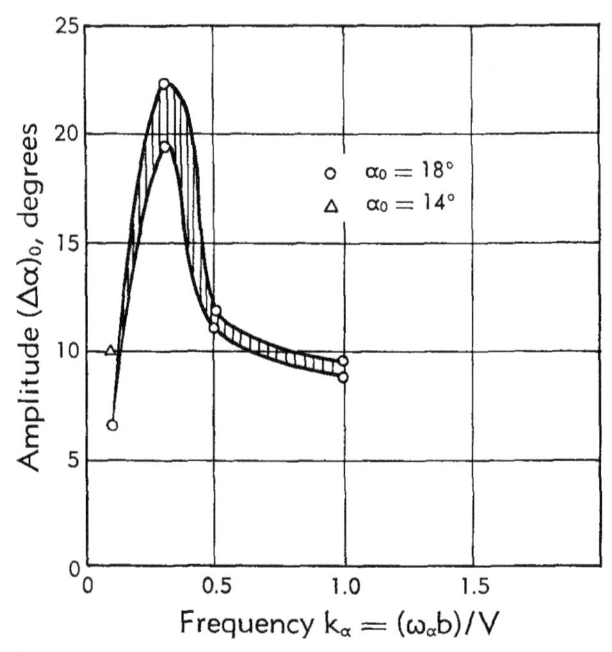


Fig. J,5e. Calculated variation of steady state limit cycle amplitude with variation of frequency $k$. Pure pitching motion, $x_p$ = 37 per cent chord, $Re = 1 \times 10^6$.

At very small $k$ values, the moment follows the same trace back and forth, even in separated flow (for the average cycle at least), and therefore no circumvented area arises, of either positive or negative work. This explains the small amplitudes found for small $k$ values in Fig. J,5e.

For very large $k$ values, there is no separation at all; a negative work is fed to the blade, which is damped rather than excited. This explains the decreasing tendency of the amplitudes at larger $k$ values in Fig. J,5e. In Fig. J,6b of Art. 6, experimental results of vibrating two-dimensional airfoils in pitching motion are shown to substantiate the general concepts derived above.

*Aerodynamic mechanism in stalled flow of single airfoil in translational motion.* Through a basically very simple kinematic consideration this

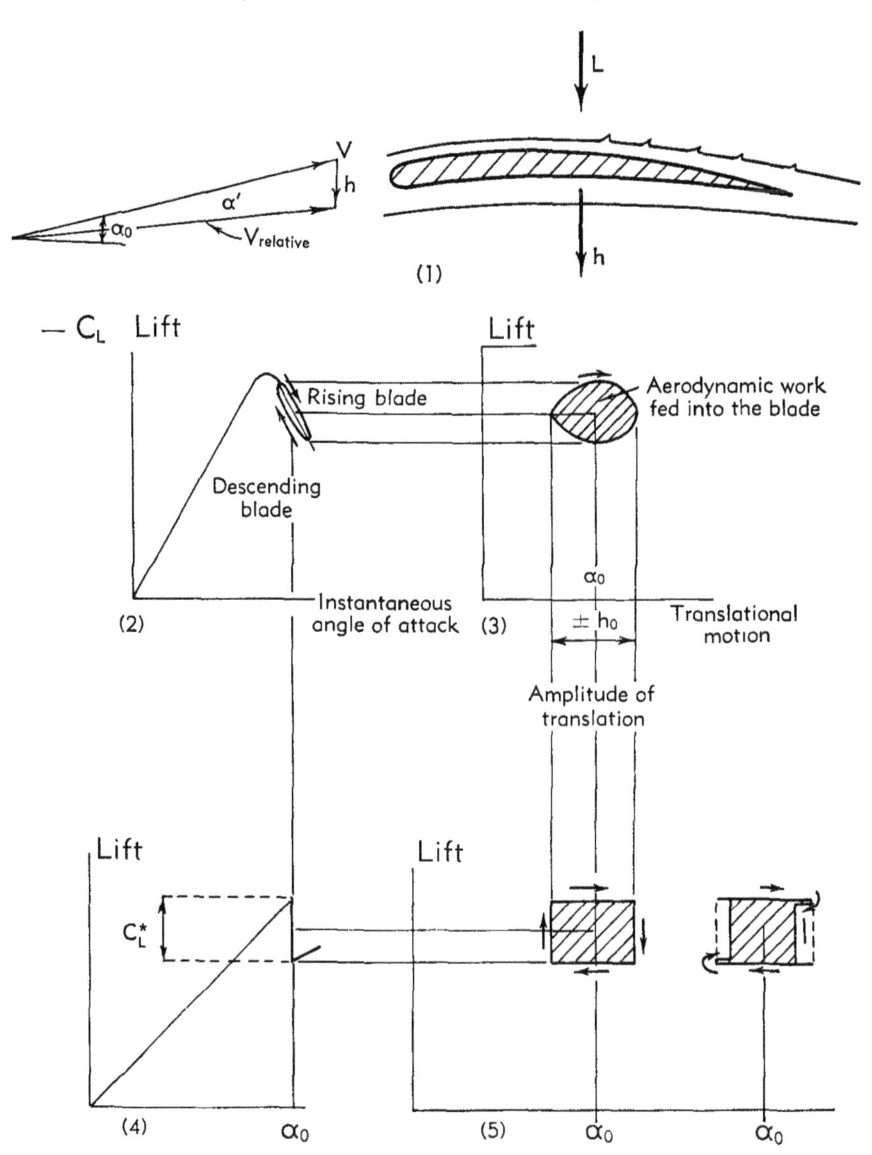


Fig. J,5f. Explanation of the aerodynamics of translational motion flutter in stalled flow.

case is traced back to the case of aerodynamic lift at varying angle of attack, as outlined in Fig. J,5f. The instantaneous angle of attack is composed of two parts, namely, the initial angle of attack $\alpha_0$ and the apparent change of angle $\alpha' = \dot{h}/V$, so that the airfoil in descending motion (positive $h$ values) experiences a decrease of angle of attack, and vice versa in rising motion (Fig. J,5f(1)). Putting $h = h_0 \sin \omega t$, one finds that in stalled motion with a lift following a descending slope indicated in

Fig. J,5f(2), a positive aerodynamic work is fed into the wing, Fig. J,5f(3), or, mathematically,

$$W_{LR} = \oint L dh > 0 \tag{5-4}$$

A limit cycle is eventually reached when essentially structural damping, dissipating this aerodynamic work, is obtained at sufficiently large amplitudes.

Somewhat different lines of thought have been followed in order to obtain a useful semiempirical analysis for the present case. In an analysis given by Sisto [*23*] the lift curve is expressed as a polynomial

$$f(\alpha) = \sum_{n=1}^{N} a_n \alpha^n \tag{5-5}$$

where $\alpha$ is the instantaneous angle of attack and $a_n$ the coefficients, adjusted to fit the mean dynamic lift curve as indicated in Fig. J,5g, top. Assuming a structural damping $\delta$ and a third degree polynomial, he solves the equation of translational motion:

$$\ddot{h} + \omega\delta\pi^{-1}\dot{h} + \omega^2 h = \frac{2bq}{m} f(\alpha) \tag{5-6}$$

It is thus supposed that the lift curve is followed regardless of the time involved for this process, that is, independent of the $k_h$ value. The result for a typical case is shown in Fig. J,5g, bottom. The expression for the steady state amplitude of the limit cycle becomes

$$\left(\frac{h_0}{b}\right)^2 = 8[f'''(\alpha_0)]^{-1}\left[-f'(\alpha_0) + \left(\frac{m}{\pi\rho b^2}\right)\left(\frac{\omega b}{V}\right)\delta\right]\left(\frac{V}{\omega b}\right)^2 \tag{5-7}$$

It is seen that the range of $h_0/b$ with respect to the initial angle of attack $\alpha_0$ is determined by the factor $[-f'(\alpha_0) + (m/\pi\rho b^2)(\omega b/V)\delta]$ where the first term represents aerodynamic excitation and the second, structural (and eventually aerodynamic) damping. The dependence of $h_0/b$ on $k_h$ value follows as a result of the relations between structural damping and dynamic pressure and the ratio of translational velocity of blade and air flow velocity $V$.

Another slightly different development was indicated by Schnittger [*16*], which gave the possibility of a unified approach to the design problem. In Fig. J,5f it is shown that the aerodynamic work fed into the blade would be reproduced in a rather similar way if the part of the lift coefficient curve with a negative slope is replaced by a concentrated jump $C_L^*$ as in Fig. J,5f(4) and J,5f(5). As in the previous case one must solve the limit cycle of the "slightly nonlinear" equation:

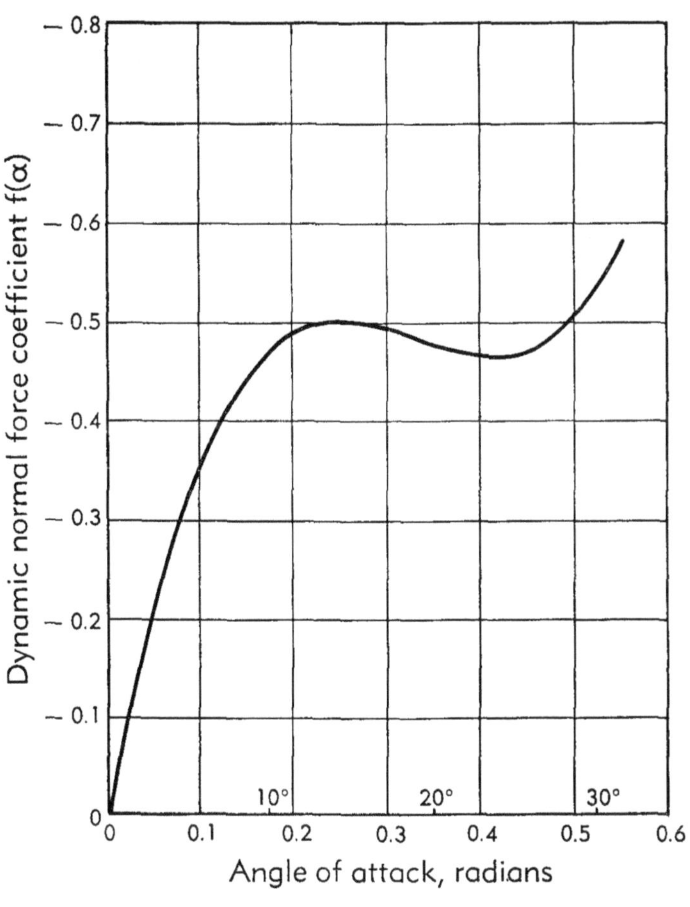


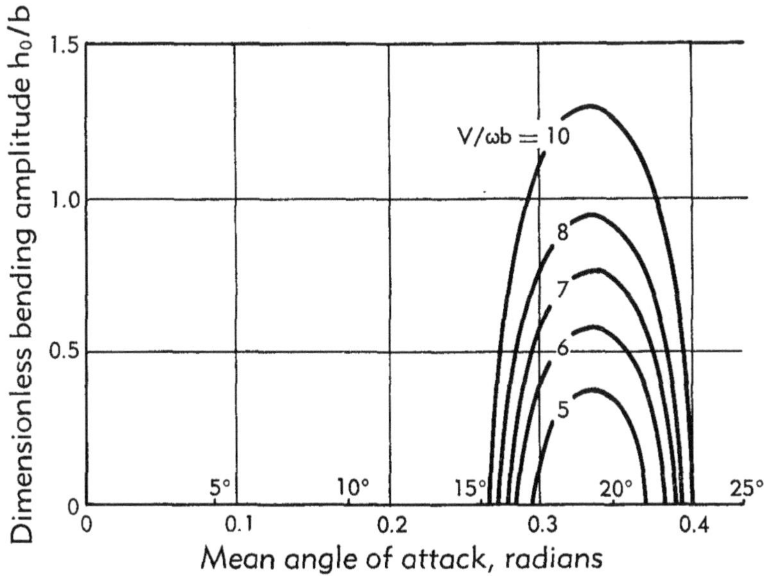


Fig. J,5g. Translational motion flutter in two-dimensional cascade. Calculated amplitudes as function of angle of attack with $V/\omega b$ as parameter [*23*].

$$\ddot{h} + \left(\omega\delta\pi^{-1} + \psi\frac{2\kappa}{k}F\right)\dot{h} + \omega^2 h = \begin{cases} \text{zero alternatively} \\ C_L^* \dfrac{4b\rho V^2}{2m} \end{cases} \tag{5-8}$$

where $\psi$ is a constant, indicating how large a percentage of the theoretical aerodynamic damping $(2\kappa/k)F$ (compare the second term of Eq. 3-1 where $F$ is the real part of the Theodorsen function) may act in case of only partially stalled blades. The jump $C_L^* 4b\rho V^2/2m$ works periodically on and off the blade. As in the above analysis, either standard methods of nonlinear mechanics or an energy balance of dissipated and supplied work per cycle gives the dimensionless amplitude

$$\frac{h_0}{b} = \frac{2\kappa C_L^*}{k^2\pi^2\left(\dfrac{\delta}{2\pi} + \psi\dfrac{\kappa F}{k}\right)} \tag{5-9}$$

As in the previous case, $h_0/b$ will be limited in range with respect to $\alpha_0$ because $C_L^*$ should be put proportional to the negative slope (derivative) of the lift curve. The dependence on $k$ is also apparent from Eq. 5-9. However, recalling the time lag appearing in the pitching motion, one should also take this lag of lift behind the motion of the airfoil into account here.

In sinusoidal motion this time lag corresponds to a phase shift which is proportional to the reduced frequency $k$. This may be conceived from the fact that

$$\frac{\omega b}{V} \text{ equals } \frac{\omega}{V/b} = 2\pi\frac{t_{\text{flow}}}{t_{\text{cycle}}}$$

where $t_{\text{flow}}$ is the time for a particle in the flow to travel with mean flow velocity across the airfoil and $t_{\text{cycle}}$ is the time for one cycle of the blade vibration. Thus a separation effect is naturally connected with $t_{\text{flow}}$ and its occurrence in the vibration cycle is proportional to the ratio of $t_{\text{flow}}$ to $t_{\text{cycle}}$. Assuming this phase shift in sinusoidal motion to be proportional to $k$ (flow particle velocity—airfoil motion kinematics), one finds that the aerodynamic work $\oint L dh$ is reduced by a factor $\cos\phi$ where $\phi$ is the phase shift in radians, because the amplitude is reduced to $h_0 \cos\phi$ until the retarded jump sets in. Putting the constant of proportionality between $\phi$ and $k$ equal to $\eta$, Eq. 5-9 is modified to

$$\frac{h_0}{b} = \frac{2\kappa\cos(\eta k)C_L^*(\alpha_0, k)}{k^2\pi^2\left(\dfrac{\delta}{2\pi} + \psi\dfrac{\kappa F}{k}\right)} \tag{5-10}$$

In Art. 6 it is shown that an actual case of bending flutter could be correlated by this formula and that both the phase shift constant $\eta$ and the magnitude of jump $C_L^*$ are in general agreement with values obtained from an interferometric investigation by the author, an example of which is shown in Plate J,5. In contrast to torsional flutter, it should be observed that the assumption of the phase shift $\phi$ is not essential in order to explain the self-excitation, but it is a function of $k$ and is part of the boundary conditions which determine the limit of $h_0/b$ with respect to $k$. Recalling the amplitudes $h_0/b$ for forced vibrations, Eq. 4-1, 4-2, and 4-3 respectively, it is seen that the amplitude of stalled flutter in translational motion has been obtained in the same general form, the amplitude being directly proportional to the magnitude of the disturbance.

## J,6. The Stress Problem of Vibrating Compressor Blades.

Generalization of the Design Problem. Stress Functions. According to C. R. Soderberg,[1] "The most characteristic feature of stall flutter is its incalculability." In the previous sections some theoretical and experimental attempts to grasp the aerodynamics of vibrating compressor blades have been reviewed, but there exists today no theory fully capable of explaining, or even predicting, the stresses that will be measured when the blades of an axial compressor start to work under actual conditions. This has recently been disappointingly demonstrated in an interesting paper by Pearson [*5*]. Instead of relating a complete survey of various test results, we shall try to make a generalization of the design problem as such and to illustrate the modest possibilities of this approach by a few examples.

Consider an elastic airfoil of uniform cross section of length $l$ supported as a cantilever beam. Let the cross section have a width $c$ corresponding to the chord and a maximum thickness $t$. Further, put $I_x$ equal to the moment of inertia of the cross section with respect to the neutral axis parallel with the chord. Let $A_x$ be the area of the same cross section and put the radius of gyration equal to $r_x = \sqrt{I_x/A_x} = Zt$. The constant $Z$ in the formula $r_x = Zt$ varies with the type of cross section. The rectangular section $ct$ has $Z = \frac{3}{6} = 0.289$ and the elliptic cross section with the axes $c$ and $t$ has $Z = \frac{1}{4} = 0.25$.

As an introduction to the study of the problems of stress of vibrating turbomachine blades, an elementary discussion of pure bending vibrations in the above airfoil will be made, since practice has shown that the bending stresses are the most important. It is first assumed that the cantilever blade is exposed to a uniform flow with velocity $V$ and an angle of attack $\alpha_0$ with respect to the chord direction. Next, suppose that the blade is subjected to the various disturbances discussed in the previous sections, and that the aerodynamic conditions at the tip of the blade are completely determining the amplitudes of the tip. It is obvious that this is only approximately true, and the integrated reactions may make more

[1] Personal communication.

correct a reference section, located somewhere at around 75 per cent of the length from the root section. However, it merely necessitates a slight change of a constant numerical value in the following discussion.

Assume a force $P$ applied at the tip of the blade. This causes a maximum stress $\sigma_b$ in the root section of the blade of length $l$:

$$\sigma_b = \frac{Pl(t/2)}{I_x} \tag{6-1}$$

The deflection at the tip section is

$$h_0 = \frac{1}{3}\frac{Pl^3}{EI_x} \tag{6-2}$$

where $E$ is the modulus of elasticity.

The elimination of the product $Pl$ gives

$$\sigma_b = \left(\frac{3}{2}\right) E \left(\frac{t}{c}\right)\left(\frac{h_0 c}{l^2}\right) \tag{6-3}$$

The natural frequency may be written

$$\omega_h = \frac{\lambda^2}{l^2}\sqrt{\frac{Eg}{\gamma}}\sqrt{\frac{I_x}{A_x}} = \frac{\lambda^2}{l^2}\sqrt{\frac{Eg}{\gamma}}\, r_x \tag{6-4}$$

$$\omega_h = \frac{\lambda^2}{l^2}\sqrt{\frac{Eg}{\gamma}}\, Zt$$

The smallest of the eigenvalues $\lambda$ equals $\lambda = 1.875$ for a straight uniform beam. The factor $\sqrt{Eg/\gamma}$ is a material constant with the dimension of a velocity, which can be shown to be the velocity of the propagation of waves in the solid material. For convenience it is put as

$$\sqrt{\frac{Eg}{\gamma}} = v \tag{6-5}$$

Thus

$$\omega_h = \lambda^2 Z v \left(\frac{t}{c}\right)\left(\frac{c}{l^2}\right) \tag{6-6}$$

The elimination of $t/l^2$ in Eq. 6-3 and 6-6 gives

$$\sigma_b = \left(\frac{3}{2}\right) E(\omega_h h_0)\left(\frac{1}{\lambda^2 Z v}\right)$$

and after a slight rearrangement and introduction of $k_h = \omega_h b/V$

$$\sigma_b = \frac{(\frac{3}{2})E}{\lambda^2 Z}\left(\frac{V}{v}\right) k_h \left(\frac{h_0}{b}\right) \tag{6-7}$$

The functions $h_0/b$ of the possible maximum amplitude, when the system is in resonance with the various external disturbances, are already known from Art. 4 and 5:

*Forced vibrations in potential flow.*

*Case 1.*

$$\frac{h_0}{b} = \sigma\alpha_0 f_\sigma(k_h) \tag{6-8}$$

*Case 2.*

$$\frac{h_0}{b} = (\Delta\alpha)_0 f_{(\Delta\alpha)}(k_h) \tag{6-9}$$

*Case 3.*

$$\frac{h_0}{b} = \left(\frac{p_0}{bq}\right) f_p(k_h) \tag{6-10}$$

For a true self-sustained vibration, there is no resonance condition in the ordinary sense. If the jump in the lift at stall is $C_L^*$, one has:

*Vibration in stalled flow.*

$$\frac{h_0}{b} = C_L^* f_{C_L^*}(k_h) \tag{6-11}$$

Put for all four response functions:

$$F_m = k_h f_m(k_h) \tag{6-12}$$

$$m = \sigma,\ \Delta\alpha,\ p, \text{ and } C_L^*$$

Then the following expressions for the maximum stress at the root section are derived:

*Forced vibrations in potential flow (wake excitation, rotor unbalance).*

*Case 1.*

$$\frac{\sigma_b}{\sigma\alpha_0} = \frac{(\frac{3}{2})E}{\lambda^2 Z}\left(\frac{V}{v}\right) F_\sigma(k_h) \tag{6-13}$$

*Case 2.*

$$\frac{\sigma_b}{(\Delta\alpha)_0} = \frac{(\frac{3}{2})E}{\lambda^2 Z}\left(\frac{V}{v}\right) F_{\Delta\alpha}(k_h) \tag{6-14}$$

*Case 3.*

$$\frac{\sigma_b}{(p_0/bq)} = \frac{(\frac{3}{2})E}{\lambda^2 Z}\left(\frac{V}{v}\right) F_p(k_h) \tag{6-15}$$

*Vibration in stalled flow.*

$$\frac{\sigma}{C_L^*} = \frac{(\frac{3}{2})E}{\lambda^2 Z}\left(\frac{V}{v}\right) F_{C_L^*}(k_h) \tag{6-16}$$

Thus the bending vibratory stress is always directly proportional to the magnitude of the actual disturbance of the wake $\sigma\alpha_0$; and $(\Delta\alpha)_0$, of the mechanical force $p_0/bq$ and the lift stall jump $C_L^*$ respectively. Further, two ratios appear, which will be shown to be rather fixed, and finally the general stress functions $F_m(k_h)$.

In design practice the problem often is to design blades for a given flow channel, where the length $l$, flow velocity $V$, and the blade material are given. Then the ratio $V/v$ is fixed and so is $E$. Usually the general shape of the blade is known, i.e. the cross-sectional constant $Z$ and the eigenvalue $\lambda$ are given ($\lambda$ is influenced by the taper of the blade). Thus the ratio $(\frac{3}{2})E/\lambda^2 Z$ is also fully determined. Furthermore, one finds from values of $E$, $\gamma$, and fatigue stress, given by Pearson [5], that little is gained by changing the blade material except in the case of titanium (Table J,6a).

*Table J,6a.* Properties of blade material.

| Metal | $E$, lb/in.$^2$ | Specific gravity | Fatigue stress $10^7$ reversals, tons/in.$^2$ | $\frac{E}{\text{Fatigue stress}}$ | $v$, ft/sec | $\frac{E/(\text{fatigue stress})}{v}$ |
|---|---|---|---|---|---|---|
| Aluminum alloy R.R. 56 | $10.3 \times 10^6$ | 2.75 | 10.7 | 430 | 5060 | 0.0846 |
| Steel S.61 | $29.7 \times 10^6$ | 7.73 | 25.0 | 530 | 5140 | 0.1030 |
| Steel S.62 | $29.7 \times 10^6$ | 7.73 | 34.5 | 384 | 5140 | 0.0745 |
| Aluminum bronze | $16.0 \times 10^6$ | 7.47 | 23.0 | 310 | 3840 | 0.0806 |
| Titanium | $15.5 \times 10^6$ | 4.50 | 30.0 | 231 | 4860 | 0.0475 |

From these values it is clear that the stress functions $F(k_h)$ and the magnitude of the disturbances are the determining factors. For the case of forced vibrations, the stress functions are given in Table J,6b.

*Table J,6b.* Stress functions of forced vibrations of cantilever blades in potential flow.

| Reduced frequency $k_h$ | 0.1 | 0.2 | 0.5 | 1.0 |
|---|---|---|---|---|
| Stress functions $F_\alpha$ | 2.20 | 2.38 | 2.65 | 2.90 |
| $F_{\Delta\alpha}$ | 1.01 | 1.00 | 0.88 | 0.70 |
| $F_p$ | 0.0945 | 0.109 | 0.130 | 0.145 |

As long as the aerodynamic damping is predominant (no structural damping was assumed in these cases of wake excitation), the interesting result is that the reduced frequency $k_h$ plays a minor role for the blade root stresses, because both excitation and damping are aerodynamic in nature. The same conclusion is not true, however, for vibrations in stalled flow when the aerodynamic damping is small and the magnitude and phase shift of the disturbance are functions of $k$.

Typical Behavior of Blades in Stalled Flow. The pitching motion (torsional mode of cantilever blades) usually has a stronger influence on the flow over the airfoil. Therefore it appears to the most dominant vibration of single airfoils. Translational (bending mode) flutter does occur, however, but in most cases in a much more limited range of angles of attack. On the other hand, the related stresses in cantilever blades for single airfoils are already of greater concern in the bending mode. Tests in a two-dimensional model indicated that the translational mode was particularly important as a "trigger device" for the pitching motion. Thus, locking of the translational mode considerably suppressed the pitching motion (see Fig. J,6a from [*24*]).

Arranging the blades in a two-dimensional cascade usually results in suppressing the vibrations for both two-dimensional stiff airfoils

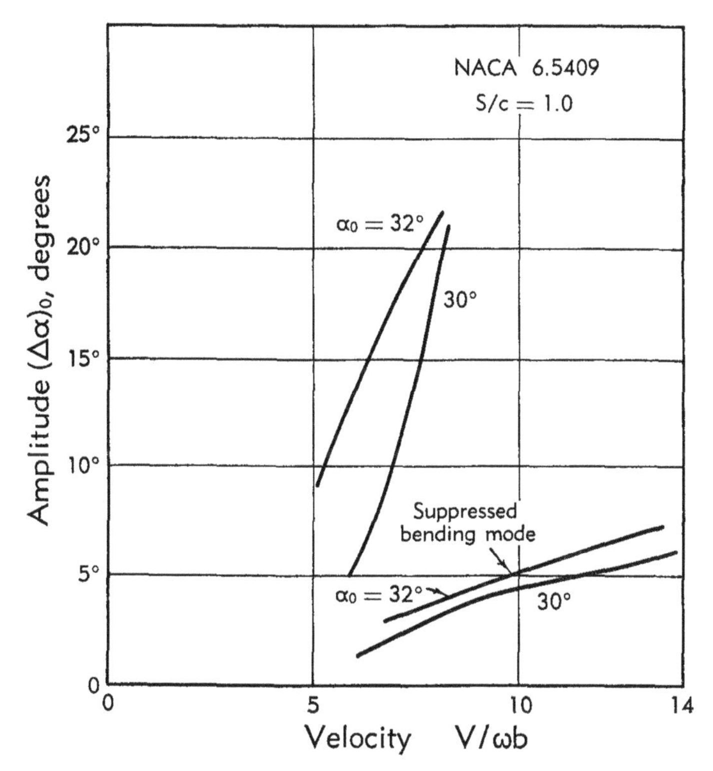


**Fig. J,6a.** Influence of suppressed bending mode on two-dimensional torsional limit cycle amplitude for cascaded airfoil NACA 6.5409 [*24*]. (One blade motion.)

with artificial springs and cantilever blades. The reduction is commonly greater for the pitching than for translational motion, so that emphasis will be further shifted to the bending mode stresses.

A common behavior of cantilever blades when subjected to a continuing increase in flow velocity (for $\alpha_0$ values where both torsional and bending flutter occur) is to start with a flexural vibration, which at a certain flow velocity does not increase any more but remains constant or even decreases. Simultaneously a strong torsional vibration begins.

In the practical application, where the blades are arranged in an annular compressor cascade, the torsional vibrations have ceased to be of any interest from the point of view of stresses, but the insight into aerodynamics of bending flutter, obtained from studies of flow in pitching motion has been of great use, as already demonstrated in Art. 5.

A typical result of torsional flutter amplitudes in two-dimensional cascades of stiff airfoils with artificial springs is demonstrated in Fig. J,6b, rearranged from [23] with $k_\alpha = (\omega_\alpha b/V)$ instead of $1/k_\alpha$ as abscissa

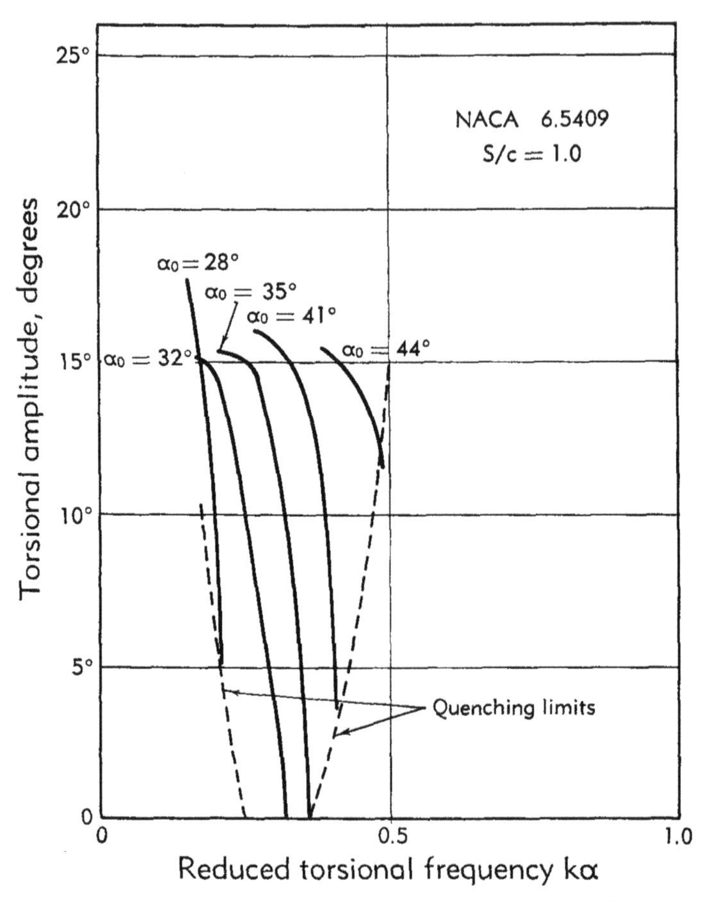


Fig. J,6b. Torsional amplitudes in an experimental two-dimensional cascade with $k_\alpha$ as abscissa and initial angle of attack as a parameter [23].

in order to demonstrate the general agreement with the concepts of aerodynamics demonstrated in Art. 5. It is seen that a rather narrow range of $k$ values was observed and quenching limits, featuring the phenomena of hard flutter, could be indicated.

The common practice of indicating a critical $k$ value (a certain stress level), where $k$ is based on torsional frequency $\omega_\alpha$, could be justified on the ground that at a certain flow velocity a torsional vibration may sometimes set in, after which the bending stresses do not increase anymore,

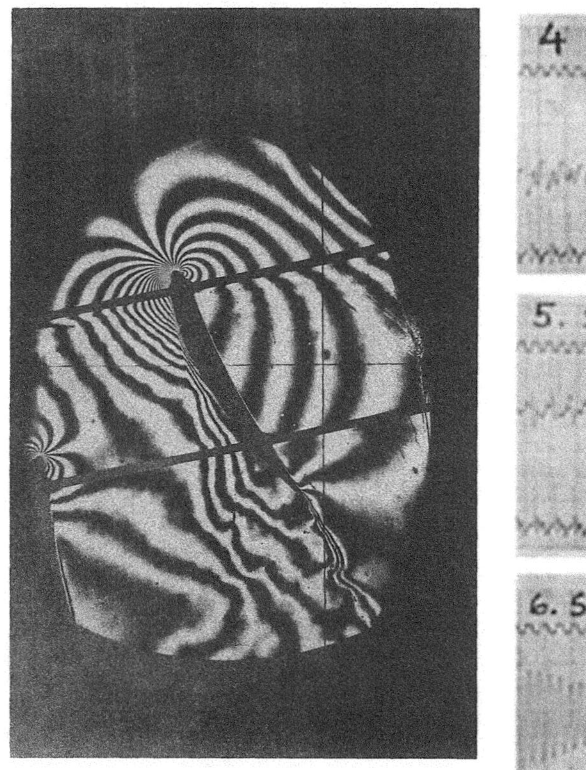

Plate J,5. Interferogram showing boundary layer separation on upper surface of compressor blade.

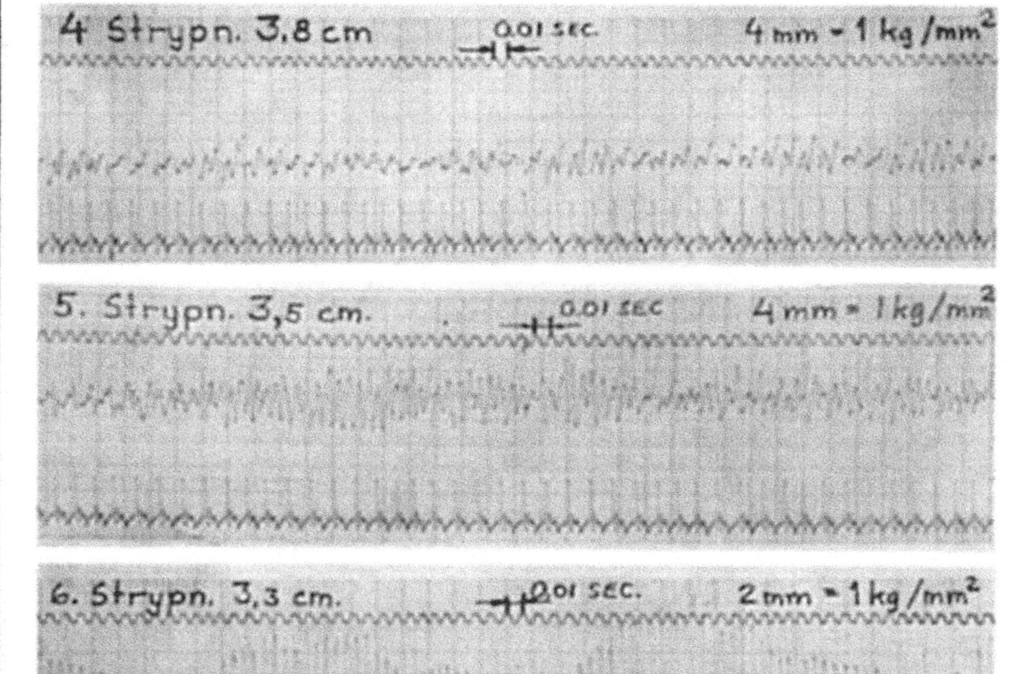


Plate J,6. Typical stress records during rotating stall.

and also on the fact that for most blades in a given design material there is a relatively close relationship between $\omega_\alpha$ and $\omega_b$. A more rational approach, however, appears to be the use of a $k$ value $k_h = (\omega_h - b)/V$, based on bending mode frequency. In the concluding part to follow, this suggestion will be followed in an example, where both cascade and compressor tests on a certain blade arrangement were undertaken.

Bending Flutter in Cascade Tests. The first three stages of an aircraft axial flow compressor were tested in a compressor test rig.[2] In a small cascade test setup of five fixed cantilever blocks of uniform cross section, the mean section of the first rotating blade row was used so that the results of the cascade test should simulate the aeroelastic behavior of the first blade row.

The ratio of measured maximum bending stresses to nominal stress $\sigma_b^*$ (Eq. 6-19) for the cascade tests versus reduced bending frequency is reproduced in Fig. J,6c, where the parameter is the equivalent to the initial angle of attack expressed as the ratio of actual turning angle $\epsilon$ to nominal turning angle $\epsilon^*$, in accordance with the terminology of Howell. The particulars of the steel blades are given in Table J,6c.

*Table J,6c.* Properties of steel blade in cascade test.

| | | | |
|---|---|---|---|
| Profile | NACA 7407 | Bending frequency $\omega_h$ | 585 rad/sec |
| Blade chord | 2.73 in. | $v = \sqrt{Eg/\gamma}$ | 16,700 ft/sec |
| Stagger angle | 53.8° | Young's modulus $E$ | $29.2 \times 10^6$ psi |
| Thickness ratio $t/c$ | 7 per cent | $\lambda^2 Z$ | 1.20 |
| Aspect ratio $l/c$ | 3.24 | Mass ratio $\kappa = \rho\pi b^2/m$ | 0.0025 |
| Space-chord ratio $s/c$ | 1.33 | Logarithmic damping $\delta$ | 0.012 |

| | | | | | | |
|---|---|---|---|---|---|---|
| Flow velocity, ft/sec | 656 | 590 | 525 | 459 | 394 | 328 |
| Reduced frequency | 0.102 | 0.113 | 0.128 | 0.146 | 0.170 | 0.204 |
| "Nominal stress" $\sigma_b$, psi $\times 10^{-3}$ | 380 | 342 | 304 | 266 | 228 | 190 |

In Art. 5, Eq. 5-10 was derived for the amplitude of bending mode flutter, which when introduced in Eq. 6-12 and 6-16 yields the expression for maximum bending stresses:

$$\sigma_b = \frac{(\frac{3}{2})E}{\lambda^2 Z}\left(\frac{V}{v}\right)\frac{2}{\pi^2}\frac{\cos(\eta k)C_L^*}{\frac{k}{\kappa}\frac{\delta}{2\pi} + \psi F} \tag{6-17}$$

In partially stalled flow as it very often occurs in annular cascades, in both tangential and radial directions, one has some aerodynamic damping and accordingly the constant $\psi > 0$. In the actual cascade tests $\psi \cong 0$ for $\epsilon/\epsilon^* > 1.0$, and Eq. 6-17 for the case of exclusively mechanical type of damping becomes

$$\sigma_b = \frac{6}{\pi}\left(\frac{E}{\lambda^2 Z}\right)\left(\frac{V}{v}\right)\left(\frac{\kappa}{\delta}\right)\frac{\cos(\eta k)}{k}\,C_L^* \tag{6-18}$$

[2] The tests reviewed here were undertaken by the author's company and permission to publish the following results is gratefully acknowledged.

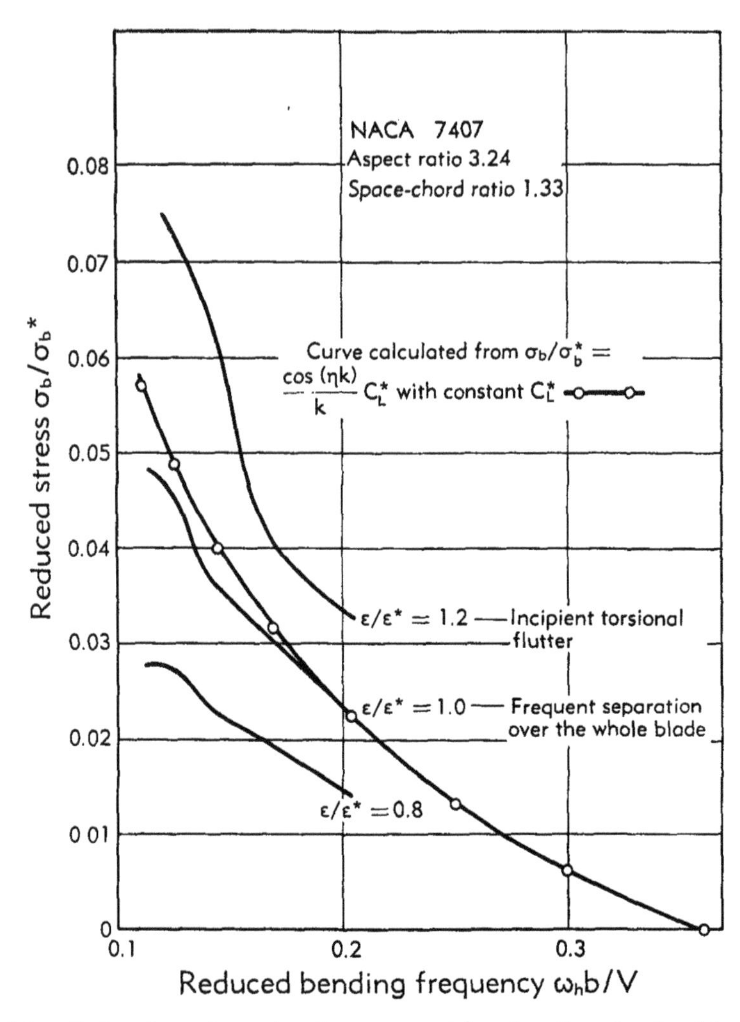


**Fig. J,6c.** Reduced maximum bending stress $\sigma_b/\sigma_b^*$ as function of reduced bending frequency $k_h$ with aerodynamic loading $\epsilon/\epsilon^*$ as a parameter for a cascade of five cantilever blades. The blades have a uniform *NACA* 7407 section, aspect ratio 3.24, and space-chord ratio 1.33.

For convenience put

$$\sigma_b^* = \frac{6}{\pi}\left(\frac{E}{\lambda^2 Z}\right)\left(\frac{V}{v}\right)\left(\frac{\kappa}{\delta}\right) \tag{6-19}$$

and finally,

$$\frac{\sigma_b}{\sigma_b^*} = \frac{\cos(\eta k)}{k} C_L^* \tag{6-20}$$

The ratios $(E/\lambda^2 Z)$, $(V/v)$, and $(\kappa/\delta)$ have already been mentioned. It remains to make a statement about the magnitude of the stall jump $C_L^*$ and the phase shift constant $\eta$.

Now it is interesting that a quite different approach, in which an interferometer was used to study the unsteady aerodynamic forces in a similar cascade [*19*], gives the right order of magnitude of $C_L^*$ and $\eta$. It was found that average $C_L^*$ values over whole cycles were of the order of magnitude of 0.01 and that a phase shift of $\pi$ radians corresponds to $k \cong 0.072$, that is, $\eta \cong \pi/0.72 = 4.36$ radians. Making the assumption that $C_L^* = 0.007$ at $\epsilon/\epsilon^* = 1.0$ and using the given $\eta$ value, one obtains at $k = 0.204$ coincidence of experimental and calculated values for Eq. 6-19. Supposing that $C_L^*$ is independent of $k$ in the studied range results in a curve, which fits the experimental values of $\sigma_b/\sigma_b^*$ very closely. One also reaches an absolute flutter limit of $k = 0.35$. The result is in general agreement with the results of Pearson, though he has the curves displaced toward somewhat higher $k$ values. Thus one cannot expect to anticipate exactly correct values, but the above example gives an idea of the order of magnitudes involved and a possible physical hypothesis to use for the purpose of the correlation of test results.

*Recent Work.* The present survey was concluded in 1955. The recent report by Carter and Kilpatrick [*25*], including clarifying communications, gives rise to further modifications of Eq. 6-19 and 6-20. Eq. 6-18 may be obtained from an energy balance between aerodynamic work $W_A$ fed into the wing and the energy dissipation $W_D$ due to mechanical damping. Pearson and Carter and Kilpatrick make the simplest possible assumption about the premises for the aerodynamic excitation, i.e. the existence of an incremental force directly proportional to the apparent change of angle of attack $\alpha^1 = \dot{h}/V$. Thus

$$F = \text{const}\,\frac{b\rho V^2}{m}\left(\frac{\dot{h}}{V}\right) \tag{6-21}$$

Performing the integration

$$W_A = \oint F dh$$

with $h = h_0 \cos \omega t$ gives with $m = \text{const}\; bt\rho_{\text{mtrl}}$

$$W_A = \text{const}\; h_0^2 \frac{\rho}{\rho_{\text{mtrl}}} \frac{V\omega}{t} \tag{6-22}$$

Since both $W_A$ and $W_D$ contain $h_0^2$, $h_0$ is canceled altogether and the energy balance produces a "velocity of divergence," above which amplitudes increase indefinitely to the point of destruction. This "velocity of divergence" takes the form:

$$V_{cr} = \text{const}\; \delta t\omega \frac{\rho_{\text{mtrl}}}{\rho} \tag{6-23}$$

The values of $h_0/b$ either from the analysis of Sisto, Eq. 5-7, or Schnittger, Eq. 5-9, represent an aerodynamic work $W_A$ proportional to a nonlinear function $f(h_0)$ and to $h_0$ respectively. Both methods result in the concept of stress level rather than that of "velocity of divergence,"

[25] now appears to confirm the applicability to present compressor experience of the stress level concept. Eq. 6-18 may generally be written as:

$$\frac{\sigma}{\sigma_{\text{fatigue}}} = \text{const}\, \frac{\rho}{\rho_{\text{mtrl}}} \frac{\sqrt{E\rho_{\text{mtrl}}}}{\sigma_{\text{fatigue}}} \frac{b}{t} \frac{V}{\delta} \frac{\cos \eta k}{k} C_L^* \tag{6-24}$$

Pearson has shown the ratio $\sqrt{E\rho_{\text{mtrl}}}/\sigma_{\text{fatigue}}$ to be rather independent of the choice of material (Table J,6a in present survey). Putting quite generally $F(k) = \cos \eta k/k \cdot C_L^*$, one obtains

$$\frac{\sigma}{\sigma_{\text{fatigue}}} = \text{const}\, \frac{\rho}{\rho_{\text{mtrl}}} \frac{b}{t} \frac{V}{\delta} F(k) \tag{6-25}$$

Carter and Kilpatrick demonstrate a reasonable agreement with respect to dependence on air velocity. They report a definite rise of stresses with air density $\rho$, but apparently there exists no linear relationship as implied by Eq. 6-23 or 6-25. Concerning material density $\rho_{\text{mtrl}}$, Pearson states that stresses are not reduced by material density. Instead, his results indicate that the ratio $\sigma/\sigma_{\text{fatigue}}$ remains relatively constant when exchanging blade material in an actual case. Thus, according to Pearson, $\rho_{\text{mtrl}}$ should not appear in Eq. 6-23 and 6-25. Carter and Kilpatrick believe that the reason for the discrepancy with respect to $\rho$ and $\rho_{\text{mtrl}}$ is partly explained by variations in mechanical damping $\delta$, particularly the important root damping. Experimental evidence is insufficient to settle the question.

The influence of the frequency parameter $k$ is still not wholly clarified. To assume that $F(k)$ of Eq. 6-25 is a constant quantity with respect to $k$ appears to be a premature conclusion from existing experimental evidence.

Bending Flutter in Compressor Tests. From the compressor tests two series of measurements on the first stage at 4000 to 6000 rpm, corresponding to 60 and 90 per cent of design speed, are illustrated in Fig. J,6d and J,6e. This stage has a midchord of 2.5 inch, an aspect ratio of 3.48, and a midspace chord ratio of 1.25. The influence of flow velocity and somewhat higher natural frequency due to thicker root section and rotation have been eliminated by calculating the nominal stress $\sigma_b^*$ from Eq. 6-19. Thus the reduced bending stress $\sigma_b/\sigma_b^*$ could be compared with the corresponding $\sigma_b/\sigma_b^*$ of the cascade tests at the same $k$ values and aerodynamic loading $\epsilon/\epsilon^*$.

In both cases it is found that the stresses of the compressor test are of the same order of magnitude as those of the cascade test, that is, the ratio of the two reduced stresses $\sigma_b/\sigma_b^*$ is sometimes a little higher than 1.0, sometimes smaller. Pearson [5] found that the stresses in his compressor were less than in the cascade test, due to recirculating flow which alleviated the stall, that is, one has partial stall and therefore some aerodynamic damping. In this case they appear to be a little higher in the compressor but the deviations are probably within the accuracy of the stress and flow measurements.

However, there is in each compressor series a spectacular increase of the stresses in the compressor, far above those of the cascade at similar conditions, and peculiarly enough at quite different $\epsilon/\epsilon^*$ values.

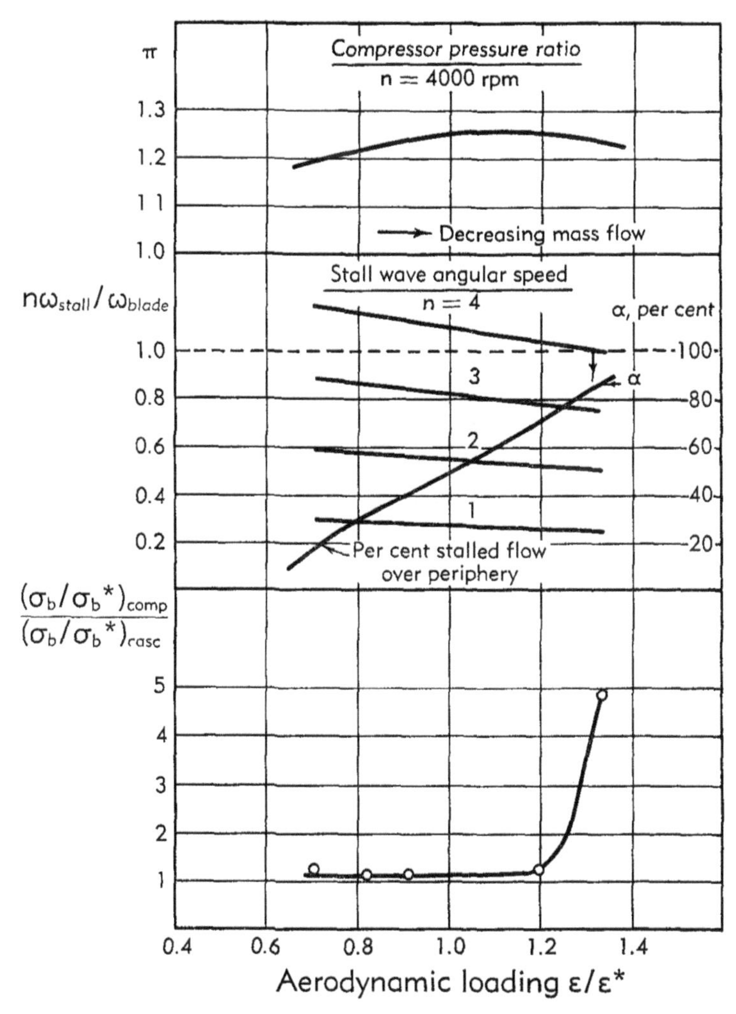


Fig. J,6d. Compressor tests at 4000 rpm. Ratio of reduced maximum bending stress of compressor blades to stress of cascade at the same $k$ value and aerodynamic loading $\epsilon/\epsilon^*$. Stall wave angular speed, per cent stalled flow over periphery. Compressor pressure ratio.

Keeping in mind the recent findings on stall propagation, the author employed Marbles's theory [*9*], which enables prediction of stall propagation speed, in reasonable agreement with the experiments of Iura and Rannie [*8*]. His expression for a cascade with simplified characteristics is

$$\frac{\omega_{\text{stall}} R}{C_x} = \frac{1}{\sin 2\beta_1} \tag{6-26}$$

where $\omega_{\text{stall}}$ is the annular speed of the stall waves with respect to the stalled rotating blade row, $C_x$ the axial velocity, $R$ the radial positions of

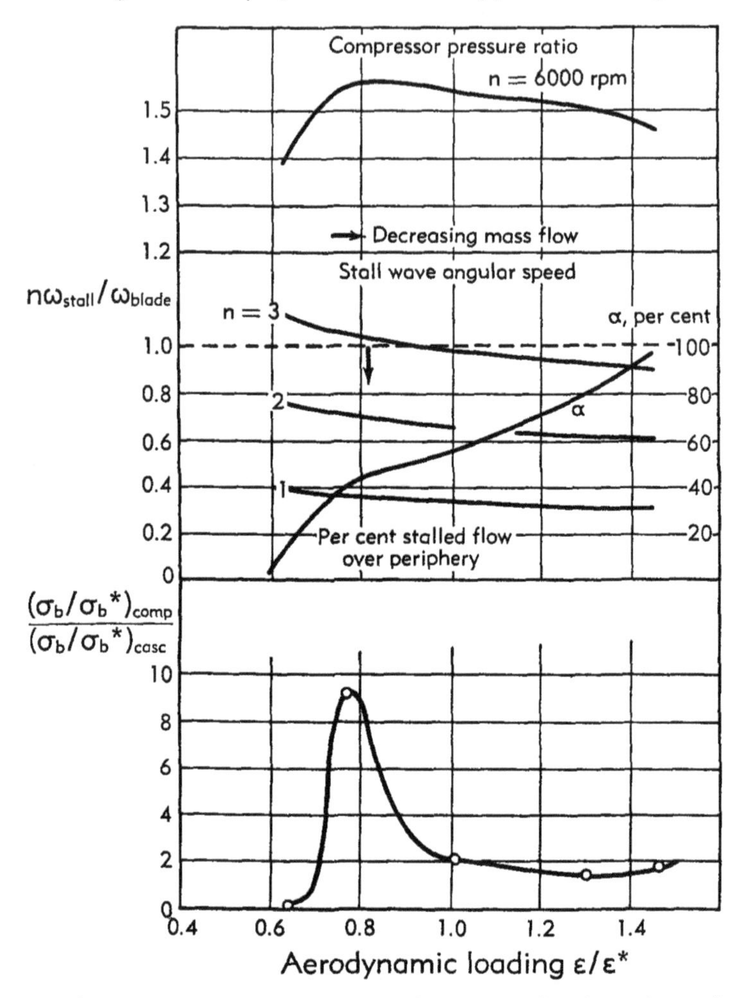


Fig. J,6e. Compressor tests at 6000 rpm. Ratio of reduced maximum bending stress of compressor blades to stress of cascade at the same $k$ value and aerodynamic loading $\epsilon/\epsilon^*$. Stall wave angular speed, per cent stalled flow over periphery. Compressor pressure ratio.

the blades from the shaft center line, and $\beta_1$ the inlet flow angle measured from the axial direction. The expression for the per cent of stalled flow in the peripheral direction is

$$\alpha = \frac{1}{2} - (\beta_1^* - \beta_1)(\csc 2\beta_1)\, 4\left(\frac{\frac{1}{2}\rho C_x^2}{\Delta p^*}\right) \tag{6-27}$$

Here $\beta_1^*$ and the pressure rise $\Delta p^*$ of the cascade correspond to the nominal loading (where $\epsilon/\epsilon^* = 1.0$).

The use of these formulas yielded some interesting results. It was found that the increase of the bending stresses, in the first compressor stage above values which could be anticipated from the cascade tests, corresponded closely in both cases to a coincidence of a multiple of the stall angular speed with the blade angular frequency. The $\alpha$ values are in general agreement with these results from the point of view that the larger this value is, the greater the number of symmetric stall waves rotating in the row. Iura, Rannie, and others found that with increasing stall one, two, three, and four waves, and eventually a still greater number of smaller waves, appear, until finally one great wave is created. Thus the fourth multiple resonance at 4000 rpm corresponds to $\alpha = 80$ per cent stalled flow and the third multiple resonance at 6000 rpm has an $\alpha$ value of $\alpha = 44$ per cent. It may be assuming a little too much, but at least it is a fair possibility to believe that one has a four-, or a three-wave pattern of rotating stall, respectively, in these cases. It must be strongly emphasized that, even if these patterns are the most stable, they may shift back and forth and the stress measurement records do not display a continuous smooth vibration. Plate J,6 gives three examples from the runs at 4000 rpm.

Examination of these records leads to the following general conclusions:

1. The stress amplitudes vary irregularly in time.
2. Superimposed on the blade frequency is a frequency roughly corresponding to the stall waves.
3. A much slower modulation or beat is recognized which at first sight could be taken for stall wave resonance. A careful check revealed, however, that this slower modulation was much too slow, with a frequency approximately corresponding to $\frac{1}{30}\omega_{\text{blade}}$ and $\frac{1}{10}\omega_{\text{stall}}$. This phenomenon, which appears to be capable of periodic, almost full extinction of stress amplitude, is probably a case of blade interference effect, also observed in two-dimensional cascade models with stiff airfoils [*12*].[3]

Two notes on similar compressor tests should be added here. In a similar compressor with $k$ values, raised from around 0.2 to 0.3 by means of larger chords, and an aspect ratio equal to 3, the stresses were much lower than in the referred case, giving some weight to the true existence of a limiting $k$ value as indicated in Fig. J,6c.

[3] After this was written in 1955 the report of Carter and Kilpatrick [*25*] and the included communication by H. Pearson appear to confirm the general validity of the concept of rotating stall cell excitation. There is also a general agreement on the irregular stress amplitudes.

In another compressor with a $k$ value of 0.725 and an aspect ratio of 2.3, flutter occurred in a very limited rotational speed range. It was nonexistent at higher angular speeds with smaller $k$ values and at smaller angular speeds with larger $k$ values. A tentative explanation of this phenomenon is offered by inspection of Eq. 6-20. It is seen that, if this equation holds, flutter does not occur for arguments of $\eta k$ which make $\cos(\eta k)$ a negative quantity. Thus, assuming the previous phase-shift constant to be valid, flutter is possible in the regions $0 < k < 0.35$ and $1.08 < k < 1.80$. The limit 1.08 is too high to check the value of $k = 0.72$, but the essential fact is the division in possible $k$ regions.

It must be emphasized that the connected concept of boundary layer time lag should not be mixed up with the time lag for stall propagation. There are two different concepts. The first is the adjustment of minor patterns of different degrees of stalled flow over the vibrating blade, which is a lag of the flow pattern behind the blade motion. The second is the time to shift the stall wave along the cascade, which is a matter of flow acceleration in accordance with classical equations of motion of fluid flow.

## J,7. Cited References and Bibliography.

*Cited References*


1. Scanlan, R. H., and Rosenbaum, R. *Introduction to the Study of Aircraft Vibration and Flutter*. Macmillan, 1951.
2. Theodorsen, T., and Garrick, J. E. Mechanism of flutter, a theoretical and experimental investigation of the flutter problem. *NACA Rept. 685*, 1940.
3. Mendelson, A. Effect of aerodynamic hysteresis on critical flutter speed at stall. *NACA Research Mem. E8B04*, 1948.
4. Studer, H. L. Experimentelle Untersuchungen über Flügelschwingungen. *Mitt. Inst. Aerodynam. E.T.H. (Zurich) 4/5*, 1936.
5. Pearson, H. The aerodynamics of compressor blade vibration. *Fourth Anglo-American Aeronaut. Conference*, 1953.
6. Emmons, H. W., Pearson, C. E., and Grant, H. P. Compressor surge and stall propagation. *Am. Soc. Mech. Engrs. 53-A-65*, 1953. Later transl., *Am. Soc. Mech. Engrs. 77*, 455 (1955).
7. Huppert, M. C., and Benser, W. A. Some stall and surge phenomena in axial-flow compressors. *J. Aeronaut. Sci. 20*, 835–845 (1953).
8. Iura, T., and Rannie, W. D. Experimental investigations of propagating stall in axial-flow compressors. *Trans. Am. Soc. Mech. Engrs. 76*, 463–471 (1954).
9. Marble, F. E. Propagation of stall in a compressor blade row. *J. Aeronaut. Sci. 22*, 541–554 (1955).
10. Ferri, A. Investigations and experiments in the Guidonia supersonic windtunnel. *NACA Tech. Mem. 901*, 1939.
11. Daley, B. N., and Humphreys, M. D. Effects of compressibility on the flow past thick airfoil sections. *NACA Tech. Note 1657*, 1948.
12. Sisto, F. *Flutter in Cascade. Sc. D. Thesis*, Mass. Inst. Technol., 1952.
13. Sisto, F. Unsteady aerodynamic reactions of airfoils in cascade. *J. Aeronaut. Sci. 22*, 297–302 (1955).
14. Mendelson, A., and Carroll, R. W. Lift and moment equations for oscillating airfoils in an infinite unstaggered cascade. *NACA Tech. Note 3263*, 1954.
15. Wang, C. T., Lane, F., and Vaccaro, R. J. An investigation of the flutter charac-

teristics of compressor and turbine blade systems. *J. Aeronaut. Sci. 23*, 335–344 (1956).
16. Schnittger, J. R. The stress problem of vibrating compressor blades. *Trans. Am. Soc. Mech. Engrs. 77*, 57–64 (1955).
17. Greenberg, J. Airfoils in sinusoidal motion in a pulsating airstream. *NACA Tech. Note 1326*, 1947.
18. Greenberg, J. Some considerations on an airfoil in an oscillating airstream. *NACA Tech. Note 1372*, 1947.
19. Schnittger, J. R. Single degree of freedom flutter of compressor blades in separated flow. *J. Aeronaut. Sci. 21*, 27–36 (1954).
20. den Hartog, J. P. *Mechanical Vibrations*, 3rd ed. McGraw-Hill, 1947.
21. Halfman, R. L. Experimental aerodynamic derivatives of a sinusoidally oscillating airfoil in two-dimensional flow. *NACA Tech. Note 2465*, 1951.
22. Halfman, R., Johnson, H. E., and Haley, S. M. Evaluation of high-angle-of-attack- aerodynamic-derivative data and stall flutter prediction technique. *NACA Tech. Note 2533*, 1951.
23. Sisto, F. Stall flutter in cascade. *J. Aeronaut. Sci. 20*, 598–604 (1953).
24. Krucicanin, S., and Martin, A. C. *Effect of Varying Parameters in Cascade Stall-Flutter. M. Sc. Thesis*, Mass. Inst. Technol., 1953.
25. Carter, A. D. S., and Kilpatrick, D. A. Self-excited vibration of axial-flow compressor blades. *Proc. Inst. Mech. Engrs. 171*, 1957.

*Bibliography*

Benser, W. A., and Finger, H. B. Compressor-stall problems in gas turbine type aircraft engines. *Soc. Automotive Engrs. Preprint 751*, 1956.

Bollay, W., and Brown, C. D. Some experimental results on wing flutter. *J. Aeronaut. Sci. 8*, 313–318 (1941).

Bratt, J. B., Wight, K. C., and Chinneck, A. Free oscillations of an airfoil about the half-chord axis at high incidences and pitching moment derivatives for decaying oscillations. *Brit. Aeronaut. Research Council Repts. and Mem. 2214*, 1940.

Chang, C. C., and Chu, W. C. Aerodynamic interference of cascade blades in synchronized oscillation. *Trans. Am. Soc. Mech. Engrs. 77*, 503–508 (1955).

Kemp, N. H., and Sears, W. R. The unsteady forces due to viscous wakes in turbomachines. *J. Aeronaut. Sci. 22*, 478–483 (1955).

Kilpatrick, D. A., and Ritchie, G. B. Compressor cascade flutter tests; 20° camber blades, medium and high stagger cascades. *Brit. Aeronaut. Research Council Cp. 187*, Dec. 1953.

Meldahl, A. Self-induced flutter of wings with one degree of freedom. *Brown Boveri Rev. 33*, 386–393 (1946).

Mendelson, A., et al. Experimental investigation of blade flutter in an annular cascade. *NACA Tech. Note 3581*, 1955.

Parry, J. F. W., and Pearson, H. Cascade blade flutter and wake excitation. *J. Roy. Aeronaut. Soc. 58*, 505–508 (1954).

Sabatiuk, A., and Sisto, F. A survey of aerodynamic excitation problems in turbomachines. *Trans. Am. Soc. Mech. Engrs. 78*, 555–563 (1956).

Sears, W. R. On asymmetric flow in axial-flow compressor stage. *J. Appl. Mech. 20*, 57-62 (1953).

Shannon, J. F. Vibration problems in gas turbines centrifugal and axial flow compressors. *Brit. Aeronaut. Research Council Repts. and Mem. 2226*, 1946.

Stenning, A. H. Stall propagation in a cascade of airfoils. *Mass. Inst. Technol. Gas Turbine Lab. Rept.*, May 1954.

Stenning, A. H., Kriebel, A. R., and Montgomery, S. R. Stall propagation in axial-flow compressors. *NACA Tech. Note 3580*, 1956.

Victory, M. Flutter at high incidences. *Brit. Aeronaut. Research Council Repts. and Mem. 2048*, 1943.

Wang, C. T., Vaccaro, R. J., and de Santo, D. F. A practical approach to the problem of stall flutter. *Trans. Am. Soc. Mech. Engrs. 78*, 565–571 (1956).

Woods, L. C. On unsteady flow through a cascade of aerofoils. *Proc. Roy. Soc. London A228*, 1955.

# PART FOUR

# *TURBINE POWER PLANTS*

# SECTION K

## *PERFORMANCE*[1]

D. H. MALLINSON

S. J. MOYES

**K,1. Basic Constant Pressure Gas Turbine Cycle.** The ideal thermodynamic cycle, on which the sequence of operations in an aircraft turbine engine is based, is the Joule or Brayton cycle. The temperature-entropy diagram, for this cycle in its ideal form, is shown in Fig. K,1, and the processes postulated for the working fluid, in order, are as follows:

1. Isentropic compression between the minimum and maximum pressures (Eq. 1-2).
2. Intake of heat at constant maximum pressure (Eq. 2-3).
3. Isentropic expansion between the maximum and minimum pressures (Eq. 3-4).
4. Rejection of heat at constant minimum pressure (Eq. 4-1).

In the practical engine, air drawn from the atmosphere is used as the working fluid, and after compression, heating, and expansion it is discharged to the atmosphere, thus forming a continuous flow machine.

Compression is effected in a rotary-type compressor with either outward radial flow of the working fluid (centrifugal compression), or axial flow (axial flow compression), or, in some engines, with mixed flow where the working fluid is compressed while flowing in both radial and axial directions.

In the aircraft turbine engine, heating of the compressed air takes place in a combustion chamber in which fuel is burned continuously in the stream of air passing through.

The first charge on the work produced by expanding the hot gas after heating is in providing the work to drive the compressor, and this is achieved by expanding the gas in a turbine with either inward radial flow or axial flow of the working fluid. The remaining part of the expansion is available to produce useful output, the form of which depends on the type of propulsive device fitted to the engine.

In practice, isentropic compression and expansion are ideals against which the processes carried out in practical machines may be compared.

[1] Acknowledgement is made to the Chief Scientist, Ministry of Supply, United Kingdom, for permission to publish the text and diagrams of this section.

In the practical engine, therefore, we may expect the compression to be achieved with some degree of inefficiency in the compressor as compared with the isentropic ideal. Likewise the expansion in the turbine can only approach isentropic expansion without ever attaining it. Again, since heat is added to the working fluid at the maximum cycle pressure while the fluid is flowing in a duct, frictional effects must cause some loss of pressure during the heating process, to which may be added a further loss arising from momentum changes during the heating process.

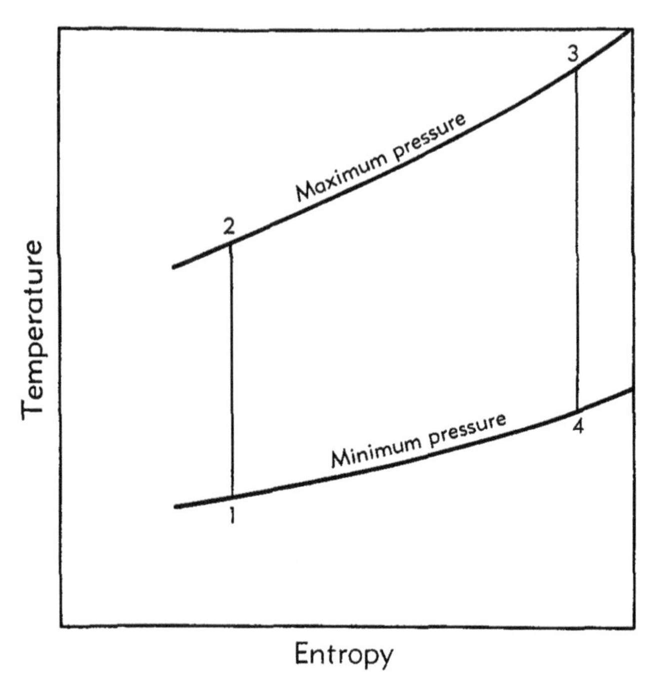


Fig. K,1. Temperature-entropy diagram for Joule or Brayton cycle.

*Application of the constant pressure gas turbine cycle to aircraft engines.* We have now modified the ideal cycle to a practical form in which (1) compression is effected in a centrifugal or axial flow compressor with some loss compared to the isentropic case, (2) heating, after compression, takes place in a combustion chamber with some pressure loss, and (3) the expansion through the turbine driving the compressor also takes place with some loss compared to the isentropic case.

Once the above processes are complete, and assuming that the expansion through the driving turbine immediately follows the heating process, we have, therefore, a flow of hot gas at exhaust from the compressor turbine which, when expanded to the minimum pressure of the cycle, is capable of producing output in the form of thrust or power.

In the pure jet engine, the main turbine exhaust referred to above is expanded to atmospheric pressure through a nozzle to form a stream of high velocity, yielding a change of momentum between the entering and exhausting streams of the working fluid. This produces a thrust on the engine which is causing the change of momentum.

In the propeller turbine engine, the main turbine exhaust is expanded through a further turbine to produce power for driving a propeller. Here, however, although the expansion is carried down to atmospheric pressure, a certain finite portion of the expansion is used to provide the kinetic energy of the residual jet velocity which the exhausting gas stream must possess in any continuous flow machine. Thus the propeller turbine engine, in fact, provides power for driving a propeller and, in addition, some thrust from the residual jet velocity. Clearly, the proportion of the expansion taken by the propeller turbine can be lessened to provide a greater residual jet velocity, and therefore greater direct thrust at the expense of propeller thrust. Depending on certain factors, there is, in fact, an optimum division of available energy in the expansion after the compressor turbine between the propeller and residual jet. Before leaving the propeller turbine engine, it should be noted that it is not necessary to employ two separate turbines, one for driving the compressor and the other for driving the propeller. Both turbines can be merged into one driving both propeller and compressor, an arrangement possessing some advantages and disadvantages compared with the scheme using two separate turbines. Likewise, with two separate turbines, it is not necessary for the propeller turbine, in expanding the working fluid, to follow after the compressor turbine. The positions could be reversed or, alternatively, the expanding gas flow could be divided to flow simultaneously into both turbines, but as will be shown later, only certain arrangements become acceptable for the aircraft turbine engine when considered from the aspects of mechanical layout and thermodynamic performance.

We turn now to the bypass or ducted fan engine, a type of engine that represents an intermediate step between the extremes of simple jet and pure propeller propulsion. A propeller can be classed as a jet propulsion device, insofar as it induces a large flow of air from the atmosphere through the blades which discharge the air so received at a slightly higher velocity to give a forward thrust. The ducted fan performs the same function for a much smaller air flow. Compared with the propeller, its blades are much smaller and of greater number, and furthermore are enclosed in a stationary duct which is continued aft of the rotating blades to form a nozzle for expanding the air passing through the fan. As such, the ducted fan may be regarded as an axial flow compressor whose function is to inspire air from the atmosphere and to pressurize it for the purpose of expelling it rearward at a higher velocity through a suitably shaped nozzle, thus creating a thrust. Power is required to drive the fan and, similar to the propeller case, this must be provided from a turbine in the engine, utilizing some of the energy made available in the expansion process. Again, similar to the propeller engine, not all the available energy in the expansion after the compressor turbine is utilized for driving the ducted fan, owing to the fact that some is required for the residual jet velocity in the engine. In fact, compared with the propeller engine, the optimum division of available expansion energy between the ducted fan turbine and the residual jet usually demands a smaller proportion of the available energy being given to the ducted fan, since its propulsive efficiency is lower than that of a propeller. The stator and rotor blades of the ducted fan can be mounted to form a second or bypass compressor around the periphery of the early stages of the main engine compressor. It is now a logical step to combine the blades of the bypass compressor and the early stages of the main compressor to form one large low pressure compressor, the outer portion of which acts as the ducted fan and the inner portion as the low pressure section of the main compressor. In this form, the engine is known as a bypass engine, so-called since the ducted fan has become part of a compressor in which a proportion of its flow bypasses the main engine and is expanded as a separate jet.

**K,2. Basic Thermodynamics of the Aircraft Turbine Engine Cycle.** Before dealing with the thermodynamics of the aircraft turbine engine cycle, we need to define compressor and turbine efficiency in precise thermodynamic terms, and to show how the thermodynamics of the cycle are modified by the engine possessing translational velocity or forward speed.

*Component efficiencies.* As shown in Vol. X, compressor and turbine efficiencies can be defined in two ways. A compressor inspiring unit air mass flow at total pressure $p^0_{i_c}$ and stagnation temperature $T^0_{i_c}$, and delivering at total pressure $p^0_{e_c}$ and stagnation temperature $T^0_{e_c}$, requires a work input of

$$c_{p_a}(T^0_{e_c} - T^0_{i_c})$$

assuming an adiabatic process. In this expression $c_{p_a}$ is the specific heat at constant pressure for air.

The ideal isentropic work of compression is given by

$$c_{p_a}T^0_{i_c}\left[\left(\frac{p^0_{e_c}}{p^0_{i_c}}\right)^{\frac{\gamma_a-1}{\gamma_a}} - 1\right]$$

The isentropic efficiency $(\eta_c)_{is}$ is defined as the ratio of the ideal to the actual work and becomes

$$(\eta_c)_{is} = \frac{\left(\frac{p^0_{e_c}}{p^0_{i_c}}\right)^{\frac{\gamma_a-1}{\gamma_a}} - 1}{\frac{T^0_{e_c}}{T^0_{i_c}} - 1} \tag{2-1}$$

It will be noticed that total head pressures and temperatures are used in this definition. Although static conditions could equally well be used, it is more convenient in dealing with the thermodynamics of the gas turbine engine to use total head conditions for efficiencies and pressure losses, and this practice will be followed throughout this section.

The concept of isentropic efficiency applied to over-all compression is more applicable to a compressor where the process is completely carried out in one stage, such as a centrifugal compressor. In the axial flow type, the compression proceeds in a number of stages, and it is a better approximation to the process to regard the compression as built up of an infinite number of stages, each operating with a certain isentropic efficiency $(\eta_c)_{\text{pol}}$ giving what is termed polytropic compression.

Applying Eq. 2-1 to an infinitesimal change, we have

$$(\eta_c)_{\text{pol}} = \frac{\dfrac{\gamma_a - 1}{\gamma_a}\dfrac{dp^0}{p^0}}{\dfrac{dT^0}{T^0}}$$

which on integration yields the result, for constant $(\eta_c)_{\text{pol}}$,

$$\frac{T^0_{e_c}}{T^0_{i_c}} = \left(\frac{p^0_{e_c}}{p^0_{i_c}}\right)^{\frac{\gamma_a - 1}{\gamma_a(\eta_c)_{\text{pol}}}}$$

Thus in polytropic compression, temperature and pressure are connected by the law

$$p^0(v^0)^{n_c} = \text{const}$$

where the exponent of compression $n_c$ is given by

$$\frac{n_c - 1}{n_c} = \frac{\gamma_a - 1}{\gamma_a(\eta_c)_{\text{pol}}} \tag{2-2}$$

Similarly, for expansion in a turbine, with work output equal to

$$c_{p_g}(T^0_{i_t} - T^0_{e_t})$$

where $c_{p_g}$ is the specific heat at constant pressure of the working gas.

The isentropic efficiency is defined as

$$(\eta_t)_{\text{is}} = \frac{1 - \dfrac{T^0_{e_t}}{T^0_{i_t}}}{1 - \left(\dfrac{p^0_{e_t}}{p^0_{i_t}}\right)^{\frac{\gamma_g - 1}{\gamma_g}}} \tag{2-3}$$

and the exponent $n_t$ of polytropic expansion is given by

$$\frac{n_t - 1}{n_t} = \frac{\gamma_g - 1}{\gamma_g}(\eta_t)_{\text{pol}} \tag{2-4}$$

where $(\eta_t)_{\text{pol}}$ is the polytropic efficiency of expansion.

It is well known that the specific heats $c_{p_a}$ and $c_{p_g}$ (and hence $\gamma_a$ and $\gamma_g$) vary with temperature and gas composition. Hence the gas properties do not remain constant during a compression, heating, or expansion process. However, it can be shown that for gas turbine performance analysis, negligible error is introduced by assuming $c_{p_a}$ and $c_{p_g}$ constant during compression and expansion respectively, provided the values used refer to the mean temperature level of the process.

*Significance of translational velocity.* If, at any plane in a duct, gas is flowing at a static pressure and temperature $p$, $T$ with velocity $V$, then total head conditions in the gas at that plane are defined as

$$T^0 = T + \frac{V^2}{2gJc_p} \tag{2-5}$$

and

$$p^0 = p\left(1 + \frac{V^2}{2gJc_pT}\right)^{\frac{\gamma}{\gamma-1}} \tag{2-6}$$

i.e. total head conditions represent the pressure and temperature that would obtain were the gas brought to rest in an ideal isentropic change. As such, therefore, total head conditions represent a measure of available energy level and are thus used in thermodynamic cycle calculations in defining component efficiencies and pressure losses. In this way, cycle calculations, for given total energy levels, are made independent of arbitrary pressure energy and kinetic energy subdivisions of total energy.

A gas turbine engine mounted in an aircraft draws in its air through a duct usually connected to a mouth or hole situated in a forward-facing part of the airframe, generally the leading edge of the wing. In forward motion, the entering air possesses a velocity relative to the engine and therefore, again relative to the engine, an energy level represented by total head conditions greater than those of the ambient air. Thus, if the forward speed of the engine is $V_\infty$ in an ambient atmosphere at pressure $p_\infty$ and temperature $T_\infty$, then ideally the total head conditions of the air entering and relative to the engine compressor are:

$$T^0_{i_c} = T_\infty + \frac{V_\infty^2}{2gJc_p} \tag{2-7}$$

and

$$p^0_{i_c} = p_\infty\left(1 + \frac{V_\infty^2}{2gJc_pT_\infty}\right)^{\frac{\gamma}{\gamma-1}} \tag{2-8}$$

Losses due to shock wave and frictional effects, however, occur when the air flows in past the lips of the air intake mouth and in passing along the ducting. Provided that the change is adiabatic, the total temperature at any plane up to the compressor entry, must equal that given by Eq. 2-7. The shock wave and frictional losses show themselves as a loss in total pressure from the ideal value given above, and though at the compressor entry we can still write

$$p_{i_c}^0 = p_{i_c}\left(1 + \frac{V_{i_c}^2}{2gJc_pT_{i_c}}\right)^{\frac{\gamma}{\gamma-1}}$$

we may also write

$$p_{i_c}^0 = \eta_i p_\infty\left(1 + \frac{V_\infty^2}{2gJc_pT_\infty}\right)^{\frac{\gamma}{\gamma-1}} \tag{2-9}$$

where $\eta_i$ is the air intake pressure recovery taken between free stream conditions and the compressor entry. The highest attainable values of air intake pressure recovery vary with forward speed, especially when the latter becomes supersonic. Under these conditions, it is of advantage in analytical work to define the forward speed in terms of Mach number $M$, i.e. the ratio of the forward speed to the local velocity of sound. Defining the latter by $a$, we have

$$a_\infty^2 = g\gamma \mathcal{R} T_\infty \quad \text{and} \quad M_\infty = \frac{V_\infty}{a_\infty}$$

where $\mathcal{R}$ is the specific gas constant. Substituting the above in Eq. 2-9 we obtain the result

$$p_{i_c}^0 = \eta_i p_\infty \left(1 + \frac{\gamma - 1}{2} M_\infty^2\right)^{\frac{\gamma}{\gamma-1}} \tag{2-10}$$

The variation of $\eta_i$ with $M_\infty$ for well-designed intakes follows a curve such as that shown in Fig. K,2a. It will be seen that over a range of flight Mach number from zero to about 1.5, the intake pressure recovery does not fall below 95 per cent, but drops fairly rapidly for values of $M_\infty$ above 1.5. Under high speed flight conditions, the difference between engine entry total head and ambient conditions against which the engine exhausts, can become considerable and will influence the thermodynamic cycle calculations in no small measure.

The increase in the total head pressure and temperature of the air relative to the engine, brought about by forward speed, is termed the ram effect, and the ram pressure ratio and temperature ratio are defined, respectively, as the ratios of the relative total head pressure and temperature of the entering air to the ambient values. When referring to air conditions within the engine system, the term "relative" is always understood and will be omitted in future discussion.

*Output and efficiency of a self-driving gas generator.* With the engine possessing forward speed, we can now modify the cycle diagram of Fig. K,1 to conform with the most general sequence of events occurring in an aircraft turbine engine. The modified cycle is shown in Fig. K,2b as a temperature-entropy diagram in which we have the following processes:

1-2 is the ram compression
2-3 is the mechanical compression
3-4 is the heating by direct combustion
4-5 is the expansion in the compressor turbine
5-7 is that portion of the total expansion remaining for producing output.

We can correlate this temperature-entropy diagram with a diagrammatic representation of the gas generator unit shown in Fig. K,2c where

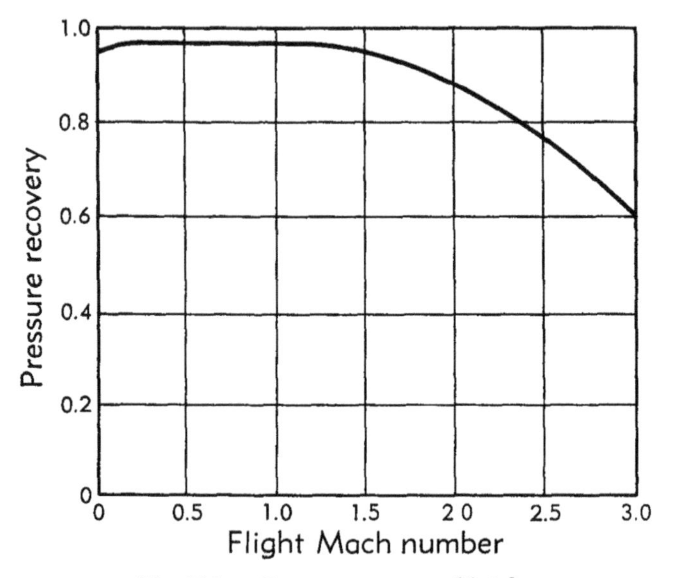


Fig. K,2a. Pressure recovery of intake.

the reference planes 1, 2, 3, etc. correspond to the end points of the processes shown on the temperature-entropy diagram. To denote the pressure, temperature, and velocity of the air at any reference plane, the appropriate symbol will take, as a subscript, the number of the plane.

If we consider the plane 1 as just ahead of the air intake to the engine, the static pressure $p_1$ is equal to the ambient value $p_\infty$. At this plane 1, the total head pressure and temperature of the air relative to the engine are $p_1^0$ and $T_1^0$ given by Eq. 2-7 and 2-8. Intake losses from 1-2 render $p_2^0$ slightly less than $p_1^0$. Thus we may take the over-all total head pressure ratio as $p_3^0/p_2^0$ and use the ram pressure ratio $p_2^0/p_\infty$.

We now derive expressions for the available energy $P_3$ contained in the hot gas at the exhaust from the compressor turbine. The processes in the cycle up to this point combine to form what may be termed a self-driving gas generator whose function is to provide hot gas which, on expansion to ambient conditions, can deliver useful output either in the

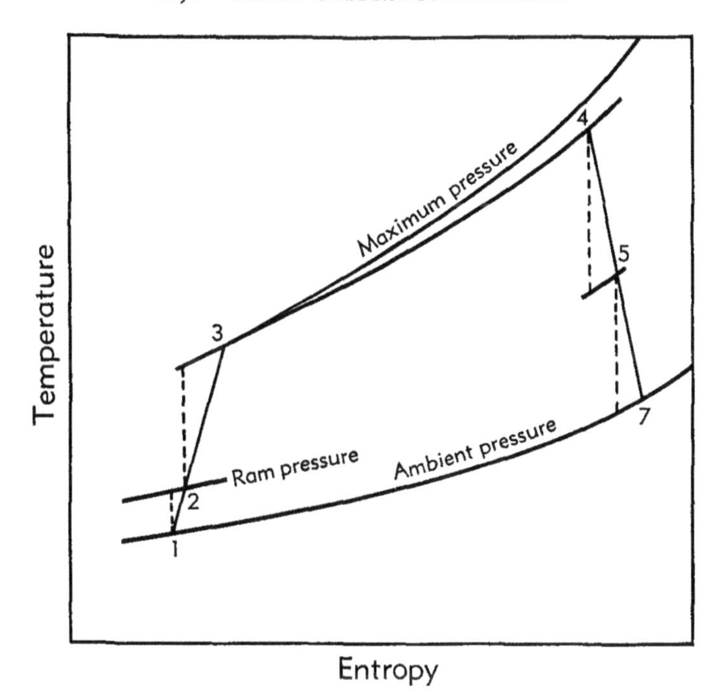


Fig. K,2b. Temperature-entropy diagram for turbine engine.

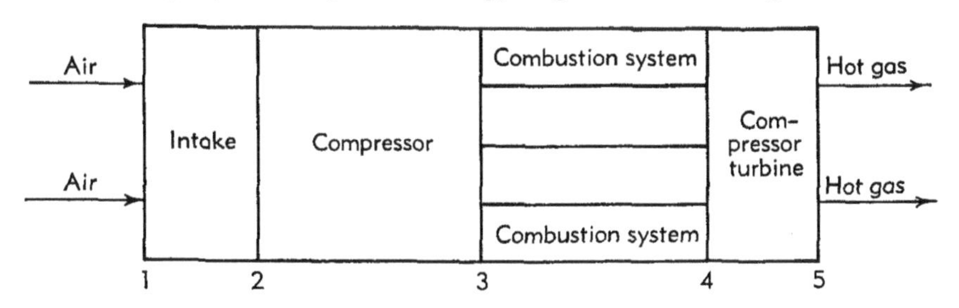


Fig. K,2c. Diagram of gas generator unit.

form of thrust or power. The expressions derived apply to unit air mass flow.

For a total head temperature rise in the compressor from $T_1^0 = T_2^0$ to $T_3^0$, the work of compression is given by

$$c_{p_a}({}_1\Delta T_3^0)$$

where ${}_1\Delta T_3^0 = T_3^0 - T_1^0$, and the compressor delivery pressure

$$p_3^0 = p_2^0 \left(1 + \frac{{}_1\Delta T_3^0}{T_1^0}\right)^{\frac{n_c}{n_c - 1}} \tag{2-11}$$

where $n_c$ is the exponent of polytropic compression of small stage efficiency $(\eta_c)_{pol}$ as defined by Eq. 2-2.

After compression, the air is heated by direct combustion of fuel in the combustion chamber, suffering a pressure loss, as described in Art. 1. If this total head pressure loss is ${}_3\Delta p_4^0$, then

$$p_3^0 - p_4^0 = {}_3\Delta p_4^0$$

By heating, the total head temperature is raised from $T_3^0$ to $T_4^0$, with the result that the heat input to the cycle is equal to

$$c_{p_b}(T_4^0 - T_3^0) \tag{2-12}$$

The obtaining of a mean value for $c_{p_b}$ is not so simple as for the compression and expansion processes since, in the heating process by direct combustion, the mass flow of the gas and its composition change as well as the temperature. For the purpose of the analysis, it is proposed to retain the use of $c_{p_b}$ which may be obtained from the heat balance equation as follows:

$$c_{p_b}(T_4^0 - T_3^0) = \left[h_4\right]_{288}^{T_4^0} - \frac{1}{1+f}\left[h_3\right]_{288}^{T_3^0}$$

where $h_4$ is the enthalpy of the products of combustion taken between the limits of $T_4^0$ and the reference temperature (288°K), $h_3$ is the enthalpy of the air entering the combustion chamber taken between the limits of $T_3^0$ and 288°K, and $f$ is the over-all fuel-air ratio.

During the compression process, air may be bled for cooling or other purposes, and during the heating process the fuel mass flow is added to the air mass flow entering the combustion chamber. The net result is that, in some engines, the mass flows of air and gas in the compressor and turbine, respectively, may differ by a small amount. The effect may be easily allowed for, but as such differences do not significantly affect the broad conclusions of a performance analysis, for the purposes of this study the air and gas mass flows in the compressor and turbine are assumed equal.

$p_4^0$ and $T_4^0$ are the total head conditions at inlet to the turbine (refer to planes 4-5) whose function is to drive the compressor. Taking $c_{p_g}$ as the specific heat of the gas during expansion, we have, equating the compressor input to the turbine output,

$$c_{p_a}({}_1\Delta T_3^0) = \eta_m c_{p_g}({}_4\Delta T_5^0) \tag{2-13}$$

where ${}_4\Delta T_5^0 = T_4^0 - T_5^0$ and $\eta_m$ is a mechanical efficiency term to allow for bearing losses and turbine and compressor disk friction losses, etc. The outlet total head pressure from the compressor turbine is given by

$$p_5^0 = p_4^0\left(1 - \frac{{}_4\Delta T_5^0}{T_4^0}\right)^{\frac{\gamma_g}{(\gamma_g - 1)(\eta_t)_{pol}}} \tag{2-14}$$

where $(\eta_t)_{pol}$ is the small stage efficiency of polytropic expansion. Also, the outlet temperature $T_5^0 = T_4^0 - {}_4\Delta T_5^0$.

The total head pressure and temperature $p_5^0$ and $T_5^0$ represent the energy level at which hot gas is generated for the production of useful output. If the expansion of this gas were carried out isentropically, then the power developed would be

$$c_{p_g}T_5^0\left[1 - \left(\frac{p_7}{p_5^0}\right)^{\frac{\gamma_g - 1}{\gamma_g}}\right] \tag{2-15}$$

where the static pressure $p_7$ against which the engine exhausts is equal to the ambient pressure $p_\infty$. This power is denoted by the symbol $(P_g)_{is}$ and we now, by means of the relationships given above, derive an expression for $(P_g)_{is}$ in terms of quantities that may be said to define the cycle.

From Eq. 2-13 and 2-14 we have

$$p_5^0 = p_4^0\left(1 - \frac{c_{p_a}({}_1\Delta T_3^0)}{c_{p_g}\eta_m T_4^0}\right)^{\frac{\gamma_g}{(\gamma_g - 1)(\eta_t)_{pol}}}$$

and

$$T_5^0 = T_4^0\left(1 - \frac{c_{p_a}({}_1\Delta T_3^0)}{c_{p_g}\eta_m T_4^0}\right)$$

Hence, substituting in Eq. 2-15 we obtain

$$(P_g)_{is} = c_{p_g}T_4^0(1 - X)\left[1 - \left(\frac{p_\infty}{p_4^0}\right)^{\frac{\gamma_g - 1}{\gamma_g}} \frac{1}{(1 - X)^{\frac{1}{(\eta_t)_{pol}}}}\right] \tag{2-16}$$

where

$$X = \frac{c_{p_a}({}_1\Delta T_3^0)}{c_{p_g}\eta_m T_4^0} = \frac{{}_4\Delta T_5^0}{T_4^0}$$

Using Eq. 2-11 and rearranging the terms yields the results:

$$p_4^0 = \frac{p_4^0}{p_3^0}p_2^0\left(1 + \frac{{}_1\Delta T_3^0}{T_1^0}\right)^{\frac{\gamma_a}{\gamma_a - 1}(\eta_c)_{pol}}$$

which, when substituted in Eq. 2-16 above, gives

$$(P_g)_{is} = c_{p_g}T_4^0(1 - X)\left[1 - \frac{\left(\frac{p_\infty}{p_2^0}\frac{p_3^0}{p_4^0}\right)^{\frac{\gamma_g - 1}{\gamma_g}}}{\left(1 + \frac{{}_1\Delta T_3^0}{T_1^0}\right)^{\frac{\gamma_a}{\gamma_g}\frac{\gamma_g - 1}{\gamma_a - 1}(\eta_c)_{pol}}(1 - X)^{\frac{1}{(\eta_t)_{pol}}}}\right] \tag{2-17}$$

It will be noticed that the pressure ratios in the numerator of the quotient inside the brackets represent, respectively, the inverse of the ram pressure ratio and the inverse of the pressure ratio over the combustion chamber.

If the energy $(P_g)_{is}$ is utilized with a polytropic efficiency of $(\eta_t)_{pol}$, then the expression becomes

$$P_g = c_{p_g}T_4^0\left[(1 - X) - \frac{\left(\dfrac{p_\infty}{p_2^0}\dfrac{p_3^0}{p_4^0}\right)^{\frac{\gamma_g - 1}{\gamma_g}(\eta_t)_{pol}}}{\left(1 + \dfrac{{}_1\Delta T_3^0}{T_1^0}\right)^{\frac{\gamma_a}{\gamma_g}\frac{\gamma_g - 1}{\gamma_a - 1}(\eta_c)_{pol}(\eta_t)_{pol}}}\right] \tag{2-18}$$

If the change of specific heat of the working fluid through the cycle is neglected and the combustion chamber pressure loss and the mechanical losses are reduced to zero, then Eq. 2-18 simplifies to

$$P_g = c_pT_4^0\left[1 - \frac{{}_1\Delta T_3^0}{T_4^0} - \frac{\left(\dfrac{p_\infty}{p_2^0}\right)^{\frac{\gamma - 1}{\gamma}(\eta_t)_{pol}}}{\left(1 + \dfrac{{}_1\Delta T_3^0}{T_1^0}\right)^{(\eta_c)_{pol}(\eta_t)_{pol}}}\right] \tag{2-19}$$

A similar analysis may be employed to derive the expressions for a cycle in which the component efficiencies are defined as over-all isentropic values, thus

$$p_3^0 = p_2^0\left(1 + \frac{(\eta_c)_{is}({}_1\Delta T_3^0)}{T_1^0}\right)^{\frac{\gamma_a}{\gamma_a - 1}} \text{ for the compression}$$

and

$$p_5^0 = p_4^0\left(1 - \frac{{}_4\Delta T_5^0}{(\eta_t)_{is}T_4^0}\right)^{\frac{\gamma_g}{\gamma_g - 1}} \qquad \text{for expansion in}$$

the compressor turbine.

With component efficiencies defined in this manner we obtain

$$(P_g)_{is} = c_{p_g}T_4^0(1 - X)\left[1 - \frac{\left(\dfrac{p_\infty}{p_2^0}\dfrac{p_3^0}{p_4^0}\right)^{\frac{\gamma_g - 1}{\gamma_g}}}{\left(1 + \dfrac{(\eta_c)_{is}({}_1\Delta T_3^0)}{T_1^0}\right)^{\frac{\gamma_a}{\gamma_g}\frac{\gamma_g - 1}{\gamma_a - 1}}\left(1 - \dfrac{X}{(\eta_t)_{is}}\right)}\right] \tag{2-20}$$

which, if utilized with an isentropic efficiency $(\eta_p)_{is}$ gives energy $P_g = (\eta_p)_{is}(P_g)_{is}$. If, however, the whole of the expansion from pressure $p_4^0$ to $p_\infty$ is carried out with an isentropic efficiency of $(\eta_e)_{is}$, then we have

$$P_g = (\eta_e)_{is}c_{p_g}T_4^0\left[1 - \left(\frac{p_\infty}{p_4^0}\right)^{\frac{\gamma_g - 1}{\gamma_g}}\right] - c_{p_g}T_4^0X$$

$$= (\eta_e)_{is}c_{p_g}T_4^0\left[1 - \frac{\left(\dfrac{p_\infty}{p_2^0}\dfrac{p_3^0}{p_4^0}\right)^{\frac{\gamma_g - 1}{\gamma_g}}}{\left(1 + \dfrac{(\eta_c)_{is}\,{}_1\Delta T_3^0}{T_1^0}\right)^{\frac{\gamma_a}{\gamma_g}\frac{\gamma_g - 1}{\gamma_a - 1}}}\right] - c_{p_g}T_4^0X \tag{2-21}$$

Again, by neglecting the change of specific heat of the working fluid through the cycle and reducing combustion chamber pressure losses and mechanical losses to zero, we can simplify the above expression to

$$P_g = (\eta_e)_{is} c_p T_4^0 \left[ 1 - \frac{\left(\frac{p_\infty}{p_2^0}\right)^{\frac{\gamma-1}{\gamma}}}{1 + \frac{(\eta_c)_{is\,1}\Delta T_3^0}{T_1^0}} \right] - c_p({}_1\Delta T_3^0) \qquad \text{(2-22)}$$

which, under zero forward speed conditions becomes

$$(P_g)_{V_\infty = 0} = c_p \frac{(\eta_e)_{is}(\eta_c)_{is} T_4^0 ({}_1\Delta T_3^0)}{T_1^0 + (\eta_c)_{is}({}_1\Delta T_3^0)} - c_p({}_1\Delta T_3^0) \qquad \text{(2-23)}$$

The derivation of the simpler expressions, as in Eq. 2-19, 2-22, and 2-23, is convenient for observing the qualitative influence, on the available energy, of altering the principal quantities that may be said to define the cycle, and of observing certain properties that in an approximate fashion hold for the more complex expressions.

Thus from Eq. 2-19 and 2-23 we may observe that, other things being equal, the available energy $P_g$ per unit air mass flow is a function more of the product of the component efficiencies than of the component efficiencies taken individually. The equations show also that increasing the maximum cycle temperature $T_4^0$ or decreasing the compressor inlet temperature $T_1^0$ (other quantities being held constant) will bring about an increase in the available energy $P_g$. The effect brought about by decreasing the compressor inlet temperature is of great significance in aircraft turbine engine performance since it means that, with other conditions held constant, the engine is able to deliver more output per unit air mass flow with increasing altitude, the latter giving rise to decreasing compressor inlet temperature.

The energy $(P_g)_{is}$ made available for useful output, if isentropic expansion could be achieved, is not actually that made available by the engine when possessed of forward speed. In this case, it must be remembered that the energy $(P_g)_{is}$ is derived by carrying the expansion down to the ambient pressure, whereas the engine receives its air at a pressure higher than the ambient value by virtue of the forward speed. Consequently, the amount of energy that the engine itself can make available is:

$$J(P_g)_{is} - \frac{V_\infty^2}{2g}$$

where $V_\infty^2/2g$ represents the kinetic energy of the incoming air relative to the engine which is traveling at a forward speed $V_\infty$.

In order to complete our analysis of the thermodynamics of the self-driving gas generator, we have to derive expressions for its thermal efficiency. This is simply the ratio of the energy that the engine has made available for useful output divided by the energy equivalent of the heat input. Returning to Eq. 2-12, we see that this latter, per unit air mass flow is equal to

$$c_{p_b}(T_4^0 - T_3^0)$$

and as previously defined, this is equal to

$$\eta_b f Q_H$$

where $Q_H$ is the net calorific value of the fuel supplies at an over-all fuel-air ratio of $f$ and burned at a combustion efficiency of $\eta_b$. Net calorific value of the fuel is defined as the heating value with all vapor products.

On the above definition, the thermal efficiency $\eta_{th}$ of the gas generator becomes:

$$\eta_{th} = \frac{\eta_b\left((P_g)_{is} - \frac{V_\infty^2}{2gJ}\right)}{c_{p_b}(T_4^0 - T_3^0)} = \frac{(P_g)_{is} - \frac{V_\infty^2}{2gJ}}{fQ_H} \tag{2-24}$$

It will be noticed that the thermal efficiency is defined with respect to $(P_g)_{is}$, i.e. the maximum possible energy that could be extracted from the gas delivered by the gas generator. This is convenient, since all losses connected with the use of the gas can be debited to the propulsive device and appear in an over-all propulsive efficiency.

An examination of Eq. 2-12 shows that, with other quantities held constant, an increase in the maximum cycle temperature $T_4^0$ or a decrease in the compressor inlet temperature $T_1^0$ (and hence compressor outlet temperature $T_3^0$), while both conferring an increase in the energy $(P_g)_{is}$, also demand a greater heat input. The efficiency equation, Eq. 2-24, however, does not reveal readily the effect that either of these changes has on the thermal efficiency and it is necessary to proceed to numerical examples to demonstrate the changes that occur in the thermal efficiency when the quantities defining the cycle are altered.

In making these cycle performance calculations, it is first necessary to establish data relating to the specific heat of the working fluid, in this case, air in the compressor and vitiated air in the combustion chamber and turbine. It will suffice here to state the well-known fact that, in common with most gases, the specific heat of air increases with temperature, with values such as those shown in Fig. K,2d. For more complete data, the reader is referred to [*1,2*]. With the specific heat data, we may now proceed to calculate the cycle performance by substituting values in the equations for the energy output and thermal efficiency given above. To show the variation of these two quantities with all the independent variables would demand unlimited space. Here it is proposed to assume values for the component efficiencies, typical of good practice, and to show how the energy output and thermal efficiency of a self-driving gas generator alter with the maximum cycle temperature $T_4^0$, the compressor temperature rise ${}_1\Delta T_3^0$, the forward speed $V_\infty$, and the altitude, the latter representing the compressor inlet temperature $T_1^0$. The curves shown in

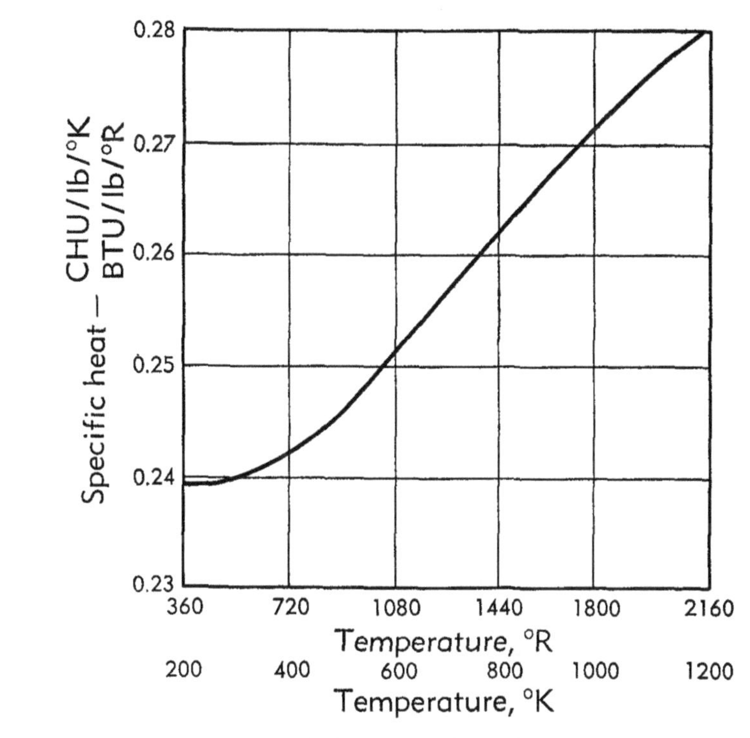


Fig. K,2d. Specific heat of air.

Fig. K,2e and K,2f are based on the following values of the component efficiencies:

| | |
|---|---|
| Compressor polytropic efficiency | = 87 per cent |
| Intake pressure recovery as in Fig. K,2a | |
| Isentropic efficiency of compressor turbine | = 87 per cent |
| Combustion efficiency | = 98 per cent |
| Mechanical efficiency | = 99 per cent |
| Fuel, kerosene; net calorific value | = 10,300 CHU/lb |

Combustion system pressure loss is taken as 5 per cent of the compressor delivery pressure $p_3^0$

In Fig. K,2e the energy per unit air mass flow, $J(P_g)_{is} - (V_\infty^2/2g)$ as defined above and termed hereafter the potential specific output, is plotted as horsepower per lb per sec of air mass flow, against the com-

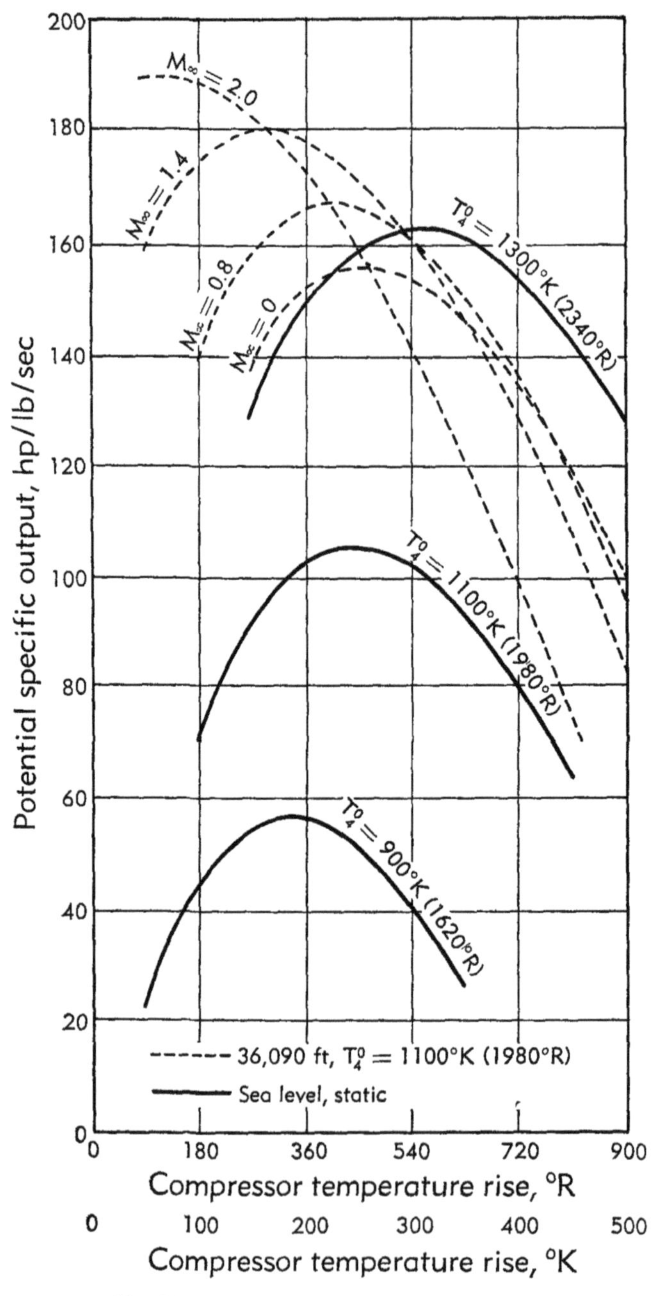


Fig. K,2e. Specific output of gas generator.

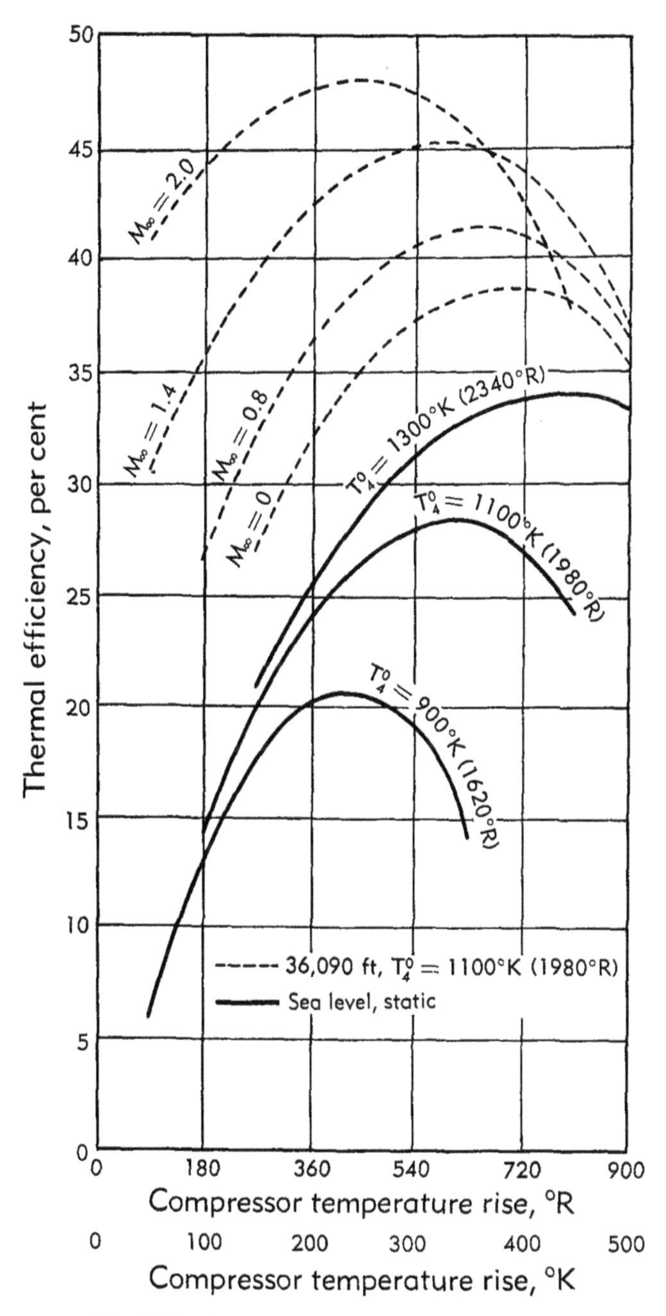


Fig. K,2f. Thermal efficiency of gas generator.

pressor temperature rise. Similarly, Fig. K,2f shows the thermal efficiency of the gas generator as defined in Eq. 2-24 plotted against the same quantity. Both Fig. K,2e and K,2f have curves plotted for the following conditions:

1. Maximum cycle temperatures of 900°K (1620°R), 1100°K (1980°R), and 1300°K (2340°R) at sea level static conditions, i.e. $p_1^0 = 14.7$ lb/in.²abs, $T_1^0 = 288$°K (518.4°R).
2. Maximum cycle temperature of 1100°K at an altitude of 36090 ft. $p_\infty = 3.283$ lb/in.² abs, $T_\infty = 216.5$°K (389.7°R) for flight Mach numbers of 0, 0.8, 1.4, and 2.0.

As shown in Eq. 2-11, a given value of the compressor temperature rise will correspond to a definite compressor pressure ratio depending on the compressor inlet temperature $T_1^0$.

Comparing the corresponding relevant curves in Fig. K,2e and K,2f, the effect deduced from the formulas of the potential specific output increasing with the increase of maximum cycle temperature or the decrease of compressor inlet temperature (other quantities being held constant) is at once shown. It will be noticed that the thermal efficiency is similarly affected but to a lesser extent. Over a small range of compressor temperature rise, a change of forward speed has little or no effect on the potential specific output. Below this range, the potential specific output increases with the forward speed, the converse occurring for the higher values of the compressor temperature rise. Over the greater part of the range of compressor temperature rise considered, the thermal efficiency, defined as above, increases with forward speed.

An important property exhibited by the curves is that optimum values occur for the compressor temperature rise or pressure ratio, which, for given ambient conditions, forward speed, and maximum cycle temperature will yield maximum values of the potential specific output and thermal efficiency. In this connection, the curves show that, in all cases, the optimum compressor temperature rise for maximum thermal efficiency is higher than that yielding maximum potential specific output for the same forward speed and temperature conditions. Thus, under sea level static conditions with a maximum cycle temperature of 1100°K, the optimum compressor temperature rise for maximum efficiency is 350°K as opposed to a value of 250°K yielding a maximum output. A further study of the curves in Fig. K,2e and K,2f shows that the optimum values of compressor temperature rise, and thus pressure ratio, fall with increase of forward speed. Under conditions of constant forward speed and compressor inlet temperature, the optimum values of the pressure ratio increase with maximum cycle temperature. With the latter held constant, then, under constant forward speed conditions, the optimum pressure ratios increase with altitude, i.e. with decreasing compressor inlet temperature.

It is worth emphasizing that all of the properties of these cycle performance curves, described above, refer to conditions in which the component efficiencies are held constant. The effect on output and efficiency of changing the component efficiencies may easily be deduced from the equations given previously. Obviously, an increase in the component efficiencies cannot but help to improve output and efficiency, but a less obvious result is that such an increase will cause the optimum pressure ratios to move to higher values.

So far, we have treated the analysis of the performance of the gas generator cycle in terms of both dimensional and nondimensional quantities. The latter relate to the component efficiencies, whereas the former include such quantities as forward speed, maximum cycle temperature, etc., which we have used, quantitatively, to define the cycle. Such treatment enables the significance of these dimensional quantities to be clearly demonstrated in their effect on absolute performance. It is possible, however, to treat the analysis in a purely nondimensional manner, though, in so doing, certain limitations are introduced which will be made clear later. Referring to Eq. 2-17 and 2-20 we have that the energy $(P_g)_{is}$ can be expressed in the form

$$\frac{(P_g)_{is}}{c_{p_g}T_1^0} \text{ is a function of } \left(\begin{matrix} \dfrac{T_4^0}{T_1^0}, \dfrac{{}_1\Delta T_3^0}{T_1^0}, \dfrac{p_2^0}{p_\infty}, c_{p_a}, c_{p_g} \\ \text{and the component efficiencies} \end{matrix}\right)$$

$p_2^0/p_\infty$ is, however, a function of $V_\infty^2/T_\infty$ which, in turn, represents the forward speed Mach number $M_\infty$. Making use of these relationships, we can therefore derive an expression for the nondimensional potential specific output, as follows:

$$\frac{(P_g)_{is} - \dfrac{V_\infty^2}{2gJ}}{c_{p_g}T_1^0} \text{ is a function of } \left(\begin{matrix} \dfrac{T_4^0}{T_1^0}, \dfrac{{}_1\Delta T_3^0}{T_1^0}, M_\infty, c_{p_a}, c_{p_g} \\ \text{and the component efficiencies} \end{matrix}\right)$$

The thermal efficiency expression given in Eq. 2-24 may be treated similarly, giving the result

$$\eta_{th} \text{ is a function of } \left(\begin{matrix} \dfrac{T_4^0}{T_1^0}, \dfrac{{}_1\Delta T_3^0}{T_1^0}, M_\infty, c_{p_a}, c_{p_b}, c_{p_g} \\ \text{and the component efficiencies} \end{matrix}\right)$$

These relationships show that the nondimensional potential specific output and thermal efficiency of the gas generator are functions of three independent variables, provided that the air and gas specific heats and the component efficiencies are held constant. On this basis, compared with the plotting in Fig. K,2e and K,2f, the variables, in plotting the gas generator performance, could have been reduced by one in number. Such simplification, however, leads to approximation in some degree, since the specific heats $c_{p_a}$, $c_{p_b}$, and $c_{p_g}$ are, as shown in Fig. K,2d, functions of the temperatures prevailing during the cycle, and they cannot be considered as constants, but will in their turn be affected by the cycle temperatures. Apart from this, the relationships given above are valuable in showing that the nondimensional performance can, to a major extent, be defined in terms of the forward speed Mach number, pressure ratio, and the ratio of the maximum cycle temperature to the compressor inlet temperature. On the whole, however, it is considered preferable at this stage to retain the dimensional form of plotting, since with but little loss of generality it facilitates the presentation of the effects of various factors on the basic cycle performance of an aircraft turbine engine. Nondimensional analysis becomes very valuable in dealing with component matching and specific engine performance, as demonstrated in the later articles of this section.

*Output and efficiency of a jet engine.* In order to convert the gas generator unit into a jet engine, it is only necessary to add a duct receiving the hot gas exhausted from the compressor turbine, for the purpose of expanding the gas to the ambient pressure. To do this, the duct must be suitably shaped to form a nozzle, and as shown by the usual thermodynamic treatment, the nozzle cross-sectional area must diminish as expansion proceeds, forming what is known as a convergent nozzle. Provided that the final velocity attained in complete expansion to ambient pressure is less or equal to the local sonic value, then the nozzle or propelling nozzle, as we shall call it, will be convergent in form. However, should the total head pressure at the turbine exhaust be sufficiently high to give a supersonic velocity when expansion of the gas is carried to ambient pressure, then, as it is well known, the propelling nozzle should be of a convergent-divergent form. Under static conditions and for subsonic flight velocities, the available expansion pressure ratio across the nozzle is usually less than 4 in most designs of jet engines. For such expansion pressure ratios, although the jet ultimately attains supersonic velocity, the flow Mach numbers are not greatly in excess of unity. Under these conditions, negligible thrust is lost by omitting the divergent portion of the nozzle and allowing the gas stream, after sonic velocity has been attained in the convergent portion, to expand freely to ambient pressure without guiding walls. Convergent nozzles are thus used in jet engines designed for subsonic flight conditions. In supersonic flight, however, the available expansion pressure ratio across the propelling nozzle can reach high values and it is then of considerable benefit to employ a correctly designed convergent-divergent nozzle.

In proceeding to the thermodynamic analysis of the jet engine, we shall need to add to the diagrammatic arrangement of the gas generator unit shown in Fig. K,2c a length of ducting and a propelling nozzle as in Fig. K,2g to form the diagrammatic arrangement of the jet engine. Reference plane 7 contains the exit from the propelling nozzle, which, in order to simplify the analysis that follows, is assumed to be of convergent or convergent-divergent form as the expansion demands. The static pressure at plane 7 is therefore equal to the ambient pressure. Returning

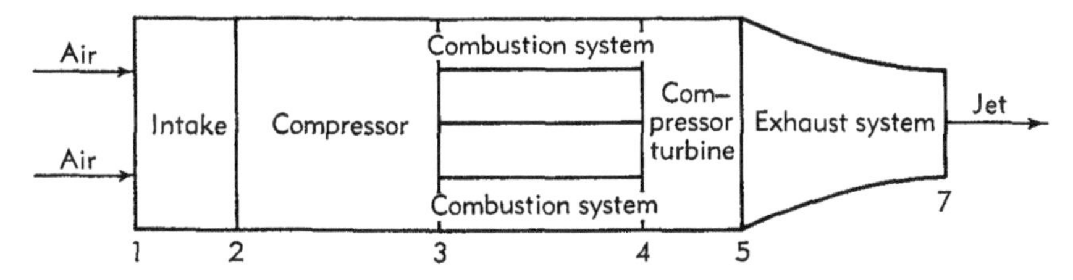


Fig. K,2g. Diagram of jet engine.

to Eq. 2-15, we defined $(P_g)_{is}$ as the energy that could be made available for useful output, if the hot gas exhausted from the compressor turbine were expanded isentropically to ambient pressure. Thus we have

$$(P_g)_{is} = c_{p_g}T_5^0\left[1 - \left(\frac{p_7}{p_5^0}\right)^{\frac{\gamma_g - 1}{\gamma_g}}\right] \qquad \text{where } p_7 = p_\infty$$

The expansion which, compared to the isentropic, will be carried out with some efficiency $\eta_j$ simply serves to produce kinetic energy in the case of the jet engine, and if $V_7$ is the final jet velocity relative to the engine, we have that

$$\frac{V_7^2}{2gJ} = \eta_j(P_g)_{is} \tag{2-25}$$

Thus, for unit mass flow of air through the engine, the thrust $(F_n)_{sp}$ exerted by the air on the engine, in being equal to the rate of change of momentum, may be written as

$$(F_n)_{sp} = \frac{V_7 - V_\infty}{g} = \frac{\sqrt{2gJ\eta_j(P_g)_{is}} - V_\infty}{g} \tag{2-26}$$

Being referred to unit air mass flow, the thrust $(F_n)_{sp}$ is usually termed the specific thrust.

The specific thrust power obtained is given by

$$(F_n)_{sp}V_\infty = \frac{V_\infty(V_7 - V_\infty)}{g} \tag{2-27}$$

Referring now to a previous argument, we saw that the amount of energy that the gas generator by itself can make available in isentropic expansion, or the potential specific output, is defined as

$$J(P_g)_{is} - \frac{V_\infty^2}{2g}$$

Consequently, the propulsive efficiency $(\eta_p)_j$ of the jet, given by the ratio of the specific thrust power to the potential specific output, is given by

$$(\eta_p)_j = \frac{2V_\infty(V_7 - V_\infty)}{2gJ(P_g)_{is} - V_\infty^2} = \frac{2\dfrac{V_\infty}{V_7}\left(1 - \dfrac{V_\infty}{V_7}\right)}{\dfrac{1}{\eta_j} - \left(\dfrac{V_\infty}{V_7}\right)^2} \tag{2-28}$$

or, for a given jet expansion efficiency $\eta_j$, the jet propulsive efficiency $(\eta_p)_j$ is a function of the ratio of the forward speed $V_\infty$ to the relative

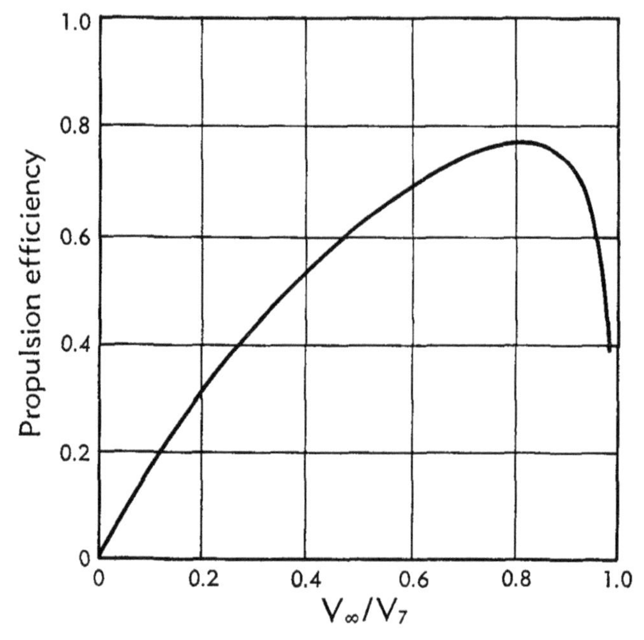


Fig. K,2h. Propulsive efficiency of simple jet.

jet velocity $V_7$. Fig. K,2h shows a curve of propulsive efficiency plotted against the parameter $V_\infty/V_7$ for a jet expansion efficiency of 0.95. Besides showing the obvious result that the propulsive efficiency is zero at zero forward speed ($V_\infty/V_7 = 0$), the curve shows that as the value of the forward speed $V_\infty$ approaches that of the jet velocity $V_7$ the propulsive efficiency almost attains 80 per cent. For this condition, however, the thrust and thrust power tend to zero, a result easily demonstrated by considering Eq. 2-26 and 2-27. When the jet expansion efficiency is made equal to unity, a condition very nearly realized in practice in which values of 95 per cent and over are realized, Eq. 2-28 can be reduced to

$$(\eta_p)_j = \frac{2\frac{V_\infty}{V_7}}{1 + \frac{V_\infty}{V_7}}$$

which gives $(\eta_p)_j = 100$ per cent when $V_\infty/V_7$ equals unity, the condition, however, yielding zero thrust. Thus, for a given forward speed, increasing the jet velocity $V_7$ by raising the potential specific output decreases the jet propulsive efficiency but increases the specific thrust. Now the over-all efficiency $\eta_0$ of the engine must equal the product of the gas generator thermal efficiency and the jet propulsive efficiency, as follows:

$$\eta_0 = \eta_{th}(\eta_p)_j = \frac{(P_g)_{is} - \frac{V_\infty^2}{2gJ}}{fQ_H} \frac{\frac{V_\infty}{g}(V_7 - V_\infty)}{J(P_g)_{is} - \frac{V_\infty^2}{2g}} \tag{2-29}$$

where

$$V_7 = \sqrt{2gJ\eta_j(P_g)_{is}}$$

Fig. K,2e and K,2f show that if the potential specific output is increased, then the thermal efficiency is also raised. However, for a given forward speed, as we have seen, although an increase of the potential specific output raises the specific thrust, it reduces the propulsive efficiency. The over-all efficiency thus becomes the product of two terms, one increasing and the other decreasing with an increase of specific thrust. Hence those effects which raise the thermal efficiency of the gas generator unit may lead to a reduction in the over-all efficiency of the complete jet engine, an important point which will be dealt with more fully when the jet engine cycle performance curves are considered.

Before passing to these cycle performance curves, it is necessary to explain a convention that is adopted for their plotting, in order to present a clearer and more readily usable set of values. We have, from Eq. 2-29 above, that the over-all efficiency $\eta_0$ is given by

$$\eta_0 = \frac{\frac{V_\infty}{g}(V_7 - V_\infty)}{JfQ_H} \sim \frac{V_\infty \times \text{specific thrust}}{\text{fuel-air ratio} \times \text{fuel calorific value}}$$

Since the specific thrust is referred to unit air mass flow, then, for a given calorific value, we have

$$\frac{1}{\eta_0} \sim \frac{\text{fuel mass flow}}{\text{thrust}} \frac{1}{V_\infty}$$

Thus, for a given forward speed, the fuel consumption per unit thrust or thrust specific fuel consumption is proportional to the inverse of the over-all engine efficiency. The specific thrust and the thrust specific fuel consumption form the two dependent performance parameters that are universally used in jet engine performance studies. The thrust specific fuel consumption has the merit that it possesses a finite value under zero forward speed conditions and, as such, forms quite a valuable criterion of performance. For this case, the over-all efficiency becomes zero and thus gives no clue as to the performance to be expected under forward speed conditions.

By means of the foregoing equations, we can use the values of the gas generator specific output and thermal efficiency to give corresponding values of specific thrust and specific fuel consumption for the derived jet engine, plotted against the same independent parameters. Such curves are shown in Fig. K,2i and K,2j where the specific thrust and specific fuel consumptions are plotted against the compressor temperature rise for the same values of maximum cycle temperature, altitude, and forward speed as were used for Fig. K,2e and K,2f, referring to the gas generator. The curves of Fig. K,2i and K,2j are based on a jet expansion efficiency of 95 per cent.

Studying these curves, we may observe that in many respects they possess properties similar to those for the gas generator unit. Thus, as may be expected, optimum values of compressor temperature rise or pressure ratio occur to give maximum specific thrust and minimum specific fuel consumption for given values of the other parameters; the optimum pressure ratio for maximum specific thrust is lower than that yielding minimum specific fuel consumption, and both optima increase with maximum cycle temperature and altitude but decrease with increase of forward speed.

Again, similar to the gas generator performance, the output of a jet engine as represented by specific thrust increases with maximum cycle temperature and altitude, but decreases with increase of forward speed. Here however, the similarity to the gas generator performance ends, and the curves of Fig. K,2i and K,2j illustrate some features of thermodynamic performance peculiar to the jet engine alone. Of these, perhaps the most important is that increasing the maximum cycle temperature does not generally confer an improvement in fuel economy as on the gas generator unit. Fig. K,2j shows that for given values of compressor temperature rise, altitude, and forward speed there is an optimum value of the maximum cycle temperature which yields minimum specific fuel consumption, the fuel economy worsening as the maximum cycle temperature is decreased or increased from the optimum value. The reason for this effect is not hard to find. We saw from Fig. K,2e and K,2f that increasing maximum cycle temperature, although it brings about an increase in thermal efficiency of the gas generator unit, also allows for greater specific output which in the derived jet engine is reflected in an increase in jet velocity. For a given forward speed, Fig. K,2h shows that this means a decreasing propulsive efficiency and while, in increasing the

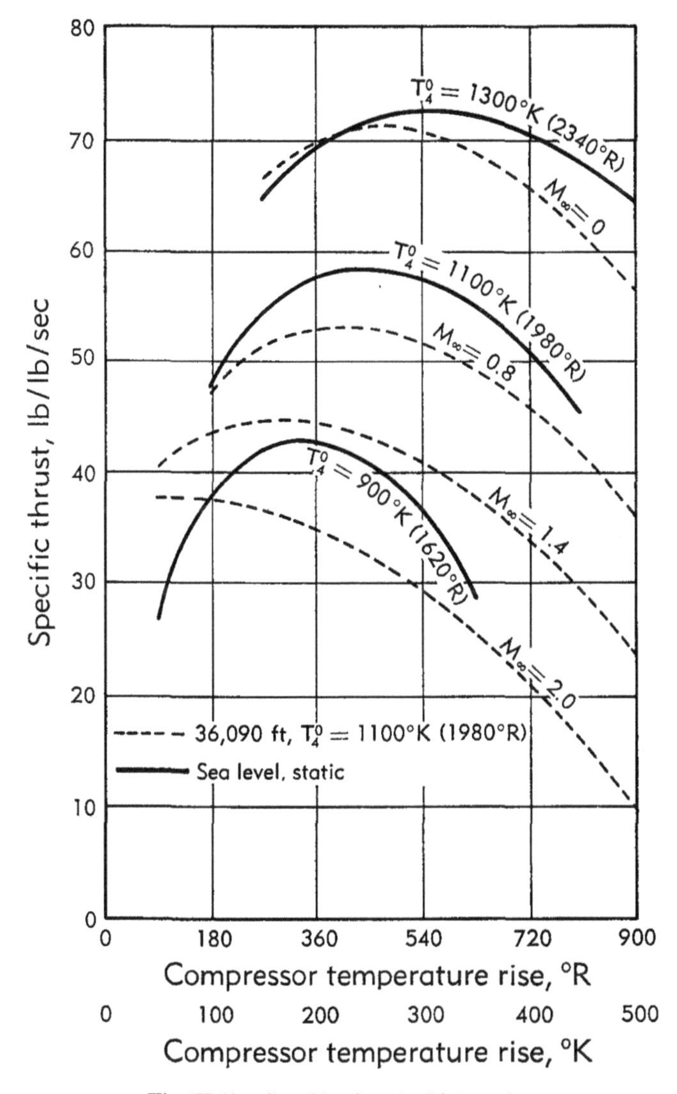


Fig. K,2i. Specific thrust of jet engine.

maximum cycle temperature on a jet engine from low values, the increasing thermal efficiency is able to counteract the decreasing propulsive efficiency, a value is soon reached where the latter effect becomes the stronger and the resulting over-all efficiency begins to decrease, which results in an increasing thrust specific fuel consumption. Comparison of

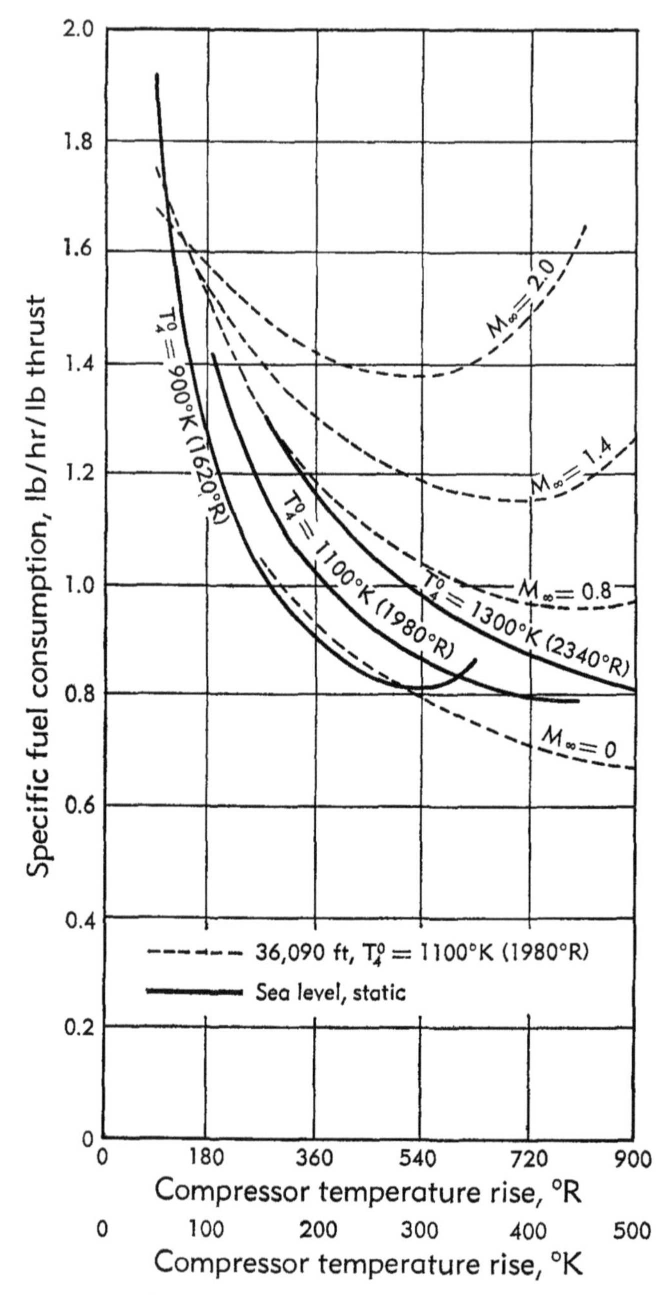


Fig. K,2j. Specific fuel consumption of jet engine.

the relevant curves shows that the optimum maximum cycle temperature for minimum specific fuel consumption increases with the compressor temperature rise and forward speed. One last point that might be mentioned in connection with these jet engine cycle performance curves is that while the specific thrust decreases with increase of forward speed, for given values of the other independent parameters, the specific thrust power, being the product of the specific thrust and forward speed, increases with the latter owing to the raising of the propulsive efficiency.

As in the case of the gas generator unit, the jet engine cycle performance curves can be plotted nondimensionally with a consequent reduction in the number of independent variables, but again it is considered preferable to retain the dimensional form of plotting to illustrate the effect of the various factors affecting the performance.

*Output and efficiency of a propeller engine.* In order to convert the gas generator unit into a propeller engine, we need to add a further turbine for the purpose of driving the propeller, followed by a duct receiving the

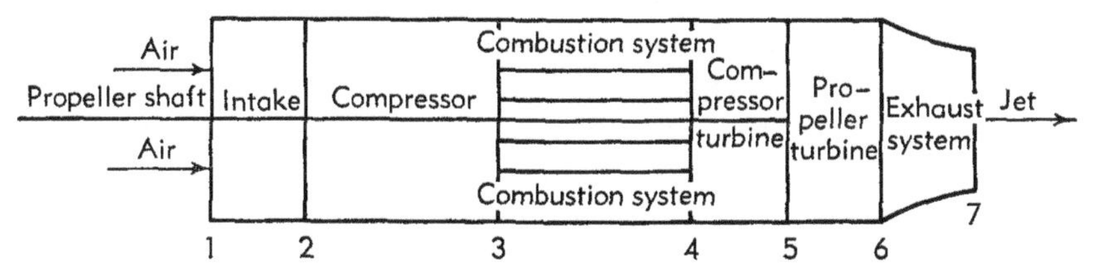


Fig. K,2k. Diagram of propeller engine.

hot gas exhausted from the propeller turbine. The static pressure of the gas emerging from this duct will be equal to the ambient pressure, and the area of the final orifice can be chosen to give any desired residual jet velocity. As mentioned before, the higher the latter quantity, the smaller is the proportion of the potential specific output that can be delivered to the propeller.

The extension of the diagrammatic arrangement of the gas generator unit to give a propeller engine is shown in Fig. K,2k. Reference plane 5, the exit from the compressor turbine, as in the gas generator unit, also becomes the inlet to the propeller turbine for the propeller engine. The exit from this latter component is at reference plane 6, from which the gas flows along the exhaust system to leave the engine at reference plane 7. The compressor and propeller turbines are shown separate here for the sake of clarity in the thermodynamic analysis, but, as mentioned earlier, they could be merged into one turbine in a practical engine or reversed in position from that shown. The thermodynamic performance analysis given below remains the same for all three cases provided that the efficiency of the over-all turbine expansion is unchanged.

Turning now to the thermodynamic performance analysis, we may see that under forward speed conditions the residual jet velocity $V_7$ must be greater than the forward speed $V_\infty$ to avoid a negative residual thrust. If $\eta_{pt}$ is the efficiency of the propeller turbine and $\eta_j$ is the efficiency of the residual jet expansion, then the power delivered to the reduction gear driving the propeller, per unit air mass flow passing through the engine, or the propeller turbine specific output, $P_{sp}$, may be written

$$P_{sp} = \eta_{pt}\left(J(P_g)_{is} - \frac{V_7^2}{\eta_j 2g}\right) \tag{2-30}$$

Also, the residual specific jet thrust $(F_n)_{sp}$ is equal to

$$(F_n)_{sp} = \frac{V_7 - V_\infty}{g}$$

Plainly, if $V_7$ is equal to $V_\infty$ and the efficiencies $\eta_{pt}$ and $\eta_j$ are unity, then the propeller turbine specific output becomes equal to the potential specific output of the gas generator unit. With $V_7^2/2g$ equal to $\eta_j J(P_g)_{is}$, the propeller input is zero and the engine becomes a pure jet engine.

Eq. 2-30 is not quite correct, owing to the fact that with $\eta_{pt} < 1$ there will be a reheating effect in the expansion which increases the available isentropic heat drop $(P_g)_{is}$ beyond the value given in one single expansion. This effect, however, is very small and, for our present purposes, can be neglected for values of $\eta_{pt}$ approaching unity as they would do in a good design.

If $\eta_p$ is the combined propeller and reduction gear efficiency, then the propeller thrust power is given by

$$\eta_p P_{sp} = \eta_p \eta_{pt}\left(J(P_g)_{is} - \frac{V_7^2}{\eta_j 2g}\right)$$

For brevity, we shall refer to $\eta_p$ as the propeller efficiency, with the understanding that the reduction gear losses are included.

The residual jet thrust power is

$$(F_n)_{sp} V_\infty = \frac{V_\infty}{g}(V_7 - V_\infty)$$

Thus the total specific thrust power may be written as

$$\eta_p P_{sp} + (F_n)_{sp} V_\infty = \eta_p \eta_{pt}\left(J(P_g)_{is} - \frac{V_7^2}{\eta_j 2g}\right) + \frac{V_\infty}{g}(V_7 - V_\infty) \tag{2-31}$$

For given values of $V_\infty$, $\eta_p$, $\eta_{pt}$, $\eta_j$, and $(P_g)_{is}$, we may differentiate the above expression with respect to $V_7$ in order to obtain the optimum residual jet velocity yielding the maximum total specific thrust power, and thus determine the optimum division of the gas generator potential output between the propeller turbine and the residual jet. Using the symbol $(P_t)_{sp}$ for total specific thrust power, we have

$$\frac{d}{dV_7}(P_t)_{sp} = -\frac{\eta_p\eta_{pt}}{\eta_j}\frac{V_7}{g} + \frac{V_\infty}{g}$$

$$= 0 \text{ for maximum } (P_t)_{sp}$$

Thus, the condition

$$\frac{V_\infty}{V_7} = \frac{\eta_p\eta_{pt}}{\eta_j} \tag{2-32}$$

is the one yielding maximum total specific thrust power, which, on substitution of Eq. 2-32 in Eq. 2-31 becomes

$$\max\ (P_t)_{sp} = \eta_p\eta_{pt}J(P_g)_{is} - \frac{V_\infty^2}{g}\left(1 - \frac{\eta_j}{2\eta_p\eta_{pt}}\right) \tag{2-33}$$

Examination of the optimum condition given in Eq. 2-32 shows that, with component efficiencies $\eta_{pt}$ and $\eta_j$ equal to or approaching unity, the residual jet velocity $V_7$ should be equal to the forward speed $V_\infty$ divided by the propeller efficiency $\eta_p$, in order to realize maximum thrust power. It will be noticed that the more inefficient the propeller turbine and/or propeller, the greater should be the residual jet velocity $V_7$ for a given forward speed $V_\infty$, i.e. the larger should be the proportion of the gas generator potential output given to creating the residual jet, and the smaller the proportion given to the propeller turbine.

The propulsive efficiency $\eta_{pr}$ of the combined propeller and jet system is given by the ratio of the total specific thrust power to the potential specific output, and under optimum conditions becomes

$$\eta_{pr} = \frac{\eta_p\eta_{pt}J(P_g)_{is} - \dfrac{V_\infty^2}{g}\left(1 - \dfrac{\eta_j}{2\eta_p\eta_{pt}}\right)}{J(P_g)_{is} - \dfrac{V_\infty^2}{2g}} \tag{2-34}$$

Under low forward speed conditions, when $V_\infty^2/2g$ becomes small compared with $J(P_g)_{is}$ the propulsive efficiency approximates, as would be expected, the product of the propeller and propeller turbine efficiencies. Thus, taking a propeller efficiency of 75 per cent and a propeller turbine efficiency of 88 per cent, the propulsive efficiency becomes 66 per cent for low forward speed conditions. Actually, the propeller efficiency cannot be assumed independent of the forward speed $V_\infty$ or rather forward speed Mach number, and in fact the propulsive efficiency $\eta_{pr}$ as given in Eq. 2-34 can only be plotted for particular sets of conditions involving a knowledge of the propeller performance to be expected.

In order to present propeller engine performance curves free of any such assumptions regarding propeller performance, it is customary to give the propeller input, the jet thrust, and the fuel consumption as separate items. It is useful, however, to use a single combined power parameter as a criterion of performance, and such a parameter can be derived which is largely independent of propeller efficiency, though not entirely so. If we divide the total specific thrust power by the propeller efficiency, we are left with the sum of the propeller turbine specific output and the quotient of the specific jet thrust power and the propeller efficiency. The sum of these two quantities is termed the specific equivalent brake power $(P_e)_{sp}$ and may be written

$$(P_e)_{sp} = P_{sp} + \frac{(F_n)_{sp}V_\infty}{\eta_p} = \eta_{pt}\left(J(P_g)_{is} - \frac{V_7^2}{\eta_j 2g}\right) + \frac{V_\infty}{\eta_p g}(V_7 - V_\infty) \quad (2\text{-}35)$$

or under optimum conditions

$$\max\,(P_e)_{sp} = \eta_{pt}J(P_g)_{is} - \frac{V_\infty^2}{\eta_p g}\left(1 - \frac{\eta_j}{2\eta_p\eta_{pt}}\right) \quad (2\text{-}36)$$

Under low forward speed conditions, it will be seen that the maximum specific equivalent brake power approximates the product of the propeller turbine efficiency and the potential specific output. When the forward speed is zero, it becomes impracticable to reduce the residual jet velocity $V_7$ to zero, since the final nozzle area at plane 7 would need to be infinite to pass a finite mass flow of gas. To keep this area within reasonable limits, a residual jet velocity of several hundred feet per second must be used. In order to form an expression similar to Eq. 2-35 above, for the specific equivalent brake power under static conditions, we can use, instead of propeller efficiency, the ratio of the thrust that a propeller delivers under static conditions to the power input to the reduction gear. Denoting this ratio by $Y$, we can, by considering the total thrust, derive that under static conditions:

$$(P_e)_{sp} = P_{sp} + \frac{(F_n)_{sp}}{Y} = \eta_{pt}\left(J(P_g)_{is} - \frac{V_7^2}{\eta_j 2g}\right) + \frac{V_7}{Yg} \quad (2\text{-}37)$$

Thus, for forward speed conditions, the product of the propeller efficiency (including reduction gear loss) and $(P_e)_{sp}$ gives the total thrust power produced by the power plant, whereas under static conditions, the product of $Y$ and $(P_e)_{sp}$ gives the total thrust. Plainly, with reasonably low values of the residual jet velocity $V_7$ in the static case and with optimum values of the ratio $V_\infty/V_7$ in the forward speed case, the specific equivalent brake power, in magnitude, approximates to the product of the potential specific output of the gas generator unit and the efficiency of the turbine driving the propeller.

Making use, therefore, of the relationships given above, we can apply these to the values of the gas generator potential specific output to give corresponding values of specific equivalent brake power for the derived propeller engine. The curves of Fig. K,21 show the specific equivalent

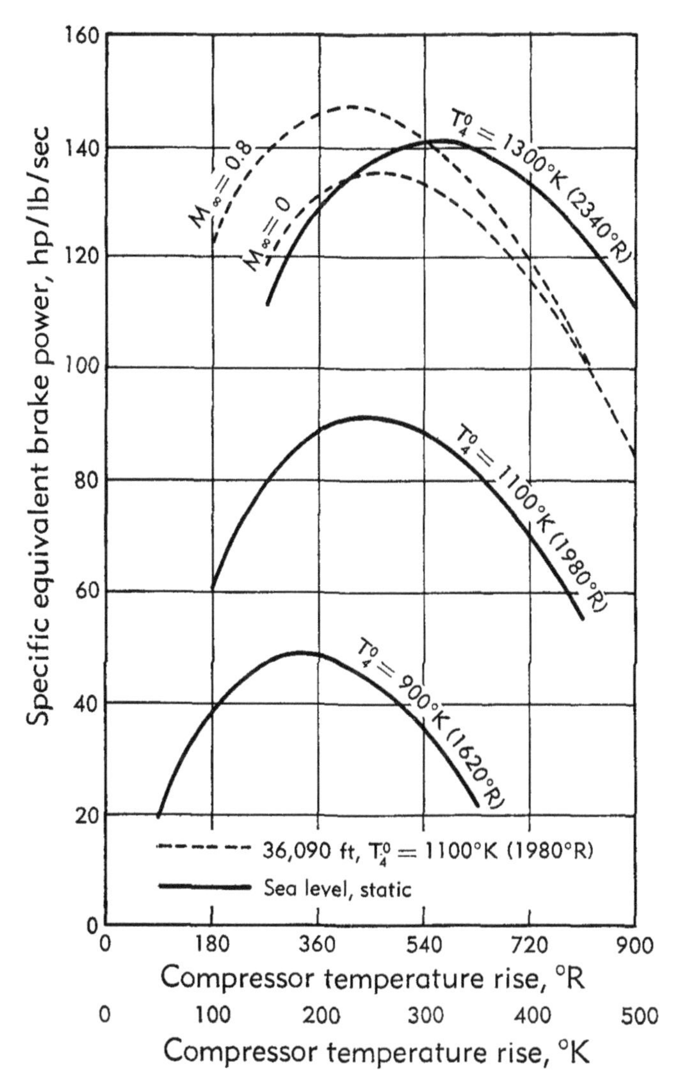


Fig. K,21. Specific power of propeller engine.

brake power plotted against the compressor temperature rise for the same values of maximum cycle temperature, altitude, and forward speed as were used for Fig. K,2e and K,2f, referring to the gas generator unit. In addition to the component efficiencies used to derive the gas generator performance, the curves of Fig. K,21 are based on the following values:

Efficiency of propeller turbine = 87 per cent

For the finite forward speed cases:

Residual jet velocity $= \dfrac{\text{forward speed}}{\text{propeller efficiency}}$, to give approximately optimum conditions.

Propeller efficiency = 78 per cent (including reduction gear loss)

For the static cases:
Residual jet velocity = 500 ft/sec
Propeller thrust = 2.5 lb/horsepower input to the reduction gear.

The curves of the specific equivalent brake power, thus obtained, being approximately proportional to those given for the gas generator potential specific output, as explained above, will naturally exhibit the same properties, for which the reader is referred to the discussion on the curves shown in Fig. K,2e.

So far as the thermal efficiency of the propeller engine is concerned, this may conveniently be referred to the specific equivalent brake power, and becomes equal to:

$$\text{gas generator thermal efficiency} \times \frac{(P_e)_{sp}}{\text{potential specific output}}$$

which is equal to

$$\frac{(P_e)_{sp}}{fQ_H}$$

Following from the above, the curves of the propeller engine thermal efficiency would thus be similar in form and exhibit similar properties to the curves of the gas generator thermal efficiency. It is, however, usual and more convenient for engine-aircraft performance studies to plot, for a given fuel calorific value, the inverse of the propeller engine thermal efficiency in the form of fuel mass flow per unit equivalent brake power which can be termed the brake power specific fuel consumption and is equal to the fuel-air ratio divided by $(P_e)_{sp}$. Curves of the brake power specific fuel consumption are plotted in Fig. K,2m and correspond to the curves of Fig. K,2l and K,2f from which they are derived, as explained above.

It is not the purpose of this volume to compare the performance of jet and propeller-type turbine engines, since to give a worthwhile comparison it is necessary to make engine-aircraft design and performance studies. It may not be out of place, however, to mention here that for low values of forward speed, say less than 400 mph, the propulsive efficiency of a propeller can be very high, of the order of 85 per cent. Fig. K,2h shows that, for a jet engine to approach this propulsive efficiency, its jet velocity in magnitude cannot be very different from the forward speed and therefore its equivalent gas generator potential specific output and thermal efficiency will be very low, giving an over-all efficiency well under that of the propeller engine. With higher forward speeds, the propeller propulsive efficiency begins to fall off and a point is reached where the jet engine with a good thermal efficiency is able to offer a higher propulsive efficiency than the propeller; hence its undoubted superiority for high speed flight.

*Output and efficiency of a bypass or ducted fan engine.* As described earlier, in a bypass or ducted fan engine a certain proportion of the energy made available by the gas generator is taken out through an additional turbine for the purpose of compressing extra air to form an air jet which supplements the hot gas jet of the engine. In energy content, the latter

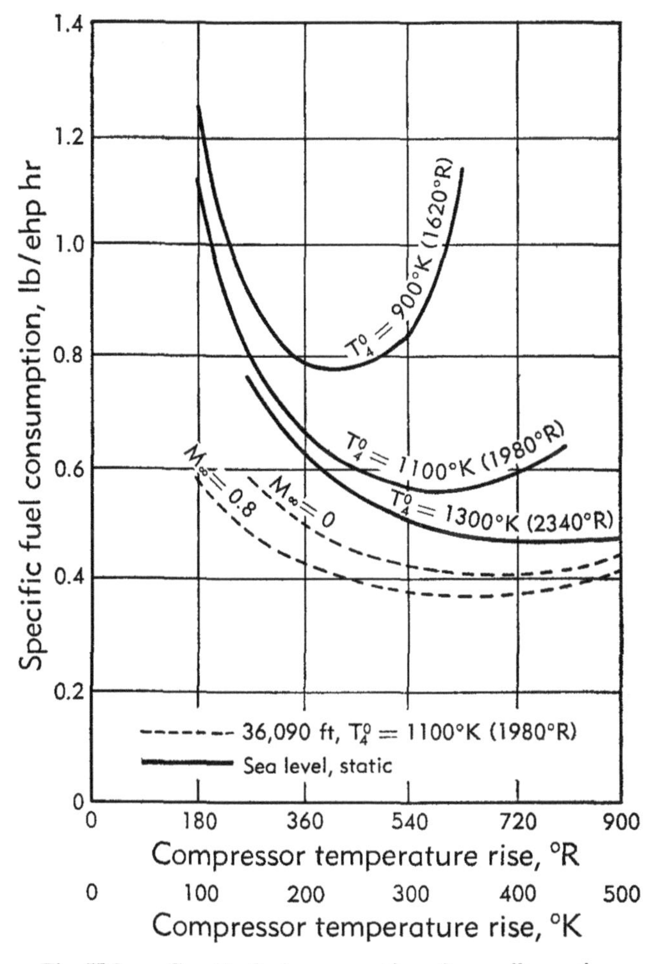


Fig. K,2m. Specific fuel consumption of propeller engine.

has less than that of the pure jet engine, since some of the gas generator potential specific output is utilized in the turbine driving the compressor element which pressurizes the additional or bypass air. Within certain limits, this form of engine has a higher propulsive efficiency than that of the corresponding pure jet engine, but obviously at the expense of simplicity and compactness. The analysis that follows shows these limits by evolving expressions for what is termed "the thrust augmentation ratio," i.e. the thrust of the bypass or ducted fan engine divided by the thrust of a pure jet engine using the same gas generator unit.

Fig. K,2n shows a diagrammatic arrangement of a bypass engine in which planes 5 and 6 denote the entry and exit of the turbine providing power for pressurizing the bypass air. From this turbine, the hot gas flows along the exhaust system to leave the engine at reference plane 7. For the sake of simplicity in the analysis, the means of pressurizing the bypass air is shown as a separate compressor placed around the main compressor drawing in air at plane 9 from the common intake and discharging it after compression at plane 10. This air flows through the annular duct bypassing the main engine components and may be expanded to ambient pressure to form an air jet discharging at plane 11. In some cases, provided that the equality of pressure between the bypass

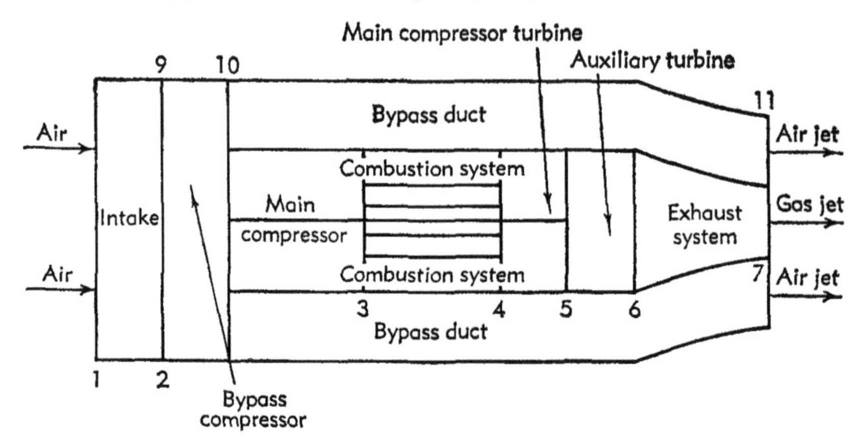


Fig. K,2n. Diagram of bypass or ducted fan engine.

air and the turbine exhaust can be achieved, the two streams may be allowed to mix and expand as a common jet. As mentioned earlier, in practical engines, the bypass air compressor is combined with the early stages of the main compressor and, similarly, the additional turbine is made part of the main turbine, thus simplifying the mechanical design to some extent.

Engines in which the parts are separated and a distinct turbine additional to that of the gas generator is used to drive an entirely separate compressor or fan pressurizing the additional air are usually termed ducted fan engines. As will be seen, however, thermodynamically there is no difference between the two forms of engine.

In the following analysis, we shall refer to the bypass compressor as a distinct element and likewise assume that some turbine stages are used solely for driving the bypass compressor. This enables the analysis to follow rather similar lines to that used for the propeller engine insofar as a proportion of the energy $(P_g)_{18}$ made available by the gas generator is used for driving the bypass compressor. For the sake of brevity, the turbine stages driving the latter will be referred to as the auxiliary turbine. Proceeding to the analysis, we have, for unit air mass flow passing through the gas generator:

$$\text{auxiliary turbine specific output} = \eta_{ft}\left(J(P_g)_{is} - \frac{V_7^2}{\eta_{gj}2g}\right) \tag{2-38}$$

where $(P_g)_{is}$ is the specific energy made available in isentropic expansion from the gas generator turbine outlet to the ambient pressure, $\eta_{gj}$ is the expansion efficiency of the residual hot gas jet, and $\eta_{ft}$ is the isentropic efficiency of the auxiliary turbine.

Similar to Eq. 2-30 for the propeller engine, Eq. 2-38 is not quite correct owing to the reheating effect of the frictional losses in the auxiliary turbine. As before, the small reheating effect will be neglected.

Denoting auxiliary turbine specific output/$J(P_g)_{is}$ by $\lambda$, which we may term the work fraction, Eq. 2-38 becomes

$$\lambda = \eta_{ft}\left(1 - \frac{V_7^2}{\eta_{gj}2gJ(P_g)_{is}}\right)$$

or

$$\frac{V_7}{\sqrt{2gJ(P_g)_{is}}} = \sqrt{\eta_{gj}\left(1 - \frac{\lambda}{\eta_{ft}}\right)} \tag{2-39}$$

The bypass compressor can be designed to pass an air mass flow of any amount. Let the bypass compressor air mass flow be $\mu$ times that of the gas generator; $\mu$ is usually termed the bypass ratio. Since the auxiliary turbine output must equal the bypass compressor input, we have

$$\lambda(P_g)_{is} = \mu c_{pa}({}_9\Delta T_{10}^0) \tag{2-40}$$

where

$${}_9\Delta T_{10}^0 = T_{10}^0 - T_9^0 \text{ or } T_{10}^0 - T_1^0$$

The pressure ratio over the bypass compressor is given by

$$\frac{p_{10}^0}{p_9^0} = \frac{p_{10}^0}{p_2^0} = \left(1 + \frac{\eta_{df}({}_9\Delta T_{10}^0)}{T_1^0}\right)^{\frac{\gamma_a}{\gamma_a-1}} \tag{2-41}$$

where $\eta_{df}$ is the isentropic efficiency of the bypass compressor. In flowing along the duct, prior to expansion in the nozzle at plane 11, the bypass air suffers a pressure loss reducing its pressure to $p_{11}^0$. Expanding through the nozzle, the jet velocity $V_{11}$ of the bypass air is given by

$$\begin{aligned} V_{11} &= \sqrt{2gJc_{pa}T_{10}^0\left[1 - \left(\frac{p_\infty}{p_{11}^0}\right)^{\frac{\gamma_a-1}{\gamma_a}}\right]} \\ &= \sqrt{2gJc_{pa}T_{10}^0\left[1 - \left(\frac{p_\infty}{p_2^0}\frac{p_2^0}{p_{10}^0}\frac{p_{10}^0}{p_{11}^0}\right)^{\frac{\gamma_a-1}{\gamma_a}}\right]} \end{aligned} \tag{2-42}$$

For unit air mass flow passing through the gas generator, the total of the air jet and hot gas jet thrusts (denoted by $(F_{tot})_{sp}$ is equal to

$$(F_{tot})_{sp} = \frac{V_7 + \mu V_{11} - (1+\mu)V_\infty}{g}$$

where $V_\infty$ is the forward speed. Substituting the relationships of Eq. 2-39 and 2-42 in this equation, we have, after some reduction

$$\frac{g(F_{tot})_{sp}}{\sqrt{2gJ(P_g)_{is}}} = \sqrt{\eta_{gj}\left(1 - \frac{\lambda}{\eta_{ft}}\right)} + \mu\sqrt{\frac{c_{p_a}T^0_{10}}{(P_g)_{is}}\left[1 - \left(\frac{p_\infty}{p^0_2}\frac{p^0_2}{p^0_{10}}\frac{p^0_{10}}{p^0_{11}}\right)^{\frac{\gamma_a - 1}{\gamma_a}}\right]} - (1+\mu)\frac{V_\infty}{\sqrt{2gJ(P_g)_{is}}} \tag{2-43}$$

For a pure jet engine, using the same gas generator unit, the specific thrust $(F_{sj})_{sp}$, according to Eq. 2-26, is given by

$$g(F_{sj})_{sp} = \sqrt{2gJ\eta_{sj}(P_g)_{is}} - V_\infty \tag{2-44}$$

where $\eta_{sj}$ is the expansion efficiency of the jet in the pure jet engine. Compared with the latter, the hot gas jet of the bypass engine will have a much smaller velocity and, in this case, the above expansion efficiency $\eta_{gj}$ mainly accounts for the losses in the exhaust system between the turbines and the propelling nozzle.

Since the specific thrusts of both the bypass engine and the corresponding pure jet engine are referred to the same gas generator air mass flow, the thrust augmentation ratio $k$ of the bypass engine is thus: $k = \frac{(F_{tot})_{sp}}{(F_{sj})_{sp}}$, which from Eq. 2-43 and 2-44 becomes

$$k = \frac{\sqrt{\eta_{gj}\left(1 - \frac{\lambda}{\eta_{ft}}\right)} + \mu\sqrt{\frac{c_{p_a}T^0_{10}}{(P_g)_{is}}\left[1 - \left(\frac{p_\infty}{p^0_2}\frac{p^0_{10}}{p^0_{11}}\right)^{\frac{\gamma_a - 1}{\gamma_a}}\left(\frac{p^0_2}{p^0_{10}}\right)^{\frac{\gamma_a - 1}{\gamma_a}}\right]} - (1+\mu)\sigma}{\sqrt{\eta_{sj}} - \sigma} \tag{2-45}$$

where

$$\text{the velocity ratio } \sigma = \frac{V_\infty}{\sqrt{2gJ(P_g)_{is}}}$$

$$\frac{c_{p_a}T^0_{10}}{(P_g)_{is}} = \frac{\lambda}{\mu} + \frac{c_{p_a}T^0_1}{(P_g)_{is}} \qquad \text{from Eq. 2-40}$$

and

$$\left(\frac{p^0_{10}}{p^0_2}\right)^{\frac{\gamma_a - 1}{\gamma_a}} = 1 + \eta_{df}\frac{\lambda}{\mu}\cdot\frac{(P_g)_{is}}{c_{p_a}T^0_1}$$

The quantity $c_{p_a}T^0_1/(P_g)_{is}$ is a function of the performance of the gas generator under given flight conditions which, for a given intake, also determines the pressure ratio $p_\infty/p_2^0$. The pressure loss in the bypass duct is represented by the pressure ratio $p_{10}^0/p_{11}^0$ and, for a well-designed component, this should be close to unity.

For a given gas generator and component efficiencies, the thrust augmentation ratio $k$ is mainly dependent on the work fraction $\lambda$, the

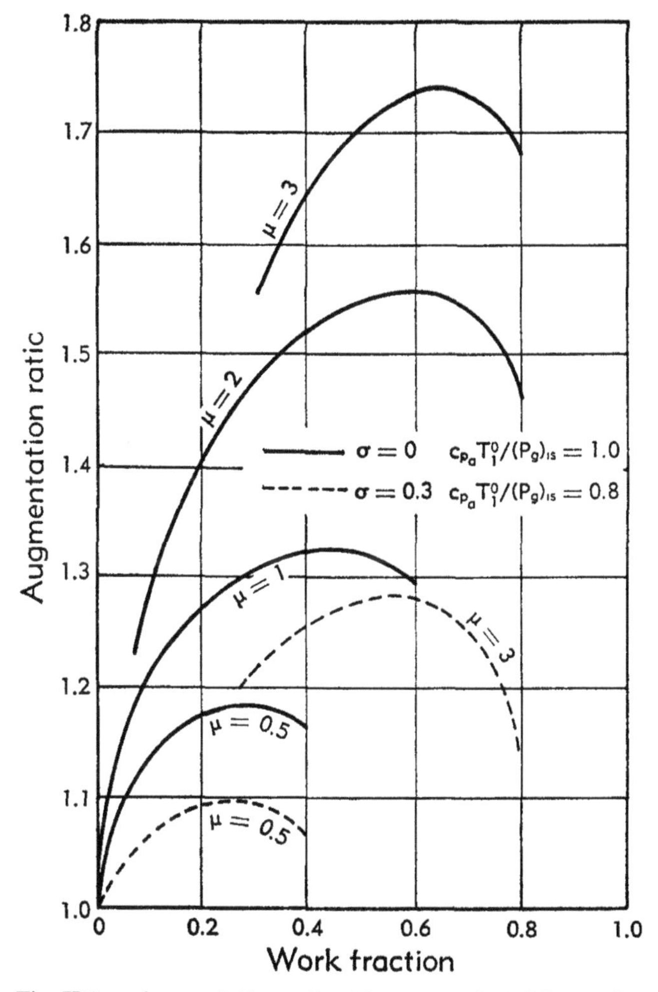


Fig. K,2o. Augmentation ratio of bypass or ducted fan engine.

bypass ratio $\mu$, and the velocity ratio $\sigma$. In Fig. K,2o, $k$ is plotted against $\lambda$ for various values of $\mu$ and for given values of the other quantities as follows:

$$\eta_{ft} = \eta_{df} = 87 \text{ per cent} \qquad \eta_{gj} = \eta_{sj} = 95 \text{ per cent} \qquad p_{11}^0/p_{10}^0 = 0.95$$

$$\sigma = 0 \text{ for } c_{p_a}T_1^0/(P_g)_{is} = 1.0 \text{ and } p_2^0/p_\infty = 0.95$$

$$\sigma = 0.3 \text{ for } c_{p_a}T_1^0/(P_g)_{is} = 0.8 \text{ and } p_2^0/p_\infty = 1.5$$

Fig. K,2o shows that for given values of the other parameters, there is an optimum value of the work fraction $\lambda$ for yielding maximum thrust augmentation. It may also be seen that the optimum value of $\lambda$ increases with the bypass ratio $\mu$, as might be expected. In the following Fig. K,2p, the maximum augmentation ratio derived as a function of $\lambda$ is plotted against the bypass ratio $\mu$ for values of the velocity ratio $\sigma$ ranging from 0 to 0.6. The curves here show that, depending on the value of $\sigma$, there is an optimum value of $\mu$ for yielding the highest augmentation ratio. Further, this optimum value of the bypass ratio decreases with increase

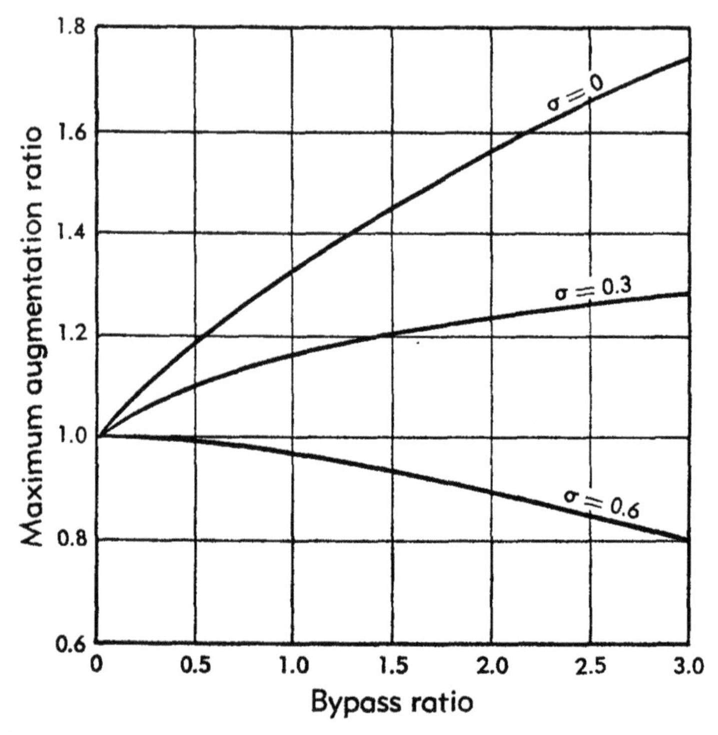


Fig. K,2p. Maximum augmentation ratio of bypass or ducted fan engine.

of the velocity ratio. Plainly for low values of the forward speed, the optimum bypass ratio is very high. Such optimum values would, however, be impracticable for use in any reasonable design, since they would demand fans and ducts of excessive diameter to pass the necessary bypass air flow and, in fact, would compare in size with an orthodox propeller.

One important fact shown by the curves of Fig. K,2p is that the thrust augmentation ratio diminishes with increase of the velocity ratio $\sigma$. Above a certain value of the latter, the augmentation falls below zero for a finite bypass ratio. In these cases, therefore, a ducted fan or bypass system fitted to a gas generator unit would yield less thrust than a simple pure jet system and would also show a corresponding disadvantage in over-all efficiency, since the fuel consumptions of both types of engine, being derived from the same gas generator unit, are the same.

We can use the curves of the potential specific output $J(P_g)_{is} - V_\infty^2/2g$ of the gas generator unit given in Fig. K,2e to plot corresponding values of the velocity ratio $\sigma = V_\infty/\sqrt{2gJ(P_g)_{is}}$ against compressor temperature rise. This is done in Fig. K,2q for the range of flight Mach numbers used in Fig. K,2e at an altitude of 36,090 feet and a maximum cycle temperature of 1100°K (1980°R). The combining of Fig. K,2p and K,2q

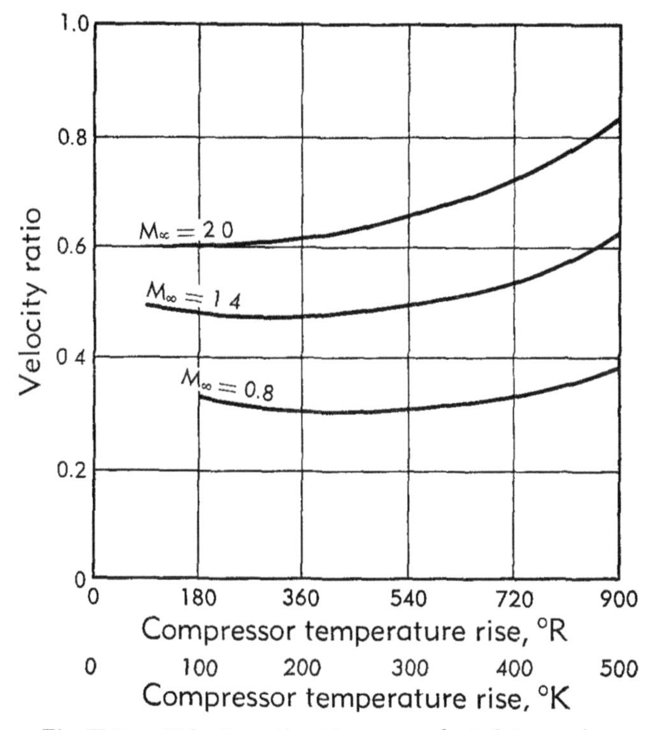


Fig. K,2q. Velocity ratio of bypass or ducted fan engine.

will give the thrust augmentation ratio to be expected under any particular set of conditions covered by the curves. This ratio applied to the pure jet engine performance curves of Fig. K,2i and K,2j gives the specific thrust and thrust specific fuel consumption of a ducted fan or bypass engine of given component efficiencies, any desired bypass ratio, and employing an optimum division of the available expansion energy between the compressed bypass air and the residual hot gas jet. It is apparent from Fig. K,2p and K,2q that, for the design quantities used, it would not pay to use a ducted fan or bypass engine in preference to a pure jet engine for flight Mach numbers much in excess of unity, since the augmentation ratio would approach and fall below unity. For lower forward speeds, where the augmentation ratio is above unity, the ducted fan engine can therefore offer an advantage over the pure jet engine so far as over-all efficiency is concerned, but now has to face a competitor in the shape of the propeller engine. In the lower forward speed range, the propeller can offer a much greater propulsive efficiency than a ducted fan of practicable dimensions. Therefore we can see that for the lower and higher forward speed ranges, the propeller and pure jet engines, respectively, offer the most efficient form of turbine engine power plant. In a comparatively narrow band between the lower and higher ranges of forward speed, the ducted fan engine can show some advantage in over-all efficiency. It needs, however, engine-aircraft performance studies to define zones of forward speed where one form of engine shows superiority over the other two. The choice, here, must necessarily introduce engine design factors, such as weight and frontal area, besides aircraft performance and operating characteristics.

Before leaving the ducted fan or bypass engine, as mentioned earlier, the bypass air and turbine exhaust can be mixed to expand as a common jet. In some cases, it can be shown theoretically that, provided mixing is complete, such a single jet can give a higher propulsive efficiency than the dual jet system assumed in the above analysis. As yet, however, it has not proved possible to realize the theoretical gains over the dual jet system, since although the bypass air and hot gas streams can be expanded through a single nozzle, mixing of the streams is usually incomplete or accompanied by pressure losses greater than predicted by the theoretical treatment.

*Methods of boosting output.* Returning once more to the gas generator unit, it is easy to see that if heat is added to the main turbine exhaust by combustion of further fuel in the hot gas stream, the energy available for useful output can thereby be increased. Thus, from Eq. 2-15, the potential specific output may be written

$$(P_g)_{is} - \frac{V_\infty^2}{2gJ} = c_{p_g} T_5^0 \left[ 1 - \left( \frac{p_\infty}{p_5^0} \right)^{\frac{\gamma_g - 1}{\gamma_g}} \right] - \frac{V_\infty^2}{2gJ}$$

The burning of fuel in the compressor turbine exhaust raises the total head temperature $T_5^0$ before subsequent expansion and thus increases the energy to be derived from that expansion.

The application to a pure jet engine follows quite simply and needs no further explanation except to remark that the burning of the extra fuel will be accompanied by a pressure loss, which must be taken into account with the normal jet expansion losses in calculating the final boosted jet velocity and thus the thrust. Such extra heating of the gas stream in a pure jet engine is termed jet pipe reheating or afterburning.

Plainly the output of a propeller engine could be boosted by introducing reheating of the gas stream between the outlet of the compressor turbine and the inlet of the propeller turbine. The equation given previously can be used directly to deal with this case, again remembering to introduce the pressure loss occasioned by the reheating of the gas stream.

In the case of the ducted fan engine, additional fuel for thrust boosting can not only be used to reheat the gas emerging from the ducted fan turbine, but may also be used to heat the air delivered by the ducted fan before its final expansion. Thus both the air and hot gas jet velocities and therefore the two jet thrusts, may be increased by this means, which in this case can provide a high degree of thrust boost.

Except at very high flight speeds, $M_\infty > 2$, this method of thrust or power boosting can result in a loss of thermal and over-all efficiency. Considering the gas generator unit, the reheating of the gas stream after the work of compression has been supplied can be shown to have an adverse effect on its thermal efficiency. Thus the over-all efficiency of the derived propeller engine will be similarly affected by using this form of power boosting. With the derived pure jet engine, however, the increased jet velocity resulting from reheating lowers the jet propulsive efficiency (see Fig. K,2h) and thus still further adversely affects the engine over-all efficiency. The same is true of the ducted fan engine except that, in this case, the additional fuel being injected after the ducted fan and its turbine adds heat to the cycle at a pressure well below the maximum cycle pressure, thus adding to the effects causing a lowering of thermal efficiency.

A further method of thrust or power boosting is by injection of a liquid into the air stream prior to compression. During compression, the air is cooled by losing heat to the liquid, ultimately vaporizing some or all of the liquid injected. So far as the air is concerned, the compression will therefore move from an adiabatic towards an isothermal process, and for a given work of compression yields a higher pressure ratio compared with the uncooled case. This, in turn, makes more energy available in the expansion process, thus providing the thrust or power boost as the case may be. In order to assess the price paid for boosting the output, the coolant flow must be added to that of the fuel in order to give the total liquid flow. Expressed per unit of thrust or power delivered, the resulting specific liquid consumption is always higher than that given by the unboosted engine.

*Methods of reducing the specific total liquid consumption.* Apart from placing the over-all pressure ratio and the maximum cycle temperature at their optimum values and improving component efficiencies, there is nothing practicable that can be done in a pure jet engine to reduce the specific liquid consumption when employing fuel of a given calorific value. In fact, in this form of engine and in the ducted fan engine, except in extreme and rather impracticable cases, the basic cycle cannot be modified to bring about any reduction in specific total liquid consumption, although it is possible, by employing cooled compression, to reduce the specific fuel consumption.

It is in the propeller engine where economies can be made in specific total liquid consumption by modifying the basic cycle and this involves the addition of a heat exchanger. In this case, the hot gas exhausting from the engine is constrained to flow through passages in a duct, giving up heat (otherwise lost) to the compressor delivery air, which, before passing to the combustion chamber, is constrained to flow through passages adjacent to those carrying the hot gas. The exhaust heat, so transferred, preheats the compressor delivery air, thus reducing the amount of fuel required to raise this air to the given maximum cycle temperature. These additional processes, however, incur pressure losses in the heat-exchanging ducts and cause reduction of the specific output. Nevertheless, provided that the heat transferred from the exhaust gas is of sufficient magnitude, the thermal efficiency of a propeller engine can be increased by this means, but naturally at the expense of lightness and simplicity.

**K,3. Nondimensional Representation of Compressor and Turbine Performance.** The study of the performance of actual gas turbine power plants benefits greatly from the advantage that the characteristic behavior of its main components can be depicted in a relatively simplified form by the use of dimensionless parameters involving the different variables which influence the component's behavior.

Thus, for a given compressor design and geometrical shape it can be said that the mass flow of air $m_a$ which can be compressed in unit time and the pressure $p_e$ to which it can be compressed will depend upon the following variable factors:

1. The pressure of the inlet air $p_i$
2. The temperature of the inlet air $T_i$
3. The rotational speed $n$
4. The power input $P$
5. The linear scale $l$

So

$$m_a = f_1(p_i, T_i, n, P, l) \tag{3-1}$$

and

$$p_e = f_2(p_i, T_i, n, P, l) \tag{3-2}$$

but, by the use of Riabouchinski's theorem, the five variables may be reduced to two dimensionless parameters defining a third such quantity and the equations become

$$\frac{m_a \sqrt{T_i}}{l^2 p_i} = f_3\left(\frac{nl}{\sqrt{T_i}}, \frac{P}{l^2 p_i \sqrt{T_i}}\right) \tag{3-3}$$

and

$$\frac{p_e}{p_i} = f_4\left(\frac{nl}{\sqrt{T_i}}, \frac{P}{l^2 p_i \sqrt{T_i}}\right) \tag{3-4}$$

The complete performance of a particular compressor wherein linear scale is no longer a variable may therefore be represented graphically by a diagram showing the variation of compressor pressure ratio, $p_e/p_i$, with the mass flow parameter, $m_a \sqrt{T_i}/p_i$, for a series of values of $n/\sqrt{T_i}$ and $P/p_i \sqrt{T_i}$, the speed and power dimensionless parameters, respectively. In this way the performance of a compressor under a variety of inlet pressure and temperature conditions can be condensed into a single diagram. Alternatively any two of the dimensionless parameters may be expressed as functions of the other two. Thus $P/p_i \sqrt{T_i}$ and $p_e/p_i$ are clearly functions of $n/\sqrt{T_i}$ and $m_a \sqrt{T_i}/p_i$. If the viscosity of the air must also be taken into account, then a further dimensionless quantity, in this case the Reynolds number, is required to define completely the performance of the machine. In the main, however, the effects of Reynolds number changes are generally not great and variations of this parameter are omitted from these generalized performance studies.

It is apparent that an exactly comparable argument to that given above may be used to demonstrate that the performance of a turbine may be shown graphically by a chart of the pressure ratio across the turbine, varying with the nondimensional mass flow, speed, and power parameters.

While the number of dimensionless parameters required to express the complete behavior of a compressor or turbine can thus be shown to be four (of which any two determine the remaining two), the actual parameters used need not necessarily be those chosen to illustrate the above discussion. Alternative, though related, parameters may be substituted and indeed it is quite common to use in place of the power parameter either the ratio of air or gas temperature across the component or the isentropic efficiency of the device. The power absorbed by a compressor or produced by a turbine is equivalent to the product of the mass of fluid passing through in unit time and the change in enthalpy per unit mass experienced by the fluid in passing through the component. If the variation in the mean specific heat of the fluid is sufficiently small to be ignored, then the enthalpy change is proportional to the change in the total head temperature and we may write

$$P = k_1 m({}_i\Delta T_e^0) \tag{3-5}$$

where ${}_i\Delta T_e^0$ = the change in the total head temperature of the fluid between the inlet and the exhaust of the component.

In terms of dimensionless parameters and assuming again that the scale factor $l$ is unity, this becomes

$$\frac{P}{p_i\sqrt{T_i}} = k_1\frac{m\sqrt{T_i}}{p_i}\frac{{}_i\Delta T_e^0}{T_i} \tag{3-6}$$

So from Eq. 3-3

$$\frac{m\sqrt{T_i}}{p_i} = f\left(\frac{n}{\sqrt{T_i}}, \frac{m\sqrt{T_i}}{p_i}\frac{{}_i\Delta T_e^0}{T_i}\right)$$

$$= f\left(\frac{n}{\sqrt{T_i}}, \frac{{}_i\Delta T_e^0}{T_i}\right) \tag{3-7}$$

and from Eq. 3-4

$$\frac{p_e}{p_i} = f\left(\frac{n}{\sqrt{T_i}}, \frac{m\sqrt{T_i}}{p_i}\frac{{}_i\Delta T_e^0}{T_i}\right)$$

$$= f\left(\frac{n}{\sqrt{T_i}}, \frac{{}_i\Delta T_e^0}{T_i}\right) \quad \text{from Eq. 3-7} \tag{3-8}$$

Thus in place of the nondimensional expression for power, the temperature ratio ${}_i\Delta T_e^0/T_i$ might be used as one of the variables defining compressor or turbine performance. Further, since

$$T_e = T_i \pm {}_i\Delta T_e^0 \tag{3-9}$$

and hence

$$\frac{T_e}{T_i} = 1 \pm \frac{{}_i\Delta T_e^0}{T_i}$$

it is clear that the temperature ratio $T_e/T_i$, or if necessary ${}_i\Delta T_e^0/T_e$, might equally well be used in place of ${}_i\Delta T_e^0/T_i$. From this it follows that the mass flow parameter, for example, may be made to relate when necessary to exhaust rather than inlet conditions. For

$$\frac{m\sqrt{T_e}}{p_e} = \frac{m\sqrt{T_i}}{p_i}\frac{p_i}{p_e}\sqrt{\frac{T_e}{T_i}} \tag{3-10}$$

and the three terms on the right-hand side of this equation are all uniquely determined by the relationships already established.

Further, since the ratios $p_e/p_i$ and ${}_i\Delta T_e^0/T_i$ are related by the isentropic efficiency of the compressor or turbine, a further modification is possible as a result of which the quantities $p_e/p_i$ and $m\sqrt{T_i}/p_i$ may be expressed as functions of $n/\sqrt{T_i}$ and the isentropic efficiency $\eta$.

Finally the temperatures and pressures used to construct these nondimensional groups need not be the static values, for they can be replaced by the appropriate total head or stagnation quantities, which are often more convenient. This may be seen to follow from a restatement of our original premises in terms of stagnation quantities or may be demonstrated from the foregoing analysis in the following manner.

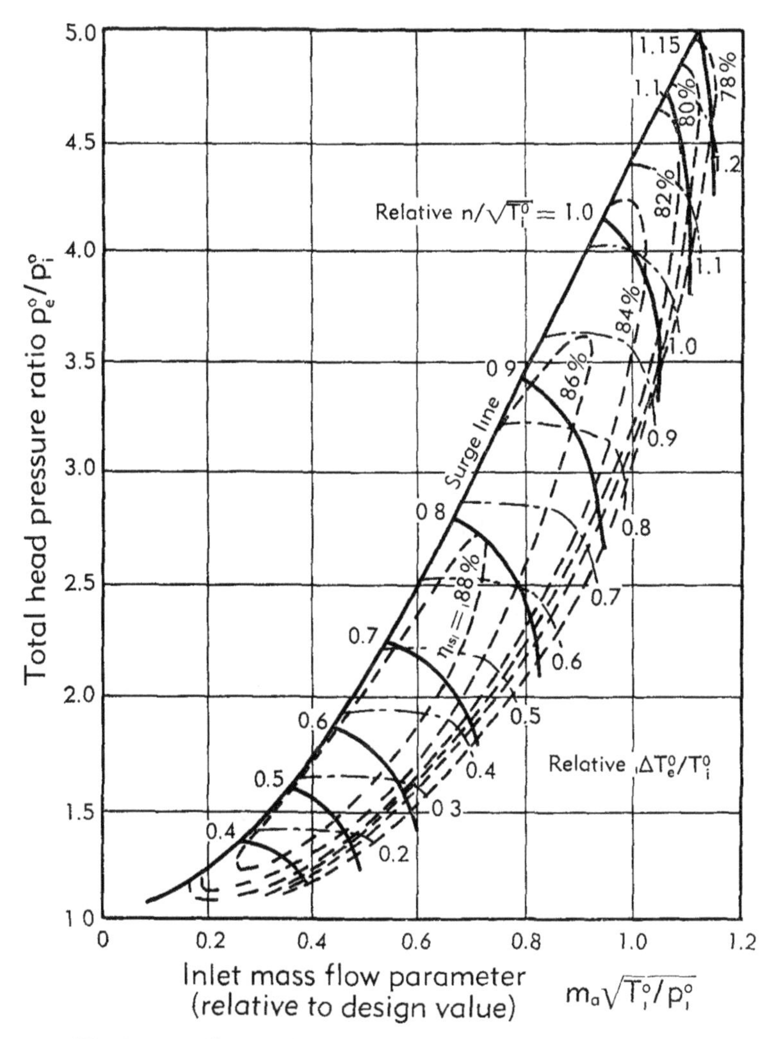


Fig. K,3a. Characteristics of a typical aero-engine compressor.

Considering the inlet plane to the compressor we may write

$$m = A_i \rho_i V_i$$

where $A_i$, the inlet cross-sectional area, is a constant. Hence for constant values of $\mathcal{R}$ and $\gamma$

$$m = \text{const}\,\frac{p_i V_i}{T_i}$$

or

$$\frac{m\sqrt{T_i}}{p_i} = \text{const}\,\frac{V_i}{\sqrt{T_i}} = \text{const}\,M_i$$

Hence, for any value of $m\sqrt{T_i}/p_i$ there is a unique value of $M_i$, the inlet Mach number. But for this value of $M_i$ there are also unique values of $T_i^0/T_i$ and $p_i^0/p_i$ so that it follows that the relationship between $m\sqrt{T_i}/p_i$ and $m\sqrt{T_i^0}/p_i^0$ is also unique.

Performance Characteristics of Compressors and Turbines.

*Compressors.* The typical form of a complete compressor characteristic is shown in Fig. K,3a plotted in the form of the variation of over-all pressure ratio with inlet mass flow parameter $m\sqrt{T_i}/p_i$ for differing values of the speed parameter, $n/\sqrt{T_i}$, with lines for constant values of the temperature ratio, ${}_i\Delta T_e^0/T_i^0$ also plotted. In addition the "contours" corresponding to constant isentropic efficiency are superimposed on the grid of constant temperature ratio lines. For present purposes it is immaterial whether the characteristic shown refers to an axial or a centrifugal compressor since the basic features are similar in each. Briefly these may be summarized as follows.

1. The total head temperature rise in the compressor varies approximately as the square of the rotational speed.
2. There is a stable and an unstable operating region separated by what is commonly called "the surge line." This line tends to be roughly parabolic in shape, its gradient increasing with increasing mass flow. The unstable region, of course, is that in which at a given pressure ratio the mass flow is less than that at the surge line. Much remains to be discovered about the phenomenon of surging. The elementary form of surge is the instability which arises when the pressure ratio of the compressor operating at controlled rotational speed and inlet conditions reaches a maximum and begins to fall as the mass flow is further reduced. Any momentary fall in pressure in a capacity situated after the compressor, instead of being compensated by an increased delivery of air from the compressor as would occur in the stable region, is, under these conditions, exaggerated by a decreased delivery of air so that the divergence from the initial conditions increases rapidly until an entirely new regime of flow is set up. In practice, however, the phenomenon is more complex and its study is complicated by the fact that instability may often be introduced before the true surge point is reached by the aerodynamic stalling of one or more rows of blades in an axial compressor or even of the diffuser cascade in a centrifugal type of compressor.

   From the present viewpoint, where we are concerned with the problems of matching a compressor with a driving turbine, however, it is sufficient to recognize the existence of this barrier to stable operation, represented by the surge line, without delving too deeply into the many factors which influence its exact form and position.

3. The region of maximum efficiency tends to occur in the stable operating zone of the compressor and to lie roughly parallel to the surge line. The position of this maximum efficiency region is, however, a function of the compressor design and in axial compressors is dependent mainly upon the mean value of the stagger angle used in the blading design.
   Low stagger angles and relatively highly cambered blades tend to cause the region of maximum efficiency to occur close to the surge line. In compressors with blades set at high stagger angles the maximum efficiency region is usually displaced down the characteristic, to lower pressure ratios than those associated with surging.
4. In aircraft engine compressors, where in order to achieve lightness the minimum possible number of compression stages is generally used, there is a tendency for the higher pressure ratios to be obtained with a lower efficiency than the optimum for the design.
   A characteristic such as that shown in Fig. K,3a might be obtained theoretically by a study of the design details supplemented by the results of cascade testing and of experience on compressors of comparable design. Alternatively it might be obtained experimentally by testing the compressor as a separate component driven by, say, an electric motor or a slave turbine.

*Turbines.* An example of a turbine characteristic plotted in the same form of pressure ratio against nondimensional mass flow parameter is shown in Fig. K,3b. The most striking feature in this diagram is the way in which the characteristic lines for different values of the nondimensional speed parameter become coincident along a line of constant $m\sqrt{T_i^0}/p_i^0$. This constant value of the inlet mass flow parameter represents the limit imposed by "choking conditions" being reached in the turbine stator blade throats. When sonic velocity conditions are attained in the throats of these passages, the mass flow which can be passed through them is fixed provided that the inlet total head temperature and pressure remain unaltered. Neither alterations in the rotational speed of the rotor blades nor changes in the exhaust pressure of the turbine can influence the gas throughput of the nozzles so long as they remain choked. Similarly if the inlet conditions to the turbine are altered then the critical or choking mass flow of the machine will vary so that the quantity $m\sqrt{T_i^0}/p_i^0$ remains constant. This is to be expected since, if the stator blade throat conditions are considered (using the subscript $t$), then

$$m = A_t \rho_t V_t$$

But $A_t$ is constant and $\rho_t = p_t/\mathcal{R}T_t$; while for choking conditions the velocity $V_t = \sqrt{g\gamma\mathcal{R}T_t}$. Hence $m\sqrt{T_t}/p_t$ is constant, if $\gamma$ is constant. Further, if there is no heat loss, since $V_t$ is sonic the temperature ratio

$T_i^0/T_t$ is equal to $(\gamma + 1)/2$ and is constant if $\gamma$ is constant. Finally, if the isentropic efficiency of the expansion in the stator blade passage does not alter, the pressure ratio $p_i^0/p_t$ will also be constant while choking is occurring in the throats. Hence for most practical purposes the parameter $m\sqrt{T_i^0}/p_i^0$ can be considered to be constant under these conditions.

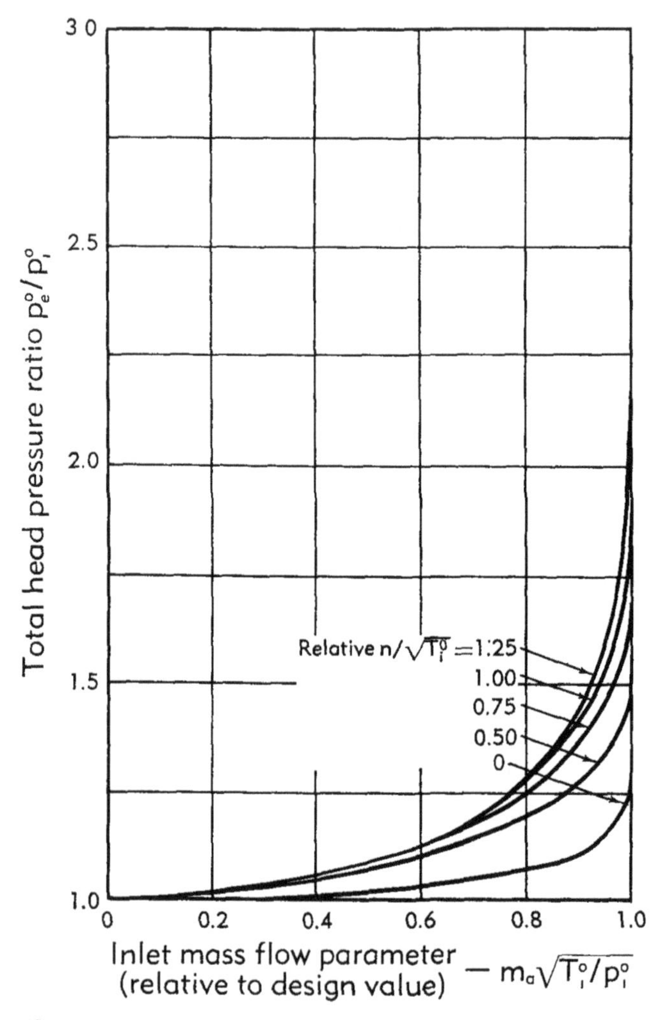


Fig. K,3b. Characteristics of a typical aero-engine single-stage turbine.

At pressure ratios not great enough to cause choking, the mass flow passing through the turbine when operating at fixed inlet conditions increases as the rotational speed decreases.

Lines to show the variation in isentropic efficiency of the turbine are not included in Fig. K,3b since they tend to confuse the diagram. Instead the variation is presented independently in Fig. K,3c where the isentropic efficiency of a single-stage turbine is plotted for different values of over-all pressure ratio against the quantity $U_m^2/2gJc_p({}_i\Delta T_e^0)$ where $U_m$ is the rotor blade speed at mean blade height. It is found that in the range over which the turbine is required to operate in driving a compressor the variation in isentropic efficiency is only of the order of 1 per cent to 2 per cent. No great change in efficiency is to be expected provided that the incidence of the gas on the various rows of blading does not depart by more than a few degrees from the values for which the blading is designed. This condition is satisfied if at all probable operating points the velocity vector triangles for the turbine are geometrically similar

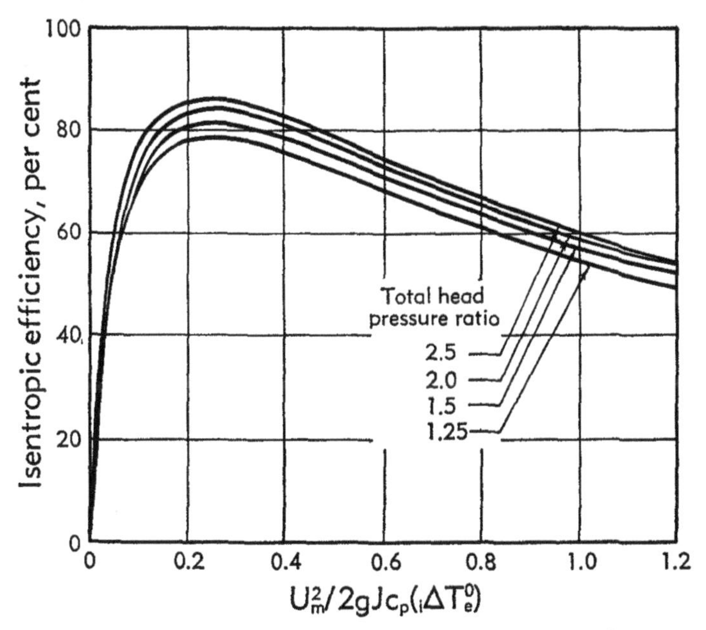


Fig. K,3c. Variation of turbine efficiency with the parameter $U_m^2/2gJc_p({}_i\Delta T_g^0)$.

to the design triangles or nearly so. Considering, for simplicity, a single-stage turbine in which the residual swirl velocity in the exhaust is negligible, which is generally the case when a propulsive nozzle follows the turbine, then the work output of the turbine per unit mass flows will be proportional to the product of the component of the stator blade outlet gas velocity in the direction of rotation of the rotor blade and the velocity of the rotor blade itself. Thus

$$ {}_i\Delta T_e^0 = \text{const } U_m V_{ax} \tan \alpha \tag{3-11}$$

where $V_{ax}$ is the gas axial velocity between the stator and rotor blade rows and $\alpha$ is the gas angle at the stator blade outlet measured from the axial direction. The divergence between the gas angle and the actual blade angle is small over a wide range of conditions so that the gas angle is virtually constant.

Since it has been seen that, to a first approximation, the total head temperature rise in a compressor varies as the square of the rotational speed, then it must follow that in the turbine which drives such a compressor by direct mechanical linkage and by using air at a similar rate, a

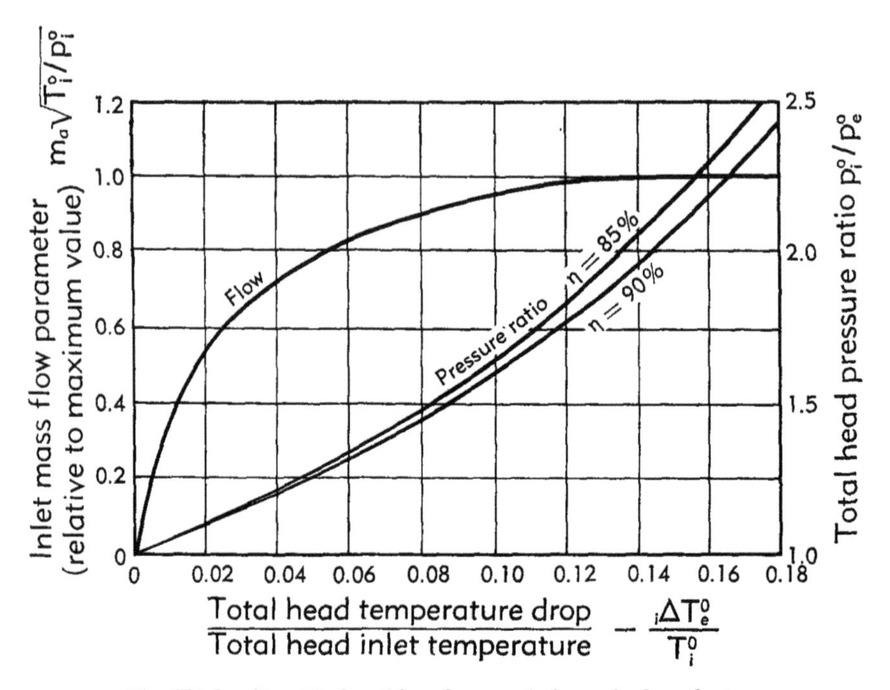


Fig. K,3d. Simplified turbine characteristic used when the turbine drives the compressor only or a fixed-pitch propeller.

comparable relationship must exist. That is, again to the first order of approximation,

$$ {}_i\Delta T_e^0 = \text{const } U_m^2 \tag{3-12}$$

Since $\alpha$ is a constant, the reconciliation of these two requirements results in $U_m$ being proportional to $V_{ax}$. The velocity vector triangle therefore maintains geometrical similarity while the output and throughput of the turbine are altered and in consequence the angle of attack of the gas on the rotor blading changes very little.

Under these conditions the operation of the turbine is found to be restricted to a narrow zone of the characteristic field, corresponding to a constant value of $U_m^2/2gJc_p({}_i\Delta T_e^0)$ and, as Fig. K,3c shows, the efficiency of the turbine varies very little until pressure ratios too low to be used in normal operation in flight are reached. In consequence it is often possible, for a small sacrifice in accuracy, to assume for the characteristic of a turbine which drives only a compressor a single line characteristic of the form shown in Fig. K,3d, combined with a constant value of isentropic or polytropic efficiency.

As the number of stages in the turbine is increased there is a tendency for this single-line approximation to the turbine's operating zone to approach the form that has for many years been taken as characteristic of multistage steam turbines, namely

$$\frac{m\sqrt{T_i}}{p_i} = \sqrt{1 - \left(\frac{p_e}{p_i}\right)^{k_2}}$$

where $k_2$ is a number only slightly different from 2.

## K,4. Determination of Equilibrium Running Conditions by Matching of Component Characteristics.

MATCHING OF COMPRESSOR AND TURBINE IN A SIMPLE SELF-DRIVING SYSTEM. It is now necessary to consider how a compressor and a turbine can be combined into a self-driving system and how the characteristics of the two components, obtained in the forms just described, may be used to estimate the factors necessary for the equilibrium of the system at the various operating points.

We may proceed as follows, using the subscripts 1, 2, 3, 4, and 5 to refer to the same stations in the engine as they do in Art. 2. For any operating point on the compressor characteristic the following quantities are defined:

$$\frac{m_2\sqrt{T_2^0}}{p_2^0}, \quad \frac{n}{\sqrt{T_2^0}}, \quad \frac{p_3^0}{p_2^0}, \quad \text{and} \quad \frac{{}_2\Delta T_3^0}{T_2}$$

By the use of the turbine characteristic we wish to find the value of $T_4^0/T_2^0$, the ratio of the turbine and compressor inlet total head temperatures, which is just adequate for the turbine to drive the compressor at the chosen conditions. Denoting this value by $Z$, then it will be necessary that the quantities given by the three following expressions shall satisfy the requirements of the turbine characteristic.

$$\frac{m_4\sqrt{T_4^0}}{p_4^0} = \left(\frac{m_2\sqrt{T_2^0}}{p_2^0}\right)\left(\frac{m_4}{m_2}\right)\left(\frac{p_2^0}{p_3^0}\right)\left(\frac{p_3^0}{p_4^0}\right)Z^{\frac{1}{2}} \tag{4-1}$$

$$\frac{n}{\sqrt{T_4^0}} = \left(\frac{n}{\sqrt{T_2^0}}\right)Z^{-\frac{1}{2}} \tag{4-2}$$

$$\frac{{}_4\Delta T_5^0}{T_4^0} = \left(\frac{{}_2\Delta T_3^0}{T_2^0}\right)\left(\frac{{}_4\Delta T_5^0}{{}_2\Delta T_3^0}\right)Z^{-1} \tag{4-3}$$

Before a value of $Z$ to satisfy the turbine requirement can be determined from the quantities already defined by the compressor operating point it is necessary to assess the size of the remaining terms,

$$\frac{m_4}{m_2}, \quad \frac{p_3^0}{p_4^0}, \quad \text{and} \quad \frac{{}_4\Delta T_5^0}{{}_2\Delta T_3^0}$$

As explained in Art. 2 the mass flow ratio $m_4/m_2$ may generally be taken as unity, the small mass flow of fuel being neglected or presumed equivalent to the mass flow of cooling air bled off from the main compressed air stream. Under these circumstances the ratio $c_{p_a}\,{}_2\Delta T_3^0/c_{p_g}\,{}_4\Delta T_5^0$ is the measure of the efficiency of the mechanical transmission of power from the turbine blading to the compressor blading. This efficiency is very high and is commonly estimated at 99 per cent. If a constant value is assumed for the ratio of the mean specific heats $c_{p_a}/c_{p_g}$ the ratio ${}_2\Delta T_3^0/{}_4\Delta T_5^0$ can then be defined.

Finally, there is the total head pressure ratio across the combustion system $p_3^0/p_4^0$ to be considered. While it is often adequate to estimate the combustion pressure loss as a small percentage of the entry total head pressure it is sometimes necessary to make a more realistic assessment based on the results of rig tests of the combustion system. In this case the pressure loss is generally expressed as a multiple of the pressure head at some reference plane, such as the station denoted by subscript 3. If at the reference plane the gas velocity is sufficiently low for the difference between static and stagnation temperatures and pressures to be small this relationship may be expressed in the form

$$1 - \frac{p_4^0}{p_3^0} = k_3\left(\frac{m\sqrt{T_3^0}}{p_3^0}\right)^2 \tag{4-4}$$

where $k_3$ is a constant, and $m\sqrt{T_3^0}/p_3^0$ may be determined from the compressor information.

For a very exact determination of combustion chamber pressure loss the factor $k_3$ should contain a term which is proportional to $T_4^0/T_3^0$, i.e. to $Z(1 + {}_2\Delta T_3^0/T_2^0)^{-1}$.

However, it will be seen that whichever expression is adopted for the combustion chamber pressure loss it is now possible to determine $Z$, the value of the temperature ratio, $T_4^0/T_2^0$, which is required for equilibrium operation of the turbocompressor unit at the particular condition initially considered. In this way the combination of the two units into a free-running system may be investigated and the necessary matching of the two component characteristics depicted by superimposing upon the compressor diagram lines for constant values of the temperature ratio $T_4^0/T_2^0$. This is done in Fig. K,4a, whence it will be seen that the new lines tend to fan out over the diagram and that at a particular mass flow the higher values of temperature ratio are associated, as is only to be expected, with the higher values of compressor pressure ratio.

When the turbine is operating with choked stator blade passages so that $m\sqrt{T_4^0}/p_4^0$ is a constant, the condition that $T_4^0/T_2^0$ is also a constant is satisfied by the pressure ratio $p_4^0/p_2^0$ being proportional to $m\sqrt{T_2^0}/p_2^0$. Since the pressure ratio $p_4^0/p_2^0$ differs from the compression ratio $p_3^0/p_2^0$ only by the effect of the small combustion pressure loss term, it is to be

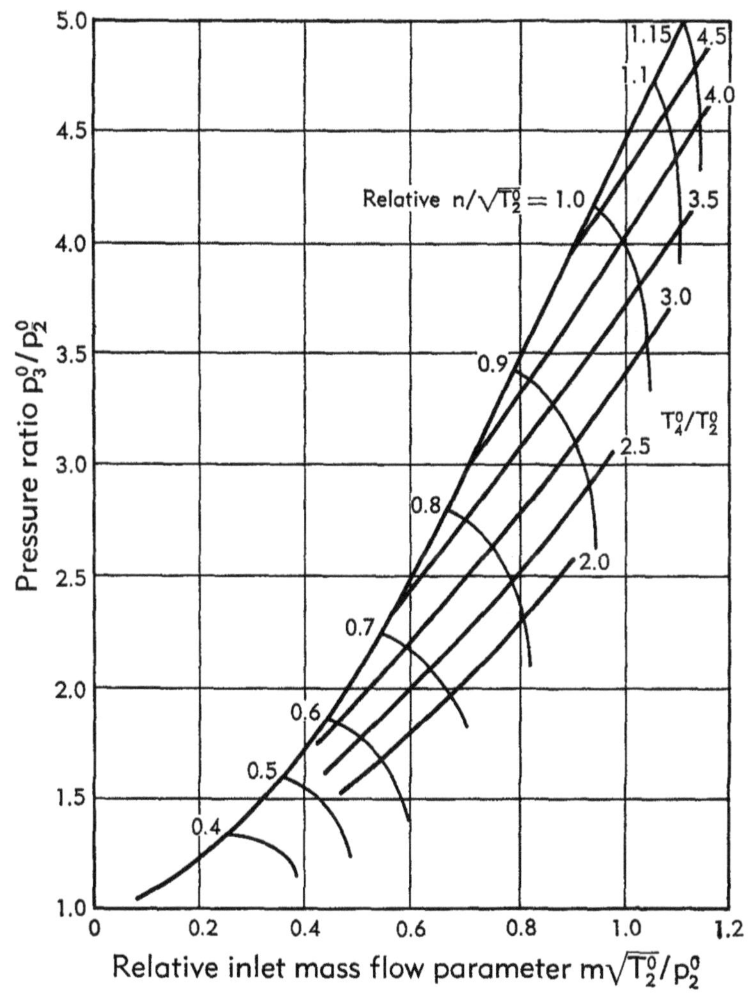


Fig. K,4a. Matching of compressor and turbine characteristics by lines of constant inlet temperature ratio, $T_4^0/T_2^0$. Value of $T_4^0/T_2^0 = 4.0$ assumed at design point.

expected that in the region where the turbine is choked the constant $T_4^0/T_2^0$ lines are virtually straight lines radiating from the origin. In the region where the turbine is not choked the lines curve round as though to pass through zero mass flow at a pressure ratio of unity.

Equilibrium Running of Simple Jet and Propeller Turbine Engines. As a result of the matching process described in the previous paragraph the conditions required to maintain a state of equilibrium between the compressor and its driving turbine can be determined for any operating point on the characteristic. Since $T_4^0/T_2^0$ has been determined and ${}_4\Delta T_5^0/T_2^0$ is also known it is possible, by the use of the turbine efficiency, to express the compressor-turbine outlet conditions of temperature and pressure, $T_5^0$ and $p_5^0$, as fractions of the inlet conditions. So for any compressor characteristic point the quantities $m\sqrt{T_2^0}/p_2^0$, $T_5^0/T_2^0$, $p_5^0/p_2^0$, and hence $m\sqrt{T_5^0}/p_5^0$ are uniquely determined and these quantities are required in order to define the gas horsepower available for use in the propulsive device. The inlet total head conditions $p_2^0$ and $T_2^0$, however, might arise from many combinations of different ambient conditions ($p_\infty$ and $T_\infty$), flight Mach number ($M_\infty$), and intake efficiency $\eta_{a2}$, while the gas horsepower available for the propulsive device is dependent upon the pressure $p_\infty$ to which the gas is free to expand. To determine a value for the nondimensional gas horsepower parameter, $(\text{hp}_g)_{is}/p_\infty\sqrt{T_\infty}$, appropriate to a chosen compressor operating point, it is therefore necessary to specify also a value of both $M_\infty$ and $\eta_{a2}$. From these, values of $T_2^0/T_\infty$ and $p_2^0/p_\infty$ may be obtained, and hence $m\sqrt{T_\infty}/p_\infty$, $T_5^0/T_\infty$, and $p_5^0/p_\infty$ can be estimated. If the potential gas horsepower $(\text{hp}_g)_{is}$ is defined as the power available in the isentropic expansion of the gas from the compressor turbine exhaust condition to atmospheric pressure, minus the free stream kinetic energy of the air entering the engine,[2] then

$$(\text{hp}_g)_{is} = m\left[c_{p_g}J({}_5\Delta T_\infty) - \frac{V_\infty^2}{2g}\right]$$

$$= m\left\{c_{p_g}JT_5^0\left[1 - \left(\frac{p_\infty}{p_5^0}\right)^{\frac{\gamma_g-1}{\gamma_g}}\right] - \frac{V_\infty^2}{2g}\right\} \tag{4-5}$$

$$\therefore \frac{(\text{hp}_g)_{is}}{p_\infty\sqrt{T_\infty}} = J\frac{m\sqrt{T_\infty}}{p_\infty}\left\{c_{p_g}\frac{T_5^0}{T_\infty}\left[1 - \left(\frac{p_\infty}{p_5^0}\right)^{\frac{\gamma_g-1}{\gamma_g}}\right] - \frac{\gamma_a - 1}{2}c_{p_a}M_\infty^2\right\} \tag{4-6}$$

or

$$\frac{(\text{hp}_g)_{is}}{p_2^0\sqrt{T_2^0}} = J\frac{m\sqrt{T_2^0}}{p_2^0}\left\{c_{p_g}\frac{T_5^0}{T_2^0}\left[1 - \left(\frac{p_\infty}{p_5^0}\right)^{\frac{\gamma_g-1}{\gamma_g}}\right] - \frac{\gamma_a - 1}{2}c_{p_a}M_\infty^2\right\}$$

This potential gas horsepower is plotted for the example considered in Fig. K,4b, for each point on the compressor characteristic, using the second version of the horsepower parameter in order to make a more compact diagram. A range of flight Mach number is considered and the

[2] The potential gas horsepower, $(\text{hp}_g)_{is}$, is thus the product of the mass flow and the potential specific output, $(P_g)_{is} - \left(\frac{V_\infty^2}{2g}\right)$, defined in Art. 2.

intake efficiency variation shown in Fig. K,2a is assumed. It will be seen that for a particular $M_\infty$ and $n/\sqrt{T_2^0}$ the maximum value of the potential gas horsepower parameter occurs at the surge line and that the output naturally falls as the pressure ratio and temperature ratio of the cycle decrease, even though there is a small increase in mass flow parameter. It should also be appreciated that although the ordinates in this diagram are the values of $m\sqrt{T_2^0}/p_2^0$ used for the compressor characteristic, the scale can easily be converted, for a particular Mach number, to one of

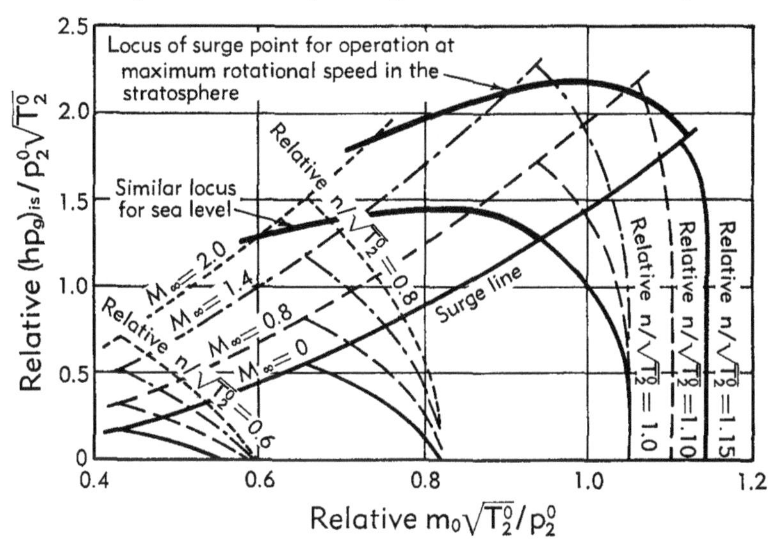


Fig. K,4b. Variation of potential gas horsepower at different compressor operating points for a range of flight Mach number. Intake efficiency variation as in Fig. K,2a. This yields the following conversion factors:

| $M_\infty$ | 0 | 0.8 | 1.4 | 2.0 |
|---|---|---|---|---|
| $p_2^0/p_\infty$ | 0.95 | 1.489 | 3.055 | 6.915 |
| $T_2^0/T_\infty$ | 1.00 | 1.128 | 1.392 | 1.800 |
| $\sqrt{T_2^0/T_\infty}$ | 1.00 | 1.062 | 1.180 | 1.342 |
| $p_2^0\sqrt{T_2^0}/p_\infty\sqrt{T_\infty}$ | 0.95 | 1.582 | 3.605 | 9.295 |
| $p_2^0\sqrt{T_\infty}/p_\infty\sqrt{T_2^0}$ | 0.95 | 1.400 | 2.588 | 5.155 |

$m\sqrt{T_\infty}/p_\infty$ by the use of the values of $T_2^0/T_\infty$ and $p_2^0/p_\infty$ appropriate to the value of $M_\infty$. The speed parameter may likewise be converted from $n/\sqrt{T_2^0}$ to $n/\sqrt{T_\infty}$. The table accompanying the figure gives all the necessary multiplying factors to make these conversions and with the use of this information the diagram enables $(\text{hp}_g)_{is}$, $m$, and $n$ to be evaluated at each operating point once the values of the ambient air pressure and temperature are specified.

We must next consider the application of this potential gas horsepower to the production of propulsive thrust, either in a pure reaction jet, or by a propeller driven by a further turbine, or further turbine stages.

*Simple jet.* If the turbine exhaust gas is permitted to expand down to the atmospheric pressure, $p_\infty$, in a propelling nozzle with an isentropic efficiency of $\eta_n$ then the nozzle velocity $V_7$ is given by

$$\frac{V_7}{\sqrt{T_\infty}} = \sqrt{2gJc_p}\sqrt{\eta_n \frac{T_5^0}{T_\infty}\left[1 - \left(\frac{p_\infty}{p_5^0}\right)^{\frac{\gamma-1}{\gamma}}\right]} \tag{4-7}$$

Hence, using the information already obtained, the gross thrust of the nozzle, $F_g$, may be determined in nondimensional form from the expression

$$\frac{F_g}{p_\infty} = \frac{1}{g}\frac{m\sqrt{T_\infty}}{p_\infty}\sqrt{2gJcp}\sqrt{\eta_n}\sqrt{\frac{T_5^0}{T_\infty}\left[1 - \left(\frac{p_\infty}{p_5^0}\right)^{\frac{\gamma-1}{\gamma}}\right]}$$

$$= \sqrt{\frac{2Jcp\eta_n}{g}}\left(\frac{m\sqrt{T_\infty}}{p_\infty}\right)\left(\frac{T_5^0}{T_\infty}\right)^{\frac{1}{2}}\left[1 - \left(\frac{p_\infty}{p_5^0}\right)^{\frac{\gamma-1}{\gamma}}\right]^{\frac{1}{2}} \tag{4-8}$$

The velocity $V_7$ is the speed of the gas relative to the nozzle, but since the nozzle has a motion through the air, $V_\infty$, corresponding to $M_\infty$, then the net thrust $F_{net}$ of the turbine engine and propulsive jet is less than the gross thrust by the amount $mV_\infty/g$, so that

$$\frac{F_{net}}{p_\infty} = \frac{F_g}{p_\infty} - \frac{m\sqrt{T_\infty}}{p_\infty}M_\infty\sqrt{\frac{\gamma\mathcal{R}}{g}} \tag{4-9}$$

The nozzle outlet area $A_7$ required to make possible the expansion considered above must be such that the equation

$$m = A_7 C_{D_7}\rho_7 V_7 \tag{4-10}$$

is satisfied. Or if the effective area $A_7'$ (equal to $A_7 C_{D_7}$) is considered

$$A_7' = \frac{m}{\rho_7 V_7} = \frac{m\mathcal{R}T_7}{p_\infty V_7} \tag{4-11}$$

But

$$T_7 = T_5^0 - \frac{V_7^2}{2gJc_{p_g}}$$

so that $A_7'$ may be expressed

$$A_7' = \frac{m\sqrt{T_\infty}}{p_\infty}\left(\frac{V_7}{\sqrt{T_\infty}}\right)^{-1}\mathcal{R}\left[\frac{T_5^0}{T_\infty} - \frac{1}{2gJc_{p_g}}\left(\frac{V_7}{\sqrt{T_\infty}}\right)^2\right] \tag{4-12}$$

The position has thus been reached where the size of propulsive nozzle required to make the compressor of a turbojet engine operate at any particular point can be determined, subject to the translational Mach number $M_\infty$ and the intake efficiency $\eta_{a2}$ being known. Furthermore the net thrust developed under these conditions may also be determined in terms of the ambient static pressure.[3]

[3] Strictly, the nondimensional group for thrust should be $F_{net}/l^2p_\infty$; it is the assumption that the scale factor $l$ may be regarded as unity when only one engine is being considered that permits the use of the quasi-nondimensional form $F_{net}/p_\infty$.

From this information it is thus possible to identify compressor operating points for which $A_7$ is constant and the locus of these points is therefore the compressor operating line for a gas turbine engine fitted with this particular size of propelling nozzle.

The above algebraic derivation of the propelling nozzle area required to make the compressor of a simple jet engine operate at a particular condition and of the thrust that will then be developed suffers from certain shortcomings, however.

Firstly it presents the temptation to assume that one particular propelling nozzle, when operating over a range of pressure ratio, will have a constant isentropic efficiency at all conditions. This assumption can lead to anomalous results. For example, at constant inlet conditions the maximum mass flow that can be passed through a nozzle of a given throat area is found on computation to occur at a pressure ratio less than that at which the Mach number of the gas flow through the throat reaches unity. Further, as the assumed efficiency is decreased the pressure ratio at which maximum mass flow through the nozzle occurs is found to become smaller and the throat Mach number at this condition becomes more and more removed from unity. These are features which cannot be substantiated by practical tests, there being no evidence to show that maximum mass flow ever occurs other than when sonic velocity is reached in the throat.

Secondly, the complete expansion of the gas in the nozzle presupposes, in many instances, the use of a convergent-divergent nozzle. The area $A_7$ which is thus calculated is therefore that to which the divergent portion of the nozzle must extend. The orifice controlling the mass flow through the engine is nevertheless the throat area of the nozzle and it is for constant values of this throat area rather than the final area that operating lines should be drawn. The foregoing analysis, therefore, requires modification so that at exhaust pressure ratios great enough to cause choking of the final nozzle, the area $A_7'$ is such that the velocity $V_7$ is equal to the local velocity of sound. $A_7'$ is then the throat area of the propelling nozzle and if the nozzle is terminated at this plane there will then be a pressure discontinuity at the exit from the nozzle since $p_7$ will be greater than $p_\infty$, and the gross thrust will be the sum of a force due to the acceleration of the gas to the exit velocity and a force due to this pressure discontinuity acting over the effective area of the jet. The addition of a divergent portion of propelling nozzle after this throat section affects the gross thrust produced in a manner to be discussed later, but it should have no influence at all on conditions upstream of the sonic throat.

In order to overcome the first of these difficulties it is often convenient, since the matching of the compressor and turbine characteristics is a graphical rather than an analytical process, to adopt for the nozzle also an empirical characteristic in graphical form. This is simply a curve of

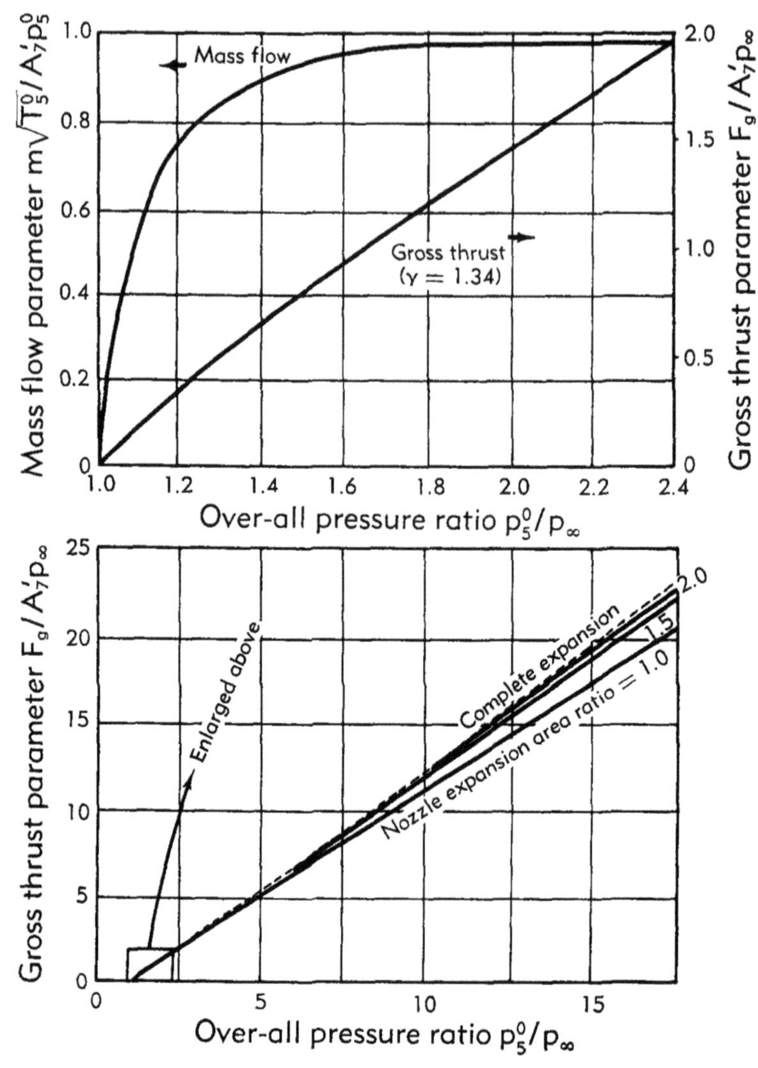


Fig. K,4c. Typical jet-pipe and nozzle characteristics. Note: Mass flow parameter is expressed relative to the value when the expansion is isentropic, e.g. 0.390 for $\gamma = 1.34$.

$m\sqrt{T_5^0}/p_5^0A_7'$ plotted against $p_5^0/p_\infty$ as shown in Fig. K,4c. For the maximum value of $m\sqrt{T_5^0}/p_5^0A_7'$ to occur when $M_7 = 1$ there must be a slight increase in isentropic efficiency with increasing pressure ratio in the unchoked region. Once the nozzle is choked the quantity $m\sqrt{T_5^0}/p_5^0A_7'$ remains constant.

Strictly speaking the curves in this form apply to a series of similar-shaped exhaust pipes and final nozzles for which the area $A_7'$ is a measure of scale, but they can also be used, without appreciable loss of accuracy, as the characteristics of a particular jet pipe to which a series of final nozzles of differing degrees of convergence is fitted.

By the use, then, of this exhaust system characteristic in place of the analytical method described above, it is possible to determine the propelling nozzle effective throat area $A_7'$ required to cause the compressor to operate at any particular condition, the flight Mach number $M_\infty$, and the intake efficiency $\eta_{a2}$ being known. From this information the equilibrium operating conditions of the components of an engine incorporating a fixed-throat-area final nozzle may easily be derived. Fig. K,4d shows the characteristics of the compressor on which the matching with the driving turbine is depicted by the lines of constant $T_4^0/T_2^0$, while the operating lines when the engine is completed by the addition of a fixed-throat-area propelling nozzle are superimposed for a range of flight Mach numbers. It is seen that the general position of these lines is roughly parallel with the surge line, although they tend to have a more rapid change of slope which causes them to approach the surge line at the upper and lower ends of the compressor pressure ratio range.

When the final nozzle is choked the compressor operating lines for different flight Mach numbers become coincident. The lines for the higher Mach numbers converge first, which would be expected since the higher the ram pressure ratio, the smaller are the compressor and turbine pressure ratios needed for a critical pressure ratio across the exhaust system to result. The coincidence of the operating lines is required by the condition that with a propelling nozzle of fixed throat area the parameter $m\sqrt{T_5^0}/p_5^0$ must remain constant when the nozzle is choked. From the matching of the turbine and the compressor it may be deduced that at any value of $n/\sqrt{T_2^0}$ there must be a unique value of $m\sqrt{T_2^0}/p_2^0$ regardless of the value of $M_\infty$, if this condition is to be fulfilled. The appropriate value of $T_4^0/T_2^0$ is also unique. A guide to the variation of $T_4^0/T_2^0$ with $n/\sqrt{T_2^0}$ along the choked nozzle operating line may also be obtained from the approximations to the turbine and compressor performance which have already been suggested. For if a single line characteristic and constant efficiency are assumed for the turbine the constancy of $m\sqrt{T_5^0}/p_5^0$ requires also the constancy of $m\sqrt{T_4^0}/p_4^0$, $p_4^0/p_5^0$, and in particular of ${}_4\Delta T_5^0/T_4^0$. This means that ${}_2\Delta T_3^0/T_4^0$ must also be approximately constant, but if the assumption that ${}_2\Delta T_3^0$ is roughly proportional to $n^2$ can be made we find that, to this order of approximation, $T_4^0$ must also vary as $n^2$, and hence $T_4^0/T_2^0$ as $(n/\sqrt{T_2^0})^2$.

At any particular altitude, for which we may assume a constant value of $T_\infty$, increasing Mach number $M_\infty$ causes an increase in $T_2^0$ and a decrease, therefore, in the value of $n/\sqrt{T_2^0}$ corresponding to operation at constant rotational speed $n$. Therefore in these circumstances there will be a movement of the compressor operating point down the equilibrium running lines. Conversely, if there is a tendency for the operating lines

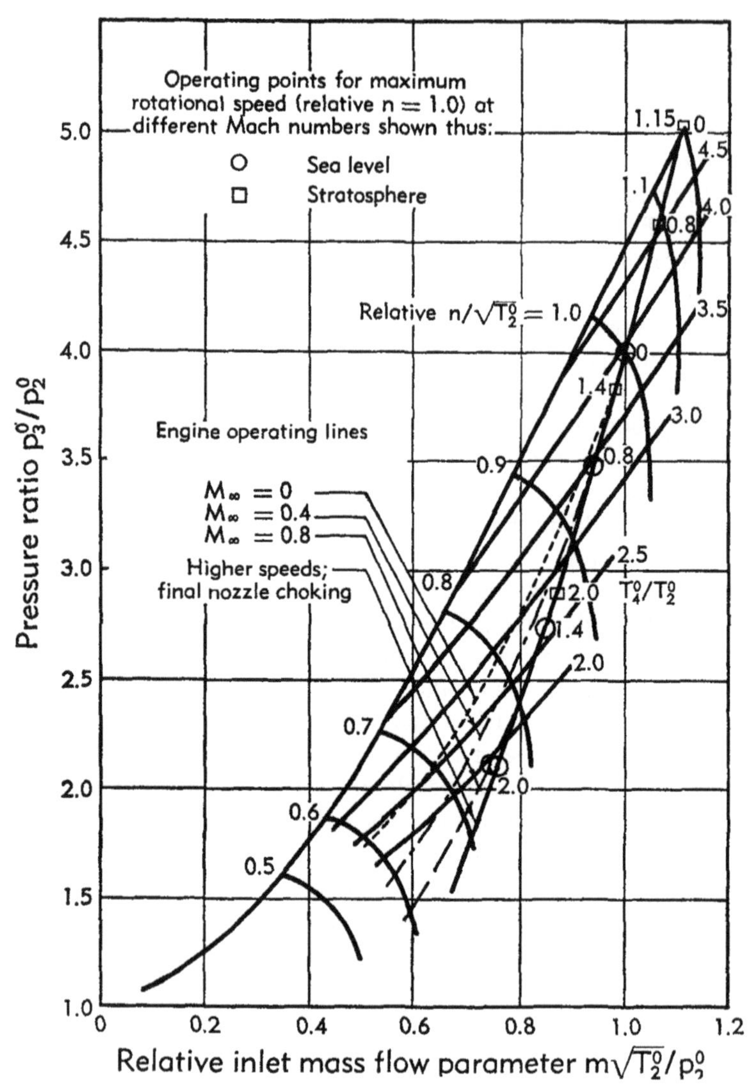


K,4d. Equilibrium operation of a simple jet engine with fixed final nozzle throat area.

to run into the surge region of the compressor at high pressure ratios, as may happen in an ill-designed engine, or if a too small propelling nozzle is fitted, then at a particular altitude and engine rotational speed, compressor surging is most likely to be encountered at the lowest flight speeds, for instance during climb, when $T_2^0$ is low and $n/\sqrt{T_2^0}$ consequently high.

Surging may also be encountered when rapid accelerations are made from idling conditions to full rotational speed. While this phenomenon clearly cannot be fully demonstrated on an equilibrium running diagram, it can nevertheless be seen that the addition of excess fuel to bring about the acceleration causes an increase in the temperature ratio $T_4^0/T_2^0$ above the value required for equilibrium running at the value of $n/\sqrt{T_2^0}$, which has been reached momentarily. Hence, if the equilibrium operating condition at idling lies fairly close to the surge line and if the rate of over-fueling is rapid, surging of the compressor can obviously result.

In addition to giving the mass flow characteristic of the propelling nozzle, Fig. K,4c also shows the variation of the gross thrust produced by the nozzle. This quantity is plotted in the form $F_g/A_7'p_\infty$ varying with $p_5^0/p_\infty$. The lowest curve labeled "nozzle expansion area ratio = 1.0" refers to a purely convergent nozzle, in which case the gross thrust at pressure ratios above the critical choking value is made up of a velocity component and a pressure component as indicated above. The figures shown by the uppermost curve relate to the assumption that a divergent portion of the nozzle is added in order to give complete expansion to atmospheric pressure. In producing the graph, losses in the nozzle have been assumed to account for 5 per cent of the available total head pressure $p_5^0$. With this assumption, it is seen that at an over-all pressure ratio of 15 the gross thrust parameter is increased by about 11 per cent by the addition of the divergent section. Smaller pressure losses in the nozzle would result of course in a slightly greater increase in gross thrust. It will be appreciated that at any particular flight condition the percentage improvement in net thrust gained by the addition of a divergent section to the propelling nozzle will be noticeably greater than the percentage increase in gross thrust.

At very high nozzle pressure ratios such as can occur at high flight Mach number the area ratio, outlet to throat, of the divergent portion required to complete the expansion of the exhaust gases to ambient pressure becomes quite large, with the result that if such a nozzle were used the outlet area could easily be the greatest cross-sectional area of the engine. The curves included in Fig. K,4c for nozzle divergence area ratios of 1.5 and 2.0 show, however, that if it is desirable to do so the divergent section can be stopped short of that required for complete expansion of the gas without incurring much loss in gross thrust.

What the diagram does not attempt to show is the loss in thrust which can occur when a nozzle of fixed divergence area ratio is operated at a pressure ratio appreciably below that which would yield an exhaust static pressure equal to the ambient pressure. Under these conditions, provided that the pressure ratio is great enough to produce sonic velocity at the throat of the nozzle, the gas flow often breaks away from the walls of the divergent section so that the nozzle no longer "runs full." At best, the thrust produced can then only equal that of a nozzle whose divergence area ratio is that appropriate to the stage at which the flow breakaway begins in the actual nozzle. In practice, however, the situation can be more serious than this, in that the jet may be deflected and tend to adhere to one wall of the nozzle or it may be unstable with a wandering region of breakaway. Considerable advantage is to be gained therefore by the use of a propelling nozzle in which the divergence area ratio can be varied to suit the changing flight conditions.

*Propeller turbine engines* (*turboprops*). The operation of a propeller turbine engine at equilibrium conditions is influenced by the greater degree of flexibility that is usually present compared with the simple jet engine, for at given flight conditions the final nozzle area is no longer sufficient to itself to determine where the turbine and compressor will operate. The power-absorption characteristics of the propeller and, to a lesser extent, the efficiency of the propeller-driving turbine or turbine stages are also important factors. Moreover, it is common practice to vary the power absorbed by a propeller at a given speed by the use of variable pitch blading and this clearly makes possible a greater freedom in selecting the turbocompressor operating points. Finally, in considering the equilibrium running conditions of propeller turbine engines it is necessary to distinguish between the behavior of an engine in which the propeller is driven from the same shaft as the compressor and that of one in which it is driven by an independent power turbine.

It should, however, be realized that, for a particular compressor, the chart of the type we have presented in Fig. K,4b which shows the horsepower ideally available in the expanding gases at the stage when sufficient work has been done to maintain the compressor at a chosen operating point, is not greatly affected by the choice between a "combined" or "series" turbine arrangement. For the positions of the lines of constant values of $T_4^0/T_2^0$ are, as we have seen, largely dependent upon the swallowing capacity of the turbine, this being the name often given to the quantity $m\sqrt{T_4^0}/p_4^0$. Now this quantity is generally determined by the choking conditions in the first turbine stator blade throats and it will in consequence be proportional to the area of these throats. This area is in turn determined by the maximum temperature that can be permitted at the design conditions. Therefore, in comparing a combined turbine arrangement with a series turbine scheme using the same compressor characteristics and the same design limitations, it will be seen that there can be little difference in the position of the $T_4^0/T_2^0$ matching lines in the two cases, provided that the turbines are designed with choking stator blade throats in the high pressure stages. The need to use a minimum number of stages to save weight in aircraft gas turbines results in this condition being met almost invariably. Hence for any particular compressor operating point the power ideally available in the gas as it reaches station 5 can only differ in the two cases as a result of the difference in the efficiency with which the work to drive the compressor has been taken out. Again there is no reason for this difference to be appreciable, unless there is a fairly large change in the quantity $U_m^2/2gJc_p({}_4\Delta T_5^0)$, discussed in Art. 3, and of which mention is again made in the present article.

The different performances to be obtained from combined and series turbine engine arrangements and from variations in propeller pitch may therefore be considered as arising from two main causes.

1. The changes in gas horsepower ideally available for useful work due to the different positioning of the final operating lines on the compressor characteristic.
2. The efficiency with which the gas horsepower ideally available is converted to useful propulsive power.

In this latter connection the over-all efficiency referred to can be considered as the product of three component efficiencies, firstly that with which the turbine extracts power from the expanding gases, secondly that of the mechanical transmission of this turbine power to the propeller, and thirdly the efficiency of the propeller itself. Since it is usual to consider the output of a propeller turbine engine in terms of the brake horsepower delivered to the propeller (with possibly some allowance being made for the small residual thrust of the exhaust gases), rather than in terms of thrust horsepower produced by the propeller, the variation of propeller efficiency scarcely influences the present discussion. Further, since the transmission efficiency can be assumed to be essentially constant, it follows that the differences in performance which result from differences in turbine arrangements and propeller controls depend mainly on the location of the operating lines on the compressor characteristic and on any variations in power turbine efficiency that may be experienced.

*Combined turbines.* Starting with perhaps the simplest propeller turbine engine of all, one with a combined compressor and power turbine and a fixed pitch propeller, it is to be expected in this case that, at a particular flight Mach number, the power absorbed by the propeller will be a function of the rotational speed. As a first approximation we may say that at constant altitude the power absorbed varies as the cube of the speed, and since the propeller rotational speed is equal to or proportional to the turbine speed and hence to the compressor speed, a relationship can clearly be established between the power and rotational speed parameters used in Fig. K,4b. For the purpose of our present approximation, we may obtain this relationship by assuming that the transmission efficiency is constant and that the power available from the turbine to drive the propeller will be a constant percentage of the gas power ideally available. Taking $p_\infty$ and $T_\infty$ as constants, an operating line can therefore be obtained from Fig. K,4b by using the fact that $(\mathrm{hp_g})_{\mathrm{is}}$ varies as $n^3$. This line is shown in Fig. K,4e plotted on the compressor characteristics and is of a form fairly similar to the operating line for a simple jet engine having the same design point though running somewhat nearer to the compressor surge line.

A more accurate determination of this running line, making use of a full propeller characteristic (which could be expressed nondimensionally, thus removing the restriction to one altitude which is imposed above)

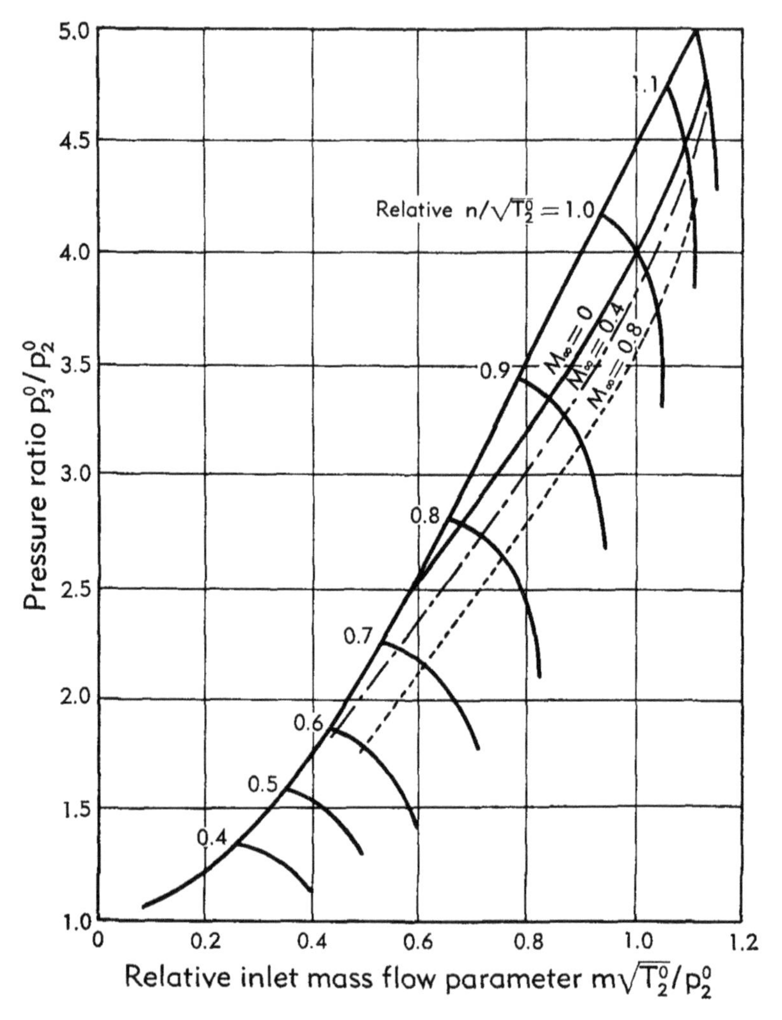


Fig. K,4e. Equilibrium operation of a single-shaft, propeller turbine engine for one setting of propeller pitch.

and a proper turbine characteristic, making allowance for the small amount of kinetic energy which must remain in order to exhaust the gases through the final nozzle, would not materially alter the shape of this operating line. A family of operating lines for a range of $M_\infty$ can in fact be drawn in the same fashion as for the simple jet engine.

If, however, the propeller is now changed for one in which the blade pitch is variable it is possible for the propeller to absorb different amounts of power while running at constant speed. The operating lines on the compressor characteristic are therefore not fixed but may move over the breadth of the rotational speed lines, actual operating points being determined by the pitch setting of the propeller. Hence a constant speed governor might be fitted and, without altering the compressor rotational speed, a wide range of propeller power might be obtained by varying the pitch setting of the propeller only. In these circumstances coarsening the propeller pitch so that more power is absorbed for a given rotational speed causes more fuel to be supplied to the engine in order for the turbine to provide the extra power. The operating point is thus displaced up the appropriate speed line to higher values of $T_4^0/T_2^0$ and so towards the compressor surge line. Conversely a change to fine pitch causes the operating point to move down the compressor speed line into the region of falling compressor isentropic efficiency and lower temperature ratios. As will be appreciated from the discussion in Art. 2 the thermal efficiency of the engine deteriorates rather quickly as a result of these two influences.

It follows that, with this type of engine and at a given flight condition, it is possible to obtain a particular fractional power in more than one way by independent alteration of the engine speed and the propeller pitch. For each particular power requirement, however, there will be one combination of rotational speed and associated pitch setting which will give optimum fuel economy, as will be more evident from the article that follows. If, as is desirable, the pilot's control of his engine is to be by the movement of a single lever it is clear that means must be provided so that when the pilot calls for a change in power, by altering the amount of fuel being supplied by the engine, the pitch setting of the propeller is automatically adjusted so that the engine speed settles to the value at which most economical use is made of the fuel being supplied. The provision of an automatic linkage such as this is clearly a problem of some magnitude, especially since the optimum operating conditions of speed, pitch, and fuel supply are likely to vary with changes in flight Mach number and aircraft altitude. A compromise solution has therefore to be accepted, but the form of the solutions which have been suggested is outside the scope of the present section.

The change of propeller pitch setting invalidates the simplifying assumptions made above to the effect that the power turbine efficiency remains sensibly constant. Since changes in propeller pitch mean that the propeller turbine can be called upon to deliver a range of power at the same rotational speed, it follows that there will be a change in the relationship between the blade speed, $U_m$, and the turbine temperature drop ${}_4\Delta T_6^0$. A change in turbine efficiency therefore results after the manner shown in Fig. K,3c. Since one turbine drives both compressor and propeller it follows that the efficiency with which the work to drive the compressor is provided will also be affected. In this instance, therefore, some variation in the value of potential gas horsepower associated with a particular compressor operating point must be expected. At the same time however, the fact that the compressor load is taken by the turbine which also takes the propeller load means that a given change in propeller load/speed relationship causes a smaller change in the mean values of $U_m^2/2gJc_p({}_4\Delta T_6^0)$ in the turbine than it would if the turbine drove only the propeller. The change in turbine efficiency is therefore smaller than in a free-running power turbine under similar conditions, but it is, of course, applied to the whole of the expansion process and not just to the portion which provides external power.

*Free power turbine.* Consideration is now given to the type of propeller-turbine engine in which the compressor and the propeller are driven by separate turbines, situated in series with respect to the flow of expanding gases but mechanically independent of each other. The salient fact which emerges is that, because of this mechanical independence, the power-absorption characteristics of the propeller, which are determined primarily by changes in pitch, have a greatly reduced influence on the compressor equilibrium conditions when comparison is made with the combined-turbine type of engine. If, in an engine with independent turbines, equilibrium conditions have been achieved and the pitch of the propeller is then coarsened, the effect will be to cause a reduction in the speed of the power turbine. The equilibrium of the turbocompressor unit will only be affected, and hence alterations made to the potential gas horsepower available for external work, if the changes in power turbine speed also cause a change in the swallowing capacity of the power turbine. Such a change would be noticed by the compressor turbine and a readjustment of equilibrium conditions would result. If the power turbine should be operating with choked turbine stator-blade throats, however, no change in swallowing capacity would result from a change in power turbine speed, and the equilibrium conditions of the gas generator part of the engine would not be affected.

Since the position of the operating line on the compressor characteristic is of such importance in determining the percentage power output and the thermal efficiency which the engine has at any particular value of compressor rotational speed, it is of interest to investigate a little further the influence which the location of the power turbine has upon the position of the compressor operating line.

In order to demonstrate this feature it is convenient to assume for the moment that the power is being absorbed by a propeller of fixed pitch, since we have seen that it is only when the compressor- and power turbines are combined that changes of propeller pitch can have a major influence on the operating line position. The pressure and temperature

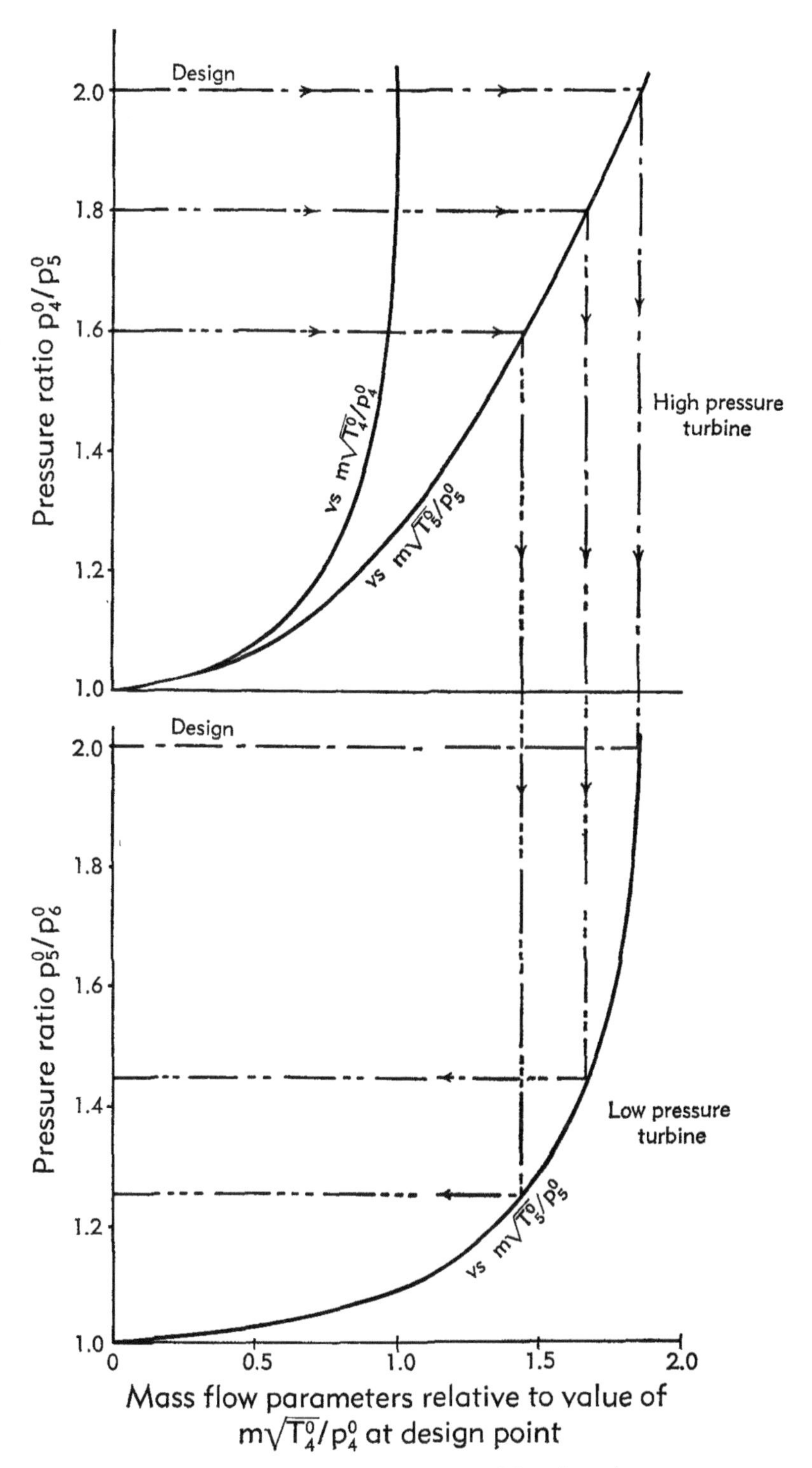


**Fig. K,4f.** Behavior of two turbines in series.

at the inlet to the compressor are also assumed constant. If the propeller is of fixed pitch it is a reasonable approximation to assume a single-line characteristic of the form mentioned in Art. 3 for the power turbine as well as for the compressor turbine. In Fig. K,4f the part-load operation of two turbines in series is demonstrated by the use of these single-line characteristics. The suffixes 4 and 5 are used to indicate the stations between which the high pressure turbine operates and the suffix 6 defines the outlet plane of the low pressure turbine. The residual jet velocity is presumed to be small. The designed pressure ratios of the two turbines are assumed, for the purpose of the illustration, to be equal and the characteristics give the variation of the turbine pressure ratio with changes in the mass flow parameter. The characteristic of the high pressure turbine, however, is plotted against both the inlet mass flow parameter, $m \sqrt{T_4^0}/p_4^0$, and also the outlet mass flow parameter, $m \sqrt{T_5^0}/p_5^0$, in the manner suggested in Art. 3. This latter quantity is, of course, the inlet flow parameter of the low pressure turbine, against which the pressure ratio variation of this turbine is plotted in the lower half of the Fig. K,4f. Corresponding operating points for the two turbines must therefore be at the same value of this common mass flow parameter. The figure shows quite clearly that, because of the shape of the characteristics, if the over-all pressure ratio across the two turbines is reduced, say by a reduction in compressor speed and delivery pressure, it is the low pressure turbine which experiences the bigger reduction in pressure ratio available to the individual turbines. Indeed, so long as the stator blade throats of the low pressure turbine remain choked there is no change in the pressure ratio of the first turbine in the series. From this it may be inferred that if this first turbine is the one that drives the compressor, now assumed to be demanding a power below its design value since its delivery pressure has fallen, the turbine temperature level must fall from the values at the design point if equilibrium between turbine and compressor is to be restored.

If, on the other hand, external power is taken from the high pressure turbine and the low pressure turbine is used to drive the compressor the position would be much different: the compressor-driving turbine then suffers a rapid fall in pressure ratio as the over-all pressure ratio of the engine is reduced, with the result that in order for the low pressure turbine to provide sufficient power to drive the compressor the temperature level of the turbines must remain high, and in some instances the turbine inlet temperature, $T_4^0$, may rise above its value at the design condition.

A comparison between two engines using similar compressors and having two turbines in series, but with power to drive the compressor taken from the first turbine in the first case and from the second turbine in the other case, would therefore be expected to show the following principal features:

1. At a particular pressure ratio of the compressor below that at the design condition of the two engines, the first engine will operate at a much lower value of the temperature, $T_4^0$, than will the second.
2. In consequence, the power output of the two engines relative to the common value at the design condition will at this particular pressure ratio be lower in the first case than in the second.
3. The operating point of the compressor of the first engine will be farther from the compressor surge line than the corresponding point for the second engine at the same pressure ratio. On the assumption that the high pressure turbine nozzles are designed well into the choking region, with the result that $m\sqrt{T_4^0}/p_4^0$ is constant over much of the probable operating range of the engines, it follows that, at a particular value of $p_4^0$, $m$ is inversely proportional to $\sqrt{T_4^0}$. Hence, if $T_4^0$ is high, as is typical of the second engine compared to the first, the mass flow at a given pressure ratio tends towards the surge region.
4. If the engines are without heat exchangers, and at the present time the weight of these components makes them unattractive for aero-engine applications, the comparative figures for over-all efficiency at a particular fractional output are not markedly different in the two cases, for at the same pressure ratio the second engine is more efficient than the first, but gives a larger fractional output.
5. If it were practicable to equip the engines with efficient heat exchangers the efficiency of the second engine would exceed that of the first when operating at a similar pressure ratio, or even at a similar fraction of the power output at design conditions. Consideration of the cycle performance shows that with a heat exchanger of high thermal ratio the advantages of keeping a high cycle temperature while reducing the cycle pressure ratio are appreciable. This is, of course, what an engine of the second class tends to do, but only at the risk of surging the compressor and of raising the temperature at inlet to the high pressure power turbine to values which are prohibited by mechanical considerations.

Most of these characteristics are illustrated by the curves of Fig. K,4g and K,4h. From Fig. K,4g it is seen that the typical compressor operating lines of the two series-turbine engines lie on opposite sides of the operating line for an engine of similar design requirements in which a single turbine is used to drive both the compressor and the constant pitch propeller. The variations of turbine temperature with varying output are also such that the combined-turbine engine has a performance which is intermediate between that of the two engines with the turbine in series.

It follows from the arguments used above, that, if the compressor and propeller-driving turbines are placed in parallel with respect to the expanding gas flow, then the compressor operating line is very similar to that of the engine with a single turbine and using a constant pitch propeller. Further, in common with the other engines with independent power turbines and in contrast to the single-turbined engine, changes in the propeller pitch will make only small alterations in the position

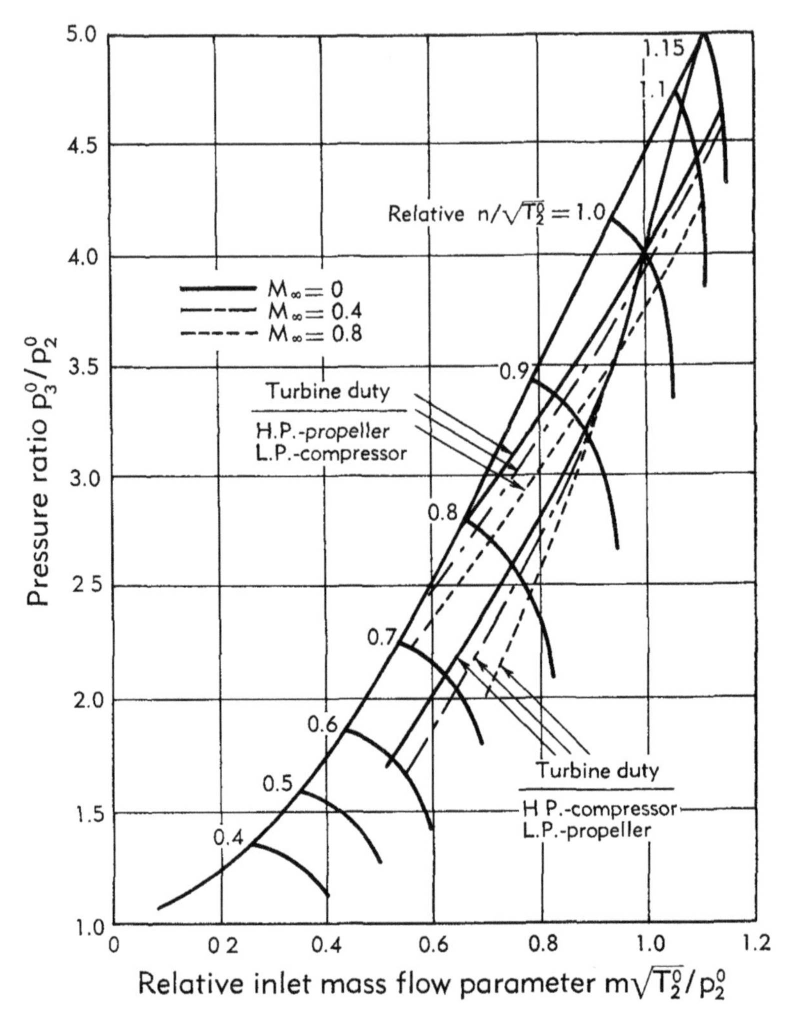


Fig. K,4g. Equilibrium operation of two otherwise similar engines having independent turbines in series, but with the compressor driven by the first turbine in one case and by the second in the other.

of the compressor operating line and consequently in the over-all thermodynamic performance of the engine.

This discussion of the use of uncoupled power turbines has been intentionally over-simplified in order to emphasize the principal effects which the position of the power turbine has upon the equilibrium running and performance of otherwise-similar engines. Although, in the foregoing discussion, consideration has been given only to operation at constant conditions of temperature and pressure at the inlet to the compressor, the same general conclusions may be reached and made applicable to varying inlet conditions by treating the component-matching problem in terms of the nondimensional performance parameters.

A further simplification has been the implication that the isentropic efficiencies of the components do not vary greatly over the range of operating conditions required. It has been presumed that at a particular pressure

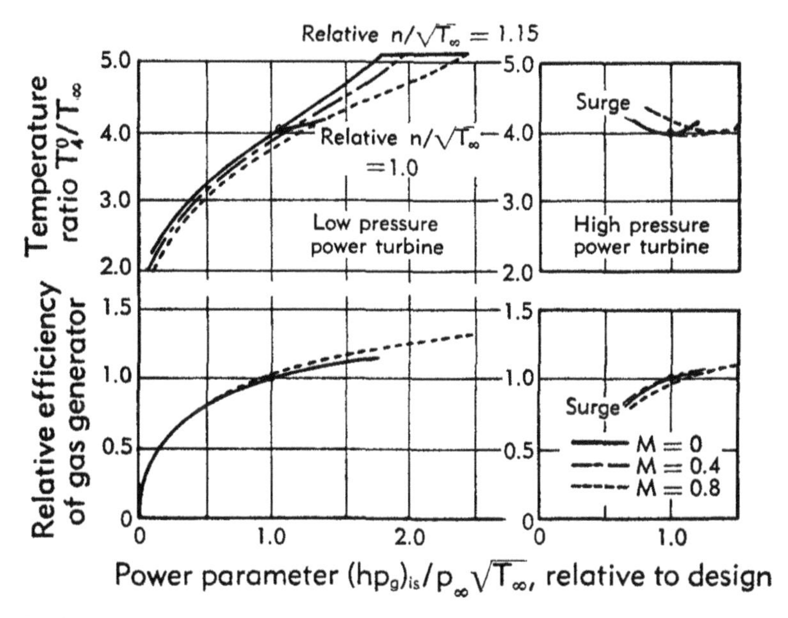


Fig. K,4h. Variation of maximum temperature and thermal efficiency with power parameter and flight Mach number in two engines with independent power turbines. Note: The variation in efficiency is discussed in Art. 5.

ratio the compressor temperature rise or the power demanded by the compressor per unit mass flow of air, will be independent of the actual mass flow through the compressor, that is of the final position of the equilibrium operating point which the argument has sought to determine. If this assumption does not hold good and, instead, there is a variation in compressor efficiency over the operating range of the compressor, the results of the preceding generalized discussion need to be modified. The nature of the modification, however, can also be assessed, qualitatively at least, by methods similar to those used in the generalized discussion.

For example, if at a particular pressure ratio a loss of compressor efficiency is experienced, then a greater amount of power will be required, per unit of mass flow, to drive the compressor at this pressure ratio. With two turbines operating in series, the pressure ratio available to each at any part-load condition has been shown to be largely independent of most factors other than the division of pressure ratio between the two at the design condition. The only way, therefore, in which the compressor-driving turbine of a particular engine can provide the extra specific power demanded of it to meet a loss in compressor efficiency is by demanding an increase in the operating temperature level in the turbines. This means that the mass flow of the engine will fall if, as is generally the case, the mass flow is determined by the swallowing capacity of the first turbine stages. Because of the rise in turbine operating temperatures the propeller turbine also gives an increased specific output power per unit mass flow, although the increase in actual output at the particular pressure ratio under consideration will be less because of the lower mass flow through the engine under the new conditions. The rise in turbine inlet temperatures also means, of course, that more fuel per unit air flow must be burned than formerly and it is inevitable that this increase in fuel consumption is greater than the increase in the specific output of the propeller turbine. The loss in compressor efficiency is thus responsible for a loss in over-all thermal efficiency. With typical aerodynamic compressors the operating line for an engine in which the low pressure turbine is used to drive the propeller tends, at partial-load operation, more towards the region of falling compressor efficiencies than does that for an engine in which propeller power is taken from the same turbine as the internal compression power. It is with the former type of engine, therefore, that most modification of performance estimates based on assumptions of constant component efficiencies would have to be made in order to obtain a more realistic assessment.

A loss in compressor efficiency from values assumed in the design of an engine may, of course, result from causes other than the position of the operating line on the compressor characteristic. Accidental damage to the blading or pollution by oil or atmospheric dirt may cause a deterioration in the compressor efficiency in service conditions.

In an engine with propeller and compressor turbines coupled together, the result of a drop in compressor efficiency is again to make the engine operate at higher temperatures to give a particular output. For in this type of engine the turbine inlet temperature corresponding to particular values of compressor pressure ratio and mass flow is determined principally by the turbine swallowing capacity. With a loss in compressor efficiency, however, more of the turbine power has to be used in driving the compressor and less is available as external power. If, therefore, the propeller power and the speed of the engine were originally such as to satisfy the requirements of the propeller in a particular pitch-setting, the external power would now be inadequate to do this. If the pitch-setting is retained the compressor equilibrium operating point must move to a new position where, principally by the use of higher turbine temperatures, the propeller power returns to a value appropriate to the engine speed. The lines of constant compressor speed, however, do not correspond to constant values of compressor pressure ratio, but rather to rising values of pressure ratio with diminishing mass flow. It follows that the new equilibrium position will be dependent upon the way in which changes in pressure ratio and engine mass flow as well as turbine temperature affect the power supplied to the propeller at the chosen speed. The shapes of the constant speed lines on the compressor characteristic are, therefore, of importance in determining the over-all effect of a loss in compressor efficiency in an engine with coupled turbines, in the same way as they have been shown earlier to be of importance in determining the variation in engine performance with changes in propeller pitch.

The shapes of these lines, as we have seen, are of little significance in determining the equilibrium operating lines for engines with free-running power turbines, but they cannot be overlooked when comparison is being made between the possible practical merits of otherwise-similar engines having external power produced by (1) a high pressure turbine, (2) a turbine in parallel or coupled to the compressor turbine, or (3) a low pressure turbine. Considered in conjunction with a particular power-speed relationship in the propeller, it has been demonstrated above that, at a particular percentage of the design pressure ratio, the third of these engines would be operating at a lower turbine temperature than the others and giving a lower percentage of the design output common to all the engines. The characteristic shapes of the compressor constant speed lines and of the typical engine operating lines, however, lead to the conclusion that the compressor speed will be higher in this third engine than in the others. The required operating range of output is therefore covered by a smaller change in the speed of the gas generator set in this engine than in the others, if operation at fixed propeller pitch-setting is considered. Assuming that the engines are designed at the maximum permissible conditions of rotational speed and turbine temperature, the use of a low pressure power turbine offers an appreciable reduction in temperature and a smaller reduction in rotational speed in covering a particular load range, while at the other extreme the use of a high pressure propeller turbine would involve a wide variation in compressor speed but little reduction, and in some cases a prohibitive increase, in turbine operating temperatures. It is true to say that up to the present time a reduction in temperature conditions at part load has been considered to be of more value than an appreciable reduction in rotational speed.

If operation with variable propeller pitch is considered, the same conclusions apply so far as the engines with independent power turbines are concerned. The engine with coupled turbines, however, might be operated at constant speed and the required reduction in output obtained by the use of lower turbine operating temperatures.

**K,5. Performance of Simple Jet and Propeller Turbine Engines.** From the preceding article it may be seen that the thrust of a simple jet engine of fixed propelling nozzle geometry may be plotted nondimensionally in a single diagram in which the net thrust parameter $F_{net}/p_\infty$ is plotted against the engine rotational speed parameter $n/\sqrt{T_\infty}$ for different values of the flight Mach number $M_\infty$.

Further the potential gas horsepower parameter $(hp_g)_{is}/p_\infty \sqrt{T_\infty}$ has been shown to be a function also of $n/\sqrt{T_\infty}$ and $M_\infty$, together with the engine mass flow parameter $m \sqrt{T_\infty}/p_\infty$. Since by the use of a power turbine characteristic and an exhaust system characteristic, both of which may be expressed nondimensionally, the operating line for an engine driving a fixed-pitch propeller may be determined, (that is a relationship established between $n/\sqrt{T_\infty}$ and $m \sqrt{T_\infty}/p_\infty$) and the efficiency of the conversion of potential gas horsepower into equivalent brake horsepower, $hp_e$, evaluated, it follows that $hp_e/p_\infty \sqrt{T_\infty}$ may be expressed as a function of $n/\sqrt{T_\infty}$ and $M_\infty$. Again, therefore, the performance of the engine may be depicted graphically in a single diagram, though further diagrams would be required to represent the performance at other propeller pitch-settings.

The nondimensional parameter for the fuel consumption of a particular engine (scale $l$ = unity) is $m_f Q_H/p_\infty \sqrt{T_\infty}$ where

$$m_f = \text{fuel consumption per unit time}$$
$$Q_H = \text{calorific value of the fuel.}$$

This leads to the parameter for the specific fuel consumption of a simple jet engine in the form $m_f Q_H/F_{net} \sqrt{T_\infty}$ and for a propeller engine, $m_f Q_H/hp_e$. Again these are functions of $M_\infty$ and $n/\sqrt{T_\infty}$. Attention must, however, be drawn to a flaw in the completely dimensionless treatment of gas turbine engine performance which consideration of the fuel consumption brings to light. It is caused by the fact that the variation in the specific heat of air with changes in temperature, that is with altitude, cannot be expressed nondimensionally. So, although by the matching of component characteristics it may be determined that at a given operating point the air mass flow is given by a value of $m_a \sqrt{T_\infty}/p_\infty$ and the temperature rise through which it has to be heated by $(T_4^0 - T_3^0)/T_\infty$, the fuel flow parameter $m_f/p_\infty \sqrt{T_\infty}$ is not uniquely determined. The fuel-air ratio, $m_f/m_a$, required to heat air through a temperature range of $T_4^0 - T_3^0$ depends on the absolute values of the two temperatures, since the specific heat of air is an independent function of temperature as Fig. K,2d shows. The relationship between $m_f/m_a T_\infty$ and $(T_4^0 - T_3^0)/T_\infty$

therefore depends to a small extent on the absolute value of $T_\infty$. If a value of $T_\infty$ corresponding to an altitude of, say 20,000 ft, is assumed, and a diagram relating the nondimensional fuel flow, flight speed, and rotational speed parameters is prepared, the error involved in using this diagram to obtain information on fuel consumption at other altitudes does not exceed 3 per cent. This is a procedure often followed, since, in practice, variations in the efficiency of the combustion process of this order may be experienced in covering this range of altitude.

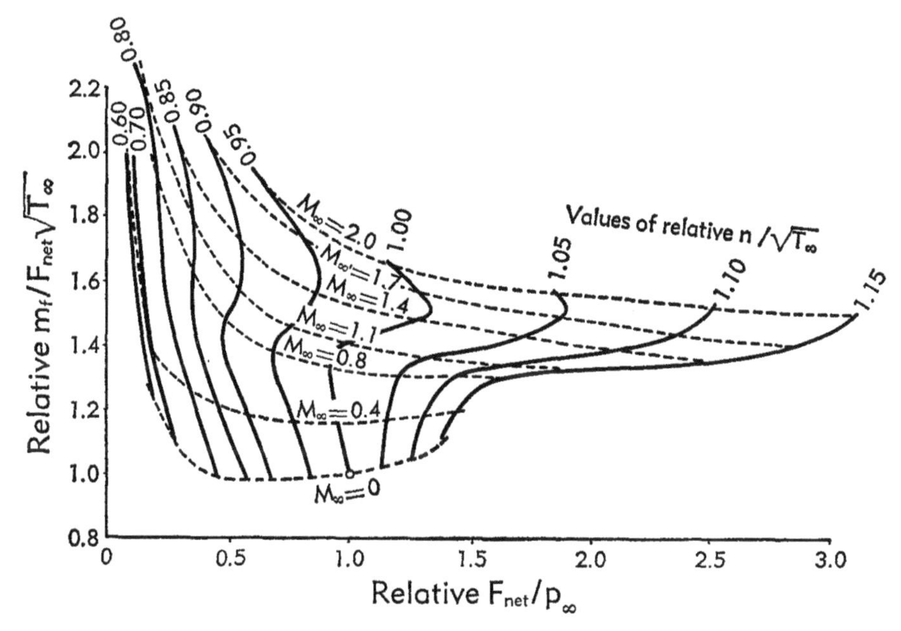


Fig. K,5a. Performance of a typical simple jet engine. Variation of thrust ($F_{net}/p_\infty$) and specific fuel consumption ($m_f/F_{net}\sqrt{T_\infty}$) with changes in $M_\infty$ and $n/\sqrt{T_\infty}$.

*Simple jet. Effect of altitude and forward speed.* Fig. K,5a depicts the variation with changes in $n/\sqrt{T_\infty}$ and $M_\infty$ of $F_{net}/p_\infty$ and of $m_f/F_{net}\sqrt{T_\infty}$ for a typical simple jet engine which, at full rotational speed under sea level static conditions, has a compressor pressure ratio of 4.

*Net thrust.* Over the range of $n/\sqrt{T_\infty}$ shown, which is sufficient to cover maximum speed and cruising speed at all altitudes, the rate of increase of $F_{net}/p_\infty$ with $n/\sqrt{T_\infty}$ at constant $M_\infty$ is very great, the thrust parameter increasing as the third or fourth power of the speed parameter. In the neighborhood of the design value of $n/\sqrt{T_\infty}$, the typical form of the curves of $F_{net}/p_\infty$ against increasing Mach number for a constant value of $n/\sqrt{T_\infty}$ is a fairly gentle decrease initially, followed by a tendency for the thrust parameter to increase again at higher values of $M_\infty$, and after reaching a peak to decrease once more. This characteristic form of curve is the result of two opposing effects which occur when the flight speed increases. Firstly, as shown in Art. 2, the thrust created per unit mass flow of air tends to decrease if the engine rotational speed is unaltered. (Compare corresponding points in Fig. K,2i for a particular compressor temperature rise.) Secondly, however, the mass flow through the engine increases as the pressure at the controlling orifices, the turbine nozzle throats, is increased as a result of the ram pressure rise of the air due to the increasing flight speed. It is when the latter effect begins to exceed the former in magnitude that the thrust parameter begins to increase again. The point at which the former effect once more predominates depends to an appreciable extent on the assumed variation of intake efficiency with flight Mach number, since this influences both the specific thrust and mass flow terms.

The typical form of curves described above is modified at low values of $n/\sqrt{T_\infty}$, in that the peak in $F_{\text{net}}/p_\infty$ becomes more and more insignificant, while at high values of $n/\sqrt{T_\infty}$, corresponding to operation at full rotational speed at the higher altitudes, the influence of the peak predominates. In the chosen example the variation of $F_{\text{net}}/p_\infty$ against $M_\infty$ at the high values of $n/\sqrt{T_\infty}$ becomes a slow increase followed by a very steep rise to a peak which occurs at a flight speed in excess of $M_\infty = 2$. The change from an initial slight fall in the values of $F_{\text{net}}/p_\infty$ with $M_\infty$ increasing from zero, which typifies the curves for relative $n/\sqrt{T_\infty}$ values of 1.0 and under, to the initial slight rise at higher $n/\sqrt{T_\infty}$ values can be shown to be logical from a study of the thermodynamic cycle, but it tends to be somewhat exaggerated in the example shown. The reason for this lies in the nature of the compressor efficiency variation and since it arises from a feature which may commonly occur in aero-gas turbines it is worth a little further comment. The performance curves of Fig. K,5a have been derived by the use of typical compressor characteristics similar to those shown in Fig. K,3a. At values of $n/\sqrt{T_2^0}$ greater than the design value it is noted that the compressor efficiency falls fairly rapidly along the line on which the simple jet engine operates. Since at a particular rpm and altitude, that is a particular value of $n/\sqrt{T_\infty}$, the value of $n/\sqrt{T_2^0}$ increases with decreasing speed, it is to be expected that it will be the low speed, high altitude performance of the engine which will suffer most from reductions in compressor efficiency. The operating points marked on Fig. K,4d help to illustrate this point.

In the illustrations of Fig. K,5b the net thrust and specific fuel consumption of the engine postulated in this example are expressed in absolute, instead of dimensionless, units and plotted against altitude for different aircraft speeds. This illustrates the effect of altitude on the engine thrust obtained at the maximum permissible rotational speed. In the stratosphere the thrust is directly proportional to the ambient air pressure since $F_{\text{net}}/p_\infty$ is then a constant for a particular flight speed. In the troposphere, however, the thrust does not decrease as rapidly

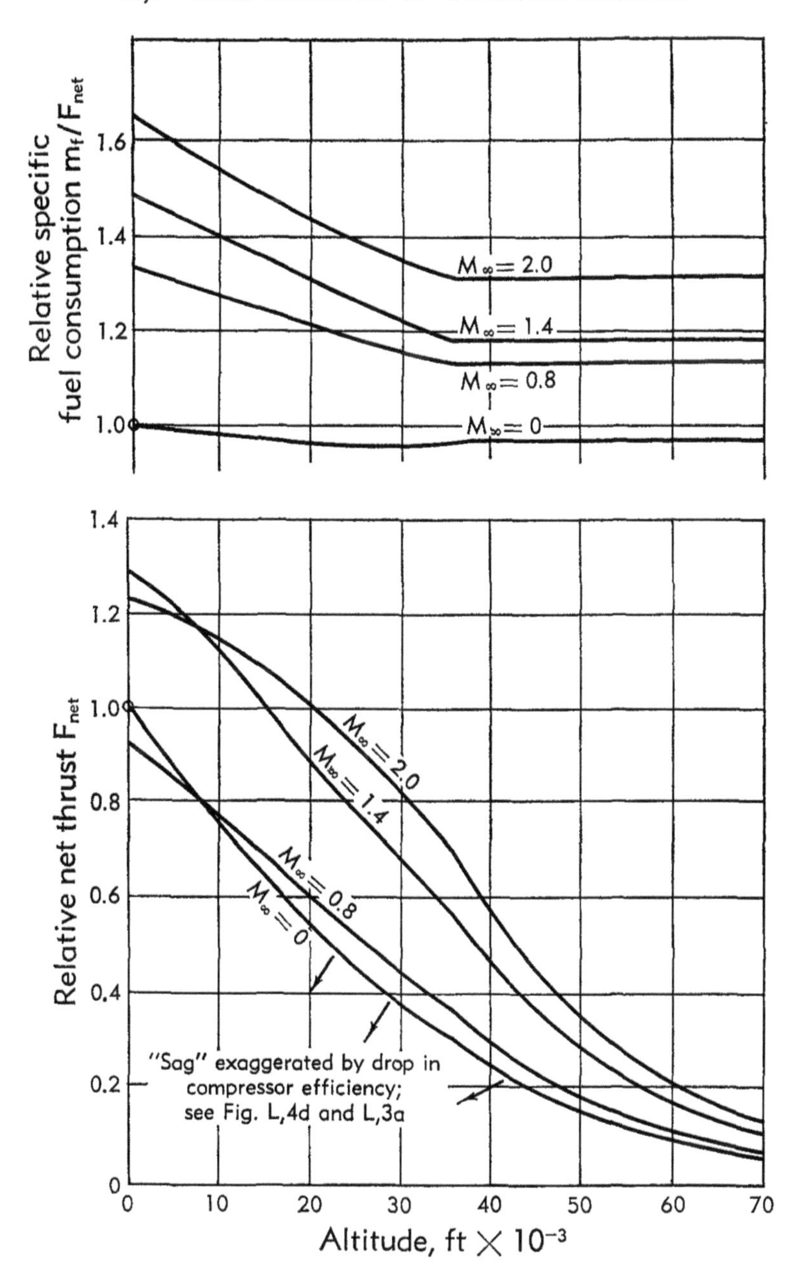


Fig. K,5b. Variation of net thrust and specific fuel consumption with altitude and Mach number.

as the pressure $p_\infty$ since the simultaneous reduction in $T_\infty$ causes an improvement in the specific thrust obtainable from the engine.

*Specific fuel consumption.* At zero flight speed the curves of Fig. K,5a show that the specific fuel consumption parameter is a minimum at about 0.90 of the design value of $n/\sqrt{T_\infty}$ and then increases with increasing $n/\sqrt{T_\infty}$, though the rate of increase is not large. In fact if the increase in $n/\sqrt{T_\infty}$ from 100 per cent to 115 per cent of the design value is assumed to be caused by moving from sea level to the stratosphere while maintaining constant rpm, the parameter $m_f/F_{net}\sqrt{T_\infty}$ does not increase by as much as 15 per cent, so that the actual specific consumption in the stratosphere is less than that at sea level. This again is in keeping with the results of the cycle performance investigation as depicted in Fig. K,2j. The curves in both cases are based on the assumption that there is no change in combustion efficiency with altitude, other than that which would compensate for the error introduced by ignoring the changes in the specific heat of air. With increase in the flight speed of the aircraft the minimum point of the specific fuel consumption curve moves to higher engine speeds, so that at a Mach number of 0.4 it occurs at about the design value of $n/\sqrt{T_\infty}$, while at $M_\infty = 2.0$ the minimum is close to a relative value of $n/\sqrt{T_\infty}$ of 1.15, i.e. to the condition of full rotational speed in the stratosphere. Naturally the exact form of the curves for specific fuel consumption, as of those for net thrust, is dependent upon the peculiarities of the particular engine and air intake design. The main trends shown in the example of Fig. K,5a however, are typical of the average aircraft gas turbine jet engine.

Both Fig. K,5a and K,5b show that the effect of increasing forward speed at a particular value of engine speed and altitude is in general to increase the specific fuel consumption of the engine. It must be remembered in this connection, however, that the specific fuel consumption is not a true measure of the over-all efficiency of the engine and propulsive device as a whole under flight conditions. This latter efficiency, as anticipated by the cycle performance discussion in Art. 2, clearly increases as the forward speed increases, at least in the range of flight conditions with which we are concerned.

*Matching the pure jet engine with a supersonic intake.* The typical performance curves presented in the preceding paragraphs incorporate two assumptions which require further comment. Firstly it has been assumed that the exhaust gases are always fully expanded in the propelling nozzle in the manner indicated by the appropriate curve of Fig. K,4c. To achieve this level of performance the divergent portion of the propelling nozzle would need to be capable of being varied to provide the correct expansion area ratios at all times. If this were not done and, for example, a purely convergent nozzle were employed, a deterioration in over-all engine performance would naturally result. In the example used as an illustration, expansion pressure ratios rise to a value of about 10 at a flight Mach number of 2.0. Fig. K,4c shows that at these pressure ratios the gross thrust of the engine is decreased nearly 10 per cent by the omission of the divergent portion of the nozzle. The effect on net thrust and hence on specific fuel consumption is more than double the percentage effect on the gross thrust parameter at these flight conditions.

Secondly, it has been assumed that at any particular flight Mach number the intake recovery achieved will be that shown in Fig. K,2a, which strictly speaking relates to a family of different intake designs and not to the performance of a particular intake over a range of operating conditions. Now in subsonic flight it is not difficult to design an air inlet which will pass a range of mass flow without suffering much variation in efficiency other than small, frictional effects. In supersonic flight, however, this is rarely the case. At any flight speed there is a maximum critical air mass flow which a fixed intake can swallow and this it can deliver to the engine either at a maximum pressure recovery level or if necessary at a reduced pressure level. If the engine cannot accept this air flow rate, difficulties arise. Except in the case of a Pitot intake, which is only effective at relatively low supersonic speeds, any attempt to reduce the air mass flow of the intake by external spillage tends to lead very quickly to the unstable flow regime known as "buzz" or else to design complications and drag penalties of an equally serious nature. It is reasonable to suppose therefore that the assumed intake recovery and hence the derived over-all engine performance would only be achieved if the air inlet were capable of being varied to capture the air required by the engine with maximum efficiency.

It would be dangerous to be dogmatic about the effects of the mismatching between inlet and engine where it is not possible to use variable geometry inlets, since many design variants would have to be considered. A few general trends may nevertheless be identified. Suppose that an engine and intake are matched so that at maximum flight Mach number the engine running at full rotational speed will accept the maximum mass flow at the inlet at peak (or critical) pressure recovery and the flight Mach number is then reduced, say by an increase in aircraft drag. Assuming no change in altitude, then at this reduced flight speed the inlet will have a reduced swallowing capacity and a reduced delivery pressure even at peak recovery. Assuming further that the engine rotational speed and propelling nozzle throat area remain unchanged, it is probable that if at the new conditions the engine were to receive air at the peak pressure recovery of the intake, it would demand more air than the intake could now provide. Consequently the pressure recovery of the intake at once drops (by the action of increased shock losses) to a level at which the fixed air flow of the inlet once again matches the requirement of the engine. The loss in intake performance naturally has an adverse influence on the performance of the propulsion unit as a whole. A remedial action would be to restore the intake operation to its most efficient condition for the new flight speed, by reducing the engine's demand for air. The most obvious way to do this is to reduce rotational speed although this still results in some loss of thrust, with possibly an attendant increase in specific fuel consumption compared with the engine using full rotational speed and a variable inlet. The engine's air consumption might alternatively be reduced by decreasing the propelling nozzle throat size. An increase in thrust could be obtained in this way but only to the extent permitted by considerations of turbine overheating and the risk of compressor surge.

By a similar argument to the above it may be seen that if at the originally postulated well-matched condition, the engine rotational speed were to be reduced by throttling the fuel supply, the air demand of the engine would drop and a fixed geometry inlet would be forced to spill air with the attendant risk of intake instability. Before engine rotational speed could safely be reduced in these circumstances, therefore, it would be desirable first to lose flight speed by such means as drag plates or by putting the aircraft into a climb.

A fixed geometry inlet designed for supersonic flight also creates a serious restriction to the air flow required by the engine at subsonic, and particularly at take-off, conditions. So serious in fact is the restriction that the provision of auxiliary inlets bypassing the fixed supersonic inlet is a virtual necessity.

*Propeller turbines. Effect of altitude and forward speed.* As in the case of the simple jet engine the effect of altitude and forward speed on the output and efficiency of a propeller turbine engine running at constant speed can be assessed qualitatively from the cycle performance variations determined in Art. 2. If the rotational speed is fixed then to a first order of approximation the compressor temperature rise can be regarded as constant and independent of changes in flight conditions. If the propeller is driven by an independent low pressure turbine which operates with choked stator blade passages, then the maximum gas temperature of the cycle will also tend to remain constant at constant compressor speed.

Referring to Fig. K,2l and K,2m and considering a value of compressor temperature rise of 200°C and a maximum cycle temperature of 1100°K, the curves of Fig. K,2l show that there is a very slight increase in specific equivalent brake power when the flight speed is increasing from 0 to 500 mph. The increase of equivalent brake power with increasing flight speed in an actual engine would therefore be expected to result from the increasing flow of the engine, which as indicated in the case of the jet engine is approximately proportional to the turbine inlet pressure. The turbine entry pressure is not directly proportional to the pressure after ram compression, $p_2^0$, as with increasing flight speed there is a reduction in compressor pressure ratio due to the increased temperature level in the compressor.

Fig. K,2l also shows that the specific power increases by about 50 per cent when the engine is operated at a height of 36,000 ft instead of at sea level. As the relative air pressure at this height is 0.225, the actual power given by the engine at the tropopause would, by these assumptions, be some 35 per cent of the power at the same Mach number at sea level. Similarly, Fig. K,2m shows that an improvement in specific fuel consumption of the engine of about 25 per cent is possible by the same change in altitude, while flight speed has only a small influence on the economy over the typical operating range of speed.

The component characteristics in an actual engine have a modifying influence upon the quantitative results deduced from the cycle performance curves. At constant engine speed, neither compressor temperature rise nor turbine entry temperature remain exactly constant with changing flight conditions. Changes in component efficiencies must also be expected. In particular the loss in compressor efficiency which is particularly noticeable at low compressor entry temperatures will, as in the jet engine example considered earlier, reduce the output and increase the specific fuel consumption of the engine at high altitudes compared with the values based on the cycle performance curves for constant component efficiencies.

The variation in equivalent brake power with changes in engine speed, if the power turbine efficiency remains fairly constant, will be very similar to the variation in potential gas horsepower. Fig. K,4b has already been used to illustrate the typical form of this latter variation. The corresponding figure showing the variation of the thermal efficiency of the gas generator set is shown in Fig. K,5c. Considering a propeller turbine engine with a combined power and compressor turbine, it is seen that the potential gas powers of Fig. K,4b and the gas generator efficiencies of Fig. K,5c can be converted to the equivalent brake power and thermal efficiency of the engine respectively, by multiplying by the value of the ratio of equivalent brake power to potential gas power appropriate to the particular operating point. The multiplying factor which might be called the efficiency of power production varies, of course, and is influenced most by the variation in turbine efficiency. Consequently on Fig. K,5c lines have been drawn showing the locus of the operating points at each value of $M_\infty$ for which the turbine efficiency is estimated to be a maximum for the pressure ratio available. These are flanked by lines showing the operating points at which the turbine efficiency is likely to have fallen by 5 per cent from the maximum value. This figure shows that with increasing flight speed the locus of optimum turbine operation moves away from the compressor surge region. It will be recalled that a

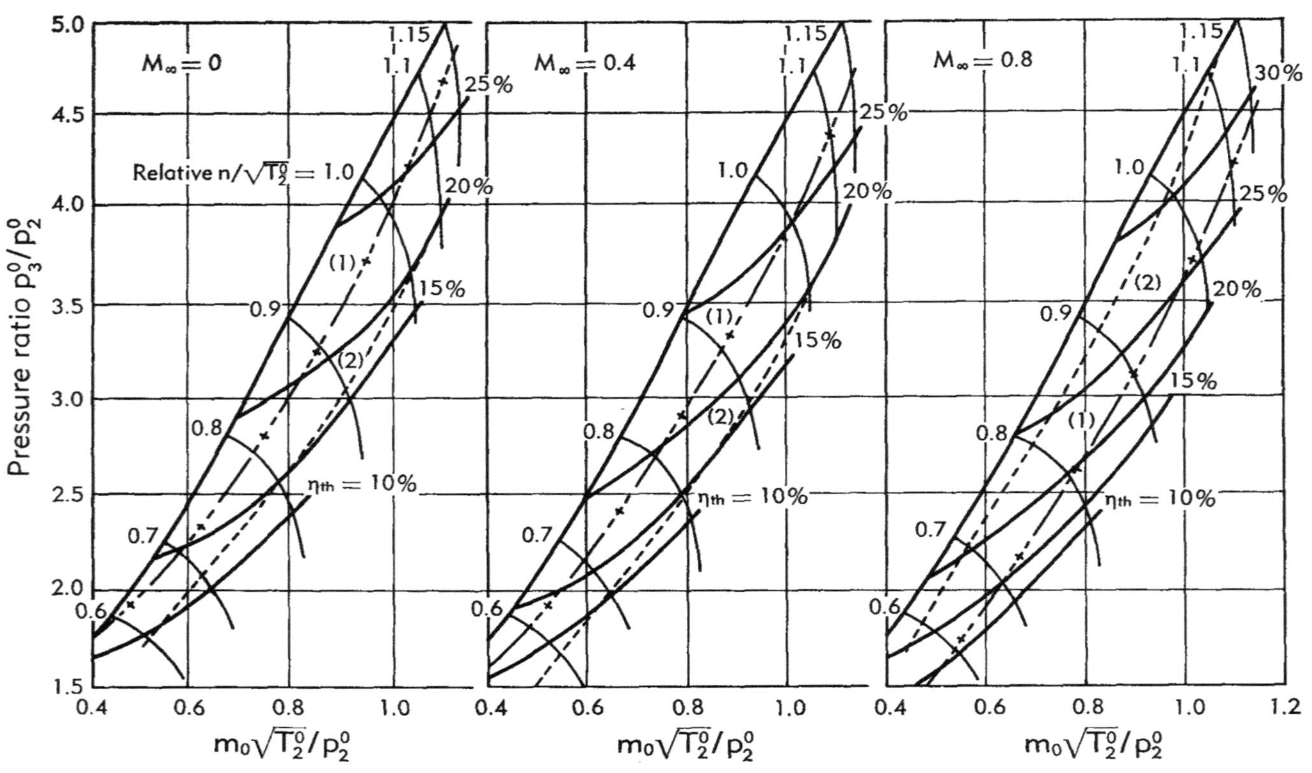


Fig. K,5c. Variation in efficiency of generation of potential gas horsepower (shown in Fig. K,4b). Additional lines show where efficiency of turbine (1) is a maximum and (2) has fallen 5 per cent from maximum.

similar movement of the compressor operating line was observed in the discussion of Fig. K,4e for the case where this combined turbine drives a fixed pitch propeller. This similarity serves to illustrate that this mode of operation tends to keep the turbine operating within a narrow range of the parameter $U_m^2/2gJc_p({}_i\Delta T_e^0)$ which, as Fig. K,3c shows, is a condition for optimum turbine efficiency.

With variable-pitch operation of the propeller, which might spread the compressor operating points over a wide band of the characteristic, the reduction in turbine efficiency is not, however, in any way catastrophic. At the lower flight speed illustrated in Fig. K,5c the loss in turbine efficiency by the time the surge line is reached is not as great as 5 per cent, and hence the appropriate line for 5 per cent efficiency loss only appears on the diagram for a flight Mach number of 0.8. On the "low-power" side of the maximum efficiency line there is again a fairly wide band of compressor operation before the condition of 5 per cent loss in turbine efficiency is reached. With movement in this direction, however, there is also a loss in the thermal efficiency of the gas generator portion of the engine and this is therefore not an attractive operating region. For a given power requirement a higher over-all efficiency would be obtained by reducing the rotational speed and controlling propeller pitch to give engine operation nearer to maximum turbine efficiency locus.

Whereas in the combined-turbine type of engine operating at constant rotational speed the variation in the efficiency of power production is applied to values of potential power and gas generator efficiency which are themselves varying, in an engine with a separate power turbine following the compressor turbine the movement of the equilibrium operating point on a particular compressor speed line is restricted. As a result the efficiency of power production is applied to values of potential power and gas generator efficiency (Fig. K,4h) which are virtually constant for that particular speed and flight condition.

Nevertheless, it follows that, in this type of propeller-turbine engine also, at a particular flight condition and compressor speed a range of actual thrust power produced by the propeller can be obtained with varying over-all efficiency by altering the propeller pitch-setting. At a somewhat different compressor speed an overlapping range of power may be produced but with different efficiency. For any particular power there must therefore be a value of compressor speed and propeller pitch-setting at which the over-all efficiency of the engine and propeller combination is at an optimum. The aim of the engine designer is, therefore, so to interrelate the controls on the compressor speed and the propeller pitch that the engine will always operate at or near these optimum conditions.

**K,6. Equilibrium Running of More Complex Engines.** The methods which have been outlined in the preceding articles, by which the equilibrium operation of relatively simple gas turbine engines can be investigated by matching the pressure, temperature, and mass flow characteristics of the individual components, can of course be applied equally readily to the study of more complex forms of engine. Fortunately, since it is almost axiomatic that in an aero-engine simplicity offers large dividends, the number of complex forms of turbine engine which merit attention is not large. Components such as heat exchangers or compressor intercoolers which offer attractive means for improving the efficiency or the power output of gas turbines for land and marine applications are generally too heavy to be considered for aircraft applications.

Consequently the only forms of complex engine which we shall consider here are those incorporating ducted fans and those employing two compressors in series. The latter types are commonly referred to as double-compound or two-spool engines.

*Ducted fan or bypass engines.* In Art. 2 it was convenient to consider the design point performance of a ducted fan engine in relation to the performance of the simple jet engine from which it may be considered to have developed. When considering the equilibrium operation of a ducted fan engine, however, it is desirable to think of it as a form of propeller-turbine engine, since the problems it presents are almost identical with those of the engine driving a more orthodox propeller.

Considering first a power unit in which the main engine and the ducted fan have separate exhaust systems we will first study the matching of the fan with its outlet jet. The fan, which takes in its air at conditions $p_9^0$ and $T_9^0$ virtually identical with those of the main compressor inlet, has characteristics which are essentially of a form similar to those of the compressor itself. The mass flow may be larger, and the pressure ratio at design conditions smaller, than in the main engine compressor with the result that the fan characteristic will be "broader" than the typical compressor characteristic of Fig. K,3a, the constant speed lines being more nearly horizontal.

The fan raises the pressure of the air in what we can term the secondary system from $p_9^0$ to $p_{10}^0$ from which pressure it expands again to $p_\infty$, the ambient static pressure, through the propelling nozzle of area $A_{11}$. At any particular value of fan pressure ratio, $p_{10}^0/p_9^0$, therefore, the pressure ratio across the fan exhaust is simply the product of this pressure ratio and the ram pressure ratio $p_9^0/p_\infty$ which depends on $M_\infty$ and the efficiency of the fan intake, $\eta_{a9}$. If $M_\infty$ and $\eta_{a9}$ are known the characteristics of the exhaust system will define a unique value of $\mu m_a \sqrt{T_{10}^0}/p_{10}^0$ (where $\mu m_a$ is the air flow in the secondary system) for the given value of $p_{10}^0/p_9^0$. There will be only one value of $\mu m_a \sqrt{T_9^0}/p_9^0$ which will satisfy the relationship between the inlet and outlet flow parameters and give the required values of $p_{10}^0/p_9^0$ and $\mu m_a \sqrt{T_{10}^0}/p_{10}^0$. Thus for a fixed nozzle area $A_{11}$ and a particular Mach number $M_\infty$, the fan operating line is uniquely determined. The lines for different values of $M_\infty$ all converge into a single operating line corresponding to choking conditions in the final nozzle, when the pressure ratio $p^0_{10}/p_\infty$ is sufficiently large.

Considering the turbine which drives the ducted fan, it follows that at a particular value of $M_\infty$ the fan demands a particular power-speed relationship comparable to that of a fixed-pitch propeller. The effect of the fan on the equilibrium operation of the main engine is therefore similar to that of a fixed-pitch propeller fitted to a propeller-turbine engine. The use of an independent power turbine to drive the fan, instead of taking power from the compressor-driving turbine for this purpose will have influences upon the operational line of the compressor similar to those already discussed in Art. 4 in relation to propeller-turbine engines.

It was suggested in the discussion of the optimum work fraction and bypass ratio for a ducted fan design in Art. 2 that the residual gas jet velocity will probably be higher in the case of a ducted fan engine than in a comparable propeller-turbine engine. Since in discussing the equilibrium operation of the latter engine the influence of the small exhaust jet has been presumed negligible, it is desirable at this point to consider the effect on the equilibrium-matching of components of using higher residual jet velocities.

From the findings in Art. 4 it is clear that if the ducted fan is being driven by an independent low pressure turbine, the engine operating lines on the compressor characteristic will scarcely be affected by a change in the division of work between the turbine and the jet at the design condition. Only changes in the swallowing capacity of the low pressure turbine can influence the operation of the gas generator set. While the fact that the power turbine, when designed at a lower pressure ratio, would possibly not operate in its "choked" condition suggests the possibility of some variation in swallowing capacity, the presence of an appreciable pressure ratio in the exhaust jet tends to preclude this occurrence. The turbine and the jet operate in series in much the same way as the two turbines illustrated in Fig. K,4f. Consequently as the over-all pressure ratio of the engine falls from the design value, it is the exhaust jet which experiences the major proportion of the reduction. The change in power turbine swallowing capacity in the course of normal operation is therefore not as great as might have been expected, and in consequence the change in the equilibrium condition line for the gas generator set is very small. There is, however, an increase in work fraction as the pressure ratio, and hence the compressor speed, is reduced.

The equilibrium operation of a ducted fan engine in which the air and gas jets are mixed before passing out through a common propelling nozzle may be determined by assuming that the static pressures in the two streams at the mixing plane are equal and that the bypass ratio is therefore a function of the relative temperatures, velocities, and flow areas of the two streams at this station. The common propelling nozzle area is the factor determining the actual air flows at any particular condition, in much the same way as the jet area determines the operating condition in a simple jet engine.

Allowance has to be made, however, for the degree of mixing which can be expected between the two streams (which are at different temperatures and moving at different velocities) and for the loss in pressure encountered in achieving this degree of mixing as was implied in Art. 2. This information is usually of a semiempirical nature. Experience suggests that, within the relatively short lengths available in a typical installation, very little mixing of the streams occurs before the nozzle is reached, unless assistance is provided in the form of mechanical mixing devices. Such devices introduce additional pressure losses, and a compromise must then be struck between the loss in thrust imposed by these pressure losses and the gain in thrust which in general accrues from obtaining a homogeneous mixture before expansion.

*Double-compound or two-spool engines.* In double-compound engines two independent compressors are used in series and are driven by separate turbines. They represent an attempt to achieve higher design pressure ratios than can be obtained in a single compressor where the difficulties of matching the initial and final stages over the necessary speed range prevent design pressure ratios much in excess of about 7 being contemplated for the present. As shown in Art. 2, these higher pressure ratios in turn enable improved fuel consumption figures to be achieved, though an increase in engine weight for a given output is also likely.

A brief study of the behavior of two compressors operating in series is a valuable guide to the main design and operational problems of double compound engines. If a compressor characteristic, such as that in Fig. K,3a, is plotted against its outlet mass flow parameter it is found that, over the majority of the pressure ratio range below the design value, the surge line occurs at an almost constant value of the outlet mass flow parameter, and the peak efficiency region occurs at a slightly higher, but again sensibly constant value, of the parameter. The outlet mass flow parameter for a low pressure compressor, however, is the inlet flow parameter to the high pressure compressor. If good operation of the former is to be achieved over a range of rpm, therefore, it is necessary that the latter must operate close to its design value of inlet mass flow parameter. Since, as Fig. K,3a shows, the efficiency of a compressor falls if its pressure ratio is decreased by any great amount while operating at a constant inlet mass flow parameter, a further requirement is that the speed parameter, and hence the pressure ratio of the high pressure compressor, must remain high over the operating range.

Comparing these requirements with what is already known concerning the operation of two turbines in series it at once becomes obvious that the only logical arrangement of the engine is for the high pressure turbine to drive the high pressure compressor and for the low pressure turbine to drive the low pressure compressor. The high pressure turbocompressor then operates over a small speed and pressure ratio range while the mass flow and the output of the engine are regulated by speed variations of the low pressure set.

The potential gas horsepower of the basic engine may be employed in any of the ways available in a single-compressor engine. A pure reaction jet may be used or a propeller or a ducted fan driven by the extraction of additional turbine power. In the latter case it is clear that many variations of series, parallel, or combined turbine arrangements are theoretically possible. Space does not permit the consideration of the relative merits of all these variations from the point of view of their off-design performance and operation. One general principle may, however, be defined, namely that the higher in the expansion process the external power is extracted the greater will be the tendency for the engine to cover a particular range of output by means of a decrease in mass flow and in the speed of the low pressure turbocompressor, while maintaining high turbine operating temperatures. This principle is, of course, simply an extension of that already noted in Art. 4 as applying to single-compressor engines.

As an illustration of the compressor-operating diagrams typical of double-compound jet engines, Fig. K,6 is presented. It is seen that the low pressure compressor operates very close to surging for much of the range of pressure ratio below the design value and that the tendency to surging is greatest at the higher flight speeds. The high pressure compressor operates over a smaller range of pressure ratio and well away from surging. These facts are of particular importance to the designer of the double-compound jet engine as the following argument will show. From the characteristic behavior of turbines operating in series it can easily be deduced that, as long as the final jet is choked, the power output of the two compressor turbines will be in a fixed ratio. The ratio of the power absorbed by the two compressors is therefore fixed, and since the same mass flow passes through each, it is only the difference in the relationship between compressor temperature rise and speed in the two cases (due to the different position of the operating lines) which prevents the compressor speeds being maintained at a fixed ratio. Should this happen, then as long as the final nozzle is choked no benefit would result from the use of two independent compressors in comparison with one single compressor of similar over-all pressure ratio, since the compressors are, as it were, "aerodynamically locked" together. At low rotational speeds, where the stage-matching of a single compressor of large design pres-

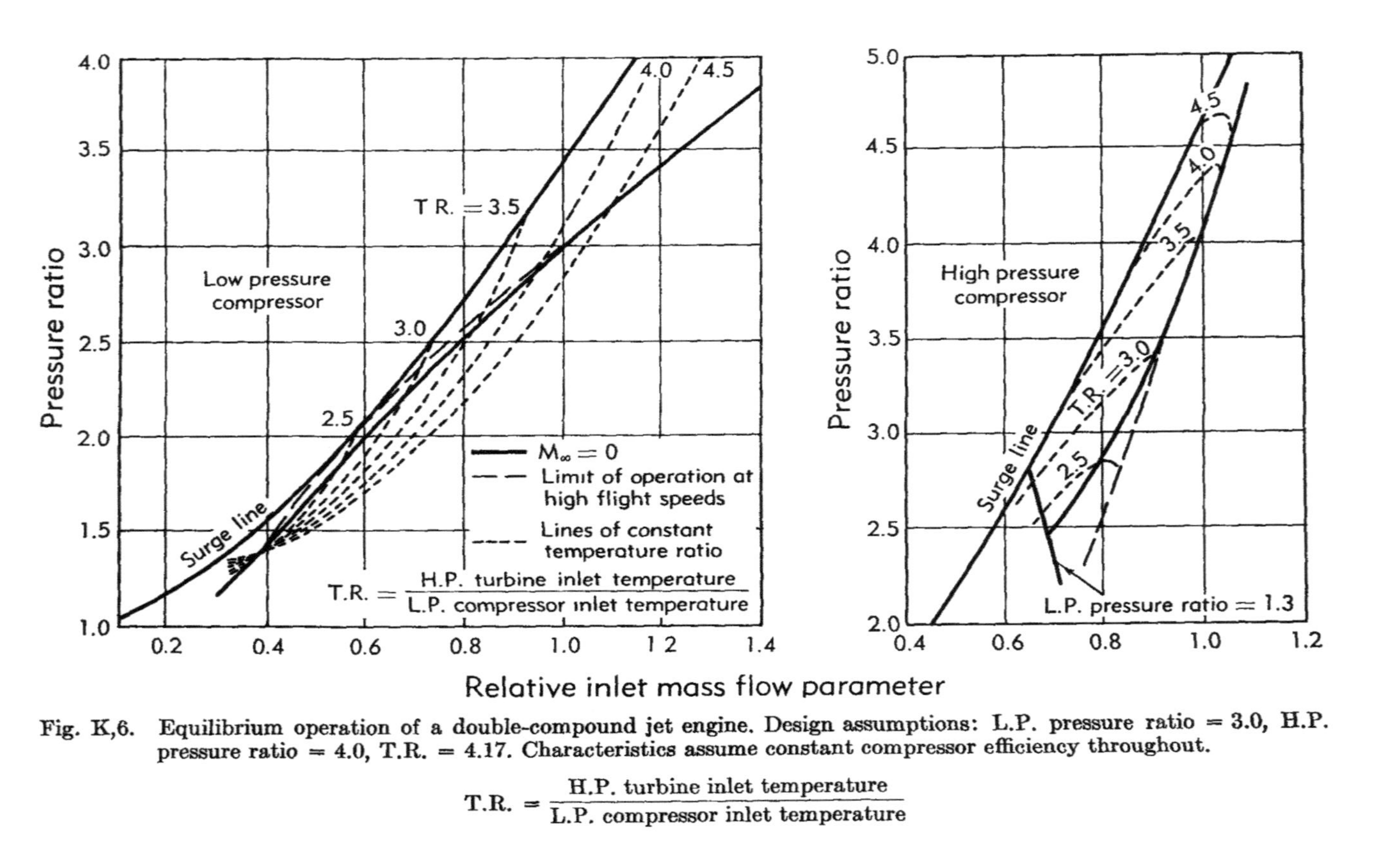


Fig. K,6. Equilibrium operation of a double-compound jet engine. Design assumptions: L.P. pressure ratio = 3.0, H.P. pressure ratio = 4.0, T.R. = 4.17. Characteristics assume constant compressor efficiency throughout.

$$\text{T.R.} = \frac{\text{H.P. turbine inlet temperature}}{\text{L.P. compressor inlet temperature}}$$

sure ratio is worst, the final nozzle is not choked and the two-spool engine operates with higher relative rotational speeds on the high pressure rotor than on the low, which considerably eases the compressor matching problem. The possibility of "aerodynamic locking" occurring is greatest in the jet engine version of the double-compound design and should not be present in versions using propellers or ducted fans unless an independent low pressure power turbine is employed.

As indicated in the opening articles of this section, a design of two-spool ducted fan engine that has proved successful in at least one instance is that in which the auxiliary fan and the low pressure compressor are combined into one unit at the front of the engine. The air flow then divides with the main engine air passing to the high pressure compressor and on through the engine, while the remaining air passes directly to the propelling nozzle. Since this air thus bypasses the bulk of the engine, the title of bypass engine is particularly apt for this design. From what has already been said concerning simpler ducted fan engines and the basic two-spool jet engine, it is seen that this design introduces no new problems in matching or control, and a study of its equilibrium operating conditions can again be made by suitable adaptations of the techniques already described.

EQUILIBRIUM RUNNING OF ENGINES EMPLOYING THRUST BOOSTING.

*Reheat.* If a large quantity of fuel is burned in the exhaust pipe of a jet engine it is necessary to increase the throat size of the final jet if the conditions within the engine are not to be adversely affected. If the turbine exhaust conditions are fixed, the density of the gas passing through the nozzle throat will decrease in proportion to the increasing temperature after reheat. The sonic velocity of the efflux, however, only increases as the square root of this temperature. An increased throat area is therefore required, and if this is not fitted, the engine, if governed at constant speed, will behave exactly as though a smaller propelling nozzle were fitted in the absence of reheat. The turbine temperature rises and the compressor tends to surge. Since reheat can only be used for a short time, provision has to be made for the propelling nozzle to be returned to its normal size when reheat is shut off, because there will otherwise be a serious loss of thrust and efficiency under normal operation.

Where a convergent-divergent nozzle is required, the problem of achieving the necessary variation in throat area is a serious one, for the nozzle must retain a satisfactory profile during the operation. In general the outlet area of the nozzle would be required to increase in proportion to the throat size, but if this would result in the outlet area exceeding the maximum frontal area of the engine this would probably not be done. Instead the nozzle in the reheated case would be allowed to operate with the gases underexpanded. Since the reheat process is bound to cause a loss of total pressure in the gas, thus reducing the pressure ratio available in the exhaust jet, the degree of underexpansion by adapting this procedure will not be quite as great as consideration merely of available flow areas would suggest.

The use of reheat between the compressor-driving and power turbines of a propeller engine, in a similar way, requires the stator blade throats of the second turbine to be enlarged in order to pass the less dense gas. Further, if the reheat process is discontinued there is a redistribution of the expansion pressure drop between the two turbines in which the compressor-turbine gains a higher proportion of the over-all drop. The result is to cause an appreciable fall in the compressor turbine operating temperature. The mechanical problems involved do not encourage the use of this technique in aircraft power plants.

*Liquid injection.* Since the injection of a liquid into the air stream of a compressor causes such far-reaching changes in the thermodynamic properties of the working fluid, it is not practicable to apply the component-characteristic-matching method to the determination of the effects that will be produced on the thrust or fuel consumption. Test experience has shown, however, that where water-methanol mixtures are employed as the coolant, it is often possible to obtain the desired amount of thrust boosting without having to make any alterations to the geometry of the engine. Further, the turbine operating temperature can in most cases be made to remain at or about the maximum permissible value by a suitable choice of the proportions of methanol and water used in the coolant.

## K,7. Cited References and Bibliography.

*Cited References*

1. Keenan, J. H., and Kaye, J. *Gas Tables.* Wiley, 1948.
2. English, R. E., and Wachtl, W. W. Charts of thermodynamic properties of air and combustion products from 300°R to 3500°R. *NACA Tech. Note 2071,* 1950.

*Bibliography*

Hooker, S. G. *J. Roy. Aeronaut. Soc. 50,* 298–322 (1946).

Mallinson, D. H., and Lewis, W. G. E. Performance calculations for a double-compound turbo-jet engine of 12/1 design compressor pressure ratio. *Brit. Aeronaut. Research Council Repts. and Mem. 2645,* 1951.

Morley, A. W. *J. Roy. Aeronaut. Soc. 52,* 305–322 (1948).

Sanders, N. D., and Behun, M. Generalization of turbo-jet engine performance in terms of pumping characteristics. *NACA Tech. Note 1927,* 1949.

Trout, A. M., and Hall, E. W. Method for determining optimum division of power between jet and propeller for maximum thrust power of a turbine-propeller engine. *NACA Tech. Note 2178,* 1950.

# INDEX

GPSR Authorized Representative: Easy Access System Europe - Mustamäe tee 50, 10621 Tallinn, Estonia, gpsr.requests@easproject.com

www.ingramcontent.com/pod-product-compliance
Ingram Content Group UK Ltd.
Pitfield, Milton Keynes, MK11 3LW, UK
UKHW021237150325
456285UK00004B/118

* 9 7 8 0 6 9 1 6 2 6 1 5 4 *